The Savant and the State

THE JOHNS HOPKINS UNIVERSITY
STUDIES IN HISTORICAL AND POLITICAL SCIENCE

130TH SERIES (2012)

1. Kurt C. Schlichting, *Grand Central's Engineer: William J. Wilgus and the Planning of Modern Manhattan*
2. Joseph November, *Biomedical Computing: Digitizing Life in the United States*
3. Robert Fox, *The Savant and the State: Science and Cultural Politics in Nineteenth-Century France*

• *The Savant and the State* •

SCIENCE AND CULTURAL POLITICS
IN NINETEENTH-CENTURY FRANCE

•

Robert Fox

The Johns Hopkins University Press
Baltimore

The Johns Hopkins University Press
2715 North Charles Street
Baltimore, Maryland 21218-4363
www.press.jhu.edu

Library of Congress Cataloging-in-Publication Data

Fox, Robert, 1938–
The savant and the state : science and cultural politics in
nineteenth-century France / Robert Fox.
p. cm.
Includes bibliographical references and index.
(The Johns Hopkins University Studies in Historical and Political Science)
ISBN 978-1-4214-0522-3 (hdbk. : alk. paper) — ISBN 1-4214-0522-9 (hdbk. : alk. paper)
1. Science and state—France—History—19th century.   2. Science—France—
History—19th century.   I. Title.
Q127.F8F694 2012
509.44′09034—dc23   2011048390

A catalog record for this book is available from the British Library.

Special discounts are available for bulk purchases of this book.
For more information, please contact Special Sales at 410-516-6936 or
specialsales@press.jhu.edu.

The Johns Hopkins University Press uses environmentally friendly
book materials, including recycled text paper that is composed of at
least 30 percent post-consumer waste, whenever possible.

CONTENTS

Illustrations follow page 126

This is a book with distant roots. My idea of writing a broadly cast cultural history of nineteenth-century French science originated in the mid-1980s, when a British Academy Readership in the Humanities allowed me three years of leave from normal duties at the University of Lancaster. Since then, work on the book has had to compete with other preoccupations, and my conception of it has evolved considerably. In this process, discussions with colleagues and students in Oxford have been important, as have opportunities for research and writing in a number of other locations in Europe and the United States.

As director of the Centre de recherche en histoire des sciences et des techniques at the Cité des sciences et de l'industrie and *directeur de recherché associé* in the CNRS between 1986 and 1988, I enjoyed an extended period of closeness to my sources. Since then a further stay in Paris, as *directeur d'études associé* in the Ecole des hautes études en sciences sociales, hosted by André Grelon, has allowed me to build on that privilege. Elsewhere, periods of intense writing at the Institute for Advanced Study, Princeton; the Bellagio Conference Centre; the Deutsches Museum, Munich; and Corpus Christi College and Clare Hall, Cambridge, came at crucial times, as did a term of additional leave from Oxford in 2005, financed by the Arts and Humanities Research Council. More recently, semesters spent as a visiting professor in the Department of History of Science and Technology, the Johns Hopkins University, and at the Department of History at East Carolina University have proved particularly fruitful.

Along the way, the Faculty of History at Oxford has always been generous in its support. Here I make special mention of Stephanie Jenkins, who with exemplary patience and attention to detail typed successive versions of a manuscript that has changed its form more than I care to remember. I am also indebted to Georgia Petrou for her assistance with an early version of the index and to the many librarians and archivists who helped to make research both efficient and a pleasure. On this score, the unfailingly helpful staff of Oxford's Bodleian and Radcliffe Science Libraries deserve especially warm thanks.

As the book neared completion, Charles Gillispie and Miriam Levin read the manuscript with a sympathetic but critical eye. The book is significantly the better for the re-working that their perceptive comments led me to undertake.

In the final stages, I have had the pleasure of working with the Johns Hopkins University Press. There I was fortunate to meet an editor, Bob Brugger, who has never wavered in his support for the project. I have also benefited from the meticulous editorial skills of Michele Callaghan and from the help I received at various stages from Kara Reiter and Josh Tong.

The Savant and the State

Introduction

• • •

Throughout the long nineteenth century that separated the Revolution of 1789 from the cataclysm of the First World War, science occupied a central place in French society and culture. In this, France resembled many other countries of Western and Central Europe and North America. But science and ways of thinking inspired by science mattered there to a degree that was arguably unmatched elsewhere. Moreover, they mattered in ways that reflected a history, political and cultural, of rare turbulence. Nineteenth-century France was scarred not only by war and violent changes of regime but also by enduring tensions between tradition and modernity and between widely divergent conceptions of what constituted the nation's best interests both domestically and in the world. In many of these tensions, science became inextricably involved. Sometimes it was seen as the symbol of an enlightened postrevolutionary order, sometimes as a dangerous Trojan horse concealing the ever-lurking menace of philosophical materialism, sometimes as a source of material well-being or national pride, and sometimes as the focus for dispute about the legitimacy of the authority claimed either by the leading figures of the academic community or by the senior Parisian administrators who controlled, or sought to control, learned culture in all its forms. On all of these issues, there were debates of an intensity that kept science consistently at the forefront of public concern. It is these debates and the various public arenas in which they were pursued that provide the focus for this book.

The Savant and the State is a study, therefore, of the public face of science—as opposed to the professional achievements of scientists or their private discoveries—in a crucial period of the history of one country. Although its coverage is broad, it does not pretend to explore this public face exhaustively. It places special emphasis on the physical and life sciences and gives only incidental consideration to the potentially no less rich fields of medicine and the human and social sciences. It begins in 1814, when the long nineteenth century, as I define it, was already well under way. This is partly to avoid duplication of the major existing studies of revolutionary and Napoleonic science. But, more important, it allows me to begin by focusing on a society making a new start, resolved that the turmoil of the preceding quarter of a century should be relegated definitively to the past. Though profoundly marked by memories of revolution and war, French society in 1814 had no choice but to face up to the shock of peace and the challenge of reconciling an age of restored Bourbon monarchy with an acceptance of changes that the years since 1789 had rendered irreversible. No amount of reactionary nostalgia for a prerevolutionary age of absolute monarchy and the powerful estates of the nobility and the Catholic Church could conceal the fact that France was now set, for good or ill, on a different path. The country was headed toward a world that promised at least some measure of democracy, a national economy increasingly adapted to urban and industrial interests, and a culture guided more by secular than by religious values.

At the other chronological extreme, the book ends in 1914. By that time, many of the transformations that had their roots in the first half of the nineteenth century had become a reality. For more than forty years, France had been a republic; economic debate and policy had come to focus to an unprecedented degree on manufacturing, not least in anticipation of the war to come; and the Catholic Church had seen its power diminished in the face of a tide of anticlericalism that had culminated in 1905 in the tempestuous separation between church and state. Over the period that I treat, the changes had been disorderly, punctuated by revolutions in 1830 and 1848, the searing defeat at the hands of Prussia in 1870, and swings of regime and political ideology that had taken the country from the clerically fired conservatism of the Bourbon Restoration (1814–30) to the more liberal July Monarchy of Louis-Philippe (1830–48), the Second Empire of Napoleon III (1852–70), and finally the determinedly secular Third Republic (declared in 1870 and a survivor until 1940).

Whatever broader analysis we adopt in trying to see our way through the thickets of the nineteenth century, it is no easy matter to identify the place of science in such a complex and troubled history. The overriding pattern was one of science's growing prominence, both in governmental policy and in public perceptions of its importance in the life of the nation. But behind that general trend, the process had its fine structure, with subsidiary trends that can essentially be reduced to three. I see these as (1) the ever closer integration of science, along with other areas of learned culture, in the bureaucracy of state; (2) the recognition of the crucial bearing of scientific education and research on the national interest; and (3) the growing prominence of philosophies and ideologies inspired by science, usually in opposition to traditional religious belief. In going on to discuss these trends individually, I do not want to imply that they each had an independent existence. On the contrary, the boundaries between them were shifting and permeable, and none went uncontested, whether by scientists and other intellectuals, politicians, or the general public.

First, then, bureaucracy and the related question of central authority. The context here is a long history of administrative and cultural centralization going back to the time of Louis XIV's minister Jean-Baptiste Colbert and reinforced under Napoleon, who created the Imperial University and the standardization and mechanisms of central authority that still have vestiges in the educational system today. Also relevant is the position of Paris. Few countries have had a capital of such sustained dominance as Paris or administrations, of whatever political stamp, so persistently intent on at least a significant measure of control, extending not only to the workings of government but also to the intellectual and cultural life of the nation. In keeping with that tradition, the ideal of centralization was as characteristic of the nineteenth century as of any other period in the history of modern France. But the nineteenth century is particularly interesting not just for the consistency with which centralization was pursued but also for the extent of the gap that, at various times, separated ideal from reality. To a degree, some such gap had always been there. The eighteenth-century Republic of Letters had been well peopled by the savants and other cultivated men and women throughout France who pursued their scientific, antiquarian, and literary interests with little concern for the authority exercised, or at least claimed, by either ministers or the national academies. What changed in the nineteenth century, especially after the creation of an autonomous Ministry of Public Instruction in 1832, was the size and capacity for intrusiveness of the "official" world of learning. For science, this meant that even savants who had no ties to the state found themselves having to adjust to decisions taken at ministerial level and, in the provinces, to deal with the ever more visible presence of academic scientists appointed from Paris to the nation's expanding network of faculties of science, fifteen of them across metropolitan France at the end of the period covered in this book. As the world of higher education grew, as it did spectacularly after 1870, a self-taught devotee simply could not hope to compete with a career academic who had succeeded in the highly competitive *agrégation* examination and worked through the sequence of the *baccalauréat*, the *licence*, and finally the essential requirement for appointment to a chair, the doctorate.[1]

Despite ministerial intervention and the drift to the professionalization of scientific life in higher education, independent savants worked hard to retain something of the role they had had since the eighteenth century. But they had always been in some measure auxiliaries in scientific agendas defined by Paris, and through the nineteenth century that trend became more marked. The ablest of them, especially those in distant provinces, might continue to provide specimens or make observations inaccessible to the disciplinary leaders trapped in their Parisian laboratories. And amateurs of all levels of competence loyally paid their subscriptions to a local society and made up audiences for lectures, exhibitions, or such peripatetic festivities of science as the annual Congrès scientifiques (beginning in 1833) or meetings of the Association française pour l'avancement des sciences (beginning in 1872). Some devotees even fashioned a national reputation: Henri Lecoq in Clermont-Ferrand, Jean-Baptiste Mougeot in Epinal, and Léon Dufour in Saint-Sever (all of them "voices on the periphery" that I discuss in chapter 2) were three provincial naturalists who did so with conspicuous success. Yet auxiliaries they remained, even the most eminent of them, and when on rare occasions they took positions at odds with beliefs approved within the ministry or by senior Parisian academicians, they had little chance of prevailing. The celebrated victory of Louis Pasteur, a quintessential man of the establishment, over the provincial museum director Félix-Archimède Pouchet on the question of spontaneous generation illustrates the point.

A particularly revealing insight into cultural centralization is the case of Arcisse de

Caumont, the most resolute of all champions of the independent world of learning against the domination of their chosen fields of science, archaeology, or the study of antiquities (especially medieval antiquities) by the approved representatives of the state. Thanks to a deep pocket and a zeal bred of legitimist politics and Catholic piety, Caumont managed to maintain his campaign for half a century until his death in 1873. His achievement, notably in mounting almost forty annual Congrès scientifiques, each in a different provincial town, from 1833, was remarkable. But the crusade that Caumont led was doomed to eventual failure, partly by the weight of ministerial opposition but also (to return to a point made above) by the steady implantation, in the provinces, of more and more formally qualified men who spoke with the authority of a chair in a faculty of the national university system. It is true that even in periods of the most aggressive attempts at centralization, Ministers of Public Instruction were happy for independent savants and scholars to pursue their work, and they offered them modest incentives and material help; starting in the 1860s and on into the Third Republic, ministers organized national congresses and prizes for such devotees and the *sociétés savantes* to which they belonged and allocated modest subsidies for their publications. But the condition for central support was that the fruits of patronage should never be used to threaten the authority of chairholders in the faculties and other seats of ministerial influence. Independent savants were meant to know their place.

The second trend, toward recognition of science as a key element in the economic and military welfare of the nation, is often hard to disentangle from the first. Again, we have to take account of a long history. In 1819, in the aftermath of the revolutionary and Napoleonic wars, the chemist Jean-Antoine Chaptal used his study of the state of French industry, *De l'industrie française*, to point to the participation of men of science during the wars as one of the outstanding contributions to France's gallant response to the Continental Blockade. And when, in the same year, the Conservatoire royal des arts et métiers in Paris created its chairs of industrial chemistry, mechanics, and industrial economy, the initiative rested on a belief in the importance of forming a workforce better informed and better adapted to the needs of modern manufacturing than one fashioned through vocational trade schools and apprenticeships. Whether the new type of scientifically oriented instruction was more effective than the existing provision appears, in retrospect, an open question. Nevertheless, a rhetoric sustained by the belief of a lobby led by the former naval engineer and rising political figure Charles Dupin in the efficacy of such science-based instruction informed educational initiatives for many years to come. Among the initiatives, the Ecole centrale des arts et manufactures, a school for training high-level industrial engineers, founded as a private enterprise in 1829, stood out. But the municipal lecture courses that existed in most large provincial towns and, in the 1850s, the attempts of Napoleon III's Minister of Public Instruction, Hippolyte Fortoul, to develop scientific courses in the faculties as a preparation for industrial and agricultural employment were part of the same movement.

As events were to show, Fortoul's strategy failed. Its failure is significant as an indicator of how difficult it was to put in place a rather academic state-controlled educational provision that employers would see as truly serving the needs of the economy. Nevertheless, in France as in all industrialized countries, the emergence of the new industries of the later nineteenth century, especially those of artificial dyestuffs (from the 1860s) and high-current electricity (from the 1880s), invested the promotion of science-based vocational education with an irresistible urgency. Quite suddenly, educational initiatives proliferated. Especially after 1880, faculties of science, in particular, were able to exploit the Ministry of Public Instruction's new policy of controlled devolution by creating institutes of applied science of which some (Nancy in chemistry and Grenoble in electrical engineering, for example) enjoyed real success, at least as measured by student numbers. The numbers, notably of students attracted from eastern Europe, were gratifying not only to the ministry but also to the many municipal authorities and local chambers of commerce that contributed to the cost of buildings and other facilities. Numbers, however, did not entirely quell debate about the appropriateness of the educational new departures. Between 1880 and 1914, discussions of what constituted a proper provision for scientific education and research continued and even assumed a new edge as France, smarting from the defeat of 1870, strove for prosperity and prestige on the battlefields of science, the arts, and the economy.

Exchanges that touched on France's economic performance and its status in the world, especially relative to Germany, inevitably became bound up with wider debates about the forces of modernity in the age of accelerating urbanization and the beginnings of mass production on something approaching its twentieth-century scale. Those who believed, like most committed partisans of the Third Republic, that science and its applications should be promoted as the supreme foundation of France's future well-being revelled in the succession of universal exhibitions in which displays of scientific and technological prowess such as the Eiffel Tower (built for the exhibition of 1889) and the huge but virtually unusable "grande lunette," a telescope constructed for the exhibition of 1900, inspired popular admiration. But, as I argue in chapter 6, that same admiration and the official shows of pride that fostered and drew on it also fired the very different response of those who saw the industrial age as one characterized by menace as well as by benefit. François Coppée's poetic tirade against the Eiffel Tower and Albert Robida's disturbing prophetic vision of a future world of polluted air and dehumanized homes and cities are just two examples of the unease provoked by the technocratic triumphalism of republican ideology. A generation that responded with enthusiasm to the excitement of the novels of Jules Verne might be fascinated by science and its applications but was also prey to profound anxiety.

Part of this anxiety was fed by concerns associated with the third of the trends that I identify. The growing prominence of secular philosophies and ideologies that drew more or less explicitly on science had important roots in the eighteenth-century Enlightenment. In the nineteenth century, however, they assumed a new character and a new vigor. From the beginning of the century until his death in 1825, the philosopher and social thinker Claude-Henri, comte de Saint-Simon, used books, pamphlets, and correspondence to sow the seeds of a tradition of secular thought and social ideals for the industrial age, and this tradition lived on, after his death, as a loosely formed but enduring body of Saint-Simonian doctrine. Saint-Simon's vision of a new society in which mathematicians, scientists, and engineers would have pride of place had its most immediate legacy in the work of Auguste Comte, his former secretary. Comte built on Saint-Simon's ideas (despite a break between the two men in 1824) with an analysis of human history in which he saw a pattern of the gradual ascendancy of scientific (in his terms, positivist) modes of thought over what he regarded as the more primitive world-views prevailing in mankind's theological and metaphysical phases. Comte's influence, like Saint-Simon's, was limited in his lifetime; this was most conspicuously the case with respect to his Religion of Humanity, with its paraphernalia of secular rites and saints. But positivist philosophy was a tenacious survivor, and beginning in the 1860s it resurfaced and drew new strength, as well as notoriety, from its easy assimilation in a congeries of controversial beliefs united in the "radical synthesis" that I discuss in chapter 4.

The radical synthesis embraced positivism, along with a number of other concepts, including evolution by natural selection, the spontaneous generation of life, and polygenist doctrines on the multiplicity of the human races. At the same time, the materialism of Ludwig Büchner, Jacob Moleschott, and Karl Vogt was filtering into France from its German-speaking heartland and beginning to take hold in circles predominantly wedded since the 1830s to the prevailing philosophical eclecticism of Victor Cousin. The conflict that the synthesis provoked in and beyond the realms of academic philosophy is a striking exemplar of the way in which science could become involved in the public sphere of debate. As "advanced" thinkers, such as Hippolyte Taine and Ernest Renan, expressed their support for, or at least their interest in, the new ideas, so the reaction to the reemergence of science at the cutting edge of philosophy became more resolute. Paul Janet at the Sorbonne articulated the discomfited response of an open-minded but at heart traditional philosopher, and the Catholic Bishop of Orléans, Félix Dupanloup, spoke out in the name of religion or at least a more spiritual perspective on the human condition. Although Janet and Dupanloup were by no means extreme in their opposition, they attacked the new wave of unorthodox ideas as the manifestation of a scientifically inspired secularism that threatened the prevailing bourgeois values, indeed the very order of the Second Empire, during which their own conservative ideas had greatest currency.

Under the Third Republic, the adequacy of a philosophy based on science grew into one of the most contested issues in French society. The conflict drew strength from a new level of Catholic assertiveness, encouraged initially by the state's abandonment (in 1850) of its monopoly in secondary education and, from 1875, by the legislation authorizing the creation of independent Catholic universities. These victories for liberty in education had also been victories for parents and others who had deplored the secular cast of public instruction since the Revolution, and they gave heart to those who aspired to a more moral, religiously founded dimension in fashioning young minds. The Catholic literary critic Ferdinand Brunetière spoke up vehemently for such people. His strategy was not that of a wholesale attack on science; rather, he insisted on science's insufficiency as a guide to conduct and the purpose of human existence. Brunetiere's assertiveness, with that of other spokesmen for the Catholic cause, helped to bolster the resolve of the faithful. But the struggle against the various manifestations of anticlerical sentiment faced formidable odds. At the turn of the twentieth century, many Catholics felt more browbeaten than ever by the mixture of secularism and scientism that underpinned republican ideology. For them, the chemist and pillar of the Third Republic Marcellin Berthelot stood as a bête noire. The same religious values that Brunetière and other Catholics (including many, like Alfred Loisy, of a relatively liberal persuasion) saw as the bedrock of a stable society were seen by Berthelot and his fellow republican ideologues as a force regressively antagonistic to the liberating march of reason. Across that divide there could be no meeting of minds.

The intensity and longevity of the conflicts surrounding materialism and the other elements of the radical synthesis are a mark of the anxiety that challenges to traditional ways, whether of thinking or of doing, provoked. The same was true of the other broad trends to which I have referred. The bureaucratic control of academic life and the alliance of powerful professors and administrators who exercised it, for example, caused resentment in the increasingly marginalized independent world of learning, including the Catholic universities. But even some of those working in the state system of higher education also had their grievances; they soon resented the heavy teaching loads and humdrum duties that came, under the Third Republic, as the price to be paid for handsome buildings and improved facilities. Some of those same critics also had reason to complain about growing ministerial preoccupation with economic priorities and the consequent diversion of funds and attention to science in its more applied aspects. The mathematician Emile Picard put the point forcefully in 1912, when he warned that if faculties continued to strengthen their association with the worlds of industry and agriculture, they would be left as little more than trade schools, with only a secondary commitment to the progress of science. "Disinterested research" needed protection, and resources had to be directed accordingly.[2]

All the debates to which I have referred were played out on a public stage, as exchanges that engaged a broad general audience as well as those immediately concerned. What could be more public, or more exciting, than the bitter four-day debate of May 1868 in the Senate about the anticlerical teaching of an unnamed professor in the Paris faculty of medicine, almost certainly the materialist Charles Robin? The debate followed a petition bearing more than two thousand signatures, of both priests and lay people, and observers extended far beyond the senators assembled in the chamber. In this fixing of science in the public sphere (in Jürgen Habermas's sense), science columns in the press (beginning in the 1830s), and books and periodicals of popular science (beginning in the midcentury) played a crucial role as settings in which the public and private faces of science could meet. It is true that by no means all popular science bore on such a big issue as materialism; much of it was straightforwardly descriptive, and most collecting in natural history and observation in astronomy was undertaken for pure amusement. But the world of popular science always had a taste for controversy, where controversy existed, and a place for an acerbic pen of the kind that the socialist popularizer Victor Meunier wielded repeatedly against the scientific establishment during the Second Empire.

Finally, I come back to my choice of the three trends underlying this discussion. The choice implies no claim to exhaustiveness. I simply believe that in nineteenth and early twentieth-century France public discussion about science was usually related to one or other of them. Collectively, too, the themes underline a position that informs the kind of cultural history of science that I offer in *The Savant and the State*. The position, put simply, is that science in my period interacted with concerns that far transcended the strictly drawn boundaries of scientists' quest for an understanding and mastery of nature. When Pasteur, in 1870, likened the French scientific community to an army condemned to fight an impossible battle without weapons, he was not only speaking about the deficiencies of academic laboratories but also joining a broader protest against a stifling bureaucracy unable, or unwilling, to comprehend the benefits, material or cultural, of intellectual endeavor. No less urgently, he was expressing a concern about France's standing in the world. The focus of his particular concern was the contrast he drew between the first quarter of the nineteenth century, when French science had been preeminent in Europe, and a present in which the rising power of Germany had disturbing implications for France, in science as in many other domains. Others at the time displayed their concern in different ways: through campaigns to modernize education, for example, or to fashion a leading role for France in the competitive world of international exhibitions and congresses. But in all such causes, a keen sense of the national interest at a time of growing challenges loomed large. Interrelatedness of this kind that bound together science and the realms of politics, the economy, and general culture makes drawing the frontiers between these various domains impossible. Hence if I leave the boundaries indistinct, I believe that, in doing so, I am reflecting the realities of the time. But I am also expressing, through the practice of writing, my more general conviction that an understanding of the public face of past science can only advance if we cast our interpretative net widely and remain as flexible as our sources allow in determining what is relevant to the questions we choose to ask of our material. It is a truism, but an important one, that the categories of today seldom survive transmission to the past.

CHAPTER ONE

Science and the New Order

• • •

When the Bourbon line was restored in the spring of 1814, in the bloated person of Louis XVIII, it was inevitable that the new regime should scrutinize with special care institutions and practices of every kind that had emerged or expanded in the preceding quarter century. Faced with this scrutiny, the scientific community had reason to be apprehensive. Since the Terror in the aftermath of the French Revolution, science and mathematics—and practical applications and ideologies founded on them—had secured the lion's share of official favor among learned pursuits and had done so to such a degree that they could all too easily be seen as disciplines that in some way encapsulated the spirit of the republic or the empire. When Napoleon referred (or is said to have referred) to the Ecole polytechnique as a hen that laid "golden eggs,"[1] he was expressing the admiration of a patron proud to associate the school's success with his reign: even if he valued the institution as "a seminary for warriors" more than as a "nursery for savants" (to use the expressions of Napoleon's general, Maximilien-Sébastien Foy[2]), he deserved at least some credit for the exceptional quality of both the *polytechniciens* and the teachers who made Polytechnique a byword for scientific and, more particularly, mathematical excellence.

In contrast, the most ambitious of all of Napoleon's educational initiatives, the Imperial University, was intellectually a mixed bag that gave less obvious reason for pride. Nevertheless, in science as in medicine, letters, law, and (to a far lesser extent) theology it too made its mark on French perceptions of learned culture. Between 1808, when the first appointments were made, and 1814, forty-seven chairs, including thirty-eight outside Paris, were created in the University's eight faculties of science alone, in addition to about 150 chairs in other faculties within the borders of France.[3] The professors who were appointed to the scientific chairs, at least in the provinces, were first and foremost representatives of the national administration charged with the humdrum task of examining secondary school pupils for the *baccalauréat-ès-sciences*. Their presence alone nevertheless had a more elevated effect in raising the profile of science and reminding the citizens in some of the country's largest cities of the emperor's solicitude for its practitioners.

Encouraged by the mixture of intellectual and utilitarian value that was attached to science, savants of the revolutionary and Napoleonic periods did not hesitate to flaunt their loyalty to those same governments which, from 1814, the new royalist order (to say nothing of many individual citizens) came to view in retrospect with distaste. With the Bourbon line restored, the service of the chemist Jean-Antoine Chaptal as Minister of the Interior between 1801 and 1804, the acceptance of senatorial rank during the empire by the mathematician Pierre-Simon Laplace and chemist Claude-Louis Berthollet could not automatically be overlooked as insignificant acts of conformity. The same was true of the naturalist Georges Cuvier's membership on the Napoleonic Council of State. The offices in question had been prominent ones, and they had often elicited a sycophancy that had gone beyond formal declarations of respect for the regime. Laplace's dedication of a volume of his *Traité de mécanique céleste* to Bonaparte in 1802 heaped unctuous praise on the First Consul as a hero who had pacified Europe (an accolade soon rendered conspicuously inappropriate by the collapse of the brief period of peace following the Treaty of Amiens).[4] Eight years later, Cuvier followed suit in even more elaborate terms. The introduction to his *Rapport historique sur les progrès des sciences naturelles depuis 1789* represented Napoleon (now emperor) as at once a hero of his nation and a man whose disinterested love of learning allowed him to soar above national rivalries and see the sciences as the possession of the whole human race.[5]

When the empire collapsed in the first few days of April 1814, such declarations were embarrassing enough. And they became even more so in the fourteen months that elapsed between Napoleon's departure from Fontainebleau for exile on Elba on 20 April 1814 and his final abdication on 22 June 1815, following the Hundred Days of his escape from imprisonment. For science, as much as for the individual scientists who had declared their allegiance to the empire, this was a precarious period of transition, and it required delicate handling. Even if the response of Laplace and Cuvier was that of outstandingly powerful savants pursuing their personal interests above all else, in both cases those interests went hand in hand with a clear-eyed recognition of the learned world's dependence on public support and hence of the necessity for an accommodation between the community and whatever political order was in the ascendant.[6] By the end of the Napoleonic Empire, in fact, the state had become, to an unprecedented degree, a prop whose resources it was essential to exploit if scientific ambitions were to be realized, whether in the world of disinterested research or in the increasingly prominent areas of science-related industry, agriculture, and warfare. As a plentiful secondary literature makes clear, savants since the 1790s had acquired a visibility and influence—a "new power," as Nicole and Jean Dhombres have analyzed it—that ensured the material support and stability necessary for their work.[7] From that hard-won position of "complementarity" between science and politics (the word is Charles Gillispie's)[8] there could be no going back.

From 1814 to 1848, first under the Bourbon and later the July Monarchy, the prominence of science in the priorities of the nation continued to grow. In these years, the interactions between government, science, and the economy multiplied, in pursuit of goals variously construed as the spread of knowledge and the pursuit of material well-being, or the less precise objectives of modernization and national prestige. All the while, advancing bureaucratization and professionalization increased the dependence of academic scientists on their ministerial paymasters and reinforced still further the roles that most of them had as biddable civil servants. This is not to say that the trend went unresisted: in this and later chapters, I have a good deal to say about savants who, as a principle or of necessity, made their reputations independently of publicly administered institutions of teaching and research. But independence was generally a recipe for a precarious existence, as is evident from the very different cases of the positivist philosopher Auguste Comte and the zealous right-wing polymath and scourge of the centralized state, Arcisse de Caumont. Scientists in the period I treat here frequently pined for greater freedom from the shackles of the nation's educational administration, which they accused variously of incomprehension, parsimony, or jobbery in the allocation of posts and influence. Few, however, achieved it, and most recognized that, on balance, assimilation in the apparatus of the state brought inestimable advantages, at least to those who knew how to make the most of the rewards that were available.

The Return of the Bourbons

Among the established leaders of French science, the degree of accommodation that Laplace and Cuvier showed in the turbulent years of 1814–15 earned them a special place in the unforgivingly censorious *Dictionnaire des girouettes.*[9] Both men were presented in the barbed vignettes of the *Dictionnaire* as *girouettes,* weathervanes ready to turn with every change in the confusing political winds that had blown during the passage from the First Empire to the Restoration. To contemporaries, their capacity for survival appeared breathtaking. Laplace's appointment to the Restoration Chamber of Peers might be regarded as a normal part of the political transition: following the closure of the imperial Senate, most of the empire's senators "survived" in this way.[10] But soon his quiver of honors was filling spectacularly. Despite his earlier allegiance to Napoleon, the restored Bourbon government recognized him as one of the glories of French culture and allowed his star to rise accordingly, notably through his election, in 1816, to the most prestigious of all the nation's academies, the Académie française (an election that would certainly have had full ministerial approval). Cuvier's passage was, if anything, even more deft and successful.[11] Like everyone who had held high imperial office, he watched with anxiety as the future of his main seats of influence—in particular, the Imperial University (where he had been a senior member of the governing council) and the even loftier Council of State (the elite political body to which he had been appointed in 1813)—was debated. Cuvier's retention, during the First Restoration of 1814–15, of virtually all the positions he had occupied under the empire was encouraging for him, but it was only after the Napoleon's brief return to power during the Hundred Days (at which time he avoided declaring himself, through the age-old solution of a Ciceronian withdrawal from the public eye) that he could breathe easily.

Although Cuvier's removal from the Council of State during the Hundred Days served to remind him of his vulnerability, it probably facilitated his rehabilitation as a member of Louis XVIII's reconstituted Council and the powerful five-man Commission of Public Instruction that was put in charge of education in August 1815. From that moment on, he had no need to hold back in his declarations of undying support for the Bourbon king and in his condemnation of the emperor he had so recently revered. Reporting retrospectively on the work of the First Class of the Institute between 1813 and 1815 (before the class reverted to its prerevolutionary title of Académie royale des sciences in March 1816), he nailed his colors to the mast. He looked back on 1815 as "yet another year of devastation and terror" to which the Bourbon line had finally restored stability.[12] In such troubled times, he wrote, it was only by taking refuge in "the depths of meditation" and the consolations of science that the nation's savants had been able to continue their lofty endeavors, unruffled by the troubles about them.[13] True science, as Cuvier presented it, was private science, a pursuit that soared above political conflict and, in doing so, promoted harmony between classes and interests.[14]

Cuvier's solution served an important function in detaching science from the revolutionary and Napoleonic past and more distantly from damaging associations with Enlightenment radicalism. But the element of self-interest in his declarations escaped no one, and the price he paid, in his exposure to public criticism for his ideological pliability, was real, though not high. For less prominent members of the scientific community, the price of pliability was barely noticeable. Such men had simply to acquiesce in loyal addresses of the kind that the president of the First Class of the Institute, the 60–year-old professor of physics at the Collège de France, Louis Lefèvre-Gineau, delivered when the king formally received members of the Institute on 11 May 1814, nine days after his first return to Paris. It is hard to imagine that Lefèvre-Gineau, a moderate partisan of the Revolution and a loyal servant of the Empire (though one who in the end had voted as deputy for the Ardennes in the national law-making assembly, the Corps législatif, for the overthrow of Napoleon[15]), genuinely experienced the extremes of "unalloyed profound joy" that he was called on to convey.[16] But the sincerity of the exchange was unimportant. The declaration of the Institute's wish to contribute to the glory of the new reign and the promise of solicitude and protection that the king offered in return had the desired effect of formalizing a new beginning. It achieved, and was expected to achieve, nothing more than that.

The cosy reciprocity of the expressions of good will could hardly be expected to survive the scare of the Hundred Days, and once Napoleon was definitively deposed, there followed a searching scrutiny of the Institute and the four national academies that composed it, as of all institutions for research and teaching. Yet even then political interference amounted to modest tinkering rather than wholesale restructuring. As part of the reorganization of the Institute in March 1816, science suffered symbolic demotion when the resurrected Académie royale des sciences, was placed third in the ranking of the four academies, behind the Académie française and the Académie des inscriptions et belles-lettres, though ahead of the Académie royale des beaux-arts.[17] But the demotion from the First Class's position of primacy did not presage significant hostility. A review of membership led to the exclusion of only two members from the new body. These were the mathematicians Gaspard Monge and Lazare Carnot, both of whom paid the price as regicides who had voted for the execution of the Louis XVI in the National Convention in 1793 and subsequently compounded their errors by unrepentant personal attachments to Napoleon. On the latter score, Carnot was especially vulnerable. For, after more than a decade of retirement from public life, he had willfully returned to the emperor's service, not just as the defender of Antwerp in May 1814 (which it might have been possible to excuse as a proper military duty) but also, during the Hundred Days, as Minister of the Interior (a show of loyalty that could not be passed off so lightly).[18]

The pattern of other punitive sanctions suggests a readiness for leniency on the part of all but the most reactionary elements of society, and there is certainly no reason to think that, among learned pursuits, science was particularly victimized. Within the Institute, in fact, purges affected the Académie des sciences rather less than the other academies: the Académie française, the Académie des inscriptions et belles-lettres, and the Académie des beaux-arts lost respectively eleven, five, and five members of the corresponding pre-Restoration classes. Moreover, Monge (aged 70 in 1816) and Carnot (at 62) were near the end of their scientific careers and for that reason cannot be accounted great losses. Age too made the removal of Louis-Bernard Guyton de Morveau from his administrative post at the Mint in 1815 and his death a few weeks before the reorganization of March 1816 a negligible loss: Guyton was already 78 in 1815 and had performed no serious chemical work for some years. Among younger men, the most notable casualties of the review of institutions and personnel were Louis-Benjamin Francoeur, in his early forties and in his prime as a mathematician, and Francoeur's slightly older contemporary Jean-Nicolas-Pierre Hachette. Francoeur may well have suffered from his friendship with Carnot. At all events, he was dismissed from the coveted position of entrance examiner at the Ecole polytechnique, while the elimination of the advanced mathematical classes in the former *lycées* (now refashioned as *collèges royaux*) deprived him of his post at the Lycée Charlemagne: in the circumstances, he was fortunate to retain his chair of algebra at the Sorbonne.[19] Hachette may also have been the victim of guilt by association, in his case with his mentor and friend Monge: Hachette's removal from the chair of descriptive geometry at Polytechnique signalled a lingering disapproval that surfaced again in 1823, when the decision of the Académie des sciences to elect him to its mechanics section was refused the necessary royal approval.[20]

Like the Académie des sciences, other bodies with prerevolutionary roots were generally treated with a light touch, amounting to little more than the obligatory addition of the tag "royal" and a minimal review of functions and expenditure. The three great research institutions survived almost intact. The Muséum d'histoire naturelle suffered a small temporary reduction in its budget,[21] and professors at the Collège de France had their salaries cut from 6,000 to 5,000 francs.[22] But the survival of the Muséum and the Collège, like that of the Observatoire de Paris, was never in doubt. The engineering schools of the ancien régime also came through virtually unscathed. The Ecole des ponts et chaussées underwent only minor trimming to take account of the loss of imperial territories outside France, while the Ecole des mines actually gained in prestige through its move back to Paris after twelve years in the mining area of Savoy.[23]

Institutions that had emerged since the Revolution, however, could expect more critical scrutiny, and two of them—the Ecole polytechnique and the former Imperial University (by this time "Royal")—received it. As both friends and critics understood, Polytechnique faced the special predicament of being at once a predominantly military school and one whose roots and reputation were immersed in the long struggle against nations that were now officially regarded as saviors. Could an institution that had had the regicide Monge as its principal founder be made to serve a monarchy to which it had no ties, of either tradition or ideology, and under which there was, quite suddenly, virtually no demand for the officers it had been so effective in fashioning? And how were the political and administrative leaders of the Restoration to regard a body of students whose Bonapartist-inspired patriotism had recently become part of Parisian folklore? As the royal allies advanced on the capital and prepared for the final attack of the last days of March 1814, the *polytechniciens* had formed themselves into armed units, conducted exercises on the courtyard of the school, and gone on to fight with legendary though vain bravery in the defense of the city.[24] During the Hundred Days, the pupils had rushed again to Napoleon's side, offering him a loyal address and an excited welcome when he visited the school, and it was only the armistice at Saint-Cloud, signed in July 1815, that prevented them from reappearing on the battlefield. Following the heroic defence of 1814, the marks of fidelity to the empire had initially been passed off as a rather admirable excess of youthful zeal, and the king had even made a conciliatory gesture by decorating three *polytechniciens* with the cross of the Legion of Honor. But in the unforgivingly reactionary climate that came to prevail from the summer of 1815, displays of admiration for Napoleon were viewed in a distinctly less favorable light.

With the will for at least some measure of reconciliation now in abeyance, Polytechnique faced the prospect of rough handling. For most of the academic year 1815–16, the school's administrators managed to maintain the appearance of normality. But nostalgia for the Napoleonic past remained strong among the pupils and provoked bouts of insubordination. These culminated, in April 1816, in an unruly demonstration against a tutor in mathematics, Louis Lefébure de Fourcy, whose arrogance and royalist sentiments had for some time earned him sustained unpopularity. As a consequence, and with memories of *polytechnicien* loyalty to the emperor also weighing heavily, the pupils were dismissed two months later, pending the report of a five-man committee appointed, under the chairmanship of Laplace, to determine the school's future.[25] Laplace proved a staunch but realistic defender of the school. He was clearly determined that Polytechnique should not only survive but also, and no less important, maintain its high standing as an elite school for the nation's ablest mathematicians and scientists. The reorganization that resulted, in September 1816, bore the marks of this elevated vision. Crucially, it protected the school against suspicions, most strongly voiced by the ultramontane Catholic priest Robert de Lamennais, that Polytechnique had become a hotbed of impiety and immorality: on that count, the dismissal of François Andrieux, the professor of literature and a committed champion of the Enlightenment, was enough to silence the critical voices. Laplace's true priorities, however, lay elsewhere, not just in the inevitable shift in the balance of studies from military to civilian subjects but more specifically in what he saw as the proper hierarchy of the sciences and the nobility of disinterested abstract inquiry. The elimination of the second-year course in architecture, a reduction by a third in the time allocated to chemistry, and a modest increase in the teaching of physics, now in the capable hands of Laplace's young protégé Alexis-Thérèse Petit, all bore his thumbprint.

Later commentators were unanimous in identifying Laplace's voice as decisive in the strengthening of the abstract, mathematical cast of the curriculum and in edging the school away from Monge's original conception of an institution dedicated, according to the circumstances, to the preparation of engineers for the military or the civil services of the state.[26] And they were correct. It is a safe assumption, for example, that Laplace was behind his committee's insistence that training for private industry, though desirable, would find a more proper home with the middle-level technical curriculum of one of the two *écoles d'arts et métiers* that had functioned at Châlons-sur-Marne (since 1806) and at Beaupréau, then Angers (since 1811).[27] It was likewise in keeping with Laplacean priorities that the committee simply ignored pressure from the Ministry of the Interior for a reduction in the mathematical requirements for entry. Slackening of that kind would have been at odds with Laplace's concern for the interests of mathematics and theoretical science.

In comparison with Polytechnique, which had the defensive advantages of scientific distinction and the potential for new functions, the Napoleonic University appeared distinctly vulnerable. It was not only a creation of the empire but also one that had performed unevenly from the start. Healthy enrolments in the faculties of law and medicine, which functioned essentially as professional schools, were no defense against the charge that—examining for the *baccalauréat* apart—the faculties of science and letters, especially those outside Paris, served little purpose. This weakness offered an inviting flank to critics, the most reactionary of whom demanded nothing less than complete closure; for them, the futility of the University was compounded by its status as a state-controlled monopoly and hence as a vestige of the despotic past. Such portrayals often went hand in hand with charges of godlessness. The abbé Claude-Rosalie Liautard, who had been (briefly) among the first *polytechniciens* in 1794 but whose early upbringing in the royal palace of Versailles had inspired him with an undying loyalty to the Bourbons, saw the institution in exactly this light.[28] As he believed, the University's divorce from the Catholic tradition had left it—and here Liautard included the pre-Napoleonic Ecole normale as well as the national network of faculties and *lycées*—an oppressive military-style institution; devoid of morality, it was a prey to deists, unbelievers, masons, divorcees, and lapsed priests who he said were swarming through it.[29] These were the men who, in Chateaubriand's equally hostile view of the educational legacy of the Empire, had sought (especially in the *lycées*) to undermine the authority of parents by imposing on them schools in which, "to the beat of a drum," their children were taught "irreligion, debauchery, contempt for domestic virtues, and blind obedience to the sovereign."[30]

Despite their virulence, such attacks would have had little effect had it not been for the reactionary turn in national politics after the Hundred Days. But, from the summer of 1815, the University needed resolute defense. It was fortunate in finding its champion in the person of Pierre-Paul Royer-Collard. By the time Royer-Collard emerged as the main champion and architect of the University under the Restoration, he had impeccable credentials as a man of the political middle ground who had abandoned his moderate revolutionary sympathies to serve the future Louis XVIII loyally during his exile. To these credentials he could add the virtue, in academic eyes, of knowing the inner workings of the University through his service as professor of the history of philosophy and dean of the faculty of letters in Paris from 1811 to 1814. During the First Restoration, with virtually the whole responsibility for planning higher and secondary education in his hands as a councillor of state and a trusted advisor to the grand master of the University, the relatively moderate comte de Fontanes, Royer-Collard came close to securing an administrative coup that would have transformed the French university system, almost certainly to its enduring benefit. His far-sighted proposal, elaborated with his young secretary and former colleague at the Sorbonne, François Guizot, was for seventeen regional universities, each with a large measure of independence, that would replace the monolithic imperial structure.[31] The necessary legislation was even signed by the king on 17 February 1815, but within days it fell victim to the disruption of the Hundred Days and was never revived.

With the University and its champions now unprecedentedly vulnerable to their detractors, Royer-Collard was fortunate to survive the passage to the Second Restoration. But, as the first president of the Commission of Public Instruction from August 1815 to December 1818 (the position of grand master now being suppressed), he was able to guide the institution through the most precarious three years of its existence. The royal decree of 15 August 1815, which granted the University "provisional" status pending a full-scale review, offered crucial breathing space, giving it time to prove itself an unthreatening body to which some measure of trust and independence could safely be restored. Eventually the trust bore fruit in the decision of June 1822 to confirm the University as a permanent institution and, two years later, to resurrect the title of grand master. The choice of Denis Frayssinous, Bishop of Hermepolis, as the new grand master bore an ambiguous message, however.[32] Could liberals be expected to see a priest known for his closeness to the most reactionary elements in Restoration society as a trustworthy defender of the University's autonomy? And, in any case, what did autonomy mean for professors whose status was now confirmed as that of obedient servants of an unremittingly clerical state?

Amid the prevailing suspicion of everything associated with the Enlightenment and its supposed consequences, the faculties of science might have been expected to fare badly in the early years of the restoration. But in the university's provisional phase, they remained relatively unscathed. Although the review of 1815 resulted in the closure of three of the empire's ten science faculties (at Besançon, Lyon, and Metz),[33] the reality was that the closed faculties (like the seventeen of the twenty-three faculties of letters that were also closed) had barely functioned. Even the chairholders in them probably observed little change in their lives: most simply reverted to what had in any case been their main occupations, usually as teachers in a *lycée* or *collège royal*. In the seven faculties of science that survived (Paris, Caen, Dijon, Grenoble, Montpellier, Strasbourg, and Toulouse), as in the six faculties of letters (Paris, Besançon, Caen, Dijon, Strasbourg, and Toulouse), the practice of holding two or more posts simultaneously—the much-criticized *cumul*—carried on much as before, with the tenure of a chair being seen as just one part of a life devoted to a greater or lesser degree of elementary examining and teaching, engagement with the economic and other needs of local elites, or (for the significant minority of professors who were ordained) priestly service.

The mix of duties also permitted research and critical inquiry. How far professors devoted themselves to intellectual activity was very much a personal choice, and levels of commitment varied accordingly. In Paris, where chairholders benefited from access to stimulating colleagues and better facilities, research and publication generally had a higher priority in professorial lives and yielded richer intellectual fruits than it did in the provinces. This is not to say that provincial professors necessarily worked in a cultural desert. Those with a will to integrate often enough found congenial, knowledgeable company in informal networks or the new and resurrected local academies and societies that had emerged, or reemerged, in all but the smallest communities since the destructive hiatus of the Reign of Terror. But assimilation in such cases could never be complete. Instances of good relations, even occasional collaborations, between professors and intellectual elites beyond the state system could not conceal the fact that administrative centralization had advanced inexorably since 1789. There was no getting away from the fact that professors were the employees of remote Parisian paymasters and that most of them were pursuing nationally rather than locally defined careers. And this set them apart. Especially in the provinces, they were seen as "new men," servants of the state with an authority that rested on a new hierarchy of formal qualifications and on the chairs they held rather than on the older academic virtues of broad cultivation and regional roots.

Patronage, Authority, and the Profession of Science

Despite the travails of the passage from Empire to Restoration, most of the leaders of French science survived, with at least the appearances of their authority intact. The careers of even the most notorious "weathervanes" of the day, Laplace and Cuvier, soon resumed their course, and Cuvier's, in particular, gained renewed momentum. He retained his chairs at the Collège de France and the Muséum d'histoire naturelle; re-emerged as a long-standing member of the Académie royale des sciences and its permanent secretary for the physical sciences;[34] followed Laplace into the Académie française in 1818; and rose quickly in the Commission of Public Instruction and its successor (from 1820) the Royal Council of Public Instruction.[35] There Cuvier exerted a powerful influence on the University, especially between 1819 (following Royer-Collard's resignation from the commission) and 1822, when the rightward lurch in politics, propelled by the comte de Villèle, led to a weakening of his position.[36] However, the formal badges of status do not tell the whole story of a transition that had its less triumphant side, even for those who survived. The select group of savants who had wielded almost untrammeled influence under Napoleon may have wanted to believe that little had changed. But as they resumed their hold on the reins of power, they found themselves functioning in a context that made authority of the kind they had enjoyed during the empire difficult to sustain. Even if the effects only became fully apparent in the 1820s and 1830s, there had been a quiet sea change in the relations between men of science and the machinery of state on whose support they depended. Dorinda Outram has analyzed what occurred as a "crisis" that was in due course to erode the system of personal patronage and authority that had been at the heart of some of the finest achievements of imperial science.[37] In the process, perceptions of the savant as a sage, with a high degree of autonomy, had given way to a recognition of the reality that men of science were, despite possibly possessing superior intellect, public servants.

In the case of Laplace, the first attacks on a savant whose standing in the physical sciences had seemed impregnable as recently as the spring of 1814 were launched early in the Restoration.[38] They were directed at a characteristically Laplacean program of physics, based on the imponderable property-bearing fluids of light, heat, electricity, and magnetism and on short-range, inter-molecular forces acting between both ponderable and imponderable matter.[39] The program had inspired Laplace's own research and that of the brilliant protégés with whom he had surrounded himself at his country house at Arcueil on the southern outskirts of Paris. And the results, in Etienne Malus's work on the polarization of light for example, were of enduring importance. But quite suddenly, as the empire fell, the program came under attack. An early criticism was voiced in a paper on the diffraction of light that Augustin Fresnel read to the Académie in October 1815. In giving convincing support to the wave theory, in opposition to the corpuscular theory favored by Laplace, Fresnel's paper signaled a turning point in optics and, more broadly, a body blow to the foundations of Laplacean physics. Immediately, the technical issues merged into a rethinking on a far broader front; in this, Petit, François Arago, and other younger physicists, played leading roles, along with the more senior Joseph Fourier, whose demonstration that thermal conduction in solids could be analyzed independently of any discussion of their cause served as a model of one possible way forward, the way of positivism.[40]

In a similar and related way, Berthollet, who was Laplace's close ally and next-door neighbor at Arcueil, witnessed a weakening of his doctrine of chemical affinities, also rooted in the notion of forces on the molecular scale; again, the assault came primarily from a rising generation, in this case from chemists who shared a growing openness to the atomic theory, which Berthollet opposed. In the life sciences, challenges took longer to manifest themselves. But the criticisms of Cuvier's conception of natural history by Etienne Geoffroy Saint-Hilaire were already simmering long before they erupted in a confrontation in the Académie des sciences in 1830, which is discussed in greater detail later in the chapter. As with the attacks on the approaches of Laplace and Berthollet, important technical questions, in this case of animal morphology, were at stake. But it was the overtones, of a clash between established authority enshrined in the work of a few powerful individuals and a new, more holistic conception of the living world, that captured public interest.

Although there was no single cause for the adjustments that occurred, an erosion of the power of the great patrons who had exercised control under the empire lies at the heart of the matter. In some instances, age played its part: by the end of the empire, Laplace and Berthollet were both in their mid-sixties, still energetic but (in conspicuous contrast with Cuvier) with limited futures ahead of them. These cases also illustrate the effect of a loss of material wealth: without the salaries they had enjoyed as senators, which contributed to total incomes of about 100,000 francs each in 1814,[41] Laplace and Berthollet could not hope to maintain what was in effect a privately financed research school at Arcueil.[42] Other factors were political: the standing of Laplace in particular seems to have suffered from a reputation (merited or not) for an undue compliance with the more reactionary policies of the Bourbon regime on the freedom of the press.[43] Others again were personal and concealed: as Outram has observed, the death of Cuvier's teenage daughter in 1827 seems to have diminished his enthusiasm for the weekly salons at which he received and influenced contemporaries of a wide variety of intellectual and political inclinations.[44]

It is too easy to represent these changing personal fortunes as part of the normal evolution of careers in their later stages. Clearly there was an element of this in the growing confidence of a younger generation ready to break with its elders. But behind the complex mosaic of interacting causes that applied in different ways in particular cases, more general trends in the tone of public debate served to make the leaders of French science (along with other figures in the cultural and political establishment) increasingly fair game for comment and challenge. The rise to prominence of a member of the scientific community as outspokenly critical of his more senior peers as François-Vincent Raspail, for example, would have been virtually inconceivable under the empire. But during the 1820s Raspail, as an independent researcher, an unqualified purveyor of medicine to the poor, and a committed republican, became the focus for a circle of mainly younger savants who did not see themselves as owing allegiance to any patron or formally designated coterie in the Académie des sciences or elsewhere.[45] The monthly journal *Annales des sciences d'observation*, which Raspail and the mathematician and kindred spirit Jacques-Frédéric Saigey launched in February 1829, stood during its brief and financially precarious life as a provocative alternative to the established journals of science.[46] Its commitment to breaking down the traditional barriers between scientific specialities and its openness to contributions that departed from academic orthodoxy (many of them by Raspail and Saigey themselves) gave the *Annales* an air of freshness. And its strategy was consistent and deliberate. Raspail knew precisely how far he was departing from the mainstream when he portrayed Lamarck as the gallant victim of adversaries who had oppressed him through the prestige of their positions and the favor of ministers, and he reveled in his campaign.[47] But repeated examples of wilful combativeness had their inevitable consequence in the enmity of the "scientific coteries" that he pilloried, and a downward spiral duly culminated in the journal's closure in June 1830.

Thereafter, through the 1830s, Raspail's depictions of "official science" became even more excoriating. A bitterly worded introduction and a twenty-three-page rant in the concluding chapter of his two-volume *Nouveau système de physiologie végétale et de botanique* (1837) pulled no punches.[48] They presented "ambition, greed, and jealousy" as the driving forces of academic career making, and the institutions of the capital as tools of the Parisian clique that exploited them. For Raspail, the Académie des sciences (with its mixture of incompetent charlatans and members whose record of good work ceased on their election), the Muséum d'histoire naturelle ("a kind of oligarchal republic"), and an officially appointed scientific élite of self-serving "pontiffs" corrupted by *cumul*, nepotism, and closeness to political power had all contributed to rendering science in France "stationary." In doing so they had contributed to silencing the opinions of the men and women of good will whom Raspail regarded as his true audience.[49]

While Raspail's attacks could be dismissed in the higher reaches of the scientific community as the fruits of politically motivated envy, they played their part in the continued undermining of the image of science as a remote, disinterested pursuit that traded only in facts and was untainted by personal rivalry. They were a sign, in fact, that science was being drawn ever more closely into the arena of broader public debate. Savants of a conservative disposition found the change and everything that contributed to it distasteful. As one such conservative, Laplace's loyal scientific disciple Jean-Baptiste Biot, maintained, once science was exposed to the gaze of the uninitiated, the routines of scientific debate would descend into a mere spectacle, and savants and science alike would become subjects for uninformed comment. And he spoke with feeling. His failure to win election as permanent secretary for the mathematical sciences in the Académie des sciences in 1822, in succession to the astronomer Jean-Baptiste Delambre, was made all the more painful by the publicity surrounding the election.[50] The tensions of his subsequent rivalry with Arago (whom he blamed for his defeat by Fourier) were likewise aggravated by their becoming common knowledge and by Arago's open encouragement of a number of younger academic scientists who broadly shared his leftist political views and disliked Biot's conformist stances in politics and religion.[51]

In a community in which by 1830 the worlds of the public and the private had become so closely entwined and in which a vigilant and relatively free press scrutinized science as it scrutinized all areas of public life, it was hard for savants to retain their old Olympian detachment. The intensifying rivalry between Cuvier and Geoffroy Saint-Hilaire after 1820, like the attacks of Raspail, showed how easily academic discussions spilled over into wider conflict about the rights of a remote elite, of which Cuvier remained a leading embodiment, to dictate doctrines and manipulate access to senior positions. The confrontation of early 1830 between the two men, conducted through a series of exchanges in the Académie, brought that conflict into sharp focus.

The purely scientific issue was straightforward enough.[52] In response to a paper on the anatomy of molluscs submitted in the previous year by two lesser-known naturalists, Saint-Hilaire advanced his ideas of an underlying unity of organization, in opposition to Cuvier's division of the animal kingdom into four distinct *embranchements,* or plans, those of vertebrates, molluscs, articulates (essentially insects and crustacea), and radiates (typified by starfish and jellyfish). But the boundaries of scientific debate were soon transcended. As in the literary "battle" provoked by Victor Hugo's play, *Hernani,* which in the same year marked a new assertiveness on behalf of Romanticism in reaction against the lingering classicism of the literary establishment, the exchanges immediately acquired overtones of a conflict between established authority (in this case Cuvier's) and the freedom to which members of the rising "Romantic" generation, including some prominent adepts of Saint-Simonian doctrines, aspired.[53] The very morality of "official" science was at issue, and, as its critics insisted, judgment had to be at the bar of lay debate. Science could no longer was regarded as the affair of experts alone.

The events of 1830 signaled the extent to which personal power in science, as in other areas of high culture, had already come to be weakened. The revolution that expelled the Bourbon line and brought Louis-Philippe of the Orleans branch of the royal family to the throne in July of that year served only to make the weakening irreversible. Of the patrons who had led French science to preeminence since the 1790s, Berthollet had died in 1821, Laplace in 1827, and Cuvier was soon to follow, a victim of cholera in the epidemic of 1832. While the transition did not entail the disappearance of savants able to exert influence, patronage after 1830 had to be exercised in a more formalized world of career-making and bureaucratic norms. In the new world, qualifications became increasingly important: one manifestation of this was the trend to implementation of the requirement (often hitherto ignored) that chairholders in the faculties should have the appropriate doctoral degree (see table 1). Even more decisive in their long-term effect were the weakening of the clerical presence in the University and a growing determination to assert the primacy of state over church in academic matters. Although these changes had roots in the restoration of the office of grand master and the additional designation of the grand master as Minister of Ecclesiastical Affairs and Public Instruction in 1824, they only became reality in 1832 with the creation of an independent Ministry of Public Instruction. There, notably under the Protestant Guizot, who was minister, with short breaks, for a total of almost four years between 1832 and 1837, educational policy took a resolutely secular turn. The consequences for Guizot and his successors were significant: through the rest of the July Monarchy ministers could never drop their guard against critics, including the comte de Montalembert and the still resolutely watchful abbé Liautard, who insisted that orderliness and morality were at risk from any administration that kept clerical interests so resolutely at a distance.[54] Yet the July Monarchy emerged as a relatively tranquil time for science. The existing institutions of education and research remained in place: administrative adjustments and some student troubles did little to change life at Polytechnique and its associated *écoles d'application*, and the Collège de France, the Muséum d'histoire naturelle, the Paris Observatoire, and the Académie des sciences likewise went their way much as before (despite one of the museum's periodic calls for urgent financial assistance in 1834).[55]

Where change did most obviously occur, albeit slowly, was in the University, now freed from the anxieties about its survival that had plagued it for much of the Restoration. The driving force lay with a succession of essentially benign Ministers of Public Instruction, especially Guizot, Abel Villemain (1839–40 and again from 1840 to 1844), Victor Cousin (for eight months between Villemain's two turns in the post), and the comte de Salvandy (1837–39 and 1845–48). More than any others in this period, these four ministers dedicated themselves to promoting the interests of the faculties and the *lycées*. A special concern for all of them was the Ecole normale, an institution that was in due course to do more than any other in raising the intellectual level and public reputation of both the secondary and the higher levels of the University. Under the July Monarchy,

TABLE I

Educational backgrounds of professors in the faculties of science on their first appointment to a chair, 1820–1879

Decade	Total number of new appointments	Number of new professors holding the *doctorat-ès-sciences*	Number of *agrégés*	Former *normaliens*	Former *polytechniciens*
1820–29	14	3	0	1	3
1830–39	38	25	4	7	4
1840–49	35	27	9	8	7
1850–59	53	48	24	22	4
1860–69	31	31	13	12	3
1870–79	73	70	32	33	2

Sources: For *doctorat-ès-sciences*: Albert Maire, *Catalogue des thèses de sciences soutenues en France de 1810 à 1890* (Paris, 1892). For *agrégation*: For the period to 1850, the list published in the *Annuaire de l'instruction publique* of 1851, pp. 283–96; for the later period, the lists published each year in the same *Annuaire*. The list in *L'Ecole normale (1810–1883)*, cited below, indicates when a former *normalien* was also an *agrégé* (which was almost always the case). For Ecole normale [supérieure]: *L'Ecole normale (1810–1883). Notice historique, liste des élèves par promotions, Travaux littéraires et scientifiques* (Paris, 1884), 83–157. For Ecole polytechnique: Charles-Philippe Marielle, *Répertoire de l'Ecole polytechnique, ou renseignements sur les élèves qui ont fait partie de l'institution depuis l'époque de sa création en 1794 jusqu'en 1853 inclusivement* (Paris, 1867) and P. Leprieur, *Répertoire de l'Ecole impériale polytechnique ou renseignements sur les élèves qui ont fait partie de l'institution depuis l'année 1854 jusqu'à l'année 1863 inclusivement. Faisant suite au répertoire de 1794 à 1853, publié par M. C.-P. Marielle* (Paris, 1867). For the identity of professors: The annual *Almanach de l'Université royale de France* (to 1848) and the *Annuaire de l'instruction publique [et des Beaux-Arts]* (from 1851), supplemented where necessary by the *Almanach royal [impérial, national]*

Note: I have excluded appointments to positions other than full chairs, e.g., to positions as *professeur adjoint* (common at various times in Paris) or *maître de conférences*. I have similarly taken no account of the temporary positions of *chargé de cours* or *suppléant*. Instances of professors who moved from one chair to another elsewhere were rare. I have counted such cases only once. This table is a revised and expanded version of table 2.3 in Robert Fox, "Science, the University, and the State in Nineteenth-Century France," in Gerald L. Geison, ed., *Professions and the French State, 1700–1900* (Philadelphia, PA, 1984), 76.

the school began its ascent to the position of eminence that it maintains to this day in the preparation of pupils for the fiercely competitive examinations of the *agrégation* and hence for entry to the top posts in both science and letters in the *lycées* and, for the ablest, in the faculties. The task was not easy: the financial resources for education were limited, and intellectual considerations, though never ignored, always had to take second place to the overriding demands of bureaucratic efficiency and political expediency. Nevertheless, dedicated administrators collaborated effectively with well-intentioned ministers in raising public instruction in the hierarchy of the state's priorities. By their concern for order and morality, they also did much to allay the fears of bourgeois families vulnerable to the siren calls of reactionary clerical forces campaigning against the Napoleonic legacy of state monopoly in education.

In their quest for public favor, ministers and administrators encouraged professors to be outward looking and, by participating in regional affairs, to demonstrate the value of faculties and *lycées* to their immediate communities. In science, as in letters, the results of the policy were a predictably mixed bag. In contrast with the faculties of medicine and law, those of science and letters remained the poor relations that they had always been. This was especially so in the provinces. Facilities in the provincial faculties of science in the 1830s and 1840s were little better than they had been during the Bourbon Restoration, and complaints about inadequate equipment and libraries, the lack of support for research, and damp cellars serving as makeshift laboratories became part and parcel of the daily correspondence that landed on the minister's desk from across France. Stories such as that of Grenoble's professors of physics, zoology, geology, and chemistry who shared a single laboratory with three assistants and the faculty's elderly *concierge* (as they did until about 1870) provided ample ammunition.[56]

Where expressions of dissatisfaction were loudest, they usually came from professors with intellectual aspirations that transcended the administration's bureaucratic perceptions of the academic role. So long as research was seen, as it was in the ministry, not as a priority but as an activity to be squeezed into a life structured around other, more important duties, there was always potential for conflict. Among those duties, examining for the *baccalauréat-ès-sciences* remained the core task. Even if the commitment in science never matched the grueling bouts of examining endured by professors in the faculties of letters, it ate into professors' time and was invariably tedious. In a thinly populated academy such as Grenoble, with a total of only 111 successful candidates for the *baccalauréat-ès-sciences* between 1811 and 1852, the load could be light.[57] But professors in the academy of Caen, where there were three ten-day sessions a year, as well as some shorter tours of duty to smaller centers, saw things differently: there, in the 1840s, between fifty and one hundred candidates were examined in science each year.[58]

Another professorial duty was teaching. By a statute of 16 February 1810, professors in the provincial faculties were required to give three lectures a week, each an hour and a half long, spread over nine months of the year, while those in Paris were to give two lectures a week over eight months.[59] The load was not excessive and became less so in 1841, when the Parisian stint of two lectures a week was made the norm for all faculties.[60] The problem was not so much the erosion of time that might have been given to more rewarding activities. For the great majority of professors in the sciences, it lay rather in the nature of the teaching that professors had to undertake. In the provinces, the number of candidates aspiring to the qualification for which the faculties were primarily intended to prepare—the *licence*, taken two or more years after the *baccalauréat*—was persistently low and showed no signs of picking up until the wholesale university reforms of the late nineteenth century: even in the relatively active academy of Toulouse an annual average of only three *licences* was awarded between 1822 and 1876.[61] The take-up for the doctorate was even smaller: in the whole of the provincial network a total of only 172 doctoral theses were written between 1811 and 1885 (compared with 548 in Paris).[62] Faced with small, often nonexistent, audiences of serious students, many professors in the provincial faculties turned to mounting lectures for the general public. Although such a choice was, in one obvious sense, a mark of failure, it aroused little concern in the Ministry of Public Instruction, where public lecturing was seen as a proper contribution to the politically important ideal of integration with the wider community beyond the university.

Precisely what profile of activities a professor chose to fashion reflected a measure of the freedom, however limited, that the July Monarchy's Ministers of Public Instruction tried to preserve for their employees, and perceptions of what constituted the academic life varied accordingly. Some professors found "decorative" lecturing rewarding enough (a point I make more fully in chapter 2). Others, though, found rhetorical displays for popular audiences hard to reconcile with their pretensions to membership of the international or even national community of science; such displays always sat uncomfortably (though not impossibly) with the requirement, expressed from the earliest days of the Imperial University, that professors should keep up to date with "the latest discoveries in science" and incorporate those discoveries in their teaching.[63] The obvious remedy was to seek to move on to the more congenial world of Paris, and advancement from a provincial to a Parisian chair remained a common professorial dream. In most cases, the dream was unrealistic. For it to be realized, a professor had to be known to the elite of the capital and both well qualified and well regarded in the ministry. And these were formidable hurdles.

The chemist Auguste Laurent learned the lesson the hard way. Six years in the chair of chemistry in the faculty at Bordeaux from 1839 to 1845 left him profoundly dejected. "I am dying of boredom here," he wrote to his friend Charles Gerhardt in June 1845. "I don't want to stay any longer. I must leave, whatever the cost, even if I have to make do with a job at 2,500 francs in Paris."[64] Two months later, in desperation (though encouraged by his recent election as a corresponding member of the Académie des sciences), he moved to the capital, initially on leave of absence from Bordeaux but from 1847 without even his faculty position to fall back on.[65] Although the life he now led on the margins of academic chemistry, until his death in 1853, was not easy, he was by no means ignored in Paris. In 1845 Claude Pouillet interceded with the Minister of Public Instruction in support of a special grant toward his research expenses, and Biot remained a steadfast champion who tried to persuade the members of the chemical section of the Académie to endorse the Collège de France's preference for Laurent as Jules Pelouze's successor in its chair of chemistry in December 1850 (following Gerhardt's withdrawal of his own candidature).[66] As it happened, the safer, more senior candidate, Antoine-Jérôme Balard, won the day and remained a powerful figure at the Collège de France for a quarter of a century.[67] Laurent saw the consolations for his dedication to chemistry as meager. The salary of 5,000 francs, simple lodgings, and the small, damp laboratory that came to him in his last years with a post at the Mint (the fruit of his intense badgering of senior figures in the Minister of Public Instruction and the Académie) fell far short of the recognition he craved.

Instances of the dissatisfaction and ambition for a Parisian post that drove Laurent are not hard to find. Laurent's friend and collaborator Gerhardt was in this respect a kindred spirit, ready to endure the financial hardship that went with a succession of leaves of absence from his chair in the faculty of science at Montpellier in order to work in the capital. The readiness of his rector and the minister to authorize his absences suggests leniency rather than harsh treatment. But Gerhardt's desire for time for his research was insatiable, and his desperation culminated, in 1851, in the hazardous decision to resign his chair and establish himself definitively in Paris, where he founded a rather unsuccessful school of practical chemistry for teaching laboratory techniques.[68] By then, a similar sense of grievance at a mixture of marginality and poor conditions had driven Armand de Quatrefages to leave his teaching position as *chargé de cours* in zoology at Toulouse in 1840 for the life of an independent naturalist that ended with his appointment as professor of natural history at the Lycée Henri IV in Paris in 1850 and to a chair at the Muséum d'histoire naturelle five years later.[69] Similarly in Rennes, fellow zoologist Félix Dujardin was seeking compensation for his own frustrations in extended periods of absence in search of a permanent move to Paris that never came his way (see chapter 2). Even those professors (the majority) who reconciled themselves to provincial faculty life and slotted into the circles of a local social elite rarely occupied the highest positions in society. Annual professorial salaries of between 4,000 and 5,000 francs left them trailing most other senior representatives of the national administration; they were less than half that of, say, a well-placed Ponts et Chaussées engineer. Hence unless a professor had family money or an auxiliary income, his comforts would be modest, and, except in the cost-free pursuits of local history or mathematics, neither his salary nor the funds available to him through his faculty would be sufficient to finance a serious engagement in research.

Even among those at the top of the academic hierarchy, time for research had to be carefully protected against competing demands and seductions. In Paris, examining and the other core duties weighed much as they did anywhere. For while there were more professors in the faculty of science at the Sorbonne (ten in 1840, plus a comparable number of adjunct professors and other teaching staff) than in the average provincial faculty (with a typical professorial complement of between five and seven), candidates for the *baccalauréat* were far more numerous. Paris, though, offered special rewards. There would always be at least a handful of serious students working for the *licence* or the *agrégation*, while the nonacademic public that made up most of the audiences in both science and letters would be larger, more assiduous, and better informed than was to be found in even the largest provincial faculty. Arriving in Paris as a young law student in the autumn of 1815, Charles de Rémusat was typical of those who flocked to hear the professorial "stars" of science, as he did for two years. Despite an education that had done little to expose him to science and inclinations and talents that drew him eventually to literature and liberal politics, Rémusat opted for the lectures of Biot and those of

Louis-Joseph Gay-Lussac and Louis-Jacques Thenard on chemistry, rather than those in the faculty of letters by philosophers or literary figures, such as Victor Cousin or Abel Villemain. And he did not regret his choice. Thenard, in particular, was captivating. As he spoke, according to Rémusat's memoirs, "It seemed . . . as though the veil of nature was being raised before my eyes. I felt that I was gazing for the first time on a world in which, until then, I had lived as though surrounded by magical mysteries or rather by senseless prodigies."[70] Nothing since his first lessons in philosophy had caused Rémusat such intellectual excitement.

Where declamatory skill was augmented by projected images or by demonstration experiments, such as the ones on electromagnetic induction that helped the physicist Claude Pouillet to attract more than eight hundred people to the Sorbonne in 1832, the excitement was ever greater.[71] Nevertheless, words mattered above all else, and science— like other intellectual activity—benefited from a taste for oratorical display that exploded during the Restoration and continued to grow under the July Monarchy. The phenomenon was one that Amaury, the hero of Charles-Augustin Sainte-Beuve's novel *Volupté*, noted on his return to France in about 1820 after an absence of three years in the New World. In Amaury's words: "What struck me most when I first returned from America . . . was that, after the Empire and the excess of military force which had been prevalent in that period, people had gone over to excess in words, to a wasteful and elaborate use of declamation, imagery, and promises, and to an equally blind confidence in these new weapons."[72] While Amaury's charge of wordy triviality could probably not be leveled at the great scientific lecturers (with the possible exception of Arago at the observatory[73]), the description evokes the tone of the performances for which the Sorbonne had become famous.

The capital also offered both the opportunities for additional earnings and an incomparable choice of intellectual peers. In Paris, *cumul* thrived, not only permitting a significant increase in income but also helping a powerful professor to extend his control over a discipline; on this point, Raspail's acerbic judgment on the practice was accurate. Gay-Lussac was one who achieved both financial gain and disciplinary power, especially the former, in this way: his mixture of professorships (at the Muséum d'histoire naturelle and the Ecole polytechnique) and administrative and advisory appointments (notably as Assay Master at the Mint) brought him an annual income of over 40,000 francs in the mid-1830s and more than 50,000 francs in the 1840s.[74] His friend and fellow chemist Thenard enjoyed similar success. A clutch of senior positions (at the Sorbonne, the Collège de France, and the Ecole polytechnique) and the title of baron that he received under Charles X and membership of the July Monarchy's newly created nonhereditary peerage earned him the status of a "grand notable," leading on in turn to prominent positions as an administrator and investor in railways and a variety of manufacturing enterprises.[75]

In any power base, the Académie des sciences was another essential element. The lure of full membership, which (in contrast with corresponding membership) was only available to those living in Paris, left its mark on a number of careers. A striking case was that of the mathematician and engineer Jean-Victor Poncelet, a savant of national standing whose ambition for a move from the provinces was fired explicitly by the prospect of election to the Académie.[76] Once in Paris, where he took up residence late in 1832 (while retaining his military rank and title of professor at the Ecole de l'artillerie et du génie in Metz), things moved quickly and on several fronts. By March 1834 he had succeeded Hachette in the mechanics section of the Académie, and over the next few years he gathered the distinctions of a major figure in Parisian science, including the newly created chair of physical and experimental mechanics at the Sorbonne (1837), the presidency of the Académie on two occasions (1840 and 1842), appointments to the rank of officer in the Legion of Honor (1837) and as examiner at the Ecole polytechnique (a position he held from 1835 to 1844), and finally (in 1848) election to a chair at the Collège de France.

The multiple attractions of Paris and its capacity to draw talent from across the country helped to confirm the city's preeminence among the world's centers for the pursuit of science. Charles Babbage's envious description of the "highly respectable, as well as profitable" situation of French savants in 1830 may have been distorted by personal resentment at his own difficulties in securing public support for his work.[77] Yet it conveyed the essential truth that the state's provision for science in France since the Revolution had provided opportunities on a scale that the British could only envy. This is not to say that opportunities were bestowed indiscriminately on all those who might have deserved them. Opportunities had to be seized, and careers had to be actively fashioned: in that respect, ambition and strategic decision making remained as important as ever if the heights of the French academic system were to be scaled. Where France differed from Britain, and differed markedly, however, was in possessing an institutional structure that offered openings to the top of the hierarchy for those with exceptional ability, commitment, and vision, while allowing lesser levels of competence to be rewarded in the form of respectable, if not glittering, academic advancement. By 1830, and a fortiori 1848, bureaucracy had taken a giant, and irreversible, leap forward. As it did so, the untidiness of career patterns in the age of personal patronage at the beginning of the century receded ever further into the mists of distant memory.

Science and the Industrial Age

The integration of scientific principles and practices in manufacturing had proceeded at such a pace since the Revolution that the economic changes accompanying the events of 1814–15 had inevitable consequences for the nation's science and engineering far beyond the confines of the academic world. The most important changes resulted from the sudden resumption of contact with Britain. Since 1792, except during the brief window of peace following the Treaty of Amiens in 1802, little information about the state of industry had crossed the Channel in either direction; in this respect, the situation was different from that for the sciences, in which some contact, albeit at a reduced pace, had been maintained.[78] But from the spring of 1814 a surge of travel and travelers' tales fed not only curiosity (on both sides) but also French anxieties about the progress of British industry in the years of war.

The laissez-faire economist Jean-Baptiste Say was among the first to see for himself, on a three-month official visit to England in the autumn of 1814.[79] His account, published in 1815 as *De l'Angleterre et des Anglais* and almost immediately in two further editions and in English and German translations, conveyed the dark as well as the light side of life in a country he had last visited in 1786–87.[80] Say found little to admire on this second visit to Britain, where crime was rampant and high prices imposed long hours on those who were able to work and acute material hardship on those who were not.[81] Nevertheless, while the prohibitive cost of books and the lack of time for reading might indeed have impoverished British intellectual life, there could be no denying the spectacular advances in the mechanization of industry that had allowed manufacturers in Britain to limit their dependence on costly labor and thereby to achieve enviable levels of profitability. Growth in the use of steam power encapsulated what had come about. In Newcastle and Leeds, Say had been amazed to see coal wagons drawn by steam-powered vehicles.[82] But it was the ubiquity of the stationary steam engine that most impressed him. Comparing what he had just seen with what he had seen a quarter of a century earlier, he wrote: "Everywhere the number of steam engines has multiplied prodigiously. Thirty years ago, there were only two or three of them in London; now there are thousands. There are hundreds of them in the large manufacturing towns, and they are even to be seen in the countryside. Industrial activity can no longer be profitably sustained without the powerful aid they give."[83]

Say was not alone in presenting technological prowess and a ruthless commercial spirit as Britain's salient characteristics. A few years later, Sadi Carnot used his study of the ideal heat engine, *Réflexions sur la puissance motrice du feu* (1824), to make a similar point about the transformation that the steam power had brought about in British industry, especially in mining. But the most remarkable of all accounts of industrial Britain were those of Charles Dupin, a brilliant product of the Ecole polytechnique (where he entered at the head of his year in 1801), a naval engineer who served during the Empire not only in French ports but also in the Ionian Islands (where he had been instrumental in setting up the Ionian Academy), and a polemicist whose voice was to be heard, on into the early days of the Third Republic, in the highest reaches of public life.[84] Dupin gained his knowledge of Britain through extended visits that he made there in 1816–17 and 1817–18, at his own expense (though with ministerial approval), and again in 1819, 1821, and 1824.[85] His first extended observations on Britain's industrial prowess, published in 1818, included an encomium of the steam engine as a main source of British prosperity that could equally have come from Say or Carnot: "It is in large measure to M. Watt that England owes her immense increase in wealth over the last century."[86] But Dupin's writings, especially his six-volume *Voyages dans la Grande-Bretagne* (1820–24), covered far more than the impact of steam power.

Even today, *Voyages* remains an invaluable source. It offered detailed accounts of Britain's main centers of industry and trade, with special emphasis on canals, roads, and trading ports, which Dupin saw as essential props of the country's success. Much of the discussion was densely statistical. But despite the book's resolutely objective tone, Dupin had to defend himself against the charge that he had overstepped the mark between detached analysis of France's manifestly successful neighbor and an unpatriotic "Anglomania." In reality, his patriotism was beyond question. Before the Royal Society in December 1816, he had shown his combative mettle by insisting that the practice of naval architecture had its most important origins in France.[87] Even more provocatively, in 1818 he reacted vehemently against the proposal of the ultra-Tory Lord Stanhope that the allied forces should extend their occupation of France.[88] Reactionary elements in the French government judged that this was a step too far, and copies of Dupin's *Réponse au discours de milord Stanhope*, first published in London and then reissued as a second edition in Paris, were confiscated, in the interests of harmony with Britain. But the *Réponse*, now an extreme rarity even in French libraries, helped to counter the charge of Anglomania that might otherwise have discredited his unwavering contention that France could only hope to recover the economic ground it had lost to Britain if it gave a far higher priority to the education of working people, at all levels from the basic three Rs to the highest reaches of science and its applications.

Dupin's belief in intellectual attainment as the backbone of a nation's strength was given free rein in 1827 in his *Forces productives et commerciales de la France*.[89] By now, as he argued, the indicators that France was well on the road to cultural as well as economic renewal were abundant. Indeed, it was the march of the mind in the years of constitutional monarchy that gave the greatest cause for pride and for hope that France might soon be ranked once again among the leading economies of Europe. A rising tide of publications—reflected in the number of printed sheets published in France in 1826 having been twice that for 1812—gave substance to Dupin's claim, the more so as the 1812 number included publications in parts of the Empire beyond France.[90] No less encouraging for Dupin, there had been signs of a shift from frivolous imaginative works and the socially undermining literature of the eighteenth century to more serious reading: a 5 percent increase in publications of a scientific nature between 1812 and 1826 gave him particular satisfaction, as a mark of the benefits of education.[91]

In his confidence in the key role of education and in the corollary of his belief in the power of science to advance industry and agriculture, Dupin was by no means alone. Outstanding among those (Say included) who broadly shared Dupin's opinion on the importance of education was Jean-Antoine Chaptal.[92] Writing from a perspective rooted in chemistry and a long association with the chemical industry of the Midi, Chaptal cast the two volumes of his *De l'Industrie française* (1819) as in essence a eulogy of the role that science had played in the nation's well-being since the Revolution.[93] It was a matter of special pride for him that science and the savants who practiced it had been the driving force of French resistance to the rigors of military blockade and isolation from crucial raw materials during the revolutionary and Napoleonic wars. While he conceded that practical ingenuity had played at least as important a role as science in advances in the mechanization of industry, it was science that had transformed agriculture and the "chemical arts." Only when chemistry had been established as a "positive science" had Berthollet, in his *Eléments de l'art de la teinture* (1791), been able to liberate a whole sector of industry from the limitations of uninformed trial and error simply by explaining the procedures of dyeing in chemical terms.[94]

Despite the encomium of science and a shared commitment to the goal of spreading enlightenment in the industrial community, Chaptal and Dupin differed on the best way of achieving that end. Chaptal's view was that of an older generation for which the way forward lay in the refinement of the age-old system of an elementary education in basic numeracy and literacy, followed by a scientifically informed but essentially vocational apprenticeship: the emphasis, as he maintained, should be on the transmission of specialized, immediately applicable knowledge. For someone who took this view, a natural model was the trade school for such skills as tailoring, shoemaking, and carpentry and the workshops for cotton spinning that the reform-minded duc de La Rochefoucauld-Liancourt had established on his estate at Liancourt in Picardy between 1788 and 1790. It was no coincidence, therefore, that early in the Restoration, Chaptal established an alliance with La Rochefoucauld-Liancourt, who after self-imposed exile in England, North America, and Germany in the 1790s had emerged as a valued member of Napoleon's advisory committees on industrial and agricultural matters (despite his rather cool relations with Napoleon himself).[95]

The setting for the alliance between Chaptal and La Rochefoucauld-Liancourt was the Conservatoire des arts et métiers. Since its foundation in 1794 and definitive installation, four years later, in the former priory of Saint-Martin-des-Champs, the Conservatoire had placed a fine collection of models, machinery, and scientific apparatus (largely made up of items confiscated after the Revolution) at the disposal of working men and offered them vocational instruction of the immediately exploitable kind that both Chaptal and La Rochefoucauld-Liancourt thought most effective.[96] When La Rochefoucauld-Liancourt was elected president of the Conservatoire's newly created administrative council in 1817, therefore, he was in his element. The institution's flagship school of technical drawing, which had survived the collapse of the empire and the inevitable ministerial scrutiny that followed, became a natural focus for his attention, and with his encouragement it flourished.[97] Elsewhere in the Conservatoire, La Rochefoucauld-Liancourt found a sympathetic, if less than dynamic, colleague in the person of Gérard-

Joseph Christian, recently appointed as director and a kindred spirit in the preparations for the national exhibition of French industry, planned for 1819. But it was Chaptal's appointment to the council in January 1819 that allowed the Conservatoire to punch its weight, even above its weight, in discussions about the role that scientific and technical knowledge should play in the national quest for economic renewal.

In these discussions, Chaptal spoke with a patriotism that had recently earned him a return to public life, despite his past closeness to Napoleon, during the short-lived liberal turn in politics under the duc de Decazes.[98] With respect to the 1819 exhibition, he also spoke with the authority of his experience as the chief organizer of the national exhibitions of 1801 and 1802, during his three years as Minister of the Interior. To achieve its aim of proclaiming France's reemergence as a major power, the 1819 exhibition depended perforce on royal and governmental approval. But, for Chaptal, it also bore a strong element of personal triumph.[99] By the time the exhibition opened, in the Louvre on 25 September, *De l'Industrie française* had been out for eight months and had come to be seen as a compelling analysis of the way forward for French industry. Although the book was not primarily about education, it left no doubt as to the author's educational priorities. Despite his general principle that governments should reduce their interventions in industry and agriculture to a minimum, Chaptal urged the creation of a network of advanced national schools for dyeing (at the Gobelins tapestry factory in France), metallurgy (under the national mining administration), and machine construction (at the Conservatoire des arts et métiers).[100]

As soon became apparent, the success of the 1819 exhibition and the personal prominence that came with the publication of *De l'Industrie française* were not enough to secure acceptance of Chaptal's proposals; the schools he advocated were never established. The explanation lies in a turn in the tide of the educational philosophy of the Conservatoire that quite suddenly raised the profile of science as the bedrock of economic success and made the traditional emphasis on highly focused trade skills appear old-fashioned. In the later months of 1819, the old guard, represented by Chaptal and La Rochefoucauld-Liancourt, lost its ascendancy to a younger generation, led by Dupin and supported by the scientifically eminent trio of Arago, Gay-Lussac, and Thenard. The new approach placed the emphasis on the scientific principles and theory underpinning technological practice, rather than on practice itself.[101] With remarkable speed, Chaptal's ideal of a workshop culture was abandoned, to his evident dissatisfaction, and the Conservatoire threw itself instead behind a program of Dupin's devising, modeled on the teaching of the Andersonian Institution in Glasgow.[102] In 1817, Dupin had been generously received at the Andersonian by the professor of chemistry and natural philosophy, Andrew Ure, and he had witnessed firsthand the benefits of the institution's cheap evening lectures for working men on geometry, mechanics, physics, and applied chemistry. The lectures had had, in Dupin's words, "astonishing" results.[103] The accompanying story of two brothers who had devoted such spare time as they could muster in their bakery to the construction of ingenious mechanical devices bore more than a whiff of fantasy. But it encapsulated Dupin's dream of what might also occur in France. At the time, however, he would have had little prospect of success. In a period still marked by political reaction, conservatives saw public lectures as insufficiently vocational and an invitation to disorder among young audiences. Also memories of a planned protest by Dupin against the exile of Napoleon's last Minister of the Interior during the Hundred Days, Lazare Carnot, in 1815,[104] followed three years later by his admiring biography of his "illustrious master" (and Carnot's fellow regicide) Gaspard Monge, still cast doubt on the depth of his loyalty to the Bourbon regime.[105] But with Decazes in power from December 1818 and moderately liberal values in the ascendant, Dupin took heart.

The window of opportunity proved brief; it ended with the right-wing reaction to the murder of the king's nephew, the duc de Berry, on 13 February 1820. Yet it was sufficient for Dupin to renew his proposals for wide-ranging lectures that, though open to all, were intended primarily for mature adults, rather than the mainly young men who attended workshop courses. With characteristically fine judgment, he secured Decazes's support for building a new lecture theater, at a projected cost of 85,000 francs, and pushed the project through to completion (see figure 1), albeit in the by now less favorable climate of late 1821.[106] As this happened, a committee appointed by Decazes (consisting of Arago, Thenard, and the industrial chemist Nicolas Clément) secured the signing of a royal decree of 25 November 1819 establishing courses in mechanics, industrial chemistry, and industrial economy (the more sensitive term "political economy" being judiciously avoided). The choice of Dupin, Clément, and Say, respectively, to teach the new courses completed the triumph of Dupin's educational philosophy.

Since the three new professors never concealed their hostility to the extremes of conservative opinion, their appointments brought with them the prospect of an uneasy existence during the six years of government by ultraroyalists (the so-called Ultras) under the comte de Villèle between December 1821 and December 1827. Throughout these years, the Conservatoire was a target for regular police surveillance, as a check on the subversive comments voiced at various times by all three professors and on the behavior of their liberally inclined audiences. However, the lectures survived. Some other areas of academic life were less fortunate. A prominent individual victim was Victor Cousin, suspended from his chair of the history of philosophy at the Sorbonne between 1820 and 1828. And collective sanctions led to a four-year closure of the Ecole normale, judged to be a dangerous seat of liberal agitation, in September 1822,[107] and four months of closure at the Paris faculty of medicine, suppressed after student disorder at the ceremony to mark the new academic year in November 1822.[108] The measures at the Ecole normale and the faculty of medicine reflected the strength of overlapping conservative and clerical influences that now prevailed in the university, even at a time when Frayssinous, in his capacity as grand master, was intended to provide the institution with some measure of immunity from political pressures.[109] The move against La Rochefoucauld-

Liancourt that led to his dismissal in July 1823 from his advisory and other positions at the Conservatoire and the *école d'arts et métiers* in Châlons makes a similar point with respect to technical education. It shows how even the moderate liberalism of a committed champion of constitutional monarchy could incur retribution.[110]

After the dark days under Villèle and a transitional period of intermittent toleration in the late 1820s, the corner was finally turned with the revolution of 1830 and the departure of Charles X. The bourgeois tone of Louis-Philippe's monarchy heralded a more tolerant, modern-minded regime in which the case in favor of scientific and technical education for the industrial world had scarcely to be made. In 1840 and again in 1848, Dupin felt obliged to respond to disturbing levels of working-class unrest in some industrial cities by dusting off earlier arguments about the value of education as a bulwark against communism and as a key to the recognition of the common interests of workers and employers.[111] But after 1830 the emphasis in public debate generally focused on the benefits rather than the dangers of an informed workforce and made such a defensive strategy unnecessary.

An immediate sign of the more supportive atmosphere for educational initiatives was the lack of governmental interference with the activities of the Association polytechnique. Formed in 1830 in the aftermath of the July Revolution, the association maintained a network of former pupils of the Ecole polytechnique who gave evening lectures in Paris on elementary mathematics, bookkeeping, and the applied aspects of the physical sciences for working men.[112] The impact of the Association polytechnique on industry is hard to assess; it was certainly limited by the loosely structured nature of the courses and, as at the Conservatoire, the absence of any diploma. But the healthy attendances at the lectures (as at those of the Association philotechnique, which emerged as a splinter group of the Association polytechnique in 1848) were proof of a growing taste for knowledge, including science and its applications, among the industrial classes, especially its younger elements.[113]

The vogue was by no means exclusively Parisian nor did it spring full-blown from the 1830 revolution. The public lectures on industrial physics that Eugène Péclet offered under municipal auspices while serving as a *lycée* teacher in Marseille (until his suspect political views led to the termination of the lectures in 1827)[114] and those on mathematics and engineering that Jean-Victor Poncelet delivered in Metz in the winters of 1827–28 and 1828–29 (in addition to his teaching as professor of mechanics at the Ecole royale de l'artillerie et du génie)[115] are just two instances of successful initiatives that laid the foundations for the expansion of technical and commercial education after 1830.

After 1830, with the Bourbon monarchs gone, the evidence of better times for industrially related instruction was plentiful. Among existing courses, those of the two *écoles d'arts et métiers* that were inherited from the empire, at Châlons-sur-Marne and Angers, had suffered during the restoration from a mismatch between the high number of young men emerging from them and the rather few outlets in the metallurgical and machine construction industries, for which they prepared particularly well.[116] In the short term, the schools could do little to adjust the imbalance, not least because of the relatively unindustrialized areas in which (in contrast with the mining school in Saint Etienne, for example) they were situated.[117] Through the 1830s, however, the slow proliferation of openings in railways, shipbuilding, and related sectors eased the problems of supply and demand and allowed an improvement in career prospects, helped significantly by a strengthening of the curriculum in mathematics and science. By 1843, things were going so well for the schools that a new one was founded at Aix, in the face of fierce competition from the rival cities of Toulouse, Nîmes, and Montpellier. As a source of expertise for regional industrial development, notably in the Paris–Lyon–Marseille (PLM) railway company, and for the technical services of the navy and merchant marine, the school soon justified its existence.[118]

The growth in the provision for scientific and technical education directed inevitable attention to the Ecole polytechnique. The school remained, in 1848, at the top of any conventional ordering of the hierarchy of academic prestige. But, by the same token, as in 1815–16, it also remained a focus for critical comment on the appropriateness of the preparation it offered for civilian life, a charge that continued to be leveled at the school well into the twentieth century. While the undiminished rigor of its entry requirements and curriculum still won awed respect, its graduates, along with those of its associated *écoles d'application*—in particular the Ecole des mines and the Ecole des ponts et chaussées—were now routinely caricatured as inept in the real world. The stereotype of the pale, impractical, but intellectually precocious *polytechnicien* who had sacrificed youth and eyesight to the struggle for admission to his school was never more tellingly exploited than in Honoré de Balzac's *Curé de village* (1839), in which the young Ponts et chaussées engineer Grégoire Gérard reflected bitterly on the effects of cramming for Polytechnique's entrance examination. In a long, pained letter to his godfather, he recalled his blighted youth in the 1820s:

> My future depended on my being admitted to the Ecole polytechnique. At the time, the work I did cultivated my intellect to excess. I almost died. I studied day and night. I perhaps subjected myself to an effort greater than my body could bear. I was resolved to pass my examination so successfully as to be certain not just of a place at the school but also of a reduction in the fees for studying there, an expense that I wanted to spare you. And I succeeded. I shudder now when I think of that appalling conscription of brains, delivered up each year to the state through family ambition. The resulting imposition of such a cruel burden of study at a time when a young man is completing his passage to manhood necessarily does untold harm. The result is the extinction, by lamplight, of precious faculties that would otherwise go on to develop and strengthen.[119]

Balzac's caricature rested on the usual mixture of truth and exaggeration. Under the Restoration (especially) and the July Monarchy, Polytechnique harbored a significant minority of students, most obviously those who helped to found the Association poly-technique, whose Saint-Simonian sympathies led them down the path of social amelio-ration and into the newer sectors of railway construction and other engineering ven-tures.[120] But such students remained a minority in a community whose educational and social mores, reinforced at the school, did little to foster a wholesale engagement with the burgeoning industrial economy. Hence Théodore Olivier, who had retained an ad-miration for Hachette and Monge since his passage through Polytechnique in the last years of the empire, was broadly correct in his observation (of 1847) that the rethinking of the curriculum in 1816 had served to marginalize the experimental sciences and tech-nical instruction in favor of an intensely specialized immersion in algebra and other areas of abstract mathematics.[121]

Despite Olivier's hostility, this abstract orientation of the curriculum would not have been seen as a failing by most *polytechniciens*, still less by their families. For once the golden age of the school's first twenty years had passed, the overriding purpose of the mathematical grind was neither fostering innovative intellectual activity (a point that Biot had already made with pained regret[122]) nor preparation for economically relevant employment. Mathematics at Polytechnique served above all to define an administra-tive and social elite. Not surprisingly, when Stendhal's eponymous hero in the novel *Lucien Leuwen* was dismissed from the school for republican agitation in 1832, his fam-ily's distress was first and foremost for his prospects of entry to the best salons.[123] A banker of the social standing of his father could only have deplored the damage that Lucien had done to his hopes of a good income (12,000 francs as a general inspector in the Corps des ponts et chaussées, for example) or the profitable marriage to which for-mer *polytechniciens* could traditionally aspire. And many of Lucien's fellow students would have shared his father's priorities. Especially after 1830, as the school's old radical-ism waned, *polytechniciens* became natural conformers.[124]

Polytechnique, in fact, was a complex, multifaceted institution, and the images of it that circulated between 1816 and 1848 lent themselves to exploitation in diverse and often contradictory ways. On the one hand, it had excelled in its contribution to the French war effort and had established a reputation for producing state engineers whose performance foreign observers from contexts as different as the Russia of Alexander I and the American military academy at West Point sought to emulate.[125] On the other hand, Polytechnique's image of cerebral otherworldliness made it a favoured butt of satire. In Balzac's hands, Grégoire Gérard's gaucheness became a way of criticizing the higher echelons of French bureaucracy, while Olivier's criticism, though accurate in its essentials, had a clear polemical purpose in the promotion of the Ecole centrale des arts et manufactures, the school for senior industrial engineers that he had founded in 1829, along with Alphonse Lavallée (a wealthy property owner who became the school's ad-ministrator), the chemist Jean-Baptiste Dumas, and (after his suspension as a lecturer in Marseille) the physicist and former *normalien* Péclet. In portraying Polytechnique as ill adapted to the needs of industrial society, Olivier was defining a gap that Centrale (where he taught for more than twenty years and was an influential director of studies from 1829 until 1836 and again in 1839–40) could fill by offering a school-based scientific and mathematical education as the foundation of the training for a new form of engineering elite.[126]

Faced with the characteristically French need to ensure that any program with an industrial orientation possessed the prestige that went with intellectual rigor and passage through an exclusive advanced school, the founders of the Ecole centrale devised a three-year course in which elements of artisanal practice and high-flown mathematical theory were brought together in a unified "industrial science." The source of both the concept and the term seems to have been Dumas.[127] For him, the otherworldly abstractions of the Polytechnique syllabus had no place at Centrale. Nevertheless, science lay at the heart of Dumas's conception: like Olivier and Lavallée, he believed that the scientific nature and breadth of the curriculum gave *centraliens* their adaptability and so distinguished them from the more specialized foremen and other industrial employees who would work under them. In practice, it was not easy to strike a balance between avoiding specialization and maintaining a common focus in a curriculum whose core disciplines ranged over physics, chemistry, mechanics, and descriptive geometry, taught by professors with loyalties to the traditional demarcations between their various subjects that were sometimes hard to break. Yet a distinctive, scientifically rooted but concrete style was aspired to and, in large measure, achieved.

The fount of the style, and the key to Centrale's originality, was an intricately structured syllabus that started with a foundation year of science and mathematics, provided in the early years by Olivier in descriptive geometry, Péclet in "industrial physics," Dumas in chemistry, and a succession of teachers—Daniel Colladon, Gustave-Gaspard Coriolis, and Joseph Liouville—in mechanics. Armed with this foundation, *centraliens* then moved on to two years of visits to factories and other industrial sites, laboratory practice (especially in chemical analysis), geological field trips, and preparation for an independent exercise in design, usually of a manufacturing installation or a piece of machinery, that formed an essential part of the final examination. The pace was tough. Olivier used a punishing curriculum, strict discipline, substantial fees (800 francs a year, in addition to a minimum of 1,200 francs for accommodation and living expenses), and a high dropout rate (far higher than at Polytechnique) to reinforce the aura of excellence to which he aspired. Such toughness, as he saw it, was indispensable if pupils with intellectual and social backgrounds on entry very similar to those of *polytechniciens* were to be transformed into doers whose attainments were measured in the machines, railways, canals, bridges, and harbors that they constructed.[128] The priorities of *centraliens* were intended to be, and were in reality, the material manifestations of technology rather than the more philosophical articulations of technocratic ideology of the kind voiced by the Saint-Simonian and former *polytechnicien* Prosper Enfantin.[129] It was characteristic of these priorities that former *centraliens* assumed a leading role in the new professional body for nonmilitary engineers, the Société des ingénieurs civils de France, from the moment the society was founded in 1848, and that they were happy to be recognized by the hundred or so medals that they won at the Exposition universelle of 1855, to which they could add no fewer than ten awards of the Legion of Honor.[130]

Such accolades do not imply that public respect for the *centralien* model of "techno-logical man" (to invoke the title of John Weiss's study of the early history of Centrale) was easily won. Lavallée still regretted, in 1847, that if industry in Britain remained more advanced than in France, the fault in his own country lay in the "false vocations given to youth by classical education."[131] In a similar vein, and in the same year, Olivier associated his denunciation of the dire influence of the Ecole polytechnique with an auxiliary com-plaint that the culture of the French educated classes since the restoration had become predominantly literary.[132] In some degree Lavallée and Olivier were right. Most children of bourgeois families at the midcentury would have gone through a secondary educa-tion sanctioned by the classically based *baccalauréat-ès-lettres*, in which long years of apprenticeship in Latin would have gone hand in hand with the acquisition of mathe-matical dexterity and an immersion in the French literary tradition, to the exclusion of any significant engagement with science. Indeed, only the minority of candidates who chose to go on to the *baccalauréat-ès-sciences* after taking the literary *baccalauréat* would have studied science at all seriously. But even if science in 1847 did not occupy a domi-nant place among the cultural attainments of bourgeois France, the place it occupied was important. Also, and more pertinent for my argument in this section, science since the end of the First Empire had gone beyond what Olivier saw as its worthy but limited function as a "recreation for the mind" or a force for the "refinement of human intelli-gence."[133] By then it also had the status of an essential foundation of material well-being and hence as a worthy pursuit for savants who had too often been seen (if seen at all) as self-indulgent spinners of intellectual webs.

In all this, Olivier and his collaborators in the Ecole centrale had not acted alone. Throughout the July Monarchy, they had received assistance from the Ministry of Commerce, in the form of financial grants and help with the organization of the en-trance examinations and the distribution of publicity about the school.[134] They were also part of the broader movement in favor of technical education to which Dupin and many other champions of the cause contributed as well. From its roots in the Restora-tion, that movement progressed to a new level in the July Monarchy. It culminated in a major report on ways of promoting vocationally oriented scientific education at the sec-ondary and primary levels[135] and a flurry of exchanges between Dumas and the Minister of Public Instruction, Salvandy, about ways in which the curricula of the Paris faculty of science might be reformed to incorporate teaching relevant to industrial and commer-cial employment. The proposal for a program of practical instruction in mechanics at the faculty, which a committee under Dumas's chairmanship submitted to the ministry in 1846, was particularly daring. But, along with recommendations for new chairs in applied subjects and a *licence* and doctorate in mechanical science, it was favorably received in the ministry and might well have been implemented but for opposition in sections of the press and from a minority of professors wedded to the traditional view that faculties should remain aloof from vocational subjects.[136] The opposition was sufficient to cause delay and, in turn, the abandonment of the scheme in the turmoil of the revolution of 1848. The failure was a decisive one that left France with important unfinished business. It meant that under the Second Empire, the scientific and industrial communities had to return yet again to the task of adapting the educational system to modern needs. The already thorny debate about how best to achieve the difficult match between education, vocational training, and employment was far from over.

A Philosophy for the Times: The Roots of Positivism

The growing prominence of science in the intellectual and economic life of postrevolutionary France attracted attention far beyond the communities immediately involved in promoting or coping with its day-to-day consequences. On the most abstract plane, analysis and commentary by philosophers conveyed the impact of the methods of science, nowhere more so than in the tradition of positivist thought that began to assume a coherent form during the Empire and continued as a staple of philosophical debate for the rest of the century and beyond. For Claude-Henri, comte de Saint-Simon, who stands as the main fount of nineteenth-century positivism in both its purely intellectual and its social sense, the only true test of reality was provided by science, and the progress of mankind consisted in subjecting all the problems of man and society to an inquiry that took no account of unobservable entities or such metaphysical notions as first and final causes. Saint-Simon's position had been honed in a tortuous intellectual itinerary pursued in parallel, first with a military career in the late 1770s and 1780s and then with several years of industrial speculation and travels in England and continental Europe. It also bore the stamp of his prolonged, and vain, attempts to break into the professorial circles of the Ecole polytechnique and the Ecole de médecine. By the time he published his first work, the *Lettres d'un habitant de Genève à ses contemporains*, in 1803, he was already in his early forties, and the essentials of his ideas, replete with the marks of his past, were in place.

In the *Lettres*, Saint-Simon voiced an extreme form of the Enlightenment's infatuation with science. For him, as for Voltaire, Newton was the secular god appropriate to the modern age. It was before Newton's tomb that Saint-Simon wished to see a subscription opened to provide support for the twenty-one "men of genius" who would make up a Council of Newton. Beneath this council there would be a hierarchy of lesser councils, each responsible for constructing a temple containing a mausoleum dedicated to Newton and a supporting structure of laboratories, libraries, and provision for education. Of the twenty-one men of genius—the "elect of humanity"—twelve were to be mathematicians and scientists, the others being men of letters, painters, and musicians.[137] The cultural balance in the plan was significant. For savants, in Saint-Simon's eyes, especially mathematicians, were superior to the rest of mankind, and they alone were worthy to wield the spiritual power that served, in his new utopia, as an essential counterweight to the temporal power vested in the property-owning class. Emphatically, they were to be the sages, not the servants, of society.

The limitations of Saint-Simon's scientific knowledge have often been commented on. But the fact that he was virtually self-taught in science and mathematics did nothing to undermine his wish to belong to the arcane world of geometers, astronomers, and physicists that, in the event, he never succeeded in entering. For this failure, the undisciplined character of his writing must take most of the blame, although his absence of tact certainly compounded his difficulties. One manifestation of this was his palpably eccentric assessment of the work of Laplace. While he accepted that the author of the *Mécanique céleste* deserved a place in his "temple of glory," he denied Laplace the highest status, that of a philosopher, because of the specialized nature of his work. Laplace's scientific achievements were flawed, too, by an excessively derivative and hence unpatriotic adherence to the principles laid down by Newton; he was left, in Saint-Simon's judgment, as possessing talent that was only "of the third order."[138] Such views, expressed at a time when Laplace's reputation was at its height both within and outside the scientific community, sat oddly, not to say suicidally, with Saint-Simon's decision to present his *Introduction aux travaux scientifiques du XIXᵉ siècle* to several members of the Bureau des longitudes, which was in effect a fiefdom of Laplace and his circle. The few polite notes of thanks that Saint-Simon received in return were small consolation, and soon financial hardship compounded a neglect of his writings that the collapse of the empire did nothing to reverse. The vigor with which he preached his ideal of a society led by savants (seconded, in his later writings, by an elite of *industriels*) remained undimmed in his last work, the *Nouveau Christianisme*, which appeared a month before his death in 1825.[139] But by then his hopes of recognition beyond his immediate circle of adepts had disappeared. Saint-Simon was, if anything, less well known than he had been fifteen years before, and his mantle as the leading exponent of scientistic philosophy in France had passed to his former secretary, Auguste Comte.

For his deeper understanding of the sciences and for the greater (if still limited) impression he made on the scientific community of his day, Comte is a more significant figure than Saint-Simon. When he entered the Ecole polytechnique in August 1814, he was ranked fourth, ahead of Jean-Marie-Constant Duhamel and Gabriel Lamé, both of whom went on to distinguished academic careers as mathematicians.[140] In fact, his first thought on arriving in Paris from his conforming Catholic home in Montpellier was that he would advance in due course to the rank of professor and stay at Polytechnique for the rest of his life. Very soon, however, the disruptions arising from the Hundred

Days and the pupils' hostility to the Bourbon regime made serious work difficult, and Comte's own career as a *polytechnicien* came to an abrupt end with the blanket dismissal of the school's pupils in the spring of 1816. As someone who had been centrally involved in the insubordination that precipitated the dismissal, Comte was a marked man, but it was above all his characteristic mixture of pride and obduracy that prevented him from returning to take the examinations after the school reopened in January 1817. With his initial ambitions in ruins, it was not long before a meeting with Saint-Simon and his appointment as Saint-Simon's secretary in August 1817 enlarged his thoughts from the highly focused Polytechnique curriculum to mathematics and science in its broader philosophical aspects.

Comte's intimacy with Saint-Simon lasted for more than six years. When he finally broke with Saint-Simon in April 1824, he did so partly because of a growing personal animosity. But the break also reflected a disenchantment with his master's emerging conception of a "new" Christianity capable of winning over the adepts of Catholicism and Protestantism. However secular the new religion might be in its tenets and rites, it represented, for Comte, the abandonment of the scientific core of the true Saint-Simonian philosophy and a flagrant retreat from the position that Saint-Simon had articulated in the *Introduction aux travaux scientifiques du XIXe siècle*. There, the constraints that religion had placed on the progress of mathematics and science had been castigated with uncompromising vehemence: "It is easy to demonstrate beyond doubt that the progress of the human mind in the mathematical and physical sciences in recent centuries is entirely due to the waning of belief in God."[141]

Looking back in 1824 on such an unqualified condemnation of the effects of religion, Comte saw the new Saint-Simonian religion as a straightforward betrayal of a high principle. Saint-Simon, however, saw only consistency. He insisted that the religious order to which he aspired would mark a break with the Christian past and that, as a distinctive new departure, it would foster rather than impede the progress of human understanding. Comte, by contrast, maintained that, if there was to be religion, its scientific foundations had first to be laid far more securely than had been done so far.

The divergence between Comte's highly intellectualist early positivism and Saint-Simon's later ideas, with their streak of mysticism and ever closer identification with social, economic, and, above all, industrial advancement, quickly widened into a chasm that neither man cared to cross. Comte had begun to articulate his own philosophy publicly in 1819, by which time he seems already to have believed that he had nothing more to learn from Saint-Simon.[142] By April 1822, he had taken a further crucial step toward independence by setting down the "fundamental" principles of positivism in the fifty or so printed copies of a "Prospectus des travaux nécessaires pour réorganiser la société."[143] The "Prospectus" contained a first statement of the fundamental Comtean notion of the three "states" through which all branches of human knowledge had necessarily to pass: the "theological" or "fictitious" state (in which natural phenomena were ascribed to the actions of supernatural beings), the "metaphysical" or "abstract" state (in which explanations invoked unobservable metaphysical entities), and the "positive" or "scientific" state (in which understanding rested exclusively on reason and observation).[144] The reference to astronomy, physics, chemistry, and physiology as sciences that had already progressed to "positive" status and divested themselves of their theological and metaphysical past was a further pointer to the full-blown hierarchy of the sciences that Comte was to elaborate in the six volumes of his *Cours de philosophie positive* between 1830 and 1842. In this later hierarchy, extending from mathematics (the most positive) to "social physics" or "sociology" (the least), the position of the sciences was not identical to that of the "Prospectus." But the criteria by which they were ordered in the "Prospectus" were essentially those of the *Cours.*

Through the 1820s, fragmentary writings and correspondence and lectures on his ideas to small but select groups of sympathizers marked the steady maturing of Comte's philosophy. His first, apparently successful course of lectures, in April 1826, was attended by three members of the Académie des sciences—the naturalist and traveler Alexandre von Humboldt, the zoologist Henri Ducrotay de Blainville, and the mathematician Louis Poinsot.[145] When he resumed the lectures in January 1829, after a period of mental instability probably brought on, and certainly aggravated, by overwork, Blainville and Poinsot were joined by the engineer Claude Navier, the physiologist François Broussais, the alienist Jean-Etienne-Dominique Esquirol, the mathematician Jacques Binet, and (the biggest catch of all) Joseph Fourier, one of the Académie's two permanent secretaries. It is far from clear whether those following the lectures could in any significant degree be regarded as adepts of the Comtean philosophy. In Fourier in particular, however, Comte saw a physicist and mathematician with the inclinations of a true positivist. Fourier had won his spurs, in Comte's eyes, through the extreme caution he had displayed with regard to the physical reality of caloric, the matter of heat, in his *Théorie analytique de la chaleur* (1822). The book's opening lines set the tone of the analysis of heat flow in a solid body that followed. In a statement that Comte can only have applauded, Fourier wrote: "First causes [*causes primordiales*] are wholly unknown to us; but they are subject to simple unvarying laws which can be discovered by observation and the study of which is the object of natural philosophy."[146]

In volume 1 (1830) and, at greater length, volume 2 (1835) of the *Cours de philosophie positive,* Comte paraded his admiration for Fourier's stance.[147] He presented what he called "physical thermology" as having long been imprisoned in its metaphysical stage by theories (of which the caloric theory was typical) founded on "illusory" assumptions about "imaginary fluids or ethers."[148] Fourier, in marked and significant contrast with Laplace, had shown how, in this area at least, physics might achieve a state of "positivité définitive" and so, by eschewing any discussion of the philosopher's traditional quest for causes, be set on the path to "the true, definitive state of human understanding."[149] Although Comte laid claim to a philosophical and even (through the lectures of 1826 and

1829) personal closeness to Fourier, he never suggested that he had influenced Fourier's position. The views of both men, in fact, can be better interpreted as facets of broader philosophical trends that gathered strength during the Bourbon Restoration. Hence, while Comte and Fourier articulated their insistence on the need for scientific statements to be subjected to the rigors of empirical testing with particular clarity, their position was far from unique. The agnosticism that Comte's contemporaries at the Ecole polytechnique, Lamé and Duhamel, maintained in their own mathematical studies of the conduction of heat in the later 1820s and 1830s is a good example of a similar stance. Lamé's use of the adjectives "positive" and "metaphysical" to describe, respectively, reliable and unsound knowledge in his *Cours de physique* in 1836 may simply reflect recognized usage: "positive" in particular had long been quite widely employed in the sense in which Lamé used it (to characterize reliable, experimentally verified knowledge). Nevertheless, the prominence of the categories of the "positive" and the "metaphysical" in Lamé's methodological credo in the preface of the *Cours de physique* suggests at least an openness, even an indebtedness to Comtean ideas, albeit an openness or indebtedness that (if this is the case) he studiously concealed.[150]

Unease about unobservable entities in physics in the 1820s and 1830s had its counterpart in chemists' attitudes to atoms. Jean-Baptiste Dumas's growing caution with regard to the existence of atoms through the early and mid-1830s can be explained by circumstances unrelated to any formal adherence to positivism. As a young man in the 1820s, Dumas was sensitive to the support that new evidence, in particular Dulong and Petit's law and the work of Berzelius and Mitscherlich, gave to the atomic theory. This was what led him to undertake measurements of the vapor densities of a number of substances, including some—phosphorus, sulfur, and mercury—that were not normally in the gaseous state.[151] At the time, Avogadro's hypothesis concerning the equal spacing of the particles of all gases at any given temperature and pressure led Dumas to expect that the vapor density of any elementary gas or vapor would be proportional to its atomic weight. Anomalies, though, abounded. Writing in 1828 he nevertheless suppressed his doubts and displayed an untroubled confidence in the atomic theory.[152] But once he returned to his vapor-density work in 1832, the anomalies became intolerable. Four years later he spoke skeptically about the whole notion of atoms in a series of lectures on the history and present state of chemistry at the Collège de France. Speculation about atoms should be regarded as "no more than a hypothesis and on this subject too many hypotheses have already been made."[153] The way ahead lay not in a wholesale rejection of hypotheses, but in the search for "reliable foundations on which to base substantial theories."[154] This was not the prescription of a positivist, and there was certainly nothing formally positivistic about this or any other aspect of Dumas's philosophy. Rather it was a manifestation of the same doubts about existing physical theories that characterized the work of Fourier, Lamé, and Duhamel.

It might have been expected that Comte's teachings would attract keen attention in decades—the 1820s, 1830s, and 1840s—in which scientific theorizing was so strongly marked by the trend to caution.[155] But while the *Cours de philosophie positive* found a prestigious scientific publisher in Blanchard and was certainly not ignored, Comte in these years never made the mark in public debate to which he aspired. In this as in other respects, the years in which Comte was at his most active and creative as a philosopher were years of frustration, aggravated by personal difficulties and a gathering sense of failure. The unhappy marriage that began in 1825 haunted him until he separated from his wife in 1842. Professional and eventually financial insecurity also bore down on him. Appointments as a teacher of mathematics in a private school, as *répétiteur* in analysis and rational mechanics (essentially a teaching assistant to Navier) at the Ecole polytechnique in 1832, and as one of the entrance examiners there in 1837 brought him a reasonable income of about 10,000 francs a year.[156] And the annual series of popular lectures on astronomy that he delivered in the town hall of the third arrondissement from 1831 to 1848 gave him the emotional satisfaction that came with large audiences.[157] As he recounted with obvious pleasure to John Stuart Mill, four hundred people sat through his three-hour inaugural lecture for the series in January 1843, and not one left the room; in the following year, when the same lecture was spread over four separate sessions, he reported the same mark of audience satisfaction.[158] But such accolades did little to assuage his sense of exclusion from the higher reaches of the scientific community. His failure on four occasions to secure the chair that he thought his right at the Ecole polytechnique weighed more heavily than any material hardship.

Comte's first disappointment at Polytechnique came in 1831, when he was a candidate for one of the two chairs of analysis and rational mechanics; a second failure followed in 1835, when he competed unsuccessfully for the vacant chair of geometry. But the greatest blows arose from his attempts to secure the chair vacated in 1836 by the unexpected death of Navier. Not only had Comte assisted Navier in this chair; he had also performed, to all appearances with some distinction, as Navier's temporary replacement, pending a new appointment.[159] When the chair was first filled, in 1837, personal friendship and a measure of qualified admiration, as well as the consolation of the position as entrance examiner (secured for him by Dulong), eased the pain of the appointment of Duhamel, rather than himself. But three years later, the death of Poisson and Duhamel's consequent promotion to succeed him as the school's permanent leaving examiner set the scene for another crisis. With the chair vacant once more, Comte advanced his candidature with conviction and, this time, with at least a show of confidence. Disdaining, characteristically, the tradition of academic visits in search of votes, and basing his claims to the chair on his record as a teacher rather than in research, he looked for no endorsement beyond that of the current generation of *polytechniciens*. But the support of this "body of noble youth," to which he laid claim in a forceful letter to the president of the Académie des sciences, Poncelet,[160] was insufficient to avert the inevitable. In an election in which the three nominated members of the Académie des sciences played a decisive advisory role, the chair went to Charles Sturm, a competent if not outstanding candidate who had the crucial advantages of membership of the Académie and of sharing the left-wing sympathies of Arago, a key figure in the appointment.

Comte's failure in 1840 stands as the clearest evidence of the weakness of his personal position, as well as of his philosophy, and he took it badly. Nothing could shake his conviction that he had been the victim of prejudice and improper procedures. That reading his letter about his candidature at a meeting of the Académie on 3 August was curtailed, in the run-up to the appointment, confirmed his resentment against the academic establishment, especially against Thenard, who seems to have been responsible.[161] The distribution of printed copies of his letter and its publication in the *Journal des débats* on the following day may have released some tension. The truth is, however, that Comte was reaping the consequences of two decades of self-destructive behavior. Throughout this period, he saw his comments on the scientific worth of certain of his contemporaries as a mark of the fearless honesty on which he prided himself. But what passed in his mind for plain speaking was seen by others as gauche handling of his personal relations even with those whom he should have courted as his closest allies.

The ease with which Comte crossed the boundaries between frankness and a destructive disregard for his own interests is nowhere better illustrated than in his erratic judgments of Duhamel and Lamé, the two men who might most effectively have advanced his cause. When Duhamel was up for election to the geometry section of the Académie des sciences in 1833, Comte endorsed his friend's candidature but did so with damningly faint praise: although he judged Duhamel to be "indisputably superior" to the other candidates (Guglielmo Libri, Sturm, and Joseph Liouville), he added (in a letter to Blainville) that "this is not saying much."[162] In the previous year, Lamé had suffered a scarcely less withering dismissal, provoked by Comte's view of himself as a possible candidate for the position of the Académie's permanent secretary for the physical sciences. With typically misplaced combativeness and arrogance (for someone who was not an academician and who worked on the fringes of the academic elite), Comte had insisted (also in a letter to Blainville) that he did not see himself as scientifically inferior to Lamé or to others intent on climbing in the scientific community.[163] In spite of this, Comte's worst fear—that the Académie might appoint Félix Savary, another near contemporary at Polytechnique, who was rumored to be running strongly for the position—proved unfounded. Dulong, a figure of indisputable eminence, was elected, though only to serve for one year before the administrative burden of the task led him to resign.[164]

A decade later Comte judged that Lamé had redeemed himself. By speaking up for Comte within the Ecole polytechnique, Lamé now had the status of an "old friend and comrade," a man distinguished, both by his personal loyalty (to Comte) and by his philosophical position, as an outstanding candidate in the forthcoming election to membership of the Académie des sciences. Writing in January 1843, Comte urged Blainville to vote for Lamé, rather than the more senior Jacques Binet (a veteran of the lectures of the 1820s):

> Among our present-day geometers, it is M. Lamé whose mind is most open to true philosophical inspiration and most capable of appreciating the scientific or logical value of studies relating to the leading branches of natural philosophy. Such merit is becoming too rare in our day for it not to be taken into serious consideration. In a word, M. Lamé appears to me to surpass on every count his estimable rival, whose mathematical work is really limited to some secondary advances in integral calculus.[165]

The election, in early March, duly went Lamé's way.

It is hard to know how far the clumsily self-interested assessments of Lamé and others that Comte conveyed in correspondence became public knowledge. But they probably did, and in any case there was much in print that conveyed the often damaging nature of his evaluations of his peers. The respect he repeatedly expressed for the phrenologist Franz Josef Gall, the "illustrious Gall" as he called him,[166] is a case in point. In two respects, Gall's ideas lent reinforcement to the model of social development that Comte elaborated in the fourth volume of the *Cours* in 1839 and then more fully, more than a decade later, in the *Système de politique positive*.[167] First, in insisting on the growing preponderance of the affective over the intellectual aspects of human nature, Gall provided an explanation of social evolution in the past and justification for believing that, in the future, human egoism would continue to lose ground to a developing social sense. Second, and far more plausibly than Lamarck, whose evolutionary doctrines Comte rejected, Gall showed how the intellectual and moral attainments of a civilization might be transmitted from generation to generation through, at least in part, a mechanism rooted in cerebral structure: even if the association between particular functions and specific structures in the brain had yet to be fully worked out, he offered the possibility of a scientific explanation of society's inexorable, if often retarded, passage through the three Comtean states.

Given such a debt, Comte's declared admiration for Gall was understandable, and there was something admirably, and characteristically, honest about it. Yet it remained an injudicious move. Gall's failure to win election as a corresponding member of the medicine and surgery section of the Académie des sciences in 1821, despite his claim to have been strongly backed by Geoffroy Saint-Hilaire, was a sign of a marginality that was only compounded by the very public manner in which he stated his case.[168] An indiscretely impassioned manifesto, in which he laid claim to a huge popular following, displayed a bravado with which Comte would easily have identified.[169] But the ability to draw audiences totaling between fifteen thousand and twenty thousand for his lectures over a period of twenty years, even if true, was something that few academicians would have seen as enhancing Gall's case. Both for the eccentricity of his views and for his populist style of self-promotion, Gall was certainly not someone to whom an ambitious careerist aspiring to enter the scientific establishment could be advised to turn. Comte could scarcely have made a worse tactical choice; of that, Cuvier's well-known judgment of Gall as a charlatan should have been warning enough.[170]

As the *Cours* moved toward completion, Comte's disappointment and sense of victimization surfaced repeatedly in correspondence and in print, and they became an all too damaging hallmark of his philosophical campaign. The immense effort that he expended in composing the last three volumes of the work, between 1839 and 1842, also left its mark in a mixture of exhaustion and depression that raised the specter, though not the reality, of a return to the mental disorder he had endured in 1826. In the circumstances, contemporaries who had long regarded Comte as an unstable eccentric found it easy to ascribe the extraordinary "Personal preface" of the final, sixth volume of the *Cours* to the overblown bitterness of a disappointed outsider. The preface took the form of an attack on the French scientific establishment as a whole and on the Ecole polytechnique in particular. A personal protest against the precariousness inherent in his dependence on the annual renewal of his positions as *répétiteur* and entrance examiner at Polytechnique led on to a withering criticism of the "pedantocracy" that had systematically done him in.[171] The chief villain and the incarnation of the rampant self-interest that Comte despised was now named publicly. It was Arago, whose pernicious influence as permanent secretary for the physical sciences in the Académie, malignly reinforced by the power he exerted at the highest levels of the Ecole polytechnique, had denied Comte his rights, especially in the election of Sturm to Duhamel's chair in 1840.

In a vituperative footnote to the preface that encapsulated his indignation, Comte leveled a charge of unambiguous ferocity: "Every informed person already knows that the irrational, oppressive measures adopted at the Ecole polytechnique over the last ten years have their origin principally in the disastrous influence of M. Arago, the loyal and willing spokesman for the passions and aberrations of the circle over which he now deplorably exercises such power."[172] It was a foolhardy attack. But Comte's determination to publish his denigration stood firm against the contrary urging of his wife (now on the point of leaving him) and what was manifestly his professional self-interest. It also ignored the request of his publisher, Blanchard, that the preface should be suppressed. Blanchard's first reaction to Comte's intransigence had been to withdraw from publishing the sixth volume altogether. With Arago's approval, however, he decided to go ahead, though with an inserted "Avis de l'éditeur" that is absent from most surviving copies of the *Cours*.[173] In this, Blanchard made public his regret at Comte's refusal to withdraw an attack on someone who was not only an eminent savant but also (though this was left unsaid) a source of patronage as the secretary of the Bureau des longitudes, for whose official publications Bachelard was responsible. Comte's victory in the subsequent lawsuit against Bachelard, on the grounds that he knew nothing of the inserted text until he received his first copy of the book, gave him fleeting satisfaction: the court ordered that Blanchard should withdraw his disclaimer from all unsold copies of the *Cours* and cover the costs of the case.

Over the next six years, victory in the affair of the preface did nothing to lessen Comte's hostility toward the elite of the French scientific community whom he believed had rejected him. In the *Discours sur l'ensemble du positivisme*, published in July 1848 in the aftermath of the revolutionary days of February, he looked back as bitterly as ever on the long history of "persecution" that had followed the publication of the *Cours* and impeded the completion of his other great work, the *Système de politique positive*, the first volume of which eventually appeared in July 1851.[174] By then, his sense of grievance could also feed on the eventual loss of his position as entrance examiner at Polytechnique in 1844.[175] What was in effect a dismissal, provoked by the school's decision to institute an annual turnover of examiners, reduced his comfortable income of over 10,000 francs by more than half and left him for the first time in real financial difficulty. Duhamel and other friends did what little they could for him against the unrelenting animosity of Arago and another Comtean bête noire, Liouville. So too did Emile Littré, a young disciple who had recently embarked on his life of loyalty to positivism. From abroad, four British sympathizers—John Stuart Mill, George Grote, Raikes Currie, and Sir William Molesworth—proffered material relief by making good the income that Comte had lost. But the relief was temporary and presaged the precariousness that blighted Comte's existence until his death in 1857.

There was much about Comte's last years that betokened a profound personal failure. Even his belated recognition that he had been unduly severe in his pillorying of Arago[176] and in his denunciation of Poinsot for reneging on an earlier promise to support him in his candidacy for the chair at the Ecole polytechnique in 1840[177] did nothing to improve his relations with the scientific elite. A man who had been relentless in his attacks on the leaders of science could expect no favors, and he received none. Nevertheless, Comte had articulated some of the pressing issues of the day with regard to the relations between science and society. His quest to promote a secular Religion of Humanity beginning in the 1840s (see chapter 4) responded to a more general questioning of the place of traditional faith in a world in which science was asserting its own claims to be the road to truth and human salvation. And he was similarly of his time in his concern that science should not remain the preserve of academicians and others with the labels of conventional academic attainment. This concern had already been apparent in his participation in the founding of the Association polytechnique in 1830. But the revolutionary climate of 1848 led him to a new sense that events might at last be running in his favor. Perhaps the time had come for him to direct his efforts more resolutely than ever to the working men whose loyalty in sitting through his three- or four-hour lectures in February, March, and April 1847 had been an inspiration to him.[178]

The resulting initiative for the creation of an Association libre pour l'instruction positive du peuple on 25 February 1848 (the day when the massed revolutionaries of Paris marched on the Hôtel de Ville to secure the overthrow of Louis Philippe) and the transformation of the association into the Société positiviste eleven days later were rich in political overtones. No contemporary could misread the naked populism of Comte's renewed ambition to break with traditional forms of instruction that depended on the passivity of the audience and, in his words, privileged "oratorical cleverness"; instead, as he insisted, his society would concentrate on the sparking of discussion and popular debate.[179] The bedrock of this new engagement, was the Bibliothèque du prolétaire du dix-neuvième siècle, an ideal positivist library whose contents Comte laid down in 1851. The 150 titles were chosen as a way of making the accumulated patrimony of humanity available to all and of thereby promoting the "social consensus" on which progress depended.[180] Divided into four loosely defined categories—sixty volumes in history, and thirty each in science, poetry, and what Comte classed as "Philosophie, morale, et religion"[181]—the library contained few surprises. Comte's own writings were prominent under both "Science" (along with those of Buffon, Broussais, Lavoisier, Condorcet, Lazare Carnot, Lagrange, Bichat, Blainville, Navier, and Poinsot, among others) and "Philosophie, morale et religion" (among works by Aristotle, Saint Augustine, Cabanis, Gall, and Barthez). And two favored authors, Thomas à Kempis and Dante, parts of whose writings he commended for daily reading in accordance with his own established practice, were singled out for special mention.

As Annie Petit has observed, the choice of books conveyed a mixture of elitism (expressed in the almost total absence of works other than recognized masterpieces in the European tradition) and a naive conviction that even the most untutored mind would benefit from direct exposure to the recommended texts without the aid of commentaries.[182] Such attitudes were Comtean to the core. And they encapsulated key reasons for the obstacles that Comte encountered in so many of his ventures. Many would have sympathized with his view of the Académie des sciences as a self-perpetuating closed world whose members could be relied on to protect any of their own or their clients' privileges that might be under threat: the Académie was indeed a seat of centralized power whose smallness (only sixty-three resident members of its eleven sections in 1848) and sense of solidarity made accommodating unorthodox ideas difficult. The caution of academicians with regard to such ideas, however, should not be confused with a refusal to engage with them. The fact is that, for all its exclusiveness, the Académie worked well and showed few of the weaknesses that led to the more justified criticisms leveled at it in the later nineteenth century. Comte, of course, saw things differently. In 1842, his resentment at his exclusion from the higher echelons of French science led him into an intemperate and quite misguided attack on the Académie's weekly printed reports on its proceedings, the *Comptes rendus*, which he dismissed as a "tedious compilation that is degenerating steadily into a routine display of our most trivial academic vanities."[183] Tragically for his reputation, to a degree that he was incapable of perceiving, that same "tedious compilation" was the envy of learned societies across the world.

If the Académie was corrupted by the presence of his sworn enemies, where else was Comte to look? One resource was the working-class public he hoped to reach through the Bibliothèque prolétaire. The other was the body of "true philosophers," those he saw as free from academic corruption. The problem was that, within this group, few of his early sympathizers were still active in 1848. Of those who had been close to him in the 1820s, only Blainville survived and he died in 1850. If the Comtean legacy was to be preserved, the flame had therefore to pass to a younger generation, and so it did. In the 1840s the mantle as Comte's main philosophical champion fell on Emile Littré. It was Littré who, from the time he encountered Comte and his ideas in 1840, did more than anyone to maintain the visibility and respectability of positivism in the face of the master's mounting eccentricity. In Littré's hands and those of Charles Robin, Georges Pouchet, and a small circle of other positivist savants, Comte's teachings were trimmed of some of their extravagance: the populism was played down, and the grandiose aspirations to a new social order were abandoned or (as in the case of the religion of humanity) redefined as admissible but not essential parts of the positivist program. To that extent, positivism was strengthened as a philosophy and guaranteed a continuing place at the heart of public debate through the Second Empire and on into the Third Republic.

It would be inappropriate to end without a comment on the bearing of the unprecedented attention that was paid to science and its applications and methods between 1814 and 1848 on the common belief that France's scientific reputation began to dip during this period. What are we to make of this belief when the institutional structure of French science remained strong and expansion (as measured by the number of chairs and other posts, for example) proceeded steadily, if unspectacularly? The most obvious development in the 1830s and 1840s was the emergence of a challenge to Parisian preeminence from other centers, notably the invigorated research-oriented universities in the German states and, in Britain, a reformed Royal Society and the modernized Mathematical Tripos in Cambridge. Challenges from abroad, however, should not be confused with faltering French achievements, which remained as impressive in many areas of the physical and life sciences as they did in France's continuing domain of excellence, mathematics. If that is so, what weight should we give to the vituperative observations of Raspail or the sense of material and intellectual privation among the more ambitious professors in the provincial faculties? Clearly the complaints smacked of special pleading bred of self-interest. But we can also see in them the first signs of a justified concern about the state of French science that was to grow in the mid and later nineteenth century. Meager funding was a reality. So too, and no less damaging, was advancing academic bureaucratization. And both told, in the long term, against the interests of science. Even if, in 1848, the effects were still limited, a powerful educational administration that valued obedience and efficiency above flair had already begun to cast its shadow. One of its greatest failings was to have fostered a reward system that judged increasingly by paper qualifications and length of biddable service. Another was to have encouraged career making and intellectual exchanges within a predominantly national context, with limited concern for the wider international community. These weaknesses will return as leading themes in future chapters.

CHAPTER TWO

Voices on the Periphery

• • •

Parallels with other countries lend an air of familiarity to the experience of French devotees who did their science, usually without formal qualifications, outside the designated institutions of research and teaching. The early marginalization of independent contributors to the mathematical and physical sciences and a continued strength in disciplines that relied on fieldwork and painstaking recording were evident in France, as they were in all the leading scientific nations. Likewise, the extensive informal networks and voluntarist organizations that are recognized to have contributed much to the observational sciences, notably though not exclusively in Britain, had plentiful counterparts in France. While such parallels exist, it is important to work beyond them. One reason is that the diffuse French realm of autonomous and largely provincial activity still tends to be obscured in a secondary literature heavily oriented to the work of well-known savants and what critical contemporaries referred to, more or less pejoratively, as *la science officielle*, an "official" science that had its being in higher education and state-sponsored research. As a result, much remains to be said not only about what went on in the independent world of learning but also about the relations between center and periphery in a country whose educational administrators always aspired to central control, even if they never wholly achieved it.

In this independent world, there were few nationally known household names, and aspirations were very different from those who held posts under the Ministry of Public Instruction or the other ministries with academic responsibilities. The highest aim of independent savants was to add to knowledge, especially in natural history, observational astronomy, and meteorology and do so as part of the loose but extensive network of like-minded practitioners that existed in even the remotest corners of France. An active participation in science or an informed appreciation of it also brought social status. It bestowed, above all, a mark of intellectual superiority that helped to justify middle-class ascendancy in a century of profound reordering in French society. To use Edmond Goblot's classic terminology, science at the level at which it was consumed and practiced by the serious devotee fitted easily into the sum of sober cultural attainment that allowed the "barrier" surrounding bourgeois status to be breached.[1] Thereby, it gave access to the "distinction" (again, the word is Goblot's) that defined the modern bourgeoisie of the nineteenth century.

So science on the periphery had both a familiar and a less familiar, peculiarly French character. While it is beyond question that science, like other forms of intellectual refinement, did much to effect or reinforce the stratification of society in many countries, the political and demographic contexts in which it fulfilled that role in France were distinctive. Most strikingly, the repeated attempts of French governments to bring the domain of learned culture under their wing provoked fierce resistance from more aggressively independent savants. Such savants saw Parisian ministries in general and the Ministry of Public Instruction in particular not as sources of support but as seats of administrative power with malign predatory aspirations. The resulting tension was a prominent leitmotif of French science for much of the nineteenth century. It was a tension that flared easily into open strife, setting the extreme champions of decentralization in the circle of the Catholic legitimist Arcisse de Caumont repeatedly at odds with the ministerial officials whom he and his allies regarded as their oppressors. The conflict turned on political and social cleavages running far deeper than the struggle for cultural authority that was their outward manifestation. The most zealous spokesmen for the provincial cause, Caumont chief among them, sought nothing less than the return to an old, prerevolutionary order. In the specific realm of learning, their ideal was an updated form of the eighteenth-century hierarchy headed by the great Parisian academies and sustained from below by local academies and societies and learned communities across France. It was an ideal profoundly at odds with the nineteenth-century pattern of advancing ministerial ambition.

It is easy to disregard such aspirations as the fruit of vain nostalgia. Certainly, they had little chance of fulfilment in the face of the centralizing tendencies of administrations of every political hue: ministerial gestures of accommodation and cooperation never allowed for the degree of autonomy that Caumont believed to be essential for provincial vitality. Nevertheless, through the nineteenth century and on into the early twentieth century, independent science and scholarship accumulated real achievements. They left a legacy not only in thousands of volumes of journals and other publications but also, more diffusely, in a nonacademic public ready to pay its subscriptions to a society, visit a museum or exhibition, or purchase one of the growing number of periodicals and books of popular science. By 1900, and a fortiori by 1914, the political ideology that fired Caumont's campaign had lost its force: it was already weakening by the midcentury and, following Caumont's death in 1873, it could scarcely hope to weather the inclement atmosphere of the resolutely modernizing, secular Third Republic. Nevertheless, the periphery remained populous and, in a limited number of activities, productive enough to remind ministers and administrators that even in the twentieth century it still had and deserved a place, however reduced, in the French scientific landscape.

Academies and Societies

The aspiration to restore the forms of academic life that had been shattered in the turmoil of the Revolution and the Terror bore its first fruit well before the end of the Napoleonic Empire. The chief beneficiaries were the great national academies, restructured and regrouped from 1795 in accordance with the moderate postrevolutionary constitution of the Year III, as constituents of the Institut national des sciences et des arts. But the wider world of learning also had its successes, drawing on deep roots in the provincial academies and other societies and in the circles of private collectors, travellers, and other *notables*. By comparison with what occurred in Paris, the successes were few and hard-won, and even in the early 1820s, after two decades of patchy resurrection of the old order of science and scholarship in the provinces, much remained to be done. With the Bourbon line definitively restored, the revival of the titles and traditions of even the most insignificant academy or society might boost morale and foster a sense of continuity with the past. Yet only the most purblind nostalgia could obscure the fact that, at a deeper level, there could be no going back. The provincial academies of the ancien régime that had struggled back into existence felt the draft of change keenly, having to reconcile themselves almost invariably to the position of impecunious postulants rather than cultural flagships. Most of them depended for their rooms and minimal facilities on municipal generosity. In return, all they could offer were gracious words and services in kind as custodians of libraries, botanical gardens, and collections of specimens, coins, or antiquities that had once belonged to them or to one of their members but that revolutionary confiscation had turned, in many cases, into public property.

The position was an undignified one, made more so by the corrosive perception among most nonmembers that such academies had had their day. Even explicit criticism of them was rare: the target was simply too easy. One rising young lawyer in Rouen, though, was moved to attack, presumably in search of a reputation for constructive acerbity. This was Gustave Rouland, who was later to become Minister of Public Instruction for seven important years under Napoleon III. Writing in the literary *Revue de Rouen* in 1835, Rouland castigated academies in general and that of Rouen in particular for their self-satisfied detachment from the modern world. In an indictment that squared with the generally progressive tone of the *Revue*, he portrayed industrialization, the popular vogue for science, and new departures in literature as having made next to no mark on the impervious world of academicians. "Poor academies! [Rouland wrote] . . . they slumber in their Utrecht velvet armchairs like Egyptian mummies deep in their granite sarcophagi. The world goes its way, human thought expends its boundless energy. All to no effect. The provincial academies are concerned only with their eternal rest."[2] Each year, as Rouland had it, academicians would don their ceremonial dress and address verse, antiquarian tidbits, and harangues in the manner of Cicero to such members of the local elite as could be persuaded to attend. Their permanent secretaries would flatter members with assurances that they and their peers were great men. And then the sanctuary would close for another year, bequeathing a volume of learned disquisitions, paid for with funds scraped together by the academy, to a posterity whose indifference would ensure the resulting publication an untroubled tranquil life.

There was much of the young Turk in Rouland's tone. But defense against his carica-ture was difficult. A nuanced, though by no means uncritical, comment by Charles Louandre (whose upbringing as the son of a librarian and archivist in Abbeville had left him with clear insights into the virtues and failings of cultural life in the provinces) could do no more than insist on the great diversity between the intellectual attain-ments and organizational capacities of different academies. As Louandre observed, the ablest of the members of the most active academies had played their part in the then gathering interest in regional history and archaeology and in the promotion of scientific studies geared to industrial and agricultural improvement.[3] To that extent, Rouland's withering judgment that the Académie royale des sciences, belles-lettres et arts de Rouen—the only one he had been able to observe at all closely (though not, at his age, as a member)—had wilfully turned its back on the nineteenth century may be judged overdrawn. The fact remained, however, that in its general thrust his characterization of the conservative proclivities of academicians and the air of lethargy that surrounded them struck home.[4]

Rouland's aim was especially accurate where literature was concerned. A decade be-fore he wrote, the academicians of Rouen had displayed a tentative sensitivity to chang-ing literary taste when they debated Romanticism's recent challenge to the classical tra-dition in which, to a man, they had been brought up.[5] Convoluted points of definition were made in search of common ground between the classical and the Romantic camps, for academicians were by nature creatures of compromise. But when the elderly judge and president of the academy, André-Nicolas-François Adam, spoke of the pain that the attacks on "age-old Parnassus" by the partisans of Romanticism had caused him,[6] his fellow academicians would at least have understood his sentiment, and most would have shared it. What grieved Adam was less the finer points of literature and art than the wider assault on conservative values dear to an aging, declining generation.

The pain that Adam experienced in the mid-1820s would have been felt even more acutely by someone of his cultural disposition by 1835, when Rouland delivered his on-slaught. By then, at least some academicians had come to acknowledge the inevitable. Speaking in 1840, the president of the Caen Academy, François-Gabriel Bertrand, could scarcely have been franker in his recognition of the diminished status of the academies in the world of learning. And in Caen his was a voice to be heeded. He was the head of provincial France's oldest formally constituted academy (dating from 1652, only seven-teen years after the Académie française) and a Norman by birth, lifelong residence, and emotional allegiance. Suspicion of authority seen as emanating from Paris and threaten-ing the individual character of his province ran in his veins. But he was also dean and professor of Greek in the town's faculty of letters, a position that gave him the authority of someone familiar with the new order of academic professionalism. Hence it was both as a citizen with deep local roots and as an employee of the national university that he evoked a seventeenth-century past that could never be retrieved. That past—of the eminent theologians and pioneers of the Caen Academy, Pierre-Daniel Huet and Samuel Bochart—had gone for ever, as Bertrand acknowledged:

> It is not, gentlemen, that I believe that any learned society, however eminent the men who compose it, can expect to emerge again with a role as important as that of the old academies, or to attract public attention in the same degree. The days of Huet and Bochart are already far behind us. . . . It is no longer in academies alone that the questions confronting the most intelligent sections of society are discussed. The affairs and ideas that forcibly engage minds are debated elsewhere.[7]

In taking a position that was realistic rather than rancorous or wholly pessimistic, Bertrand was accepting with some dignity the pattern of inexorable decline. Nevertheless, he still claimed for his own academy and others like it a significant if localized role in promoting the public good. At the very least, as he argued, academies provided the elevated neutral terrain on which differences could be conducted with a courtesy and impartiality that would at once counter the exaggerated claims of utopian system builders and disarm the old order's refusal to countenance change of any kind.[8] The claim elaborated a familiar conservative position, founded on an ideal of harmony that would permit controlled social amelioration while holding at bay the diverse counterthreats of governmental inertia (or misguided zeal), popular ignorance, and the gathering tide of populist radical politics. As a good academician, Bertrand was also extending the notion of utility from the conventional terrain of material advancement to embrace the refinement of taste and morality that came with the pursuit of science, literature, and art. It was a high-minded patrician vision that would have pleased the elderly audience of the Caen Academy but left Rouland and others of his modern-minded persuasion unmoved in their skepticism.

In science, as in other areas of learned culture, the most prominent challenge to the provincial academies came from the University. In the six provincial faculties of science that emerged from the reorganization in the aftermath of the Empire (at Caen, Dijon, Grenoble, Montpellier, Strasbourg, and Toulouse), there were, in 1816, a total of twenty-seven chairs. Thereafter, under the Restoration and the July Monarchy, new or revived faculties were opened in Lyon (1833), Bordeaux (1838), Rennes (1840), and Besançon (1845), and expansion there and in the existing faculties brought the number of scientific chairs in the provinces to sixty by 1848 (see table 2).[9] Where the professors appointed to the faculties also taught in a *collège royal* (as more than one-third of them still did in the early years of the Restoration[10]) or were otherwise already established in regional society, their assimilation was relatively painless. Local sensibilities were flattered rather than ruffled, for example, by the appointment of Jacques-Armand Eudes-Deslongchamps to the chair of natural history in the faculty at Caen in 1825. A native of Caen who had become immersed in zoology while serving as a ship's surgeon in the last years of the empire, Eudes-Deslongchamps had practiced medicine in the town since

TABLE 2

Chairs in faculties of science in Paris and the provinces, 1816–1914

Year	Number of faculties	Number of chairs			Percentage	
		Paris	Provinces	Total	Paris	Provinces
1816	7	9	27	36	25	75
1827	7	9	27	36	25	75
1852	11	18	63	81	22	78
1877	15	19	98	117	16	84
1902	15	25	129	154	16	84
1914	15	36	165	201	18	82

Sources: For 1816: *Almanach royal pour l'année bissextile M.DCCC.XVI* (Paris, 1816). For 1827: *Almanach de l'Université royale de France. Année 1827* (Paris, 1827). For 1852, 1877, 1902, and 1914: *Annuaire de l'instruction publique [et des beaux-arts]* for the relevant years. This table is a revised and expanded version of table 2.3 in Robert Fox, "Science, the University, and the State in Nineteenth-Century France," in Gerald L. Geison, ed., *Professions and the French State, 1700–1900* (Philadelphia, PA, 1984), 70.

qualifying as a doctor in 1818.[11] Once he was appointed to the chair, he proceeded to combine his professorial duties not only with teaching in Caen's *collège royal* but also, for fourteen years, with the directorship of the municipal botanic garden and, from 1845, with that of a newly formed public museum of natural history. The museum harbored one of the most important of all collections outside Paris, bringing together the collections assembled by Félix Lamouroux[12] and Eudes-Deslongchamps and important fruits of the Pacific voyages of the explorer and naval officer Jules Dumont d'Urville, as well as the chaotic though not unproductive legacy of the doctor and first holder of the Caen chair, Henri-François-Anne de Roussel.[13] It bore striking witness to the capacity of an ambitious provincial town to build on the legacy of the ancien régime through a continuing policy of opportunist acquisition.

An immersion of such centrality in the life of Caen, reinforced by a sustained involvement in the area's *sociétés savantes*, made Eudes-Deslongchamps a leading figure in his corner of the world of learning, where he enjoyed celebrity as the leader of natural history in Normandy, a true "the Norman Cuvier." A correspondence with Geoffroy Saint-Hilaire and his election as a corresponding member of the Académie des sciences in 1849 demonstrate that his work was known and respected on a larger stage as well; certainly his publications (mainly on the fossils of the Calvados) and his contacts beyond Normandy set him far above Gustave Flaubert's comic enthusiasts Bouvard and Pécuchet, with their undiscriminating passion for collecting, or even Homais, the pompous pharmacist in *Madame Bovary*, who proudly bedecked himself, as many others did, with the label of "member of several learned societies."[14] The fact remains, however, that Eudes-Deslongchamps's roots and profile of activities in the life sciences remained those of a zealous and productive auxiliary rather than a national leader. His world was busy but irremediably provincial. As he put it in a letter to Geoffroy Saint-

Hilaire in 1833: "I remain immersed in my mollusks from that other world. Withdrawn in my shell, I study, draw, and describe the ancient organic remains which, despite the accumulation of centuries since they were laid down, remain almost entirely unpublished and hence unknown to naturalists."[15] His modesty was genuine but so too was the sense of his worth, however limited, in the grander scheme of natural history.

By the midcentury, naturalists with Eudes-Deslongchamps's primary commitment to the institutions of learned culture in his own region were becoming, if not less numerous, at least less influential, as chairs went increasingly to outsiders holding national qualifications, in particular the doctorate (see table 1 in chapter 1).[16] To such "new" men, with their primary loyalty to distant employers in the Ministry of Public Instruction, membership in a local academy would typically appear as an agreeable honor, though one that demanded little beyond occasional attendance at a meeting or the even more occasional offer of a paper. The Académie des sciences in Bordeaux, for example, held no attraction for Auguste Laurent in the six unhappy years that he spent as professor of chemistry in the city's faculty of science between 1839 and 1845, and most new arrivals in provincial chairs seem to have felt as he did. A rare exception, as it happened, was Laurent's successor, Alexandre Baudrimont. Though a stranger to the region and as someone who had originally hoped to make his career in Paris, Baudrimont was distinctly more amenable to provincial ways, to the point of being one of the very few professional academic scientists who accepted membership of the Institut des provinces, the resolutely independent counterpart to the Institut de France that Arcisse de Caumont launched in 1839.[17] Even in this case, though, it was only when his highly original views on molecular structure (including an exploration of the implications of Avogadro's gas hypothesis) had been published in the more usual outlets, such as the *Comptes rendus* of the Académie des sciences, that Baudrimont began to make use of local societies and their publications. Once he did so, however, he immersed himself in the intellectual life of Bordeaux with unreserved commitment. He became a regular contributor to the Académie des sciences, belles-lettres et arts of Bordeaux and a pillar of what began in 1849 as a student society but soon, as the Société des sciences physiques et naturelles de Bordeaux, assumed real importance as a meeting place for both professors and students from the faculty of science.[18]

The contrast between professors who saw themselves as birds of passage and those who were ready to put down provincial roots had a striking counterpart in the mid-1850s in the newly founded faculty of science of Lille. Two of the four professors who were appointed when the faculty opened there in 1854—Gabriel Mahistre (in pure and applied mathematics) and Auguste Lamy (in physics)—published regularly in the *Mémoires* of the Société des sciences, de l'agriculture, et des arts de Lille. In the strength of this local allegiance, both men resembled Eudes-Deslongchamps; they were from the region, had been promoted to the faculty from posts in *lycées*, and were without realistic hopes of further advancement in the national system. Their colleagues Louis Pasteur (in chemistry) and Henri de Lacaze-Duthiers (in natural history), in contrast, were outsiders to the area who saw themselves as en route to higher things. Their engagement with the society was never close: Pasteur, who left for the Ecole normale supérieure in 1857, published one paper in the *Mémoires* in two and a half years, while Lacaze-Duthiers published two papers in the eight years he spent in the faculty before also moving to a post at the Ecole normale supérieure in 1863. It is true that Pasteur's one paper was his pathbreaking "Mémoire sur la fermentation appelée lactique."[19] But his perception of the society's marginality to the world of serious science is clear: by the time the paper appeared, it had already been summarized in the *Comptes rendus* of the Académie des sciences and was earmarked for publication in the *Annales de chimie et de physique.*[20]

At the same time as the faculties were growing in number and size, so too were new *sociétés savantes*. The threat that these societies offered to the local preeminence of the provincial academies lay in the nature of their objectives, which were essentially modernized forms of those that academicians had pursued since the eighteenth century. Like the academies, the new societies, which began to proliferate from the 1820s and even more conspicuously from the 1830s, sought the status that came with displays of seriousness (in particular through publication) and of high cultural endeavor, routinely enhanced by declarations of the value of their work to the local community.[21] They distanced themselves from the academies, however, by making much of their fitness for the more democratic age in which they functioned. To this end, while carefully fostering the cultivated male bourgeoisie's appetite for distinction, they also made a virtue of their controlled openness: membership could be bought by anyone willing to pay the annual subscription of 20 or 30 francs. For a chairholder in a provincial faculty, with a typical basic salary at this time of 4,000 francs a year, such a sum was by no means negligible, but it was affordable; for a teacher in a primary school, however, with a salary that was unlikely to exceed 1,000 francs, it represented a virtually impossible investment. In that sense, claims about the societies' openness conveyed a partial truth rather than an unqualified reality.

No less important in fashioning the new societies' self-image was their abandonment of the encyclopedic intellectual aspirations of the academies. In practice, their visions of themselves as assemblies of authoritative specialists were difficult to realize. Yet the best of them could call on a core of competent leaders supported by a larger body of amateurs of varying degrees of commitment and knowledge but with directed rather than universal interests. In both the growth and the institutional fragmentation of what had once been a unified cultural elite, Lyon was typical. Here, by the time of the Revolution, the sixty-five-year-old Académie des sciences, belles-lettres et arts de Lyon had already acquired two rivals: the Société royale d'agriculture (founded in 1761) and the Société de médecine de Lyon (1789). Under the Empire, there were further additions: a Société de pharmacie (1806) and the precursor of the future Société littéraire, historique et archéologique de Lyon (1807). And thereafter the rhythm of new foundations quickened.

A successful Société linnéenne (1822), a Société académique d'architecture (1830), and societies for the study of meteorology (1843), geography (1871), botany (1872), political economy (1876), anthropology (1881), and astronomy (1883) all reflected the evolution of educated taste and contributed to the figure of twenty-six societies in the city that were listed in the survey of the nation's *sociétés savantes* that the Ministry of Public Instruction undertook in 1886.[22]

The decision of the newcomers to stake out circumscribed disciplinary territory makes them sensitive indicators both of the learned enthusiasms of the nineteenth century and of changing forms of bourgeois sociability. Among the culturally more advanced provincial cities, Rouen and Bordeaux, with almost twenty such societies each by 1886, and Marseille, with more than a dozen, all displayed a pattern of growth comparable with that of Lyon.[23] Inevitably such larger cities made the running, but similar trends can be observed in far smaller communities: by 1846, Angoulême, Chalon-sur-Saône, Béziers, Saint-Brieuc, Saintes, Langres, and Sens all boasted societies specializing in archaeology and local history, founded as part of the vogue for historical study that blossomed from the mid-1830s.[24] By comparison, specialization in natural history came about ploddingly, although the creation of societies for the subject at La Rochelle, Versailles, and Metz in the 1830s and 1840s indicates that by the midcentury collectors, specimen hunters, and those who wanted simply to read or hear about the natural world were seeking, and finding, congenial outlets for their interests within communities that had at least some measure of disciplinary focus.[25]

During the July Monarchy, the Second Empire, and the early Third Republic, the divergence between the fortunes of the new societies and those of the old academies, most of which remained stubbornly loyal to their ceremonies, limited memberships, and all-embracing vision of culture, became steadily more marked. In 1846, the Académie des sciences, arts et belles-lettres in Caen had to rely on a trifling income of 700 francs a year, all but 100 francs of which came as a "grace and favor" payment by the General Council of the Calvados.[26] The consequences were to be seen in a chronically low pace of activity. Between 1825 and 1850, only seven rather slender volumes of *Mémoires* were published, and in 1851 the academy had to take the once unthinkable step of asking its members to pay a subscription.[27] The policy, though humiliating, seems to have worked, to judge by the volumes of some four or five hundred pages that thereafter appeared each year until the First World War. Even this heightened level of publication, however, did not restore the Caen academy to the position of cultural primacy that it had once enjoyed in the town. Its *Mémoires* had to jostle for recognition with the more focused journals of the Société linnéenne du Calvados (later the Société linnéenne de Normandie), the Société des antiquaires de Normandie, and the Association normande (a society pursuing both scientific and nonscientific regional interests), all of which had acquired substantial followings in Caen and more broadly in Normandy since their foundation in the 1820s and early 1830s.

While academies everywhere faced hardship, not all of them endured quite such severe financial constraints as those that trammeled the academicians of Caen. Things were somewhat better in Dijon, where in the mid-1840s an annual grant of 1,500 francs from the general council of the department of the Côte d'Or ensured the solvency, though little more than that, of the town's Académie des sciences, arts et belles-lettres.[28] And they were better still in Bordeaux and Toulouse, where the academies received annual grants from municipal or departmental funds totaling 3,500 and 3,000 francs, respectively.[29] Nevertheless memories of what these two academies had once been provoked sad nostalgia. In Bordeaux, the nostalgia was for a past in which the academy had maintained elegant premises in the center of Bordeaux, a fine library, a botanical garden, and a museum of natural history. In Toulouse the academy still possessed a library of 1,200 volumes. This, though, was a pathetically small vestige of a once rich prerevolutionary resource, while the gracious building in which eighteenth-century academicians had held their meetings, and the academy's botanical garden, observatory, and accumulated private donations of money and objects had all been confiscated in 1793 and were now beyond recovery.[30]

In Toulouse, as elsewhere, academicians had to adjust to a local cultural hierarchy in which their institution was just one of several societies, most of which displayed a vitality that they could not match.[31] A comparison between the ministerial surveys of 1846 and 1886 gives statistical substance to the changes that were under way. One hundred thirty-five provincial societies, including thirteen prerevolutionary academies, were listed in 1846, with another forty-three in Paris.[32] Forty years later, the academies were lost numerically in a total that had risen to 513 (excluding horticultural and agricultural societies), with another 142 (many of them with some claim to national status) in Paris, and 12 in Algeria and the other colonies.[33] The growing number of society publications too had the effect of swamping the efforts of the academies. The survey of 1886 recorded that the nation's *sociétés savantes*, Parisian and provincial, had already put out fifteen thousand volumes of *Mémoires, Bulletins*, and other publications, the overwhelming majority of them having appeared since the 1820s.[34] By the 1880s, the societies were publishing at an unprecedented rate of five hundred volumes a year, only a tiny proportion of which came from the old provincial academies.

The 1886 survey demonstrated beyond question the undiminished attraction of belonging to a society: the opportunities for cultivated sociability and the sense of pride in belonging to an exclusive body, albeit one that offered nothing like the exclusiveness of an academy, were sufficient in most cases to maintain healthy membership lists. But the number and vigor of societies at the end of the nineteenth century should not be interpreted as indicators of social aspirations alone. An ambitious bibliography, completed in six volumes and covering societies' publications in history and archaeology up to 1900, reflected the undimmed seriousness of voluntarist participation in areas of provincial research in which independent scholars had traditionally been strong.[35]

Although the strengths, as measured by published work, were increasingly oriented to history, archaeology, and antiquarianism, all with a local focus, science was by no means unrepresented. While the boundaries between scientific and nonscientific societies are hard to draw, roughly half of the 655 societies surveyed in 1886 flagged science in some form as at least one of their interests.

The survey also showed how effectively a well-led society with clearly defined purposes could cater for new disciplinary interests. One such interest with a scientific dimension was geography, a discipline that rose to unprecedented prominence in response to the colonial ventures of the early Third Republic. Between 1871 and 1884, no fewer than twenty-seven geographical societies were established, including societies of commercial geography in Paris and the ports of Bordeaux, Le Havre, and Nantes. With the vigor of youth, the new organizations prospered: they almost invariably published, and their number multiplied, reaching thirty in 1894 and forty-one by 1913.[36] In these societies, academic career making, public policy, and popular interest in foreign lands found ample common ground. There at least, the dusty elitism that had provoked Rouland's contempt was emphatically a thing of the past. They were modern societies with modern objectives.

The Devotee: Nature, Learning, and Locality

Voices on the periphery were by no means always the collective ones of an academy or a society. Members and nonmembers alike also spoke for themselves and pursued personal intellectual trajectories that owed little or nothing to an institutional affiliation. While in science, such voices tended to be discrete, they did not necessarily go unheard. Provincial settings allowed naturalists in particular to fashion a local, sometimes even a national reputation, usually based on fieldwork of a kind that the Parisian elite found it hard to conduct. Except in a minority of cases—of those who combined deep provincial roots with holding a chair in a faculty and who therefore worked with at least one eye on their ministerial paymasters—such work was done without thought of material recompense or career advancement. Its rewards took different forms, sometimes the satisfaction of a publication or an acknowledgement in print for information or specimens provided to the Muséum d'histoire naturelle, sometimes the more discrete gratification that came with a letter of thanks from a great Parisian savant, or, for those with oratorical talent, the appreciative response of the large audiences that a good lecturer was able to assemble even in a small town. The leading independent devotees, in fact, occupied an important middle ground. They served the interests of the national leaders of zoology, botany, or geology for the unique expertise that the best of them could offer in the field, while being admired by their immediate community, which placed them in the ranks of the region's *notables*.

From this key position, provincial savants were an indispensable element, as both consumers and practitioners, in the gathering vogue for the collecting and study of specimens—zoological, botanical, and geological—that marked educated taste in France from the first quarter of the nineteenth century. Natural history, as Georges Cuvier saw it as early as 1804, had become a leading cultural fashion of the day, one to which the planned multivolume *Dictionnaire des sciences naturelles* (in which Cuvier made his observation) and its contemporary rival, the *Nouveau dictionnaire d'histoire naturelle appliquée aux arts* under its main promoter Julien-Joseph Virey, were responding.[37] The standing of natural history rested on an appeal that Cuvier saw as transcending social divisions between the powerful and the weak and providing cultural common ground for both men and women.[38] Most important of all, it turned the thoughts of the finest minds from the dangerous futility of abstract speculation to the realities of the world garnered by sense experience. Cuvier's endorsement was eloquent:

> By studying it the powerful have sought a diversion from the anxieties that go with greatness, and the unfortunate have tried to forget the injustices of fate; even the fair sex, along with those men on whom fortune has bestowed independence and who have had the wisdom to preserve their freedom against the lures of ambition and vain glory, have made it the delight of their leisure hours. Finally, and this is unquestionably its supreme triumph, the superior men of genius for whom meditation is a necessity and who tire of the futility of abstract speculation have abandoned the heights of a too general philosophy and instead sought nature's true laws through the scrutiny of her works, preferring the study of the real world to the fabrication of an imaginary one. In short, it is from natural history that they have drawn confirmation of their doctrines or material for their experiments.[39]

In fashioning enthusiasms so inclusive and socially therapeutic, the impact of the national leaders of natural history was out of proportion to their modest numbers: Cuvier, Geoffroy Saint-Hilaire, and Lamarck were public figures whose influence was heightened by the accessibility of most of their writings, their celebrity in the cultivated world of the salons or (in Cuvier's case) political life, and even their appearances in fiction (a distinction that Balzac bestowed with particular frequency on Cuvier[40]). But the power of the Parisian elite to broaden horizons and stimulate imaginations was outstripped by the impact of the objects and reports brought back from the ever more ambitious voyages of the eighteenth and early nineteenth centuries. The expedition to Egypt in 1798 raised the taste for exotics to new heights, fostering an interest in the Orient that resurfaced almost thirty years later in the extraordinary public fascination with the arrival of a giraffe—the first living specimen to be seen in France—at the Muséum d'histoire naturelle.

The episode of the giraffe arose from the convergence of the ambition of Charles X to enrich the collection of the museum's menagerie and the eagerness of the eccentrically entrepreneurial French consul in Alexandria, Bernardino Drovetti, to render a service to the king.[41] It was Drovetti who persuaded the pacha of Egypt to make a gift of the giraffe, a female that eventually disembarked at Marseille in October 1826, along with its Arab attendants and three cows to provide it with milk. Housed for the winter in the courtyard of the Prefecture, where it was studied by members of the city's Académie des sciences, lettres et arts and drawn (for the benefit of the professors of the Muséum d'histoire naturelle, who feared the animal might die before reaching Paris), the giraffe was viewed first by schoolchildren and other privileged groups and then by the crowds who watched its daily two- or three-hour walk in Marseille. By March 1827, it had been decided, on the advice of Cuvier and his Parisian colleagues, that the journey to Paris should be made by road,[42] and soon afterward no less a figure than Geoffroy Saint-Hilaire arrived to take charge of the preparations on behalf of the museum. On 20 May 1827 the journey of more than 800 kilometers to Paris began.

A stop at Aix after the first day of walking confirmed the level of popular interest. Thousands turned out, making a noise and causing a disorder that persuaded an anxious Geoffroy Saint-Hilaire that access should after all be restricted to the better, more "discrete" elements of bourgeois society. This social discrimination, however, was never achieved. Throughout the seventeen days that it took to cover the 350 kilometers to Lyon, including four rest days, the giraffe remained an object of sustained excitement: in Lyon crowds were fascinated by its strange way of bending its neck to drink and its liking for the leaves of the lime trees of the place Bellecour, which it stripped. And thereafter, for the rest of the journey, which ended in the menagerie of the museum on 30 June 1827, "giraffe mania" never waned. In Paris, in the first six months, six hundred thousand visitors paid to see the animal in the rotunda of the menagerie, where it was to remain for eighteen years (apart from an excursion to Saint-Cloud, in which the king graciously received his gift on 9 July 1827). For three years consumer goods inspired by the giraffe, from cakes and toys to wallpaper and fashion accessories, reflected and fed the craze, and it was only from 1830 that interest flagged, leaving the giraffe to become, according to Balzac, a spectacle for imbeciles, maids desperate to amuse the children in their care, and "backward provincials."[43] The novelty had faded, definitively and quite suddenly, with consequences that ministers would have done well to heed: they too, even the most senior of them, might one day recede from public favour, just as the giraffe had done. When death claimed the giraffe in 1845, the animal's only distinction was to be dissected and then stuffed by a new technique before being displayed in the galleries of the Muséum d'histoire naturelle and finally, in 1936, ignominiously transferred to its present resting place in the Muséum Lafaille in La Rochelle.

While surges of interest in natural history could be stimulated on the national scale by the appearance of a strange animal or a highly publicized event, such as the confrontation of 1830 between Cuvier and Geoffroy Saint-Hilaire, traditions with local roots tended to be more enduring and at least as intellectually productive. Virtually every sizable town had its pocket of serious practitioners, of greater or less eminence, who served as sustained sources of leadership and advice. Such men (and they were almost invariably men) reveled in their freedom and independence. Or they professed to do so. For in reality few had the means to dedicate themselves wholly to research and writing and so to escape from a sense of the enormity of the tasks to be done in time snatched from busy lives. In postrevolutionary France, the leisure for serious science and scholarship that came with wealth, inherited or otherwise, was an even rarer commodity than it had ever been. But the exceptions, such as the botanist Benjamin Delessert, show what could be achieved by someone whose circumstances permitted a richly resourced, full-time dedication to science.

Delessert was not only exceptional in his wealth and intellectual commitment but also unusual among naturalists in being a lifelong resident of Paris. A financier, political figure, and philanthropist who had risen to national prominence during the First Empire,[44] he was secure in the material independence he had achieved and in the scientific standing that was recognized and consolidated by his election as a *membre libre* (a category of members unattached to the disciplinary sections) of the Académie des sciences in 1816, when he was in his early forties. Under the Restoration and July Monarchy, he moved with ease in the intellectual as well as the social elite of the capital. His passport was not just an expertise in botany, displayed most sumptuously in the five magnificently illustrated volumes of Augustin Pyramus de Candolle's *Icones selectae plantarum* that he edited between 1820 and 1846;[45] it lay also in the unique research facility that he placed freely at the disposal of savants and the serious public at his house in Passy, then on the western outskirts of the capital. There he could offer a herbarium whose 86,000 species, expertly assembled over fifty years and represented by a quarter of a million specimens, placed it second only to that of de Candolle and his son Alphonse; a collection of one hundred fifty thousand shells (including Lamarck's collection) representing twenty-five thousand species;[46] and a library of four thousand botanical books that, in the opinion of Flourens, no other botanic library could match.[47] All this made Delessert a man to be courted, even in the highest echelons of French science. He retained his prominence for more than thirty years, until his death in 1847, and for more than two decades thereafter his collection remained a resource for naturalists under the supervision of his younger brother, François-Marie.

Two other independent naturalists who achieved high levels of freedom in fashioning reputations that stood as high in the academic world as they did among their fellow devotees were Adolphe, vicomte d'Archiac, and Gaston, marquis de Saporta. Born in 1802 and in receipt of a modest pension from 1830, when his rejection of the July Monarchy led to his departure from the army, d'Archiac lived the life of an extraordinarily well-read and dedicated geologist and paleontologist, distracted by neither family nor academic duties, until, at the age of almost 60, he took the chair of paleontology at the Muséum d'histoire naturelle.[48] A generation later, Saporta worked with equal dedication once he turned to natural history in search of consolation following the death of his wife in 1850. He was 27 at the time and, for the rest of his life (he died in 1895), he used his leisured existence to advance the nascent science of paleobotany and to spark the occasional vocation in the Midi, including that of the 12–year-old Antoine-Fortuné Marion, who went on (from humble beginnings as a *préparateur* in natural history in the Marseille faculty of science) to become professor of zoology, head of the faculty's purpose-built marine zoology laboratory, and a leader in the fight against phylloxera, which was destroying grapevines throughout France and elsewhere.[49] As a paleobotanist, Saporta made contributions of genuine originality, notably through his work on Jurassic flora, and established himself as one of the few independent natural historians who, even in the more professional world of the later nineteenth century, could still compete with contemporaries holding academic posts.[50] His election to the Académie des sciences, as a corresponding member of the botanical section in 1876, was well earned and certainly not just an acknowledgment of his aristocratic status.

The degree of freedom that other, less privileged naturalists enjoyed can be mapped on a broad canvas. For half a century Jean-Baptiste Mougeot carved out time from his work as a surgeon and physician in Bruyères in the Vosges for the study of the botany of his remote region (as well as its meteorology, geology, and medieval antiquities).[51] Léon Dufour, of the same generation (the two men were born in 1776 and 1780, respectively), fashioned his reputation as one of the most notable entomologists and botanists of the first half of the nineteenth century in a similar way; like Mougeot, he was a doctor who made a virtue of physical isolation (in his case in Saint-Sever in the Landes in southwestern France) and of his mastery of a region, extending into and beyond the Pyrenees, that was not easily accessible to Parisians.[52] An even more famous example of research pursued amid other preoccupations is that of Jacques Boucher de Perthes, whose work on the fossil origins of the human race dated from the later part of a professional life in charge of the collection of customs in his native town of Abbeville in Picardy.[53] Freedom, as Mougeot, Dufour, and Boucher de Perthes knew, never came easily to any but the most fortunate provincial naturalists, and tales of the ingenuity, resolution, and sacrifice that were part and parcel of an independent naturalist's life are legion.

Among the saddest cases was that of Félix Dujardin, who famously struggled for more than a decade to sustain his personal research, travels, and family commitments by a frenetic engagement in money-making ventures. His initial strategy was based on a mixture of private tutoring, lectures on geometry and applied chemistry (paid for by the municipal authorities of Tours), journalism (including a brief period as editor of the *Echo du monde savant*), and popular writing. But this yielded insufficient support, and Dujardin finally had to seek full-time employment in the faculty system, first as professor of mineralogy at Toulouse (1839–40) and then in the chair of zoology and botany in the recently founded faculty of science at Rennes, where he spent twenty years as professor from 1840 until his death in 1860.[54] His years in Rennes were wretched ones. The delay between the creation of the faculty in 1840 and its installation in a desperately needed Palais des facultés from the mid-1850s only heightened his perception of himself

as the undeserving victim of impediments that had eaten cruelly into the time he had
been able to devote to research and reflection.[55] A difficult marriage, some cantankerous
colleagues, and the hostility toward him that Henri Milne Edwards directed from his
powerful position as professor of entomology at the Muséum had also played their part
in eroding Dujardin's productivity. Apologizing for what he perceived as a modest harvest from the seven to eight thousand hours that he had devoted to his great study of
intestinal worms, he laced the preface to his *Histoire naturelle des helminthes* with a
declaration of undisguised bitterness in 1844: "I believe I could have achieved still more
if, instead of the persecution I experienced in the course of my work, I had found the
support due to a professor; if I had not been reduced to depending entirely on my own
resources and obliged to spend time on dissections and research of a tedious nature,
time cruelly lost for science."[56]

During the first half of the century, an increasingly common solution to the peren-
nial problem of securing the time and material conditions that the serious pursuit of
natural history required was through a municipal lectureship. Where such lectureships
were available, the consequences, for both the refinement of expertise and the care of
collections, were important. For a municipal appointment not only allowed a naturalist
to pursue a life that accommodated at least some personal research; it also helped to
ensure the survival of the rich patrimony of natural history inherited from the ancien
régime. The case of Nantes was typical. There, the nucleus of what eventually became
the town's museum of natural history had been assembled from the 1770s by a self-
made pharmacist, François-René Dubuisson, who had opened his personal cabinet of
natural history specimens first to friends and then, from 1799 and for a small charge, to
the public, before making an agreement with the Department of Loire-Inférieure for its
deposit in the departmental *école centrale* in 1802. When, in the following year, the
school was closed and the teaching of natural history that had been an essential part of
its curriculum was abandoned, the collection passed by a circuitous route to the municipality of Nantes. Its future, like that of many other collections that landed on municipal laps following the closure of the *écoles centrales* and their replacement by the less
scientifically oriented *lycées*, remained uncertain. Fortunately, temporary accommodation soon gave way to a more satisfactory installation in the abandoned premises of the
town's school of surgery, and there Dubuisson, as "directeur-professeur" (beginning in
1806), was able to use what had once been his own collection for his research and in due
course for the public lectures on mineralogy that he undertook for a quarter of a century until his death in 1836.[57]

Dubuisson was unusual in the quality of the personal collection that he assembled.
But appointments of the kind he held were common, and they provided an important
context in which leading authorities on a city or region could combine research with
lecturing and the management of a collection or garden that had become municipal
property in much the same way as occurred in Nantes. Some municipalities found the

task too much for them: in Perpignan, the natural history specimens that had passed to the town on the closure of the *école centrale* of the department of Pyrénées-Orientales in 1803 fell into neglect and were not displayed again until 1840, when the municipal council appointed a curator for its newly formed natural history museum.[58] Most civic authorities, though, took their pedagogical and cultural responsibilities seriously, none more so than Marseille, which had been a trendsetter since before the restoration. In 1813, the municipality already sponsored a course of three weekly public lectures on botany and plant physiology, free of charge.[59] And the lectures continued after 1815, with steadily mounting success. In the 1820s audiences of between thirty and sixty were the norm, and in 1829 they almost reached two hundred.[60]

In Marseille, as in many other large towns, the botanical garden was the core of an integrated provision for a public culture of science on a broad front. The opening of a civic museum of natural history in 1819 added a resource that helped to make the most of the complementary talents of the museum's energetic first curator, Polydore Roux, and the lecturing skills of the garden's director, Gouffé de la Cour.[61] It is a mark of the public favor for the museum that in 1823 the city council allocated 6,000 francs for the purchase of ten thousand natural history specimens from the archaeologist and anti-quarian Félix Lajard,[62] setting in motion a tradition of municipal patronage that culminated in the museum's installation in the grandiose premises of the Palais de Longchamp in 1869. There, the inauguration of an observatory in 1872, again with municipal support, created something akin to a science center that satisfied public interest while also producing serious astronomical work, chiefly on the recording of comets.

Success in securing the safe passage of scientific collections and gardens through the revolution and empire and on into the restoration and beyond almost invariably depended on partnerships between dedicated individual naturalists conscious of the risks of vandalism or indifference and public authorities committed to providing formal instruction or improving entertainment. Finding common ground was not always easy. In Lille, in the difficult period following the replacement of the *écoles centrales* by the Napoleonic *lycées*, the botanical garden was closed, and it took a combination of a sympathetic prefect and a municipal investment in support of a variety of scientific lecture courses to bring it back into use: by 1809 up to forty people, including horticulturalists and students aspiring to medical careers, were attending botanical lectures.[63] Even in a town with as strong an eighteenth-century tradition of public lecturing as Rouen, it proved hard to maintain the momentum through the years of war and political turmoil. Bertrand Pinard, a long-serving senior doctor at the Hôtel-Dieu, was the last in a line of prerevolutionary professors of botany with salaried positions under the auspices of the Rouen academy. Following his refusal to take the required republican oath in 1793 and his consequent retirement, there followed several years of vulnerability. At this point, the botanical garden on which Pinard and his predecessors drew for their lectures might well have disappeared had it not been for a devoted head gardener, Jacques Varin, who continued to give practical instruction even in the darkest days.[64] It was Varin who replanted and supervised the garden during its brief attachment to the *école centrale* of the department of Seine-Inférieure and then tended it after its redesignation as a municipal facility, following the school's closure in 1803.

The reinvigoration of provincial intellectual life during the relative calm of the Consulate and early empire made its mark in Rouen as it did in many other towns. It did so in a context of cooperativeness between private and public interests exemplified in the career of Jean-Baptiste Vitalis. As municipal lecturer in chemistry (a position he held in parallel with teaching positions in the *école centrale* and then, until 1817, the *lycée*), Vitalis became a leading authority for students and employers of the region with interests in dyeing, calico printing, beet sugar, fertilizers, and the coal deposits of the Cotentin peninsula. And he acquired national as well as local fame through a successful *Manuel du teinturier* (1810) and *Cours élémentaire de teinture* (1823), based on the lectures on dyeing that he had given since 1805.[65] By the time declining health led him to return to Paris, where he resumed his prerevolutionary vocation as a priest in 1822, the civic lectureship that he held was firmly established in the cultural life of Rouen and it remained, with other lectureships on other subjects, a valued resource through the Restoration and July Monarchy and on into the Second Empire and beyond.[66] An account dating from 1827 gives an impression of the range of instruction that was available. In that year, lecture courses were offered, free of charge, on applied chemistry, botany, drawing in both its artistic and its industrial aspects, and geometry and mechanics. By the mid-1830s courses on physics had been added (including one for working men, taught on Sunday afternoons at one o'clock). Independent initiatives flourished too. Among them were new offerings by the Société d'émulation de Rouen, whose Sunday lectures and other courses in subjects relevant to local needs and interests helped to make Rouen a model, though by no means an exception, in the variety of opportunities for intellectual and vocational improvement available to citizens of virtually all classes.[67]

One of the greatest merits of municipal and other nongovernmental lectureships, for those who held them and for their audiences, lay in their independence of state control. The absence of the treadmill of examinations and nationally imposed syllabuses allowed lectures to be finely adapted to the economic circumstances and cultivated tastes of the audiences concerned. In Rouen, where the lecture programs and the city's Muséum d'histoire naturelle and botanical garden received particularly generous civic support from two mayors dedicated to reform, Henry Barbet and Charles-Amédée Verdrel, the circumstances were ideal for a scientific opportunist with the drive and imagination to take advantage of them. None of Rouen's lecturers exploited the circumstances with greater success than Félix-Archimède Pouchet. Refashioning himself after an initial training in medicine, Pouchet used his appointments as municipal lecturer in zoology and botany and the first director of the museum from 1828 to carve out a career enriched by high degrees of scientific autonomy and personal gratification (albeit a gratification tempered by a keen awareness of the handicaps of anyone seeking to perform serious scientific work in the provinces[68]). The foundation of Pouchet's celebrity was his annual series of winter lectures. Devoted successively to one of the main classes of the animal kingdom, these consistently drew audiences of two hundred, including the members of the medical and legal elites of Rouen and students engaged in initial training in medicine and pharmacy at the Ecole préparatoire de médecine et de pharmacie.[69] Subject to none of the constraints that weighed on *universitaires*, Pouchet was at liberty to broach "transcendant" questions,[70] including those bearing on his research on reproduction and spontaneous generation (see chapter 4), and there can be no doubt that he exploited the freedom to the full.

In Clermont-Ferrand, at about the same time, Henri Lecoq made a career of equal distinction, in his case on the foundations of a training in pharmacy and a municipal appointment as the guardian of a botanical garden and collections with a familiar history of precariousness following their use by the departmental *école centrale*. Lecoq had a broader spectrum of activities and less taste for controversy than Pouchet. Following an appointment that is said, in some accounts, to have resulted from the misdirection of a letter from the mayor of Clermont-Ferrand offering the municipal chair of natural history to another person of the same name, Lecoq pursued an idiosyncratic trajectory eased by his rapid accumulation of a significant personal fortune.[71] Had he been dependent solely on the income from the civic appointment, which he took up in 1826, his life might well have been one of modest, contented obscurity. But his pharmaceutical business, his production of a coffee substitute, and his collaboration in the commercial promotion of the waters of Vichy, coupled with his presidency of the city's Chamber of Commerce, brought prominence and prosperity sufficient for him to build a fine house (effectively a private museum, now the Muséum Henri-Lecoq), to acquire the land that eventually became a public garden, and to make substantial gifts and bequests to the town. Clearly a man who could leave a total of 150,000 francs for the municipality to provide new greenhouses for the botanical garden, a covered market, and an installation for the supply of water over a distance of 3 kilometers from Royat could take a somewhat detached view of regular paid employment. While he accepted the chair of natural history in Clermont-Ferrand's new faculty of science in 1854, it is unlikely that his annual salary as professor, which stood at 6,000 francs at the end of his career, was of much consequence to him.[72]

Complaints about the intrusiveness of the demands of either national or local administrations recur frequently in naturalists' accounts of their lives, especially from about the mid-nineteenth century. These reflect the growth in educational bureaucracy that occurred under the July Monarchy and the Second Empire. Even in a municipal lectureship, where the reins tended, by national standards, to be loose, Dubuisson in Nantes was answerable to a supervisory committee established to oversee his work.[73] In Rouen, the expansion of the national university system threatened a greater and more systematic intrusion when Pouchet's lecture course in zoology was incorporated in the two-year program of the city's newly established Ecole préparatoire à l'enseignement supérieur des sciences et des lettres in 1854. The need to teach to an imposed and largely humdrum university syllabus gravely cramped Pouchet's style and almost reduced his audience by half. It was an ironic consequence for someone who had been, and who remained, among the leaders of a long campaign to secure a faculty of science for Rouen that ended successfully, though not until 1965.[74]

From 1854, Lecoq too labored under the constraints that a professorial appointment suddenly imposed on a previously unfettered intellectual life. First, in order to qualify for the chair, he had hastily to obtain a doctorate, despite now being in his fifties; this he did in Lyon with a main thesis on the colors of flowers and their distribution by latitude and altitude and a subsidiary thesis (as was required at the time) on alluvial sediments.[75] Then there was the straitjacket of the nationally imposed syllabus that he was required, at least in principle, to follow. This was presumably one of the pressures that led him, in 1863, to seek permission to give three weekly lectures in the winter and none in the summer, an adjustment that would maximize his time for fieldwork.[76] The unfavorable reaction of the rector of the academy (who commented of Lecoq that "he believes he is entitled to some kind of exemption"[77]) reflected an irritation aggravated by Lecoq's sense of independence and his idiosyncratic and wholly unbureaucratic relationship with his public, on the one hand, and with the officials of the Ministry of Public Instruction, on the other. Periods of leave lasting a matter of days, which he had already enjoyed from time to time, were clearly insufficient for Lecoq, and in February 1864 he returned to the fray with a request for a temporary replacement to undertake his summer teaching in both that and the following year. He made the case on the grounds of ill health and the need to push ahead with his geological work, and this time the rector acquiesced. Perhaps simple resignation led to the decision, as it almost certainly did in the later 1860s, when Lecoq was granted three years of unpaid leave, apparently without difficulty.

The recognition that provincial naturalists received was as diverse as the strategies they employed in quest of it. While they were always perceived, and they represented themselves, as dedicated and self-effacing, few in reality were without ambition of some kind. Dujardin's aspiration to establish himself in Paris at almost any cost was overt and especially corrosive; it accounts for his hermit-like existence in Rennes and his strained relations with colleagues and the rector of his academy caused by his prolonged absences in the capital whenever he could get away.[78] Of the others, the most eminent and ambitious set their sights on the difficult target of election as a corresponding member of the Académie des sciences and reconciled themselves to the investment of energy that such a goal entailed if a sufficient record of publication was to be achieved and the necessary contacts with the Parisian elite were to be cultivated. When the investment paid off, the gratification was intense. Léon Dufour's election as a corresponding member of the section for anatomy and zoology in 1830 prompted him to write effusively to Académie's permanent secretary for the physical sciences, Cuvier, acknowledging an "unhoped for" honor that would fire him to continue his work.[79] Even more gushing were the letters that Eudes-Deslongchamps sent, when he was elected to corresponding membership of the same section in 1849, to Jean-Baptiste Boussingault and Pierre Flourens, respectively the president and permanent secretary of the Académie: the "signal honor" had fulfilled the wildest desires of "a poor provincial naturalist, more zealous than learned, who will redouble his efforts to ensure that the Académie will have no reason to regret its choice."[80] The honor had come, however, when Eudes-Deslongchamps, like Dufour, was well into his mature years; on election he was 55, while Dufour was 50. Though immensely gratifying, it stood as a final but essentially decorative accolade for a life of science in which day-to-day satisfaction lay first and foremost in the simple pleasures of observation, the occasional discovery, and the transmission of knowledge at whatever level would draw an audience or earn some other mark of appreciation.

However little appeal such pleasures had for someone with Dujardin's restlessly tempestuous spirit, they amply sustained less exigent peers such as Dufour and Eudes-Deslongchamps, as they also sustained Lecoq, who never lost his enthusiasm for fieldwork and communication. Even after he entered the more circumscribed world of the Clermont-Ferrand faculty, Lecoq continued to give priority in his teaching to his admiring nonacademic public. His success is reflected in audiences that for most of the 1850s and 1860s comfortably topped 150 and often, in the winter months, reached three hundred.[81] No other professor in his faculty could remotely match his drawing power. What Lecoq would have seen as a supreme mark of success, however, was viewed very differently in the Ministry of Public Instruction. In the ministerial scheme of things, his lectures (though open to the public) were first and foremost a requirement for students in Clermont-Ferrand's Ecole préparatoire de médecine et de pharmacie. But it was clear that, for this examination-focused purpose, they left much to be desired. The style of Lecoq's lectures, as of his books, was colorful, rich in anthropomorphic metaphors, and *galant*. In characteristically extravagant language, he presented the beauties of flowers and plants as part and parcel of the coquetry of the vegetable kingdom and dedicated the last chapter of his *Vie des fleurs* to the ladies in his audience, "the flowers that speak" to whom he always liked to address himself.[82] A proclivity for rhetoric and fine language in the romantic manner and a reluctance to adhere to an imposed syllabus remained at the heart of Lecoq's perception of his role, a point that was not lost on administrators.[83] A report by the rector of his regional academy in 1855 spoke of a delivery "of unquenchable fluency and facility," while by 1863 his lectures were described as "facile" and flowing "like warm water from a tap": the content, "very easy, very elegant, even a little too much so," was "accessible" but, fatally, "lacking great elevation."[84] This was seen to be especially the case in zoology, Lecoq's coverage of which was judged to be far more superficial than his treatment of geology, mineralogy, or botany, although even in these sciences too his lack of a formal academic grounding was cited as a handicap.

Ministerial perceptions of Lecoq as a populist and showman are an indication of the disparagement to which even a productive naturalist and corresponding member of the Académie des sciences (a rank he achieved in 1859 at the age of 57, after personally lobbying Adolphe Brongniart[85]) could be subjected. They contrast markedly and significantly with the respect that Lecoq enjoyed, at the highest level, in other areas of his life. His local expertise was an essential resource for the eminent naturalists—including Charles Lyell and Roderick Murchison in 1828 and Hugh Strickland in 1835—who visited the Auvergne;[86] and he was feted at national scientific gatherings in Clermont-Ferrand, such as the Congrès scientifique de France in 1838 (see figure 2) and meetings of the Société géologique de France (1833), the Société botanique de France (1856), and the Société entomologique de France (1859). Charles Darwin too respected his work, to the point of citing him a number of times in later editions of the *Origin of Species*.[87] Yet in the Académie as in the Ministry his reputation never rose above that of an auxiliary, to be gilded with faint praise and drawn on when the savants of the capital needed information, specimens, or a guided tour. Perceptions of Jean-Baptiste Mougeot, whom Augustin Pyramus de Candolle placed, in 1810, among the most important contributors to the study of French flora and whose herbarium came to contain some fifty thousand specimens, were similar. Despite a relentless correspondence with many of the greatest names in European natural history, Mougeot's place remained firmly on the Vosgean periphery. To his death in 1858 at the age of 82, the question of his election to the Académie des sciences seems never to have been raised.[88]

So while a provincial location did not necessarily imply obscurity, distance from Paris and an absence from the main seats of academic power limited the formal recognition that naturalists in the provinces could hope to achieve. It also meant that if recognition came at all, it did so in later life. The Auxerre lawyer and prolific authority on echinoids, Gustave Cotteau entered the Académie when he was in his late sixties, while Saporta, like Lecoq and Eudes-Deslongchamps, did so in his mid-fifties. In all these cases, the difficulty of maintaining regular contact with the leaders in their various specialities weighed heavily. Visits to Paris, in particular, were rare events, and were to be savored, as Dufour savored the eight occasions on which he ventured to the capital between 1818 and 1864.[89] Moreover, a provincial location did more than color the work that a naturalist chose to undertake. Crucially, it also meant that, unable to fulfil the requirement of residence in Paris, he could never aspire to full membership of the Académie and hence to a place at the high table of science.

Science and Decentralization

Most of the naturalists who pursued their studies in the provinces recognized remoteness from the disciplinary elite of the capital as an irritant but one that they accepted with resignation, if not in all cases equanimity. For some provincial savants, however, the divide between center and periphery was more than a facet of intellectual life that they had to live with and turn to such advantage as they could glean from it. These were men whose intellectual life was inextricably bound up with a more profound struggle for administrative decentralization. The struggle, driven by conservative values in politics and religion and a nostalgia for a pre-Jacobin France made up of provinces with distinctive cultural traditions and at least the trappings of autonomy, gave their endeavors an intensity far transcending the excitement of research and discovery.

In science and scholarship, the campaign to turn such nostalgia into reality began immediately with the return of the Bourbons. Although it built on foundations laid in the resurrection of academies and societies during the Consulate and the First Empire, the events of 1814–15 gave it unprecedented strength. The vehicle was a surge of enthusiasm for local studies. These included mapping regional topography, examining local geology, flora, and fauna, and describing antiquities. Within this range, favored focuses were the medieval buildings that had suffered destruction, appropriation for mundane secular uses, or, at best, neglect since 1789. On this front, there was much to do. The lamentable state of France's monastic and ecclesiastical heritage had already aroused the indignation of the traditionalists whose values Chateaubriand articulated in his *Génie du Christianisme* (1802).[90] But, with the Bourbons restored, the sense of loss and wanton destruction assumed a new vehemence, fanned by the writings of the travelers who flocked to France on the resumption of peace. The Norfolk antiquarian Dawson Turner, who traveled extensively in northern France in the early years of the Restoration, with his companion and artist John Sell Cotman, had reason to feel particular sadness at the condition of the relics of the Middle Ages that he saw in Normandy. He would certainly have shared the dismay of an anonymous contemporary who referred, in the *Quarterly Review*, to the parlous condition of buildings "doomed to neglect and destruction by the disgraceful sloth and ignorance of the French."[91] Yet his familiarity with the monuments, terrain, and people of the province would have consoled him with the hope of improvement, a hope that lay, for the French as for the British, in the dedication of the savants and antiquarians of Normandy who led erudite visitors on their tours.

As the Rouen librarian and academician Théodore Licquet observed, with a pique inspired by British disparagement of the kind voiced anonymously in the *Quarterly Review*, the assistance that his contemporaries offered to Turner and Cotman was both expert and plentiful.[92] Turner's exploration was helped, in particular, by one of the most forthcoming and distinguished members of the learned community of Normandy, Charles Duhérissier de Gerville, a resident of Valognes, a town of eighteenth-century elegance that encapsulated the prerevolutionary past of the Cotentin. As someone whose family had suffered material loss in the Terror and who had himself spent several years in his twenties and early thirties as an *émigré* in England, Gerville took a natural place among the provincial scholars who, starting in 1815, turned Chateaubriand's anti-Jacobin rhetoric into a sustained program of inquiry, recording, and reeducation.[93]

Similar figures, usually fired in their erudition (as Gerville was to an exemplary, though not exceptional, degree) by a nostalgia for the ancien régime and a loathing of the Revolution and its legacy, were to be found throughout France. But no region matched Normandy in the number of scholars of comparable stature, with backgrounds and interests similar to Gerville's. Auguste Le Prévost and Eustache-Hyacinthe Langlois of Rouen, Jean-Achille Deville of Rouen and, later, Alençon, and the abbé Gervais De La Rue of Caen all had the zeal, time, and competence (in tasks demanding painstaking accuracy rather than daring interpretative flights) that helped to make their province preeminent in the broader national movement to rehabilitate the traditions of local culture.

The resulting surge of learned publications and its role in the fashioning of a Norman regional identity would not have deserved extended discussion in this book had it not been for the shared context of science and erudition from which it emerged. As the publications and correspondence associated with the movement show abundantly, work on the history of a medieval abbey or some little-known diocesan archive was commonly stimulated not only by an intellectual admiration for the golden age of kingship and Christian piety but also by a regional loyalty that spilled over into a commitment to the broader cause of decentralization.[94] Such sentiments could, and often did, find equal expression in the study of the natural world. It was rare, in fact, for the first generation of the men who turned to local studies in this spirit to regard themselves exclusively as antiquarians or specialists in any other single intellectual endeavor. The case of Gerville, who was an authority on the geology (especially the fossil shells) and the botany of his region, as well as on its archaeology and church architecture, makes the point.[95] But the range of interests that Gerville cultivated was far from unique. Indeed, it was more nearly the norm than the exception in the provincial learned communities he frequented. In this world, to be a specialist was, if anything, the mark of an inferior mind.

The outstanding characteristic that united the descriptive study of buildings and ancient sites and topography with natural history was the emphasis on place and all that this implied for the struggle against the tentacles of central control. For those committed to the wider ideal, a savant with Gerville's credentials provided both an intellectual and an ideological model: his erudition bestowed unquestioned superiority in his immediate society, and his unconcealed sympathy for the traditions of the old provinces, with their Intendants and large measure of independence and regional identity, stamped him as an enemy of everything that smacked of Jacobin *dirigisme* and the new order of Napoleonic departments and prefects. Yet the very starkness of the contrast between the pre- and postrevolutionary worlds left the movement he represented vulnerable to the passage of time. By 1820, most of those who had had firsthand experience of the depredations of revolutionary excess were aged at least 50, and memories were fading. It had therefore become a matter of urgency to find some way of transmitting to a younger generation the sense of vocation—learned, political, and religious—that fired Gerville and his fellow conservatives.

As things turned out, the transmission was achieved more rapidly and effectively than the older generation could ever have hoped. Beginning in the 1820s, a swell of provincial opinion in favor of a diffused, independent world of learning emerged that survived in parallel with the institutions under public control and, against all odds, for half a century to come. During the Restoration, with clerical and reactionary interests well represented in the higher reaches of the educational hierarchy, such views did not provoke significant conflict: relations between the "official" and the independent sectors were almost uniformly harmonious. But under the July Monarchy, the gulf began to widen between those who held posts in higher or secondary education and those who pursued science and scholarship outside the structures of the state. It was consequently in this period and in reaction against heightened ministerial aspirations to control that the cause of intellectual decentralization acquired a new sharp edge.

Although support for decentralization fed on sentiments shared by local cultural elites across the nation, it owed its main vigor to one man: Arcisse de Caumont.[96] Born in 1801, Caumont had not lived through the Revolution. But he was brought up among elders who harbored bitter memories of the Terror. As members of an old Norman family, Caumont's ancestors had acquired land and moved easily into positions of authority in the courts and governmental administration in and around Bayeux. In this respect, his grandfather, Jacques-François Caumont, was typical. On the eve of the Revolution, he enjoyed the status of a *notable* as a senior legal official in the area of Bayeux, and it was this office that exposed him in the dark days of 1793–94. In truth, the Terror in Bayeux was a relatively tame affair. But even if it was "singularly bloodless," as Olwen Hufton has shown,[97] the arrests were numerous, especially between November 1793 and January 1794, and Jacques-François Caumont and his wife were among those who were imprisoned in the old convent of La Charité. By then, one of their two sons had fled and died in exile in Germany, but the other, François Caumont, shared imprisonment and the threat of death with his parents. Although Jacques-François Caumont emerged from prison, he was dead within months of his release, broken, as family tradition had it, by his ordeal. Bereavement and disruption on this scale were by no means exceptional, but the family's sense of loss and of regret for the old order ran deep, and it was as keen as ever when François Caumont's wife gave birth to Arcisse in Bayeux in 1801.

The return of the Bourbons and the ennobling of François Caumont (who became François *de* Caumont) in 1815 occurred at a time when his son, now aged 14, was especially responsive to the traditions handed down within the family:[98] in keeping with those traditions, Arcisse remained until his death a practicing Catholic and, at heart, a legitimist, though one who adjusted without undue difficulty to the political realities of the July Monarchy and the Second Empire. The years 1815 and 1816 were also decisive for his education. He spent them at the highly successful college of Falaise, where he was taught by the head of the college, Jean-Louis-François Hervieu, a notable polymath, former *émigré*, and now honorary canon of the cathedrals of Bayeux and Séez.[99] In this way, in Falaise as at home, Caumont was exposed to a profile of cultural attainment, including science and medievalism in the tradition of Chateaubriand, that was to mark him for the rest of his life.

By the time he enrolled as a law student in 1820, therefore, Caumont was already a formed, if young, scholar, steeped not only in Hervieu's main scientific interests of physics and geology but also in the study of gothic architecture. Such taste as he had for law soon faded amid the ferment of ideas that made Caen the center of what Jean Laspougeas has described as Normandy's "cultural renaissance."[100] The range of opportunities in Caen matched, and even extended, that of Caumont's interests. He was captivated above all by the teaching of Félix Lamouroux, a marine biologist, corresponding member of the anatomy and zoology section of the Académie des sciences, and from 1812 until his premature death in 1825 head of the Caen botanic garden and professor of natural history in the faculty of science.[101] No one did more than Lamouroux for the promotion of natural history in Normandy, and, as a mentor and quickly as a friend, he played a decisive role in reinforcing Caumont's interest to the subject.[102] In 1823 Caumont, at the age of 21, was appointed secretary of Lamouroux's brainchild, the newly founded Société linnéenne du Calvados.

In this exhilarating environment, Caumont found his scientific interests encouraged not only by Hervieu and Lamouroux of the older, local generation but also by occasional visitors from Paris. The most influential of these was Jules Desnoyers, Caumont's almost exact contemporary and former companion in Falaise, who was well launched in the capital on his way to national celebrity as a founder of the Société géologique de France, librarian of the Muséum d'histoire naturelle for over half a century, and a notable bibliophile and antiquarian whose work earned him membership of the Académie des inscriptions et belles-lettres.[103] Regular contacts, though, were more important, none more so than the emerging Norman network of younger naturalists, chief among them René Lenormand, a lawyer and authority on seaweeds from Vire, and Alphonse de Brébisson, a botanist from Falaise. Lenormand and de Brébisson had more specialized interests than were common among the older polymaths. But in all other respects they were typical of the voluntarist tradition in their independence of the academic institutions of the state and in being self-taught (though helped in de Brébisson's case by a father who was a capable entomologist and the author of an important *Flore de la Normandie*).[104]

From the rich menu of intellectual options that Caen offered, Caumont fixed on geology, and he embarked on journeys of more than 3,000 kilometers in preparation for a book-length text on the geognostic topography of the Calvados that he published in the *Mémoires* of the Société linnéenne in 1825.[105] It was a pioneering study of enduring value, accompanied by a geological map of the department (dedicated to Gerville) that was still thought worthy of reprinting more than forty years later.[106] Caumont went on publishing articles and maps treating the geology of the departments of the Calvados and the Manche until the mid-1830s.[107] But already his other abiding interest—in the region's antiquities—was eating into the time he could devote to science. By 1830, his teaching young people who chose to accompany him on his geological and antiquarian perambulations had given way to the more formal public lectures in Caen that formed the foundation of his learned six-volume *Cours d'antiquités monumentales* on the antiquities of Normandy and adjacent regions.[108] Five years later he resigned the secretaryship of the Société linnéenne to devote himself more fully to antiquarian studies and agricultural improvement.

In Caumont's mind, the intellectual displacement in the passage from geology to his new interests would have been a modest one. He had never doubted the practical value of geology for the region's agriculture, and ever since he had begun his travels in Normandy he had studied man-made monuments at the same time as he had classified the "monuments of nature" in the mineralogical realm. Like geology, Caumont's antiquarian and agricultural pursuits were entirely in keeping with his family background and sense of responsibility for his modest estate at Vaux-sur-Laizon (near Caen), devoted chiefly to the production of cider. Moreover, as domains in which the vitality of provincial life could be demonstrated and in which he could address with immediacy the men and women of every region, they provided ideal contexts for his ever more vocal commitment to the cause of decentralization. Local archaeology, antiquities, agriculture, and natural history were all elements in the almost sacred domain of independent cultural life that Caumont saw as the bulwark against ministerial intrusion.

Then, as throughout his life, Caumont remained loyal to his base in Normandy, an undifferentiated province in his eyes, despite its postrevolutionary sundering into five departments. It was entirely consistent with this nostalgic view (and almost certainly at his instigation) that the Société linnéenne du Calvados was renamed, soon after its foundation, the Société linnéenne de Normandie. Likewise, the antiquarian society that he helped to establish in Caen in 1824 and went on to serve, as assistant secretary and then secretary, for seventeen years was given the title of the Société des antiquaires de Normandie.[109] A few years later, the same regional objectives were enshrined in the short-lived *Revue normande*, which he founded in 1830, and the far more successful Association normande, a society that grew from the *Revue* in 1832 chiefly as a vehicle for promoting agricultural improvements through annual congresses, shows, and prizes.[110]

The common aim of all of Caumont's Norman initiatives was to foster not just a sense of regional identity but also a spirit of association that would result in a more systematic approach to innovation and help, in turn, to remedy Normandy's supposed intellectual and economic backwardness.[111] Even when limited to Normandy, it was a grandiose enterprise and, for Caumont's financial welfare, potentially ruinous. In its first year, the *Revue normande* attracted only thirty-four subscriptions, leaving him to find nearly 1,600 francs to meet the cost of production. Already, however, a solution was at hand. In January 1832, he married Aglaé-Louise Rioult de Villaunay, the daughter of an old Norman family steeped in the traditions of nobility, and from this moment, his ambitions, far from being inhibited, were enlarged to a coordinating mission embracing the provinces throughout France. Flushed with material security, he now embarked on a program of cultural devolution on a scale that allowed antiquarianism, science, and regional economic interests (in particular the interests of agriculture) to be pursued in a common context. Within three years of his marriage, he had founded another regional flagship, the highly successful *Annuaire des cinq départements de l'ancienne Normandie*[112] and inaugurated a national society for the protection of medieval and premedieval historic monuments—the Société française pour la conservation et la description des monuments historiques (the future Société française d'archéologie pour la conservation des monuments historiques).[113] He had also launched the greatest of all his enterprises, the Congrès scientifiques and Congrès archéologiques, peripatetic congresses that met each year in a different provincial town (though never, of course, in Paris).

Of the two congresses, the Congrès archéologiques have proved the more durable: in a barely modified form, though with a degree of ministerial involvement that would have horrified Caumont, they have been mounted regularly since the first session was held in Caen in 1834 and they remain important events in archaeology and antiquarian studies. In contrast, the Congrès scientifiques, the first of which took place (also in Caen) in 1833, lost momentum rapidly from the 1870s, following Caumont's death. Nevertheless, in the forty-seven years of their existence, they made a central contribution to fashioning the public face of science in the provinces. The core of their contribution lay in the intellectual diet of learned papers and lectures that were offered year by year in whatever town was selected for the annual meeting. But the impact of the congresses also depended on the theatricality with which they were mounted. Caumont's flair for publicity and the zeal he evoked in others ensured that the formal sessions always went hand in hand with a festive program of receptions, concerts, recitations, and dramatic performances that were more than a mere relaxation from the rigors of a packed program. Conviviality, in fact, was an indispensable way of integrating with a region's social elite and thereby of establishing a direct bond that cut out the capital. To that end, no congress would be complete without guided tours to collections and curiosities, and few congresses passed without an inaugural mass celebrated, wherever possible by an archbishop, after a solemn procession to whatever was deemed the most prestigious place of Catholic worship. Local regimental bands, too, were much in demand to provide entertainment during a firework display or the congress dinner, at which extravagant toasts and light-hearted revelry exposed the whole enterprise to ridicule. A cruelly satirical article with its accompanying cartoon (see figure 2) on the congress of 1842 in Strasbourg caught the tone.[114] Being able to read was said to be unnecessary as a requirement for participation in the congress: the mere intention to learn to read was enough. And, much as in figure 2, the inevitable wine-laden banquet was caricatured as a scientific investigation of the qualities of the flesh of chickens, ducks, and turkeys. When the gastronomic merits of the three birds had been examined and confirmed, as the article put it, "the results were henceforth acquired for science."

Certain very special excursions earned an indelible place in congress lore. One was the outing, during the Lyon congress of 1841, that took the participants in two boats from Lyon to Vienne, 30 kilometers away, to the sound of cannon and accompanying bands and choirs and amid such popular curiosity that crowds of over ten thousand in Vienne and twenty thousand in Lyon were said to have witnessed the event.[115] Although the outing was described as "a scientific excursion," the visits to the archaeological, scientific, and industrial curiosities of Vienne were certainly overshadowed by speeches and an open-air banquet for eight hundred persons. Three years later, at Nîmes, a specially chartered train carried members the more than 80 kilometers to Alais. On a line of the PLM railway company that had been inaugurated as recently as 1840, the journey itself was a novelty. But the highlight was the banquet that awaited in Alais. In the best traditions of the congresses, those attending the banquet were treated to recitations of verse: a monologue of Cornelian grandeur on Julius Caesar and two poems specially composed to honor "the dual power of intelligence and labor." Enthusiastic applause and a patriotic toast followed, before everyone made off for visits to the iron and steel works of Tamarys and the coal mines of la Grande Combe.[116]

Intellectually, the most aggressively proclaimed feature of the congresses was their emphasis on the oneness of learning. It was essential to Caumont's scheme that they should accommodate the whole range of learned activities and that participants should not regard themselves as exclusively attached to any one discipline. For most of their history, some changes of nomenclature apart, the programs of the congresses were organized in six sections: mathematical and physical science, agriculture, medicine, archaeology (a category that embraced all antiquarian studies), natural history, and a combined section for literature, *beaux-arts*, and philosophy. While the various sections exercised a measure of independence in the conduct of their business, they were never allowed to forget Caumont's guiding principle of intellectual unity, which made it essential to reduce clashes of timing to a minimum. Parallel sessions would have been unthinkable, and the sections duly met one after the other throughout the morning, beginning at seven o'clock, sometimes earlier. For the minority of zealots who followed Caumont's example by attending all or most of the sessions, as well as the plenary events that occupied the afternoons, the pace was grueling.

In their organization and peripatetic character, the Congrès scientifiques resembled two foreign precursors, the annual Versammlung deutscher Naturforscher und Aerzte that had circulated in the German states since 1822, and the British Association for the Advancement of Science, or BAAS, founded in York in 1831. In their mixture of plenary sessions and meetings of specialized sections, the Congrès scientifiques followed the German model (which Caumont particularly admired, though without ever having attended a congress in Germany[117]), and the pattern of congresses lasting between a week and ten days was the one already established in both Germany and Britain. In size, too, the ventures in the three countries were not wildly dissimilar. No peripatetic meeting anywhere could match the exceptional case of the Newcastle meeting of the BAAS in 1838, attended by more than 2,400.[118] But typical attendances of between 200 and 250 at the Congrès scientifiques in the 1830s were probably scarcely less than those at most meetings of the Versammlung der deutscher Naturforscher und Aertze.[119] Even the exceptionally well-attended meeting of the German body in Vienna in 1832, when more than 1,100 were present, had its counterparts in the Congrès scientifiques of 1841 and 1842, in Lyon and Strasbourg respectively, where the attendances were of the order of a thousand, larger than those for the BAAS meetings in those years.

Despite the outward appearance of similarity to their foreign precursors, however, the Congrès scientifiques bore the distinctive stamp of Caumont's vision. For him, the congresses were first and foremost a means of cementing a conservative union between a diversity of landowning, judicial, clerical, and industrial "establishments." As part of that objective, the kind of culture that Caumont promoted fulfilled the binding role of (in his words) a modern "truce of God." For such a calm consensus to be achieved, it was essential for contentious issues to be avoided and for the emphasis to be placed resolutely on the critical examination of facts rather than on speculation. The importance of the principle for Caumont emerged from the alarm that he and some of his closest associates voiced in an early running debate on the desirability of having a section devoted to social, political, and economic issues. Such a section had not been envisaged in the initial plans for the congresses. But during the first congress, in Caen in 1833, the decision to create an additional, sixth section of "social economy" was taken in response to the advocacy of Marc-Antoine Jullien, a former Jacobin turned liberal and one of the tiny handful of those attending who gave their address as Paris.[120]

Jullien's background marked him as an outsider to Caumont's circle, by age, political disposition, and profession. As the founder and, from 1818 to 1830, editor of the liberal periodical, the *Revue encyclopédique*, he had little natural sympathy for the religiously inspired conservatism that bound so many of those who gathered in Caen, and his dogmatic manner soon provoked further misgivings, which duly became public at the second congress, in Poitiers in 1834. The controversy sprang to life when the secretary for the congress, a local lawyer and antiquarian in the Caumont mold, Armand-Désiré de la Fontenelle de Vaudoré, warned his audience that discussions in the sixth section opened the door to the kind of political and religious debate that risked fomenting divisions among the "men of good faith but diverse political opinions."[121] For a while, the consensus in favor of the section survived through an uneasy alliance between those among the rank and file who believed de Vaudoré's disquiet to be exaggerated and a handful of conservatives among the leadership who believed that conflict could best be muted by embracing the potentially contentious disciplines: what mattered was that debate should rest on the sober "factual" discussion that was the congresses' stock in trade. But, with Jullien a continuing force at Douai in 1835 and Blois in 1836, the policy of conquest through assimilation failed. Sessions of the sixth section consistently attracted audiences larger than for any other section, and (most worryingly of all) they aroused precisely the public excitement that de Vaudoré predicted. A change of name to the section for "moral, economic, and legislative sciences" offered a thin cloak of protection. But in years of mounting bourgeois alarm at the kind of unrest that had erupted in the workers' riots in Lyon as recently as 1834 the cloak was inadequate.

It is not hard to imagine Caumont's relief when the departmental prefect delivered the coup de grâce, in the run-up to the Metz congress of 1837, by issuing a decree banning the section. Indignation at the decree was still sufficient to provoke a public outburst by François Chatelain, an independent "man of letters" and an ally of the congresses who insisted on their inalienable role as settings for the exercise of free speech and for anything but politically motivated dissension.[122] But Chatelain's compromise proposal that "social economy" should be added to the responsibilities of one of the other sections, or alternatively that the brief of the fifth section (labeled at the time "philology, literature, fine arts, philosophy") should be extended to embrace statistics as well, was rejected by a vote of sixty-nine to forty-five.[123] The dying embers of the debate flared into life briefly in the following year during the congress in Clermont-Ferrand, when Jullien returned to the fray. The opening session must have been a turbulent occasion. There, Caumont defended the suppression of the sixth section on the grounds that its debates had engendered personal conflicts and that in any case political economy and related subjects had from time to time been admitted as topics for discussion within other, appropriate sections.[124] Jullien did not admit defeat easily. Accusing Caumont of having abused his position as president of the congress in order to impose his will, he insisted on another formal vote, this time on the creation of a sixth section ("moral and social sciences, economics, and legislation") devoted to such matters as education, the organization of labor, and the condition of the poor and of women. The vote went heavily in Caumont's favor, and Jullien left, evidently enraged.

Surprisingly, Jullien retained a residual loyalty that led him to attend some later congresses and even to publish a poem in honor of the congress in Lyon and its Italian counterpart in Florence in 1841, a third of the series of congresses of "scienziati italiani.[125] But the question that he had pursued with such vigor was never raised again. In resisting his campaign, the Congrès scientifiques had turned their face against controversy and opted for the calm certainties of science, agricultural improvement, and local history. The decision was a telling one, reinforcing the suspicion of overtly ideological debate and a confidence in the power of indisputable evidence that characterized the congresses to the end, just as it characterized Caumont's own work.

Caumont's geology epitomized his approach. In his typically prim way, he declared it a mark of seriousness to avoid attempts at explanation. As he put it in his first report as secretary of the Société linnéenne du Calvados in 1824: "Modern geologists, learning from the errors of their forbears, restrict themselves to observing facts and to drawing only such conclusions as can be rigorously deduced from them, rather than seeking to explain their causes."[126] Similarly, in his antiquarian work, he insisted that "archaeology must be a positive science, as well founded as the physical sciences that derive knowledge from observation": only "facts whose authenticity is acknowledged in our own day" merited attention, and whereas the dating of styles had its place as solid knowledge, speculations about the origins of those styles and the pattern of their migration between regions and cultures remained too uncertain to deserve more than a noncommittal mention.[127] In the face of Caumont's conception of what constituted "positive" knowledge, scientific deviance could expect, and it received, short shrift. Characteristically, the report on a discussion of homeopathy at Lyon in 1841 made much of the openness of the congresses to all opinions. But, no less characteristically, it proceeded to record, with obvious satisfaction, the rout of homeopathic medicine by arguments founded on "cold, severe logic."[128]

Wherever science with religious or other nonscientific implications was in play, Caumont perceived the seeds of destructive conflict. A paper at the Clermont-Ferrand congress in 1838 in which Henri Lecoq identified thermal and other springs as the carriers of the organic matter that may have constituted the food of the earliest forms of life caused an all too predictable rumpus. What made it worse, in Caumont's eyes, was the accompanying suggestion that the waters might have been, and might still be, the setting for processes of spontaneous generation.[129] Caumont must have quaked as Lecoq, for whom the reality of spontaneous generation was essentially beyond question, elaborated his idea of life emerging at different points across the globe. It seemed that worse was to come when the abbé Jean-Baptiste Croizet, a knowledgeable geologist, rose to answer Lecoq. In fact, Croizet did no more than comment mildly on the difficulties that Lecoq's theories might raise for certain religious teachings. More disturbing, as it transpired, was the reaction of the right-wing mayor of Aurillac, Jean-Hippolyte Esquirou de Parieu, another typical member of Caumont's circle, who denounced Lecoq's views as leading on to the doctrines of Geoffroy Saint-Hilaire and hence to pantheism.[130] In replying that Saint-Hilaire's doctrines about the modifications of life since the creation (though unproven) deserved further examination, Lecoq risked an explosive reaction in such a predominantly conservative assembly. But there—mercifully for the peace of the congress—debate about the history of life rested, and not even the *Origin of Species*, which passed unnoticed by the congresses of the 1860s, could fan it back into life.

The Triumph of the Center

The decentralizing ideology of the congresses provoked tensions with the Ministry of Public Instruction that resulted in decades of more or less bitter conflict and reciprocal mistrust. To Chatelain and a number of other champions of the provincial cause,

Guizot, as the head of the newly independent Ministry of Public Instruction since 1832, was the villain of the piece.[131] It is not entirely clear where Chatelain believed Guizot to have failed. Although his criticism drew on a well of political animosity, it was intensified by memories of Guizot's inauguration, in 1830, of a national system of inspection for ancient monuments, which seven years later became the Ministry of the Interior's Commission des monuments historiques.[132] As Chatelain almost certainly believed, the centralizing tendencies of the commission, especially under its first inspector, Ludovic Vitet, ran counter to the spirit of provincial rehabilitation. Caumont, for his part, would have weighed that threat against the laudable commitment of Guizot, Vitet, and the best-known and longest-serving of the inspectors, Prosper Mérimée (1834–52), to the protection and restoration of France's patrimony of medieval architecture. The calculation was a delicate one. But Caumont's unquenchable aspirations to national as well as local eminence meant that he could never be blind to the need to maintain a modus vivendi that might one day earn him some kind of official role as a nationally recognized spokesman for provincial learning. And he fashioned his strategy accordingly.

Basic diplomacy and personal self-interest converged in Caumont's advocacy of Guizot's election as honorary president of the Congrès scientifiques of 1833 and 1835.[133] It was an astute move. But, once Guizot had finally left the ministry in 1837, the rising levels of public support for the congresses and of the Congrès archéologiques served only to multiply the possibilities of confrontation with ministerial ambitions. When attendances at the Congrès scientifiques reached their peak, as they did in 1841 and 1842 (in Lyon and Strasbourg), the Ministry of Public Instruction could no longer ignore Caumont's decentralizing agenda as an innocuous pinprick. Caumont was speaking, uncontrolled, to a dangerously large and influential constituency of sympathizers, and it soon became a matter for concern in Paris that by the early 1840s municipal councils, after several years of indifference, were beginning to make significant grants to support his initiatives. The council of Lyon led the way in 1841 with a grant of 12,000 francs;[134] thereafter, Strasbourg (1842), Angers (1843), Nîmes (1844), Reims (1845), and Marseille (1846) all allocated between 2,000 and 10,000 francs, to which the general council of the department concerned would sometimes add a contribution as well. In winning local patronage, personal contacts were crucial, never more so than in Marseille. There Pierre-Martin Roux, a prominent doctor and one of Caumont's closest associates, helped to secure not only 10,000 francs from the municipal council and 2,000 francs from the General Council of the Bouches du Rhône but also symbolic shows of support including the mayor's appearance in full uniform and a mass celebrated by the Bishop of Marseille.[135]

Another mark of public receptiveness to the call to decentralization was the proliferation both of local journals modelled on Caumont's *Revue normande* and of regional associations with functions similar to those of the Association normande. By the mid-1840s, four of these associations—the Association bretonne, the Association du Nord, the Association de l'Ouest, and the Association du Centre—had followed the Association normande by holding congresses, organizing exhibitions of agriculture and rural industry, and fostering studies of history, archaeology, and art, always with a strong insistence on the distinctiveness of their various regions.[136] Despite their individually modest nature, such ventures stood collectively as a reminder to central government of the recalcitrance of provincial traditions in culture and the lure that those traditions had for broad swaths of the aristocracy and the educated bourgeoisie. Just how broad the swaths were was brought home by the ability of the decentralizing movement, until well into the Second Empire, to call repeatedly on new men to repair the ravages of age. It might have been expected, for example, that the loss of such early leaders as Thomas Cauvin of Le Mans (a former Oratorian priest and the national doyen of heraldic scholars), Henri de Magneville of Caen (a landowner, agriculturalist, antiquarian, and pioneer of geology in the Calvados), and Charles Richelet of Le Mans (the most ardent of Caumont's allies in the cause of decentralization), all of whom died between 1846 and 1850, would lessen the momentum. But the men who replaced them in the leadership— Roux in Marseille, Charles Des Moulins in Bordeaux, and the distinguished lawyers Victor Simon in Metz and Ambroise Challe in Auxerre—were no less committed to the religious, political, and intellectual conservatism of their predecessors. All were self-taught in their scientific and scholarly specialities and, in varying degrees, polymaths with active interests in both the sciences and the humanities. In these respects, they were worthy successors of the founding generation.

While at this time as throughout the nineteenth century the vigor of the independent world of learning elicited words of praise from successive Ministers of Public Instruction, no minister could allow the vigor to grow unchecked, especially when it went hand in hand with pretensions to autonomy and an ideology that promised contention rather than obedience. By the mid-1840s, even a Minister of Public Instruction as sympathetic to the nation's *sociétés savantes* as the comte de Salvandy (now in his second period of office) was moved to act. In August 1846, as part of a review of the societies that resulted in the densely informative 1,035–page *Annuaire des sociétés savantes*, he made a conciliatory gesture by appointing Caumont as an unpaid "delegate general of the Ministry of Public Instruction to the *sociétés savantes* of the kingdom," with a roving mission to foster links between the ministry and the societies throughout France.[137] Caumont's initial reaction was one of delight at what he saw as a longed-for stamp of official approval, but soon his exaggerated conception of his role led to a break. Caumont, on the one hand, saw the appointment as a means of advancing his position as a largely independent leader of provincial intellectual life; Salvandy, on the other, regarded Caumont as no more than a biddable instrument that he might use in the usual ministerial quest for control and administrative tidiness. The two perspectives were irreconcilable, and the correspondence that Caumont and Salvandy exchanged between October 1846 and July 1847 became steadily more acrimonious. Caumont's response to

Salvandy's failure to clarify the duties of his "delegate" and to provide appropriate funding was one of hectoring indignation, while Salvandy became more obdurate.

Matters came to a head when Salvandy rejected Caumont's proposal for a 120–man national committee of representatives of the *sociétés savantes*, the grandly titled General Council of the Academies of the Kingdom. Caumont fumed at the emptiness of the ministry's commitment, and his indignation boiled over in a letter to Salvandy: "I soon recognized that the title was meaningless and that the employees of the ministry wanted to do nothing."[138] Such language was calculated to infuriate, and the letter was read accordingly. The conviction within the ministry was that Caumont had displayed an unwarranted presumptuousness and that his activities, far from being encouraged, should be curbed. The main target for attack became the Institut des provinces, a body that Caumont had founded in 1839 on the model of the national Institut de France. On the face of it, the Institut was innocuous enough: a vehicle for honoring, by election, two hundred of the most distinguished scientists and scholars working in the provinces, its activities were few and obscure.[139] It met no more than twice a year, always in a provincial town: one meeting was held either at Le Mans under the aegis of Cauvin or (after Cauvin's death) at Caen, while the other usually took place during the annual Congrès scientifique or Congrès archéologique. And its limited funds, which came largely from Caumont's pocket, allowed it to produce little more than occasional medals, a short-lived series of *Mémoires*, and a modest *Annuaire* listing its members and containing an account of its doings.[140] In the absence of significant material resources, its one strong suit was rhetoric. Self-styled as "the most eminent academy of the kingdom" (after the Institut de France, which was graciously excepted), it declared itself to be nothing less than "the peerage of the men of letters and savants of the provinces."[141] In this description, "peerage" was given special force by a membership liberally adorned with names drawn from the nobility as well as from the provincial *notables* of the Church and the learned professions.

At one level, the pretensions of the Institut des provinces simply reinforced a sense of the harmless posturing that most contemporaries saw as characteristic of all of Caumont's initiatives. During his first period of office as Minister of Public Instruction, between 1837 and 1839, Salvandy seems to have regarded the initiatives in precisely this light. But after his return to the ministry in 1845, his attitude hardened. Frosty politeness quickly descended into a state of open warfare, aggravated in the early months of 1847 by the increasing incivility of Caumont's letters to Salvandy about his roving commission with regard to the provincial societies. By now, ministerial patience was exhausted. In a move that he evidently saw as decisive, Salvandy instructed prefects throughout France to forbid the *sociétés savantes* of their various departments to associate with the Institut des provinces.[142] Such firmness, if implemented, would certainly have trammeled Caumont's campaign and achieved Salvandy's ultimate aim of a total ban on the institute. But the instruction was soon overtaken by the tumultuous end of the July Monarchy in February 1848. With Salvandy swept from office, it was now left for the Second Empire's first two Ministers of Public Instruction—Hippolyte Fortoul (1851–56) and Gustave Rouland (1856–63)—to begin the struggle for control afresh.

The champions of educational and intellectual decentralization with whom Fortoul had to deal emerged in good heart from the uncertainties about France's political destiny that marked the years from the Revolution of February 1848 to Louis-Napoléon's coup d'état in December 1851 and the declaration of the Empire twelve months later. The freedom that had been granted to primary education (by the Guizot law of 1833) and then to secondary education (by the Falloux law of 1850) meant that hopes of a successful campaign against what remained of the Ministry of Public Instruction's monopoly ran high, the more so as action could now be focused on a narrower front. Although the leading goal was the very specific one of extending the principle of freedom to higher education, that objective was inseparable from ambitions of a more general kind for a weakening of the tentacles of ministerial power. It was predictable, therefore, that many of the conservative spokesmen who led the campaign against the hated *"monopole universitaire"* were close, or at least sympathetic, to Caumont.

Among the most prominent of the spokesmen was Charles Forbes, comte de Montalembert, a member of the Institut des provinces since its earliest days, a long-standing enemy of state influence in religion as well as in education, and beginning in 1851 a member of the Académie française. An historian of strong, albeit liberal, Catholic inclinations (which led him to oppose Littré's entry into the academy), Montalembert moved easily into an alliance with some of the most resolute clerical champions of the freedom of education, including the Bishop of Orléans, Félix Dupanloup, an articulate and reflective enemy of the lingering traditions of Enlightenment thought (though not of science), and the ultramontane Bishop of Arras, Pierre-Louis Parisis. As members of the Institut des provinces, Dupanloup and Parisis were glad to endorse Caumont's ideals by assuming prominent roles when the Congrès scientifiques of 1851 and 1853 met in their dioceses. For Dupanloup, the opening ceremony in Orléans was the ideal occasion for an eloquent declaration of the Catholic Church's solicitude for the progress of science and learning. That progress, as he insisted in words that had long been common currency in Catholic circles, could only be achieved by the true savant, one who recognized (as François-Victor Rivet, the Bishop of Dijon and president of the Dijon congress of 1854, put it) that God was "the first and the last word in all things.[143]

The continued momentum of Caumont's enterprises in the early 1850s brought him into inevitable conflict with a Ministry of Public Instruction that he found to be as unyielding as ever. In some respects, in fact, the hostility of Fortoul and (less markedly) Rouland toward Caumont outdid Salvandy's. The main reason for this hardening of attitudes lay in ministers' anxiety at the growing support, in many sectors of the church, for a distancing from the imperial regime. As ministers sought to deal with this broader concern, Caumont and the movement he represented came inevitably under fire. The easiest target was the irregularity of the title of the Institut des provinces, which violated the law of 1 May 1802 (11 floréal year X), reserving the title "Institut" exclusively for the Institut de France.[144] Formal prohibition, however, proved difficult since the legislation had stipulated no sanction against an offending institution.[145] Frustrated by this legal nicety, Fortoul and Rouland resorted to harassment, most provocatively by marshaling their resources so as to duplicate initiatives already taken by Caumont.

Fortoul was pursuing precisely this policy when, in 1854, he launched his *Bulletin des sociétés savantes*, a periodical that two years later became the *Revue des sociétés savantes de la France et de l'étranger*, a large monthly digest and bibliographical review of the work of societies throughout France and, in less detail, of those in other countries. Faced with a rival sustained by the virtually unlimited resources of the ministry, the modest *Bulletin bibliographique des sociétés savantes des départements* that the Institut des provinces had published since 1851 had no hope of survival, and it quickly died. The strained politeness of the editorial note by Caumont's collaborator and leading light of the Association bretonne, Armand Du Chatellier, in the last issue of the *Bulletin* in 1854, did nothing to conceal the ill-feeling that the ministry's sudden and unannounced launching of the *Revue* had caused.[146] A rather poorly printed thirty-two-page digest appearing every two months could simply not compete with the official publication, which soon settled into a rhythm of publishing that offered more than six hundred pages a year for an annual subscription of 9 francs (compared with 5 francs a year for the *Bulletin bibliographique*).

Even more blatant, and crushing, was Rouland's inauguration, in 1861, of a ministerial Congrès des sociétés savantes, held each year in one of the greatest of France's academic shrines, the Sorbonne. The congress provided representatives of the *sociétés savantes* of both Paris and the provinces with an attractive forum in which they could present the best of their work before the élite of French science and scholarship and receive the accolades of prizes and encouragement from the minister himself. The existing independent counterpart, the annual Congrès des délégués des sociétés savantes des départements, organized in Paris under the auspices of the Institut des provinces, was an altogether more humdrum affair that never had the *éclat* of the ministerial gatherings. Its first three sessions, in 1848, 1850, and 1851, got off to a promising start: each lasting several days, they had been held in the gracious surroundings of the Palais du Luxembourg. But after 1852 the congresses found themselves relegated to the less prestigious quarters of the Société d'encouragement pour l'industrie nationale in the rue de Rennes.[147] For a gathering whose delegates came from the provincial (as opposed to Parisian) societies and whose most notable dignitaries were the leading lights of the Institut des provinces, there could be no question of official recognition, still less of the presence of the minister. Even attendance by a leading professor was a rare event. Among scientists of national standing, only Quatrefages and Alexandre Baudrimont, both of whom knew the constraints of life in a provincial faculty, were willing to jeopardize their hopes of professional advancement by attending regularly. For most employees of the ministry, the experiences of three professors in the local faculty of science who rashly agreed to serve as officers at the Congrès scientifique in Toulouse in 1852 and then withdrew, following a formal ministerial rebuke, were warning enough.[148]

Such episodes, however, served only to bolster the resolve of the inner circle of decentralizers, who maintained an unwavering rhetoric in support of activity, initiative, and honesty as antidotes to the idleness, passivity, and duplicity that resulted when the provinces succumbed to the deadening hand of Paris. The language of the dichotomy was, of course, Caumont's, and its institutional embodiment lay in the seat of "true" progress, the Institut des provinces.[149] Throughout the Second Empire, neither the ideals nor Caumont's sensitivity to what he was always ready to interpret as ministerial aggravation changed. In its undiminished virulence, his address to the Congrès scientifique in Chartres in 1869 was typical. In an emotive portrayal of a provincial intellectual life prey to the stifling effect of centralization, Caumont called yet again for action if the provinces were not to die:

> To regain activity and momentum, in fact, we need to free the provinces from the bonds that shackle them. We need true decentralization. There are those, I know, who cannot understand what it means to have *variety in unity*, as if variety did not exist in nature, as if the north were no different from the south, or Brittany from Franche-Comté, or Poitou from Dauphiné. Such people, not content with the inexorable stultifying unity that demoralizes us with regard to intellectual matters, would like to tighten the bonds still further, so as to make the unity more complete. They go out of their way to find new means to that end and keep repeating "Moisten the rope, to shorten it." We for our part say "Avoid moistening it. Cut it rather, and cut it quickly, for paralysis is spreading, the provinces are stifling, and they will die."[150]

By the late 1860s, however, Caumont's vehemence owed as much to accumulated frustration as it did to any sense of imminent achievement. The Ministry of Public Instruction's flagrant duplication of his initiatives in the 1850s and early 1860s had delivered a telling blow, and the Ministry of the Interior's decision to dissolve the Association bretonne in 1859 on the grounds of its "political tendencies" (which the minister deemed to be "mauvaises") had presaged the same cold wind.[151] The arrival of Duruy in the Ministry of Public Instruction in 1863 had brought with it a measure of controlled liberalization, though of a kind that did nothing to advance the cause of intellectual decentralization in the sense in which Caumont understood it. Under Duruy, in fact, the odds came to be stacked more heavily than ever against the autonomy of the independent world of learning. Rail travel (which made journeys to Paris incomparably easier), the easy availability of popular books and periodicals, and a proliferation of contexts (exhibitions, museums, and national societies) in which science at the highest level appeared on the public stage all gave the claims of Caumont and his circle an air of faded hopelessness. The deliberations and publications of the academies that stood at the head of Caumont's hierarchy appeared dry when set beside the exciting, imaginative science of such masters of popularization as a Camille Flammarion or a Louis Figuier or a public lecture on spontaneous generation by Pasteur before a packed audience at the Sorbonne (see chapter 5 for more on the popularization of science). And what provincial savant would choose to present his reports on storms or his readings of sunshine, wind, and rainfall on the parochial stage of a local society when by the 1860s Le Verrier was offering the Paris Observatoire as an eager recipient of meteorological observations from across the country?

If these trends were not damaging enough, there soon followed the humiliation of the Franco-Prussian war and the transition to a régime that had no time for Caumont's nostalgic conservatism. Caumont himself did not live to see the collapse of his cause. In August 1870, he was in Moulins for a Congrès scientifique held despite the war. But in the following year he was incapacitated by a stroke, and in 1873 he died. His foresight in arranging for some devolution of responsibility from the late 1860s, reinforced by the pockets of intense personal loyalty he always managed to inspire, ensured that all impetus was not lost. The Congrès scientifiques and the Parisian Congrès des délégués des sociétés savantes des départements limped on for a few years, sustained by the organizers' undimmed commitment but with signs of administrative disorder. Congrès scientifiques in Pau (1873) and Rodez (1874) proceeded much as usual, but the postponement to 1876 of both the meeting originally planned for 1875 in Autun and a hastily arranged replacement meeting in Périgueux marked a fatal loss of rhythm. Thereafter, through the later 1870s, the national network that Caumont had maintained with such skill and energy disintegrated, leaving the annual Congrès archéologiques as the sole reminder of a campaign that had marked provincial intellectual life for half a century.

In the last years of the Congrès scientifiques, the secretary and effective leader of the Institut des provinces, the independent Bordeaux naturalist J.-E. Druilhet-Lafargue, secretary of the Société linnéenne de Bordeaux and a prominent Catholic, did what he could to fill Caumont's shoes. The acrimony he displayed in his relations with the Ministry of Public Instruction would have done credit to Caumont himself. But his resolve to retain the title and, more important, the independence of the Institut des provinces smacked of fatal obduracy. When the minister offered a measure of official recognition if only the institute would change its name, Druilhet-Lafargue peremptorily dismissed the compromise.[152] This was the final straw, and no further olive branches came his way. At the last of the Congrès scientifiques, in Nice in January 1878, the battle cry of intellectual decentralization was still heard, and the two volumes of congress proceedings were among the most ambitious and best produced that had appeared for some years.[153] But fewer than 140 attended the congress, and although a high proportion of doctors, other professional men, and landowners lent the gathering an air of familiarity, the absence of Catholic dignitaries spoke eloquently of the changed climate. By this time, senior representatives of the church in France had more important battlefields on which to fight in the defence of faith and tradition against secularization and modernity: a Congrès scientifique was no longer a setting in which they could effectively make their case. So the congresses died. There was no formal winding down; the congress planned for Brest in 1880 just did not take place. The same was true of the Parisian congresses, the last of which was held in 1880, and the Institut des provinces, which simply ceased publishing and became inactive.

The collapse of these institutional settings marked the end of a grandiose conception of how an independent world of learning might be organized. Yet it left some of the core pursuits of that world intact. Regional antiquarianism continued to flourish, and in France as everywhere sciences rooted in collecting and dispersed local observations, such as oceanography for the leisured and wealthy, and ornithology and certain types of botany and zoology for a larger public, went their way much as they always had done. At an unusually high intellectual and organizational level, the founding of the Société mycologique de France in 1884 illustrates the vigor of a pursuit that allowed doctors and pharmacists with access to remote sources of fungi, such as those in the department of the Vosges, to carve out an intellectual niche in which the professional academic contingent was small.[154] But domains in which the devotee could continue to excel and secure a measure of national visibility were now significantly reduced. Two of the most important national societies that were founded shortly after the Franco-Prussian war, the Société française de physique and the Société mathématique de France, reflected the inevitable marginalization of amateur enthusiasts in disciplines to which by now the self-taught could not hope to contribute (see chapter 6). Catering as the new societies did for the expanding communities of formally qualified physicists, mathematicians, and engineers regardless of occupation, they could preach the virtues of openness. But the reality in both cases was a strong orientation to the academic world, with a parallel opening, in the case of the Société française de physique (whose publications and exhibitions always reserved an important place for applied physics), toward industry.

With academic professionalization advancing at an accelerating pace, even those national societies whose intellectual territories were accessible to the lay public had to make adjustments. The Société zoologique de France was typical.[155] At its foundation in 1876, the society had the ambience of a naturalists' club in which amateurs felt at ease. It began as the brainchild of Aimé Bouvier, a veteran of the French scientific expedition to Mexico in the mid-1860s and now a dealer in specimens with a slightly dubious reputation and an eye on the popular market by which he lived.[156] Its public soon grew, and within less than a decade it had become a popular meeting place for zoologists from the bourgeoning Parisian world of higher education and research. In this transition, the arrival of a young academic histologist, Raphaël Blanchard, as secretary in place of Bouvier in 1880 marked a decisive new departure. For while Blanchard was successful in deploying a variety of means—his personal friendliness, festive annual banquets (see figure 2), and a program of excursions—to maintain a place for seriously interested collectors, he also witnessed (with obvious regret) the tendency for leading roles in the society's meetings and publications to be taken by his academic peers in the discipline.

So long as the discussions and publications did not become too specialized or esoteric, the kind of balancing act that Blanchard attempted was possible: not only the Société zoologique but also other national societies with a strong dimension of collecting and fieldwork—the Société géologique, the Société entomologique, and the Société botanique—all worked hard to retain a wider public of members who were not professionally engaged in the discipline concerned. It was an uphill struggle: a dip in the average attendance at meetings of the Société entomologique, from twenty-seven in the late 1860s to eighteen people twenty years later reflected the difficulty of maintaining even a tiny audience.[157] Nevertheless, the societies survived, and the hard core of committed members on which all of them depended continued to use publications and a mixture of Parisian and provincial lectures and excursions in pursuit of a balance between the interests of the academic specialist and those of the serious amateur. On this count, Alfred Lacroix, professor at the Muséum d'histoire naturelle and member of the Académie des sciences, conveyed the spirit of openness when he used his presidential address to the Société géologique in 1910 to praise amateurs as essential to the life of any scientific society.[158] This was no mere window dressing.

In a similar way, though on a far broader scientific front, a characteristic product of the early Third Republic, the Association française pour l'avancement des sciences, made a principle of fostering the bonds between professional and nonprofessional communities through the annual congresses it organized in towns around France from its foundation in 1872 (see chapter 6). The declared aim of the AFAS—to encourage scientific activity beyond the confines of Paris—sat well with its auxiliary objective of opening science to the educated lay public. Superficially, the association perpetuated Caumont's model, with its migratory annual congresses held each year in a different provincial town: wherever it went, the AFAS drew on local knowledge and expertise both for organizational skills and for the all-important task of assembling an audience, just as successive Congrès scientifiques had done. But the profile of the support that was sought and proffered for a meeting of the AFAS was very different. From the start, its leaders came from the academic elite and hence from circles in which an overriding priority was the promotion of science in the faculties and other institutions of the state: with the provincial faculties now harboring a new generation of professors and lecturers whose growing prominence and (in some notable cases) indifference to the attractions of Paris have been well analyzed by Mary Jo Nye,[159] the resource was a rich one. Neither among the leaders of the AFAS nor among the rank and file subscribers was there any counterpart either to the sense of resentment against the capital or to the more specific suspicion of the university that inspired Caumont and his immediate circle. The watchword of the AFAS was cooperation between center and periphery, not confrontation.

By the end of the nineteenth century, there had been a sea change in the world of independent academies and societies. It is important to stress again that the world remained a populous one: my own estimate is that in 1914 more than two hundred thousand Frenchmen (though still strikingly few women) saw fit to pay their subscriptions to a society. Some of them, especially in the national disciplinary societies, subscribed in pursuit of professional advancement in academic careers that engendered a growing demand for opportunities for the discussion and publication of research. On the nonacademic periphery with which this chapter has been mainly concerned, however, priorities were quite different. Here, the members of an academy or society were more than ever the consumers, and not the creators, of the knowledge about which they read, heard, or talked. Most of them, in fact, were a subset of the growing body of overwhelmingly bourgeois visitors to museums, galleries, and universal exhibitions and readers of books and periodicals destined for the lay public. And their aspirations were limited accordingly. One thing to which such members and their societies did not aspire was anything approaching a vision of autonomous intellectual excellence of the kind that Caumont harbored until his death. Indeed, even Caumont's name was one that they would scarcely know. This is not to say that the voices on the periphery were silenced. But the voices spoke far more discretely and were heeded by far fewer in the educated classes than had been the case half a century before. In that sense, the triumph of the center, exemplified in the now uncontested authority of the savants of the universities and national research institutions, had been crushing and irreversible.

CHAPTER THREE

Science, Bureaucracy, and the Empire

•••

The Second Empire has had a bad press as a period in French intellectual history. Its dominant culture has generally been regarded as superficial and lacking seriousness; its treatment of the professoriate of the University has been seen as repressive; and its values have been characterized as shot through with the unadventurous bourgeois conformity of a society in which maintaining order, after the fright of the revolutionary troubles of 1848, was the overriding priority. All those perceptions have a substantial germ of truth that no amount of revisionism could dispel. The empire was a difficult time for the critics of moral and intellectual orthodoxy: the promoters of secular thought and the savants who found themselves, for one reason or another, at odds with the entrenched leaders of the academic establishment in the 1850s and 1860s learned that lesson to their cost, as eventually did Victor Duruy, the cautiously reforming Minister of Public Instruction during the later, more liberal years of the empire. This is not to say that new ideas were suppressed, still less that the cause of educational modernization (in the realm of vocational instruction as in higher education) was ignored. It was in this period, after all, that the atomic theory and the new energy physics won at least a measure of acceptance in France; Claude Bernard was in his prime as the pioneer of a new experimental medicine; and liberal intellectuals reveled in the heady mix of ideas, especially those of materialism and the higher biblical criticism emanating from Germany, that precipitated Ernest Renan's suspension and eventual removal, in 1864, from his chair of Hebrew, Chaldaean, and Syriac at the Collège de France.[1] Yet novelty of a kind that proffered even mildly disturbing moral implications was almost invariably achieved against the odds.

In the years that separated Louis-Napoléon's seizure of dictatorial power in December 1851 from the collapse of the Second Empire in September 1870 following the ignominious defeat at the hands of the Prussians at Sedan, the desire for orderliness and discipline in the academic world remained a consistent characteristic. Any account of intellectual life in the period has therefore to rest on a perspective rooted in, though not limited by, the perceptions and policies of the administrators who fashioned the empire's provision for science and learned culture. The leitmotif that emerges is of a succession of Ministers of Public Instruction—Hippolyte Fortoul (1851–56), Gustave Rouland (1856–63), and Duruy (1863–69)—who sought meticulous control of their domain in order to protect it from the very different threats from free thought, on the one hand, and the clerical challenge to its uncompromisingly secular character, on the other. Control and protection, however, always went hand in hand with fostering certain initiatives. Even the most conservative of the three ministers, Fortoul, resolutely pushed through a program of measures aimed at making the faculties of science more responsive to the needs of their various local economies. And throughout the empire, the introduction of modern curricula appropriate to the industrial age was a high priority at both the secondary and the higher levels of education. Hence, this was not a period devoid of innovations, far from it. But the innovations bore the thumbprint of an administration that identified overwhelmingly with orderly middle-class values and nationally defined priorities rather than those of the international world of learning. The kind of research that received public support might be utilitarian and painstakingly exact (as in Victor Regnault's experimental studies of the thermal properties of gases, sponsored by the Ministry of Public Works through its Central Steam Engine Committee), descriptive (as in the initiatives for scientific, topographical, and archaeological compilations, conceived as exhaustive inventories of the nation), or patriotic (as in the emperor's own efforts to identify the site of Alesia, where the Gaul Vercingetorix made his valiant last stand against the might of Rome). Such ventures all broke new ground, as did the moves, instigated by Urbain Le Verrier, to build a meteorological service using data from amateur observers throughout France. New ground, however, does not imply dangerous ground, and it was that conforming and heavily bureaucratic conception of intellectual life, rather than inactivity, that characterized the empire.

The Trials of Academic Science

During the July Monarchy, the professors of the University's dispersed network of faculties had passed from the status of a privileged, if circumscribed, elite to that of just one contingent in the nation's growing army of educational administrators. The change had been effected through the Ministry of Public Instruction, instituted as an independent administration in 1832 and ruled throughout the monarchy by a succession of ministers who, with varying degrees of commitment, saw higher education as a mechanism for satisfying the bourgeois taste for cultural distinction, safe morality, and career making. In the process, scientific or scholarly originality and intellectual freedom, though routinely trumpeted as ideals in ministerial literature, had always played second fiddle to a bureaucratic preoccupation with qualifications, honorable conduct, and efficient service to region and state.

The Second Empire saw the apotheosis of this trend. In the faculties of science, one indicator of bureaucratization lay in the regularization of the formal requirements for the holding of chairs; whereas two-thirds—fifty-two out of seventy-three—of the new professors appointed between 1830 and 1849 held the *doctorat-ès-sciences*, the proportion rose to 90 percent between 1850 and 1859 and to 100 percent between 1860 and

1869 (see table 1 in chapter 1). Even more telling for the professoriate's collective self-esteem were the relations that professors had with the ministers and lesser officials to whom they were irrevocably answerable. Only under Victor Duruy, whose ministerial reign encapsulated the ideals of the relatively liberal empire of the 1860s, did *universitaires* feel that they were serving a minister who understood their problems and aspirations. In contrast, under the two ministers who served in the empire's earlier, authoritarian phase—Hippolyte Fourtoul and Gustave Rouland—professors generally found themselves struggling against, rather than with, their administrative masters for at least some measure of the autonomy to which most of them aspired.

The sense of frustration attendant on the struggle was especially keen under Fortoul. Appointed in December 1851, Fourtoul came to the ministry as part of the new political wave that began in that month with the coup d'état and culminated a year later in the declaration of the Second Empire. His main inheritance was the choppy wake that followed the passage of the conservatively inspired Falloux law of March 1850.[2] By the law, the state finally yielded to many years of pressure to abandon its monopoly in secondary education. Henceforth, a secondary school could be established by anyone over the age of 25 and of French nationality who held the *baccalauréat* and had been a teacher or teaching assistant at secondary level for a minimum of five years. As mastermind of the legislation and Minister of Public Instruction, Alfred de Falloux intended that the measure promote above all the interests of Catholic education as an alternative to the state's diet of secular instruction, which conservative opinion saw as a primary cause of the revolutionary troubles of 1848. The reorganization of the administrative structure that went with the law pushed in the same direction. Catholic interests were significantly, and deliberately, advanced by the decision to increase the number of academies from twenty to eighty-six, one for every department, each headed by a rector, who presided over a locally appointed academic council made up of religious, legal, and civil dignitaries, with only a minimal presence of serving *universitaires*.[3]

The controlled devolution of power that was enshrined in the Falloux law made life uncomfortable for those teaching in the University, especially in the various faculties. The exercise of administrative authority on the local scale by nonacademic elements in society was bad enough. But the impact was heightened by a deliberate marginalization of the national committee in charge of education, the Higher Council of Public Instruction (Conseil supérieur de l'instruction publique). Especially in its earlier guises, as the Royal Council of Public Instruction and Council of the University (see appendix A for more on how the French educational system was organized in the nineteenth century), the council had spoken effectively for the professional interests of teachers in the University, albeit under a watchful ministerial eye. But it now found itself turned into little more than a minor consultative body. Most damaging, Falloux's restructuring of the council left the once powerful "permanent section" of eight members (appointed by the president of the council from the senior administrators of the University and professors in the faculties) outweighed by the fourteen clergy and political and legal figures, three representatives of the independent sector of *enseignement libre*, and the minister himself as chairman. The intention, as a report on the Falloux law by the liberal Catholic parliamentarian Auguste-Arthur Beugnot stated, was to give control of education to society, though by obvious implication to society in its most conformist manifestations.[4]

This structure, which Fourtoul encountered on his appointment in December 1851, was well suited to his immediate political priority of allaying the anti-Jacobin fears of the conservative families whose votes had brought Louis-Napoléon to power. A common perception of the University among these families was of a dangerously self-indulgent and poorly controlled corporation all too open to radical ideas. One response to such a perception might have been to attack, by subjecting the University to systematic hostility and a vindictive budgetary policy. Fortoul's strategy, however, was different: it was aimed rather at removing some of the most obvious targets for public criticism and enhancing the strict discipline exercised by his ministry. An early decision to do away altogether with the Higher Council's already emasculated permanent section enshrined his vision.[5] In the retrospective judgment of Charles Jourdain, an eminent literary scholar and administrator in the Ministry of Public Instruction throughout the empire, the measure was the most damaging of all those that Fortoul initiated in what were increasingly seen as the dark days of the early and mid-1850s.[6] Yet it can equally be interpreted as a mark of timely pragmatism. Arguably Fortoul had no choice but to trim the autonomy of higher education if he was to defend himself against the charge of presiding over an unruly and unnecessarily costly domain. Control and order were certainly main objectives in the decision, taken in the spring of 1854, to abandon the fragmented structure of the eighty-six departmentally based academies and replace them with sixteen more easily administered academies, each centered on a town with actual or potential strength in higher education.[7]

The other great reform of Fourtoul's administration, and the one that earned him most notoriety, was likewise intended to contribute to the fashioning of a national system of secondary and higher education that would enhance the Bonapartist regime's trademark virtues of modernity and efficiency, while standing firm against socially disruptive secular doctrines, on the one hand, and religiously inspired reaction, with its ideals of devolved power, on the other. This was the separation, dating from April 1852, between the scientific and literary streams in secondary education, which now became parallel tracks of equal status.[8] With this measure, always referred to as the *bifurcation*, went an associated remodeling of higher education in which the six existing *agrégations* (including that in philosophy, to which Fortoul was especially opposed[9]) were abolished and replaced by just two: one in science and one in letters, each with a vast but, as critics insisted, shallow syllabus.[10] The aim was, quite explicitly, to break with a system of qualifications that fostered improper intellectual pretensions. Fortoul's principle was the traditional one that *agrégés* should see themselves as, first and foremost, the transmitters of received knowledge rather than the unbridled demagogues that some of them aspired to become. In his words, lessons were to be "dogmatic and purely elementary" and those who taught them were, at all costs, to resist the temptation to yield to "ideas of the moment and the caprices of fashion."[11]

The unmistakable message that self-effacement, moral rectitude, and political conformity were now the order of the day provoked resentment in many quarters, nowhere more so than in what had long been one of Fortoul's blackest bêtes noires, the Ecole normale supérieure. There, the *bifurcation* and the associated measures were seen, rightly, as a threat to the school's intellectual independence. Preparation for the practice of teaching—the kind of "harsh novitiate" that Fortoul thought appropriate—assumed further prominence through new regulations that focused instruction on material relevant to the secondary curriculum and only allowed *normaliens* to attempt their chosen *agrégation* once they had served for three years in a *lycée* after leaving the school.[12] In the same spirit, the purely linguistic aspects of Latin and Greek were enhanced, at the expense of philosophy and history, both of which were reduced to a level of safe, uncontroversial superficiality. And a new requirement that pupils should take the *licence*, in science or letters, after two years rather than one intruded deliberately on a second year that had previously been a time of relative intellectual freedom for *normaliens*.

The yoke that Fortoul imposed on academic life weighed heavily in the humanities and thereby earned itself a place in the folklore of hardship that philosophers, historians, and literary scholars rehearsed (and arguably exaggerated) for the rest of the century.[13] The academic philosopher Louis Liard, writing in the very different atmosphere of the Third Republic, was categorical: "The early years of the Second Empire were, by general agreement, the most wretched ones that public instruction has had to endure this century: no other period has left such bad memories."[14] As one of the republic's most distinguished educational administrators, Liard spoke with authority. But his perspective was very much that of the humanities. In the sciences, the effect of Fortoul's reforms was less obviously damaging, and protests at the time were rare. Indeed, both the astronomer Urbain Le Verrier and the chemist Jean-Baptiste Dumas seem to have had a significant hand in fashioning the whole system of *bifurcation*, and it was a committee chaired by Thenard and reported on favorably by Dumas that formulated the scientific curricula.[15] The attractions of the new system, in scientists' eyes, are not hard to see. With the new *baccalauréat-ès-sciences* established as an alternative, rather than a sequel, to the *baccalauréat-ès-lettres,* pupils were free to give priority to the scientific disciplines at the age of 14 (i.e., in the *classe de troisième*) and thereafter to reduce significantly the time they devoted to the classical languages.[16] Despite cries of alarm at the weakened position of the humanities, many seized the opportunity of an earlier concentration on science and mathematics. The number of those taking the *baccalauréat-ès lettres* fell by half between 1852 and 1853, while the corresponding figure for the *baccalauréat-ès-sciences* soared, as did both the number of science teachers and the income generated for examining by the faculties of science.[17] Supported by an educational philosophy rooted in the utilitarian Saint-Simonian inclinations of Fortoul and Michel Chevalier, the sciences had scored a signal victory.[18]

In higher education, too, it could be argued that the sciences suffered little and that, in some respects, they benefited. One explanation is that most scientific study could be seen as ideologically neutral. Certainly both Fortoul and his successor Rouland saw whole areas of science in this way: astronomy (provided it was treated descriptively), natural history (purged of speculation about the origin and history of life), and aspects of physics and chemistry that focused on measurement and observation ran no risk comparable with the threat from deviant opinions or satire (such as Victor de Laprade's poem "Les Muses de l'état") voiced by professors in the humanities.[19] It is no coincidence, therefore, that a number of scientific activities enjoyed a new prosperity. One conspicuous beneficiary was the science section of the Ecole normale supérieure, whose facilities in the rue d'Ulm Fortoul enhanced by installing the research laboratory of the Sorbonne there in February 1855.[20] Later in the year, as part of a modest easing of the constraints on the intellectual life of *normaliens* and their teachers, both the sciences and the humanities benefited from the introduction of a program of advanced studies.[21] To qualify for entry to the new "upper division" twelve of the ablest students, six in science and six in letters, would attempt the *agrégation* at the end of the normal three years of the curriculum. If successful, they would be allowed to stay on for a further two years. In this time, they would have the provisional status of *agrégés* and a stipend of 600 francs a year plus room and board. In return, they were to pursue carefully supervised doctoral research that would in due course qualify them for a chair in a faculty or, failing that, for an accelerated passage into a *lycée*.[22]

Significant though it was, this injection of vigor into the intellectual life of the Ecole normale supérieure did not seriously compromise the school's conformist ethos. Advanced students were subjected to a system of biannual reporting and the constant threat of expulsion for either poor performance or bad behavior. Nevertheless, the breath of somewhat fresher air was important in preparing the way for the change in the intellectual tone of the school that followed Rouland's arrival in the ministry in 1856. The main catalyst of this transformation, especially in the sciences, was Louis Pasteur, who moved to Paris from Lille in 1857 as part of the major reorganization that brought him in as the school's administrator and director of scientific studies. Under the overall directorship of the classical and literary scholar Désiré Nisard, Pasteur rethought the privileges of the upper division with a view to further promoting research among the best of the students in his care. After 1858, his reinvigoration of the rank of *agrégé-préparateur* (to accommodate which he now sacrificed the upper division) advanced the process by giving a small number of advanced *normaliens* both elite status and three (rather than two) years of relative freedom in which to prepare a doctoral thesis, in return for a part-time engagement as laboratory and teaching assistants.[23] As Pasteur intended, the measure, reinforced in 1864 by the launching of the prestigious house journal, *Annales scientifiques de l'Ecole normale supérieure*, sowed an early seed of aspiration to a research-oriented career in science.[24]

In his ten years in the rue d'Ulm, Pasteur transformed the expectations and academic standing of the most gifted scientific *normaliens*, encouraging them into career paths that diverged increasingly from the *lycée*-oriented trajectories of even the ablest students in letters. Whereas previously the majority of candidates whose success in the entrance examinations for the science section left them with the choice between Normale and Polytechnique chose the latter, the decision thereafter became far less clear-cut. When one such candidate, the future dean at the Sorbonne and permanent secretary of the Académie des sciences, Gaston Darboux, opted to become a *normalien* rather than a *polytechnicien* in 1861, the event was reported in the national press. Nisard saw Darboux's preference, which was that of five out of the eight candidates who had been offered places at both Normale and Polytechnique in that year, as a triumph for his school.[25] And the choice that Darboux made was soon to become by no means unusual.[26] *Normaliens* entering the science section were now seen as a hand-picked elite whose status was enhanced by their small number: about fifteen a year were admitted during the Second Empire, compared with almost ten times that number who entered Polytechnique. They were, moreover, an elite in whose selection, welfare, and future careers no less a figure than Pasteur took a personal interest.

In contrast with the rapidly changing lives of the new generations entering the Ecole normale supérieure, the existence of most professors in the faculties of science under Fortoul and Rouland went its familiar, rather disadvantageous way. Some, especially in the provinces, displayed their discontent with miserable conditions, paltry student numbers, and the lack of either incentives or facilities for research (see chapter 2). But most adjusted to the realities of their role. If students working for the advanced qualifications of the University—the *licence* and the doctorate—remained rare, there was always a bourgeois public eager for accessible well-presented public lectures of the kind that in the later 1850s regularly drew audiences of over a hundred (on occasions exceeding 250) to hear professors in the Bordeaux faculty speak on sciences ranging from the ever-popular astronomy to the more austere disciplines of physics and chemistry.[27] Such lectures reaped the rewards of both personal gratification and the ministerial approval that always shone on initiatives directed at integration with local elites.

Where professors sought integration—and not all did[28]—industrial or agricultural utility provided the most obvious common ground with the wider public. This was conspicuously true at the faculty of science in Nancy in northeastern France, created by Fortoul in 1854, where the dean and professor of natural history, Alexandre Godron, and the professor of pure and applied mathematics, Hervé Faye, implemented a vision of the scientific professoriate as a resource for manufacturing and commerce in the region. By the end of the Second Empire, the faculty's highly successful evening lectures on the applications of science had given the lie to any suggestion that the University was a refuge for otherworldly self-indulgence.[29] More parochially, their popularity had also provided effective ammunition against the insistent arguments of the municipality of Metz that the faculty, always seen as Lorraine's faculty, should be moved there from Nancy.[30] At another of the new faculties of the 1850s, in Lille, Pasteur, as professor of chemistry and dean, promoted a similar policy of involvement in local life.[31] He organized visits to factories for his pupils and encouraged his professorial colleagues to do likewise. As professor of pure and applied mathematics, Gabriel Alcippe Mahistre showed a particular commitment to Pasteur's utilitarian objectives. Of the twenty-five papers that he published in the *Mémoires* of the Société impériale des sciences, de l'agriculture et des arts de Lille between his appointment in 1854 and his death in 1860, a high proportion treated applied subjects, mainly mechanics and power technology.[32] The choices that Mahistre made in the allocation of his time would certainly have received the active approval of Pasteur as dean, and Pasteur in turn earned the plaudits of Fortoul for leading the Lille faculty in the direction of utility and integration with the region.[33]

The leadership that Pasteur provided at Lille was far from unique. Everywhere, in fact, the personality of individual deans was crucial in setting the tone of a faculty. Some deans slumbered, satisfied if the irreducible bureaucratic obligation of examining for the *baccalauréat* was fulfilled adequately. Most, however, adopted more serious intellectual agendas and, in teaching and research, drove their colleagues to greater efforts, whether in the pure or the more utilitarian aspects of their disciplines. An unsung hero in this respect was Jean-Joseph-Benoît Abria, who served for almost fifty years as professor of physics at Bordeaux, from the foundation of the faculty in 1838, and, for more than forty of those years, from 1845 to 1886, as dean.[34] Abria's education and early appointments illustrate a route to a scientific chair that became common as career patterns assumed their classic nineteenth-century pattern under the July Monarchy. After entering the École normale in 1831 (at the age of 20), he became an *agrégé* in 1834 and worked for the doctorate while teaching at the *collège royal* in Limoges and then at the Collège royal Henri IV in Paris. Appointed to an interim appointment as *chargé de cours* in the newly created faculty of science in Bordeaux in November 1838, he took his doctorate in the following month, so qualifying himself for his promotion to the rank of professor in October 1839. Abria reconciled the competing demands on a professor's time with no apparent difficulty. His first priority was purchasing apparatus for lecture demonstrations, including his own very successful lectures to nonstudent audiences. But gradually a combination of careful husbandry and a modest flow of funds from the municipal council, keen to promote the cultural life of the city, allowed equipment to be accumulated for professorial research. Abria himself published fifty papers. While these were mostly rather minor experimental investigations using the kind of basic optical apparatus that any well-run faculty would have possessed, they reflected a zeal for research that he maintained into old age.

Abria was by no means the only professor in the Bordeaux faculty to allocate his time in this way. Of the six professors in post in the early years of the Second Empire, all but one (Constant Rollier, the professor of astronomy and rational mechanics from 1841 to 1858) were conspicuously active in research and publishing. Alexandre Baudrimont, professor of chemistry from 1848 until his death in 1880, published 125 papers in his thirty-two years in the chair. Almost as prolific was Victor-Amédée Lebesgue, who as professor of pure mathematics from 1838 to 1858 published sixty-eight papers. In the natural history sciences too, there were signs of sustained activity, with Victor Raulin (professor of botany, mineralogy, and geology from 1846 to 1876 and of botany alone from 1876 to 1885) and François-Aman Bazin (professor of zoology from 1839 to 1865) publishing ninety-two and twenty-nine papers, respectively. And even Rollier's rather somnolent chair became a setting for significant activity under his successor Frédéric-Gaston Lespiault, who published forty-seven papers between 1858 and his retirement in 1893. Such a record does not suggest either lethargy or incompetence, any more than the average length of tenure of the chairs (of very nearly thirty years) suggests the kind of discontent that might have led professors to seek more satisfying positions elsewhere.

The cases of Nancy, Lille, and Bordeaux show that, under the Second Empire as during the July Monarchy, professors who could reconcile themselves to their provincial location and muster the necessary resolve were able to fashion satisfying academic lives enlivened by opportunities and at least some stimulus to engage in research. But functionaries they remained, and with that status and its attendant burden of routine activities came discipline and subjection to the centralizing hand of Paris. Fortoul's arrogation to himself of the right to fire any of his employees, without passing through the Higher Council of Public Instruction (as he had previously had to do), was just one aspect of the legislation of 1852 that conveyed his vision of a minister's total authority over professors and other teaching staff of the University.[35] The closeness of the control, initially under Fortoul but then under Rouland and, later, Duruy, was made possible by meticulous reporting. Through the 1850s and 1860s, the deans of all the faculties—letters, law, and medicine, as well as science—submitted regular, often weekly, reports to the rector of the academy to which they were attached, and the rector in turn reported to the minister. In this way, accounts of the conduct of examinations, attendances, proposed programs of lectures, and the personal misdemeanors and failings of professors all fell with relentless regularity under ministerial gaze, and frequently they elicited a brisk response.

Reports on professors pulled no punches. Signs of aging, a failure to keep up with the discipline, poor delivery, or a shambling appearance could all attract an unfavorable comment. And while the faculties of letters offered the greatest opportunities for the voicing of unorthodox opinions, professors in the faculties of science too were watched carefully for signs of deviant flights of interpretation. The fact that such professors spent so much of their time lecturing to lay audiences whose interests were not bound by the formal syllabuses of the University only added to the temptation for them to stray, and this made ministerial vigilance an even higher priority. Despite the watchfulness, however, the instances of professors in the sciences who fell foul of censorship were few. Criticisms were usually limited to a professor's tendency to treat subjects in a trivial or unduly eye-catching way, and such failings were easily condoned if lectures achieved the higher goals of the spread of science within polite culture and the acceptance by local elites of the importance of a faculty for their region's economy and reputation.

Education, Industry, and the Imperial State

Except in the passing disruption that it caused to trade and business confidence, the revolution of 1848 had little impact on French industry. By 1849, the mounting of the eleventh in the series of national exhibitions that had begun over half a century before had already helped to give industrial activity an air of normality. And thereafter, as the uncertainty of the Second Republic gave way to at least the superficial stability of the empire after December 1852, the sense of continuity was put beyond doubt. One menacing aspect of the new age was that employers and workers now viewed one another with a greater suspicion than before. The revolution had reinforced the sense of a disquieting gulf between them that had been widening since the labor riots in Lyon and elsewhere in the early and mid-1830s, and during the Empire the typical paternalist response of investing in the construction of *cités ouvrières,* company housing projects for employees, as was done in Mulhouse and Le Creusot, may, if anything, have aggravated the problem by bringing workers into closer association with one another and accelerating, rather than containing, the growth of working-class radicalism.[36] These social and political matters, however, are not my concern here. For my purpose, it is important merely to stress that in the structure of industry and in methods of production, 1848 was not a significant watershed. What occurred was less a change of direction than a reinforcement of trends that were already afoot, in particular in the move toward greater mechanization.

The years of prosperity and at least the veneer of social harmony under Napoleon III provided ideal conditions for the evolution in the practices of French industry that gathered pace through the 1850s and 1860s. The first phase took the form of a pronounced, if uneven, move to more advanced production methods in the traditional sectors. Generally, the employers' aim of enhancing profits by at once increasing output and reducing costs was achieved by the introduction of machinery that had already proved its worth abroad, most commonly in Britain. The pattern of imitative development was nowhere more evident than in the cotton industry, where the use of steam power and of faster devices for combing, spinning, and weaving was greatly extended. It was in 1852 that the self-acting mule, for spinning cotton, made its first appearance in mills in the Haut-Rhin and in the late fifties that it and Hübner's advanced circular comber became widely used, not only in Alsace but also (albeit to a lesser extent) in the other main cotton areas of the Seine-Inférieure (chiefly Rouen) and the triangle of northern textile towns, Lille, Roubaix, and Tourcoing.[37] It was in the same decade too that the balance began to swing rapidly in favor of the power loom. Again the Haut-Rhin, though a follower with respect to the most advanced regions of Britain, led the way within France. In 1844, hand looms in the department had still outnumbered power looms by 19,000 to 12,000, but by 1856 the numbers were respectively 8,657 and 18,139.[38] Eight years later, the ratio was of the order of one to seven, with the three or four thousand remaining hand looms being used only for trial runs or specialty cloth. What happened in Alsace was nothing less than a fundamental retooling of the cotton industry, begun in the 1850s and then accelerated after the Cobden–Chevalier free-trade treaty of 1860 had dramatically reduced the duties on British imports and so exposed French manufacturers to the full blast of competition from across the Channel.

Where new machinery was introduced, it usually called for a workforce more systematically trained than would have been possible through the traditional route of experience gained on the job. Most employers recognized the challenge, and where this was so, their response was rapid and well directed. In Lille in 1858, the city's Société des sciences, de l'agriculture, et des arts (in which textile manufacturers were prominent) set up a specialized school to provide short courses for the operators of steam engines, nearly two thousand of which were now in use in the Department of the Nord.[39] In Mulhouse, too, employers took a leading role in the promotion of vocational schools, in this case for weaving and spinning, founded with the aid of a public subscription in 1861 and 1864, respectively.[40] Examples of similar new departures in vocational education can be multiplied without difficulty. Taken together, they demonstrate the extent of entrepreneurial resolve and pedagogical flexibility in the face of new circumstances. They certainly do not suggest a supine or even half-hearted response to technological change.

Within the relevant Parisian ministries—Public Instruction and Commerce—the surge of industrial and educational renewal in the more thrusting manufacturing areas of the provinces and in and around the capital did not pass unnoticed. Quite the opposite. The trend to mass production and to the larger factories that came with the increased investment in plant, machinery, and skills raised economic and social challenges to which no government could remain indifferent. Nevertheless, the imperial administration's response was always hampered by the divide that separated its own perceptions of the appropriate remedy from those of the dispersed communities of industrialists.[41] The point is illustrated in Fortoul's priorities as Minister of Public Instruction in the early and mid-1850s. His response could never be direct or straightforward, if only because of the multiplicity of political pressures on him and the precariousness of the middle ground he tried to occupy between modernizers (with whom he broadly sympathized) and conservatives. The challenge for Fortoul was formidable: he had at once to win bourgeois support, curb insatiable clerical aspirations to the independent control of education, and gain the ascendancy over the Ministry of Commerce's rival provision for vocational education. In pursuit of these diverse ends, he struggled to present his ministry as a fount of policies that would maintain the essentials of the French educational system while responding sensitively to the needs of society and the economy. The result was a succession of reforms in which secondary schools and faculties became targets for a committed, if not always well-directed zeal, with especially important consequences for the sciences.

The furor generated by the divisive issue of the *bifurcation* has made it all too easy to disregard Fortoul's imaginative attempts to promote the teaching of applied science and other industrially related subjects within the university system. This unbalanced perspective has distorted the secondary literature ever since, leaving Fortoul portrayed as a myopic bureaucrat who sought discipline and centralization at the expense of intellectual vigor and the true interests of education in a changing world. The reality is that Fortoul fully recognized the need to provide a more serious preparation for careers in industry and commerce, and he gave that need the importance it deserved. In this, he displayed the unmistakable marks of his immersion in Saint-Simonian principles during the 1830s and 1840s, when (as a law student and journalist in Paris and, from 1842 to 1848, a professor of literature in the faculties of letters in Toulouse and Aix) he had been intimate with Hippolyte Carnot, Edouard Charton, Jean Reynaud, and others in the movement.[42] In accordance with these principles, he saw vocationally relevant education as a leading priority if manufacturing was to liberate itself from the weight of stubborn routine and the vagaries of chance.

Fourtoul's main instrument, outlined in legislation in 1854 and made a reality in the following year, was a new diploma in applied science, the *certificat de capacité pour les sciences appliquées*. Teaching for the diploma was to be widely available, either in the grossly underused faculties of science (five more of which Fortoul now created, partly to help in the task, in Nancy, Lille, Clermont-Ferrand, Marseille, and Poitiers) or in a network of new municipally funded schools. These schools, the *écoles préparatoires à l'enseignement supérieur des sciences et des lettres*,[43] were to provide instruction in mathematics, mechanics, physics, chemistry, natural history, French literature and history, geography, and drawing at a level corresponding more or less to the first two years of the faculties of science and letters. A curriculum that had no place for the classical languages and the more theoretical parts of the various subjects marked a bold new departure in higher education. And Fortoul was under no illusion as to the difficulty of securing its acceptance.

It would have come as no surprise to him that interest in the new curriculum was strongest in towns with a keen industrial spirit and where the authorities already provided at least some appropriate instruction and were willing to foot the bill for further expansion. To a degree unmatched elsewhere, Mulhouse met these requirements, thanks to its thriving cotton industry and a supportive municipal council and local demand for scientifically trained young men with a practical bent that had sustained a successful school of chemistry since the early 1820s and was to continue to do so into the twentieth century.[44] It was in Mulhouse that the most ambitious and best-equipped of the *écoles préparatoires* was opened in November 1855.[45] A well-appointed chemical laboratory and the arrival of the young Paul Schützenberger as professor of chemistry in 1855 (after medical studies in Strasbourg and a brief period as assistant to the color chemist Jean-François Persoz in Paris) signaled the new school's special strength in the science and techniques of dyeing and dye making and helped to give it a distinction that none of the other schools could equal.

Mulhouse, in fact, had an institution very nearly the equivalent of a faculty of applied science. It was a heartening start, and Fortoul proceeded to champion his priorities of utility and relevance in the memoranda and personal letters he directed to faculties and interested local authorities. The rectors of the University's sixteen regional academies bore the heaviest burden of Fortoul's campaign. He gave them no peace. In one of his briskest memoranda, dated 30 November 1855, he delivered a typically stern reminder of the dual role of the faculties in an industrializing society. Above all, he stressed the importance of curbing the professoriate's tendency to esoteric intellectualism, driven by what he called "personal whim."[46] Whereas it was proper that teaching in the *lycées* should continue to privilege intellectual rigor and not treat the applications of science, the faculties had a more demanding mission on both counts. Their courses, unlike those at the secondary level, had to embrace science in its applied as well as its theoretical aspects; and science, for its part, had to serve the interests of both national prestige and the economy.

> Let us never forget [Fortoul told his rectors] that long before the industrial sciences made their contribution to the fame and power of our country, we had achieved indisputable pre-eminence in the study of pure science. This was a glorious tradition, and it is one that we must strive to maintain. Yet today it is just as necessary that we should direct attention to the applications of science and that the young should be able to acquire in our institutions the knowledge they need in order to participate in endeavors that equally honor the human spirit. For us, this is an obligation all the more pressing since the scientific instruction in our *lycées*, while laying a foundation for those embarking on industrial careers, is insufficient because of its inevitably general character. Secondary education must have its culmination in the faculties, whether this is in the pure sciences or in the applications of science.[47]

Homilies in this vein, bolstered by the specific proposals that accompanied them and a torrent of correspondence directed at prefects and rectors throughout France, accurately conveyed the measure of Fourtoul's commitment. The goal that Fourtoul conceived for the faculties was a locally oriented but ordered diversity. Around the carefully prescribed core of traditional academic study, there were to be departures into whatever applications were seen as appropriate to each region. It was suggested that at Marseille, for example, there should be additional lectures on the manufacture and use of soda; at Lille, the emphasis should be on the industrial aspects of sugar; for Besançon, Fortoul proposed special instruction in horology; at Lyon, there should be lectures on industrial machinery, and so on.[48] The watchword was flexibility, and the aim was a reorientation of the faculties that would make them sources of (for industry) practitioners who were better informed and (for the liberal professions) men who were learned in their specialty but not remote.[49] If the aim was to be achieved and if misconceptions about the unduly theoretical character of teaching in the faculties was to be allayed, it was essential for rectors and professors to do everything possible to persuade a skeptical lay public of the value of formal instruction in the utilitarian aspects of science. To that end, Fortoul approved the principle that lectures should normally be held in the evening so as to facilitate attendance by working men, a principle that the Lille faculty adopted from its foundation.

Despite the mix of Fortoul's determination and his attention to detail, the successes were few. Even at Lille, where Pasteur (as professor and dean) entered fully into the spirit of the new teaching, only four *certificats de capacité* were awarded between 1857 and 1860.[50] Three other faculties (Bordeaux, Nancy, and Toulouse) awarded a mere twelve between them in the same period, and the rest either failed even to begin preparing candidates for the new qualifications or, as happened at Dijon, abandoned their efforts at an early stage, in the face of negligible student interest. After 1860, no faculty awarded a single certificate. The achievements of the *écoles préparatoires à l'enseignement supérieur des sciences et des lettres* were more encouraging but still modest. One of the schools (destined, ambitiously, for the small Auvergnat town of Moulins) never opened, and, of the five that did, two (at Chambéry and Nantes) awarded no certificates at all, and another (at Angers) gave a total of three in the seventeen years between 1857 and 1874. Only Rouen (with twenty-eight certificates in the same period) displayed a response even remotely comparable with that in Mulhouse, whose eighty-seven certificates made it the one true success story. And even in Rouen and Mulhouse the flow of candidates all but ceased after 1866, while no certificates were awarded anywhere after 1874.[51]

Although respectable attendances at certain courses of lectures intended for candidates for the certificate offered some consolation, Fortoul's plan, with its formal structures and qualifications, must be accounted a failure. The main explanation lies in public apathy. This was itself a consequence of the mismatch between the multidisciplinary and (despite the ministerial rhetoric) still rather academic courses of study that led to the certificate, and the specialized, immediately applicable vocational instruction that most industrial employers regarded as their greatest need. Except in Mulhouse, where the town's *école préparatoire* responded to a niche demand in textile chemistry, it was hard to present the *certificat* as a saleable commodity on the job market. It fell firmly between two stools. On the one hand, parents and pupils shared the employers' view that the syllabus was too diffuse to serve as a preparation for any particular form of industrial or commercial career. But, on the other hand, they knew that the certificate bestowed none of the advantages that went with the *enseignement classique*: it offered neither a first step on the road to one of the liberal professions nor the more nebulous benefits of distinction and refinement. On the French scale of middle-class values, a program of study such as that for the certificate, which did not make the *baccalauréat* a prerequisite for entry, had little prospect of success with the aspiring families whose support Fortoul was seeking.

Fortoul had overestimated his capacity to remold attitudes. It is conceivable that his zeal might, in time, have borne richer fruit. But in July 1856 he died, and thereafter, under Rouland and then Duruy, the cause of educational modernization, though by no means abandoned, took other forms. The new initiatives preserved much of the utilitarian spirit and boldness of Fortoul's vision. The most promising of them, dating from the mid-1860s, was Duruy's program of "special" secondary education, the *enseignement secondaire spécial*.[52] The main administrative originality of this program and its most fundamental departure from existing practices lay in its principle of parity of esteem between the special and the classical curriculum. But the content of the four-year syllabus, for boys between the ages of 12 and 16, also broke important new ground in the emphasis it laid on science, mathematics, and at least a modest exposure to technical studies. The component of *culture générale* (an essential ingredient for a secondary curriculum) similarly paraded the stamp of change through a mix of French, modern foreign languages, economics, history, and geography that imposed no exposure to either Greek or Latin. Initially, the new program—supported by its own *agrégation* and its own status-enhancing training college for teachers in the abandoned abbey of Cluny—secured a significant following. Within six months of the legislation of 21 June 1865 that established the program, 16,882 students were enrolled, and so long as Duruy was at the ministry, the success was sustained. A new school devoted entirely to the special curriculum was inaugurated with great ceremony at Mont-de-Marsan (in the Landes) in October 1866,[53] and two years later the total number of enrolments, in some three hundred schools that offered at least part of the syllabus (most of them municipal *collèges communaux*, but some of them *collèges impériaux*, as the *lycées* were now called), had risen to 18,463. The figure represented about a quarter of the pupils in secondary education throughout France.[54]

It was an encouraging start. But after 1870 the special program limped in the shadow of its classical alternative. With Duruy gone and the optional extension of primary education, the *enseignement primaire supérieur*, offering unhelpful competition, support for the program's air of vocationally oriented modernity struggled to maintain its numbers, let alone increase them. The last remnants of it, subsumed within a distant successor, the Latin-less *enseignement secondaire moderne* of 1891, disappeared in the radical reform of secondary education in 1902.[55] The fact remains, however, that for a few years it seemed that Duruy might succeed in combating antiutilitarian cultural prejudice where Fortoul had failed. What is less clear is whether the early healthy level of enrolments in the program reflected the successful diffusion in French society of modern attitudes, including an appreciation of the importance of science and technology in the industrial age. Some such diffusion certainly did occur, notably in the families that enrolled sons in the *enseignement secondaire spécial* in Lyon, Marseille, Rouen, Strasbourg, and Colmar, in all of which numbers outstripped those at the largest center in Paris, at the Lycée Charlemagne. As a cultural indictor, though, the take-up for the *enseignement secondaire spécial* tells us little.

One reason for this is that, despite the program's localized successes, the total numbers opting for any form of secondary education—whether the special or the classical track—were small when set against the five million pupils in primary education in the mid-1860s. Moreover, a high proportion of the children of the farmers, artisans, skilled workmen, foremen, and technicians who made up roughly half of those who embarked on the special program in the 1860s withdrew after only two of the four years and went on to junior appointments in offices and government administration.[56] By abandoning their studies prematurely in this way, students were avoiding the more demanding parts of the course, including most of the instruction in applied science and the technical subjects. The quest for a soft option must have played a part in their choice. But those who dropped out could also be seen as making a perfectly rational career choice at a time when a modest office job offered a security, as well as a social status, that most industrial employment could not match. At all events, probably no more than a third of those entering the *enseignement secondaire spécial* ever embarked on industrial careers, so that the effect of the program was, if anything, to drain young men *away* from manufacturing toward bureaucracy. The Ministry of Public Instruction's attempts to adjust to the industrial age from Fortoul onward would certainly have fared better if employers had had a greater involvement in fashioning the curricula. Duruy's initiative, like Fortoul's measures, showed a sensible openness to the diversity of employment needs by supplementing the core syllabus with variants adapted to local demand. Yet neither under Duruy nor under any other nineteenth-century minister did Public Instruction manage to establish sustained common ground with the world of industry. As the economist and historian Henri Pigeonneau observed in an astute retrospective judgment, following the passage to the new *enseignement secondaire moderne* in 1891, what the *enseignement secondaire spécial* offered was too much like a watered-down form of the *enseignement classique*.[57]

In this respect, Public Instruction was no match for the Ministry of Commerce, whose mission in education allowed it to respond more immediately to the interests of manufacturers and to do so in a context less constrained by cultural traditions and older bourgeois values. Commerce exploited its advantages energetically. The scope for ministerial ambition and the extent of that ambition were made particularly plain in a policy of extending control over schools that had previously been either wholly or largely independent of the state. The policy had its most notable success in the takeover of the hitherto autonomous Ecole centrale des arts et manufactures in 1857,[58] a move that gave the Ministry of Agriculture and Commerce (as it was called at this time) its first advanced *école spéciale*. Lower down the academic scale, the ministry also assimilated a number of middle-level technical and trade schools, such as those for mining at Alais[59] and for horology and precision instrument making at Cluses and Besançon, all of which had been under private or municipal control.[60]

As a complement to this policy of expansion, the Ministry of Commerce also paid attention to the improvement of the schools already under its purview. The most significant beneficiary was the network of *écoles d'arts et métiers*. Three of these, at Châlons-sur-Marne, Angers, and Aix-en-Provence, existed before 1848, and all three enjoyed a success both then and throughout the Empire that spoke eloquently for the appropriateness of the intensely vocational training they offered. The inspector of the schools was probably not exaggerating when he reported in 1856 that the demand for *gadzarts,* or *gars des arts*, as the graduates were called, could not always be satisfied.[61] Employers liked *gadzarts,* comparing them favorably with the graduates of the Ecole polytechnique who found their way into industrial employment. In contrast with *polytechniciens*, *gadzarts* emerged from their schools not only with considerable manual dexterity, developed through solid doses of workshop practice during the three-year course, but also (and no less importantly) with a helpful modesty in their career aspirations. Employers liked it this way, and it was certainly not they who led the campaign for the slow process of "scientification" of the schools that gathered strength after the middle of the century. The pressure for tougher admission requirements and the increased rigor of the curriculum came rather from the association of former pupils intent on raising the standing of the schools and the qualifications they offered.[62] It was a familiar status-seeking strategy, and, for the *gadzarts* themselves, it eventually paid off as the ablest of them successfully colonized the higher levels of a succession of new industries, including in the early twentieth-century automobile construction and aviation. Career advancement of that order would have been beyond them had it not been for the strengthening of their academic as well as their practical attainments.

The fact that industrially and commercially oriented instruction remained a preoccupation of successive ministers of both Public Instruction and Commerce and a focus for rivalry between them helped to keep vocational education at the forefront of national concern. At the regional level equally, it helped to fire the debates that engaged virtually all local authorities, as well as chambers of commerce and individual employers. Education for the modern world was seen to matter, and it was discussed accordingly, everywhere and at all levels. The result was a flow of new ventures and reforms of existing provision such that no person—no young man at least—could complain of a lack of opportunity for occupationally relevant education or training of some kind. The quality and appropriateness of what was offered, though, was another matter. At one extreme, municipal lecture courses, which had often to meet the very different expectations of audiences in search of cultivated diversion as well as of those aspiring to an industrial, agricultural, or commercial career, tended to suffer from a diffuseness of purpose. At the other extreme, however, the still widespread system of apprenticeship offered instruction that was, if anything, too focused; moreover, it was frequently abused by employers, many of whom saw apprentices as merely a source of cheap, biddable labor. Above all, what the young and their parents knew very well was that the various levels of opportunity, far from being open to all, were rigidly stratified by social status and previous educational experience. Success in the *enseignement classique* (with all that this implied in terms of social selection) remained a prerequisite for entry to the Ecole polytechnique and virtually so (despite early attempts to break the elitist mold) for candidates aspiring to the Ecole centrale des arts et manufactures. And even in the very different and more flexible context of Mulhouse, the mixture of trade school and on-the-job preparation for color chemists seldom led on to the highest reaches of the profession, which remained, in practice, the preserve of the powerful families of the region.

So while vocational education under the empire was plentiful, the provision suffered from a lack of coordination and left all parties dissatisfied. The unsatisfactory state of affairs was pinpointed in a wide-ranging inquiry into the state of professional education that the powerful champion of the Bonapartist cause and Minister of Commerce, Eugène Rouher, set in motion in 1863 in response to alarm about France's showing at the previous year's International Exhibition in London.[63] The resulting report, by Arthur-Jules Morin and Henri-Edouard Tresca, respectively director and deputy director of the Conservatoire impérial des arts et métiers, laid bare the divergent but equally pressing concerns of parents (anxious about the absence of any moral dimension to apprenticeship and about the system's cost to families of modest means) and of employers (most of them wedded to their traditional ideal of a locally controlled method of inculcating immediately usable skills).[64] Recommendations abounded, most notably for a standing committee on technical education to advise the Ministry of Commerce and for an expansion of the network of *écoles d'arts et métiers*. While this last recommendation was left unanswered until the creation of new schools in the network at Cluny (1891), Lille (1900), and Paris (1912), some immediate steps were taken. Duruy's *enseignement secondaire spécial* had its roots in the sense of urgency that the inquiry engendered. So too did the establishment of a number of successful municipal vocational schools, including the future Ecole Diderot, founded as a school for apprentices in Paris in 1873.[65]

Despite the patchwork of individual local successes, no amount of ministerial and municipal interest could conceal the limitations resulting from the absence of a single directing influence in technical and other vocational education. Morin and Tresca were correct in asserting that the flow of well-trained recruits emerging from the educational system in the early 1860s was not at the level or of the kind required by an expanding and changing science-based economy. Identifying the deficiency immediately after the 1862 exhibition in London, they pleaded cogently for greater state investment in technical education and a more coordinated hierarchy of institutions that would broaden access while filling gaps at all levels in current provision.[66] But attempts at the fundamental structural reform that might have effected significant improvement had little chance of success in the face of persistently unhelpful interministerial rivalry, Parliament's lack of resolution, and the short-termism of employers, all of which tempered investment and prevented anything approaching a radical solution. The most that can be said is that by the end of the Second Empire two decades of rather disparate attention to the problem had left vocational instruction of some kind, however variable in quality and appropriateness, financially and geographically within the reach of virtually all young Frenchmen. And that was no small achievement.

The Bureaucracy of Learning

During the empire, the Collège de France, the Muséum d'histoire naturelle, and the Paris Observatoire remained, as they had always been, a privileged if never prosperous sector of the world of learning. As a matter of general policy, the Ministry of Public Instruction tried to give all three institutions a measure of liberty that left them relatively free of bureaucratic intrusion. But, time and again, what might have been expected to be a tranquil backwater of the ministry's domain came to the forefront of public debate and hence, necessarily, of ministerial attention. The spat surrounding Renan's removal from his chair at the Collège de France in 1864 was just one instance. In this notorious case, an administration as jealously protective of its authority as Duruy's could not remain indifferent (whatever the minister's personal view may have been) to the sniping by religious conformists who interpreted any sign of ministerial laxity as an invitation for liberals to challenge decency and good order. The same was true of a number of other controversial issues that taxed both Duruy and his less liberal predecessors. Perceived failings within the institutions of learned culture, ranging from abuses of academic freedom such as Renan was thought to have perpetrated to cases of flagrant maladministration at the museum and the observatory, were guaranteed to attract unfavorable public comment and hence to demand the attention of ministers and bureaucrats.

Of the three great research institutions, the Muséum d'histoire naturelle was the one most commonly seen as being in need of a watchful ministerial eye, not least because of an expenditure far exceeding that on the Collège de France and the Observatoire de Paris.[67] What Camille Limoges has characterized as the four decades of the museum's golden age—an age characterized by generous funding and the accumulation of a total of more than a dozen chairs over a broad range of sciences—had come to an end by about 1840.[68] For the next thirty years, with often violent public attacks emanating from left and right, usually directed at the professors' high degree of autonomy and, as critics believed, uneven performance, the museum endured a period of anxiety. Inquiries in 1850–51 (inspired by Hyacinthe Corne's critical comments in his report on the budget for Public Instruction for 1849) and 1858 reflected the seriousness with which the museum's failings were regarded. Both inquiries had eminent chairmen (the military and political figure Louis-Etienne-François Héricart de Thury in 1851 and the prominent general and Councilor of State General Nelzir Allard seven years later) and both exposed much the same inventory of defects: neglected and overcrowded collections, professors with little interest in the displays (a problem compounded by their poor payment and, until 1853, the lack of any system of professorial retirements), profoundly defective inventories, insalubrious laboratories, inadequate greenhouses, and a menagerie whose squalor and high rate of animal mortality set it way behind the Zoological Gardens in London.[69]

In a manner consistent with the authoritarian cast of the early empire, the inquiries diagnosed the root cause of the museum's difficulties as administrative failings that made it essential for the institution to be brought under closer control. The Ministry of Public Instruction's bureaucratic notion of control, however, was precisely what the professors did not want, and in the name of a cherished scientific freedom, they responded to the two reports by publishing detailed rejections of virtually all the charges made against them.[70] It was therefore a very real achievement on the ministry's part when a plan for the revitalization of the museum was completed in December 1863.[71] As the report on which the plan drew made clear, the existing practice whereby the professors elected one of their number to serve as director for only one year opened the door to a free-floating Jacobinism incompatible with efficiency or any sense of sustained responsibility. In view of the state of the museum, the remedy that was adopted, of having a director who would serve for five years and be approved by the Minister of Public Instruction and formally appointed by the emperor, could hardly be regarded as an unreasonable intrusion. Yet a mixture of professorial prevarication and outright opposition emasculated even this moderate reform. Successive possibilities that an outsider might be brought in to fill the post—the names of Jean-Baptiste Dumas and Louis Agassiz were among those proposed in this context—came to nothing, leaving vested interests to triumph. In a move that precluded any possibility of significant change, the assembly of professors, having acquiesced in the principle of the new arrangement, secured the appointment of the 77-year-old Michel-Eugène Chevreul, one of their own and a resolute defender of the museum against ministerial interference (see figure 4).[72]

For the rest of the empire, the scene was set for tense relations between the museum's professoriate and its ministerial paymaster. The incentive for Duruy to bring to heel a body of men who combined such recalcitrance with a record of underperformance was great, and he must initially have harbored hopes of succeeding where Fortoul and Rouland had failed. Intellectual authority at the top of the institution was certainly not lacking: among the museum's still impressive galaxy of great names, those of Pierre Flourens (in the chair of comparative physiology) and Antoine-César Becquerel (physics applied to the natural sciences) stood out by any academic standards. Yet, for a variety of reasons (including age, marginality to the institution, and a preference for alternative settings for their work), these potentially powerful professors failed to develop the museum as a major location for research in their various fields. More promising foundations for change, both in the quality of research and in the training of the younger generation, were the chairs occupied by Henri Milne Edwards (entomology, then zoology), Emile Blanchard (natural history), and Quatrefages (anthropology). But although these chairs might collectively have fostered a new look at some of the classic issues in the history of life to which Cuvier, Geoffroy Saint-Hilaire, and other French life scientists had contributed so much, their occupants had a less benign influence. Their combined effect, as I argue in chapter 5, was actually to impede, rather than advance, French acceptance of evolutionary theory.

The chemist Edme Fremy was another professor who might have breathed fresh air into the museum. His commitment to research was beyond question, and his pleas on behalf of science and scientists, elaborated in a series of polemical broadsides and in the personal contacts he had with Duruy and even the emperor, were forthright, especially where the plight of the younger generation was concerned.[73] Research, for Fremy, required more than better laboratories, necessary though they were.[74] It called, no less importantly, for recognized career tracks that would spare men of exceptional promise the stultifying burden of heavy routine teaching. Although in its general thrust Fremy's cause squared with Duruy's own struggles to encourage research, it did so with some distinctive glosses, including his hope that independent patronage of the kind that was helping the United States to prominence as a scientific international force would complement the traditional resource of support by the state.

Despite Fremy's seniority in the museum, his impact within the institution was slight, and almost twenty years after his campaign of the 1860s he found himself still having to make much the same case for the creation of a new type of research career distinct from those of the few brilliant *normaliens* who went on to chairs in the faculties and hence to positions that, as he saw it, risked extinguishing the "sacred flame" of a passion for science.[75] Fremy's advocacy was dulled in some measure by a dispersion of effort that came with his participation in the very system of *cumul* that he identified as one of the blights on French scientific life: he simultaneously held a chair at the Ecole polytechnique as well as at the museum, in addition to pursuing other time-consuming activities, including his charitable work, through the Société des amis des sciences, on behalf of savants and their families who had fallen on hard times.[76] At all events, Fremy's chair did not emerge as a focus for change. Despite his efforts and a long history of ministerial concern, the profile of the museum remained in 1870 what it had been since the 1840s. The combination of a stubborn defense of the administrative status quo with an unadventurous conception of natural history rooted in the collections (rather than in experiment) had condemned the institution to stagnation. The result, in Limoges's assessment, had been three decades of "decline."[77]

By comparison with the Muséum d'histoire naturelle, the Observatoire de Paris was viewed initially with favor within the ministry. At the dawn of the empire, François Arago had served for many years as the "director of observations," appointed annually by the Bureau des longitudes but de facto the permanent head of the institution. He brought to the post the aura of a long and distinguished scientific career and an outgoing personality to set against his republicanism and what Libri, Biot, and other critics saw as an unhealthy taste for the acclaim he enjoyed as a writer, lecturer, and politician (see chapter 5). After his death in 1853, his successor Urbain Le Verrier offered comparable eminence and was politically a much safer bet. Taking up his duties early in February 1854 as the observatory's first formally appointed director, Le Verrier proved to be anything but a showman, much as he enjoyed the celebrity that had come with his accurate prediction of the existence of Neptune in 1846.[78] He was first and foremost a conformist, a man of piety and political flexibility more than willing to adopt the mantle of a faithful servant of the empire. This allegiance only aggravated the bad start to his directorship, at least as viewed from within the astronomical community. Many in the community took it amiss that no professional advice on the appointment was sought and that (albeit quite legally) Le Verrier was simply parachuted in by a decree of the Minister of Public Instruction. Matters were only made worse by the legislation, which removed the last vestiges of control that the Bureau des longitudes exercised over Le Verrier's activities as head of the observatory.[79] What his enemies saw as a bad start, however, did nothing to diminish his standing in administrative circles. There, at least under Fortoul, he was regarded with unqualified favor. The emperor's resolve that no observatory in the world should surpass that of Paris squared perfectly with Le Verrier's private ambitions, and within less than a year of his arrival at the Observatoire, Le Verrier had pointed the way forward in a detailed report to the government on the state of the institution and the investment and organizational reforms he thought necessary.[80]

The opportunism of the report was flagrant. Le Verrier seized on the emperor's desire for national glory and some real evidence of neglect in the observatory as levers for his personal objectives, which ranged from essential material improvements in equipment and the lodgings of the resident astronomers and other scientific staff to much more ambitious plans for creating a "bureau des calculs," for the formidable task of reducing past and current observations, and for establishing a meteorological service under his control. As it soon transpired, Le Verrier's determination to reinvigorate meteorology, which had previously been pursued as a desultory sideline of the observatory's activity, was not only the most adventurous but also the most contentious part of the scheme.[81] Critics were quick to note that the meteorological initiative, as Le Verrier pursued it, distorted what most saw as the primary functions of the observatory. But, to those who appointed Le Verrier, the new provision for meteorology had overwhelming merits, scientific and political. Complete with a system of electrical telegraphy and a structure for exchanging information with French and foreign observatories (chiefly to help in giving warnings about storms), it offered the prospect of allowing France to compete on equal terms with Britain. Here the greatest prize in the long term lay in the study of terrestrial magnetism, an international activity in which since the late 1830s the French had become distinctly marginal.[82]

With most of his requests granted and the highly regarded, though emotionally fragile, Léon Foucault in the new position of physicist from 1855, Le Verrier had every reason to congratulate himself on his first year or two in office. Even events in the Crimea played into his hands, when he was able to show that much of the damage that had been caused to the allied Black Sea fleet in the violent weather of 14 November 1854 might have been avoided if only there had been telegraphic communication and hence a rapid means of transmitting a warning about the advancing storm, coming from the direction of Vienna.[83] Although Le Verrier's autocratic style was evident from the start, its consequences were masked initially by his celebrity and a number of significant improvements that he made in the performance of the observatory. Outstanding among his early achievements was launching the *Annales de l'Observatoire impérial de Paris*. More than any other product of the observatory, the *Annales* signaled Le Verrier's resolute engagement with the task of reducing and publishing a huge backlog amounting to more than a half century of raw observational data. The *Annales* eventually ran to ten handsome quarto volumes, in addition to a series of memoirs and more or less annual volumes of new observations.[84]

In meteorology and terrestrial magnetism, the start was less spectacular but no less solid. In March 1854, the appointment of Emmanuel Liais, whose meteorological observations—made in his private observatory in Cherbourg—had won the admiration of Arago,[85] gave Le Verrier the dedicated expertise that he needed in the area. The next four years, though, were difficult ones for Liais. Despite advancing quickly from the rank of *élève-astronome* to that of *astronome titulaire* and, in effect, assistant director of the observatory, he was the regular butt of Le Verrier's irascible behavior. Eventually, in December 1857, Liais left in acrimonious circumstances, having been effectively (though not formally) fired, and in the following year he sailed for Brazil. There, on the first of three extended stays in the country, he observed a total eclipse of the sun and accepted commissions from the Emperor Dom Pedro d'Alcantara that led on to important work in the newly founded Olinda observatory near Recife and seven years (1874–81) in Rio de Janeiro as director of the imperial observatory.[86] Despite his lengthy absences from France, his feelings remained as strong as they were on the eve of his first departure for Brazil, when he used an article in a popular magazine to express his mixture of bitterness and frustration at Le Verrier's management of the Observatoire de Paris.[87]

Despite his unhappiness, Liais left a significant legacy, doing much to advance Le

Verrier's ambitions for control over French meteorology by making improvements at all levels in the accumulation and distribution of meteorological information.[88] The key innovation, implemented under Le Verrier's supervision, was the recruitment of telegraph clerks throughout France as part-time observers using instruments (including a barometer specially designed by Liais) and procedures communicated from Paris. The criticisms by Victor Regnault and Biot, who saw the accumulation of meteorological readings as pointless so long as there was no underlying theory, would have piqued Le Verrier.[89] But Le Verrier countered them with a smothering rhetoric of utility (for rather vaguely defined ends in agriculture, shipping, and warfare) and by the evidence that techniques of forecasting were becoming, by the mid-1850s, a focus for international attention and rivalry. Here, he was correct. Any in France who doubted the international significance of meteorology had only to look across the Channel, where Robert Fitz Roy's brief but widely read *Barometer and Weather Guide* (1858) and *Notes on Meteorology* (1859) significantly raised the profile of British contributions. Further evidence lay in the plan for a network of colonial observatories for magnetic and meteorological work that was presented, with the endorsement of the Prince Consort, at the Aberdeen meeting of the British Association for the Advancement of Science in 1859.[90] Such marks of confidence in the value and accuracy of forecasting techniques abroad meant that the French could not stand idly by, and Le Verrier was glad to exploit to the full the evidence of heightened activity in other countries.

Le Verrier's sympathy for the ideals of the empire, especially in its early, authoritarian phase, and the firmness with which he had taken the observatory in hand made him a natural recruit to the Ministry of Public Instruction's closest circle of advisers. Even the simple act of setting telegraph clerks to routine meteorological work advanced the ministry's centralizing ambitions and did so through the economical and trouble-free expedient of complementing an orderly nationwide network of paid observers with far larger numbers of volunteers at all levels of ambition and attainment. The vehicle that Le Verrier wielded most effectively in pursuit of this union of formally recognized expertise and amateur enthusiasm was his own creation, the astonishingly successful Association scientifique de France. Within a few months of its foundation in 1864, the association had 3,500 members in Paris and the provinces and had already distributed 21,000 francs to support scientific work, mainly in meteorology and astronomy.[91] But its greatest achievement was in providing a framework for the engagement of amateurs in making meteorological observations. That achievement squared perfectly with the Ministry of Public Instruction's conception of a broadly based but unified structure for learned culture in which the elites of the Collège de France, the Muséum d'histoire naturelle, the Observatoire de Paris, and the Institut de France would preside over the efforts not only of professors in the faculties of science and letters but also the members of *sociétés savantes* and even the humblest *curés* or teachers in primary schools who painstakingly accumulated their readings of sunshine and rainfall.

Winning the loyal support of the more modest contributors to the meteorological enterprise was not difficult. But Le Verrier's aspirations were boundless. At the height of his zeal, in the mid-1860s, they came to embrace the *écoles normales primaires*, the institutions across France in which future teachers were trained for service in primary education. Le Verrier targeted the schools as a new source of ideal observers, and trainee teachers and their professors duly proved eager helpers.[92] To the end of the Second Empire and on into the 1870s, pupils and professors were happy to see their schools incorporated in a program of meteorological observation that engaged them and other responsible citizens in storm watching and taking simple readings under the supervision of local committees, one for each department.[93] The appointment of a senior public figure—a departmental civil engineer, a member of the state mining corps, or a prefect, for example—to chair such a committee only heightened the zeal of the amateurs who did the legwork. Indeed, it had the desired dual effect of at once lending them status in their regions and binding them to Le Verrier's enterprise and, through the prefects who were required to coordinate efforts in the various departments, to the central administration. And it worked so effectively that the departmental committees and local networks of observers remained essential to meteorology even after the establishment of a reorganized meteorological service, under the Bureau central météorologique, in 1878.

This dual allegiance to local and national interests reflected Le Verrier's own mixed objectives. For the populist and largely provincial character of the initiatives in meteorology always went hand in hand with a primary loyalty to the ministry and the senior advisers who fashioned ministerial policy on intellectual matters, of whom he was one. Along with Dumas, Adolphe Brongniart, Milne Edwards, and Emile Blanchard, Le Verrier had, and enjoyed, the authority to speak for the sciences, just as Désiré Nisard and the ancient historian and classical scholar Amédée Thierry did for the humanities. In support of their power, Le Verrier, Dumas, and Brongniart could also flaunt their titles of inspector of higher education (positions to which they had been appointed in 1852).[94] And all three of them had an important additional seat of influence in the prestigious Comité de la langue, de l'histoire et des arts de la France, a committee that, especially after its renaming as the Comité des travaux historiques et des sociétés savantes in 1858, did much to fashion the official high culture of the empire.[95]

Founded by Guizot in 1834 to oversee the publication of documents of French history, the Comité devoted the lion's share of its efforts during the July Monarchy to a program of recording and study in which independent scholars, many of them gifted and hard-working, collaborated unobtrusively but effectively with the highly trained archivists, museum curators, and librarians emerging from the prestigious Ecole des chartes; by 1852, about a hundred substantial volumes of historical documents, the Collection des documents inédits de l'histoire de France, had appeared, a number that had risen to 258 by 1874.[96] From the start, Guizot had intended that the sciences and medicine should be treated as part of the nation's cultural heritage and that savants should take their place alongside political and literary figures, philosophers, artists, architects, and the other "actors" of history and prehistory as contributors to France's glorious past.[97] That perception had been taken further in 1837, when, as Minister of Public Instruction, the comte de Salvandy had established a separate committee for science, within a structure that now distributed the work between five constituent committees corresponding to the five academies of the Institut de France.[98] But, in reality, the reform, like other structural amendments that followed, did little for science. The pattern of reports on meetings of the Comité at which no business of scientific relevance had been conducted was disheartening and, to all appearances, irreversible.

With the scientific disciplines occupying such a precarious position, Victor Cousin's reform of 1840, in which Salvandy's scientific committee was merged in a larger Comité pour la publication des documents écrits de l'histoire de France, cannot be seen as an attempt to diminish the status of science within the pantheon of French culture. But it did reflect the difficulty of determining the role that science might have in a structure in which Guizot's historical, document-based conception of the Comité's set the tone for good scholarly endeavor.[99] In fact, the place of science in the Comité was a matter of genuine concern to Cousin, as it was to his successors in the Ministry of Public Instruction. Concern, however, did not equate to a solution, and the marginality and false starts continued. At last, in 1858, Rouland's reorganization of the Comité seemed to promise science a prominence that its advocates had long hoped to secure for it.[100] The new title of Comité des travaux historiques et des sociétés savantes in itself signaled Rouland's intention of bringing science definitively into the fold of the ministerial patronage of culture. Still more important in this respect was its division into three sections: history and philology, archaeology, and science. With regard to science, it was stated that the brief of the revamped Comité should extend to the "riches" of the nation's scientific tradition, a declaration that indicated a break with the pervasive philological emphasis, especially marked under Fourtoul, on songs, poems, and other manifestations of popular culture. Rouland's will to advance science in the cultural hierarchy was beyond question. Yet the fruits of the new prominence that was to be given to France's scientific heritage were few, and even they only materialized in the 1860s, under Duruy, when the multivolume complete works of Antoine Lavoisier, Augustin Fresnel, and Joseph-Louis Lagrange began to appear.[101] Handsome and prestigious though such publications were, they represented a thin haul for the sciences after so many declarations of good intent.

The most disappointing of Rouland's well-intentioned initiatives in favor of science was a multivolume *Description scientifique* of France. The work was conceived as a detailed account of the geology, botany, zoology, and meteorology of each department.[102] But the anticipated battalions of independent contributors eager to pursue the task did not materialize. Only one volume of the *Description*, for the Bas-Rhin, was published, and in the early 1860s the Comité tacitly acknowledged the difficulty of promoting coordinated scientific activities in the provinces by abandoning the plan. Significantly, a parallel series of departmental inventories of archaeological remains, the *Répertoire archéologique de la France*, fared somewhat better. And a *Dictionnaire topographique de la France* unquestionably succeeded, albeit slowly; between 1861 and the First World War, topographical descriptions of most departments, largely written by painstaking amateurs, limped into print.[103] Both in topography and, to a lesser extent, in archaeology, the manpower and zeal for a vast project on the national scale were there and waiting to be mobilized. In the sciences, with their already well-established reward systems founded on societies, journals, and extended informal networks, they were not. Meteorology alone, in the manner in which Le Verrier conceived it, provided the exception.

Despite the failure of the *Description scientifique*, the empire's circle of favored advisers on science worked assiduously in promoting both their various disciplines and ministerial authority. They were natural allies of the successive ministers who pursued their running battles against Arcisse de Caumont's aspirations for a structure for provincial science and scholarship independent of the state. They would also have sympathized with Fortoul's attempts to counter independence of a very different and subtler form within the Institut de France. Of the Institute's constituent bodies, the Académie française was the most troublesome. Here the elections of the legitimist lawyer Antoine Berryer (in 1852), Félix Dupanloup, Bishop of Orléans (in 1854), and the Orleanist duc Victor de Broglie (in 1855)—all three of them declared critics of the empire with royalist sympathies of one sort or another—were flaunted, to Fortoul's fury, as famous victories for independence.[104] Other academies too notched up their successes. The election of the prominent Orleanist politician Odilon Barrot to the Académie des sciences morales et politiques in 1855 gave members all the more pleasure since Fortoul had tried clandestinely to thwart Barrot's chances, while the Académie des inscriptions et belles-lettres drew its mite of satisfaction from rejecting Fortoul (narrowly) in 1854 before he was finally elected in the following year.[105]

In a manner characteristic of the early Second Empire, the Académie des sciences seems to have had tranquil, if not warm, relations with Fourtoul's administration. Rouland's arrival in the Ministry of Public Instruction in 1856 heralded even better times for science as for all areas of learned culture, and the academy, like others in the Institute, looked forward to having a more receptive ministerial ear. But, despite his slightly looser rein, Rouland retained a keen sense of what might best advance his undimmed quest for the allegiance of the world of learning, both independent and "official." A conspicuous success, and one in which Rouland's scientific advisers played a prominent role, was the annual Congrès des sociétés savantes. Held each year from 1861 in a self-congratulatory atmosphere of patriotism and benevolence, the congresses allowed ministerial officials, scientific grandees, and the leading lights of the independent world of science to bask in a shared sense of achievement. In the sessions of the science section, Le Verrier (as the section's president) and Henri Milne Edwards (as its vice president) would honor the work of amateur scientists by showering compliments and contributing to an imposing sense of occasion. Their interventions and those of other leading savants were not solely, or even primarily, motivated by a disinterested dedication to the independent world of learning. The proximity to the pinnacle of educational power, displayed at the congresses on a very public stage and under the minister's eye, offered rewards to those who bestowed favors as well as to those who received them.

No one knew this better than Le Verrier, whose zeal for the objectives of ministerial policy led on to still headier contacts, ascending even to the emperor himself. Who else but Le Verrier was Napoleon III to go to for advice on the astronomical matters that arose in his writing of the life of Julius Caesar? A specific question such as whether the time of the full moon really was three o'clock in the morning on the day of Caesar's landing in England in 55 BC, as *De bello gallico* records, was easy meat for Le Verrier. But his complete concordance between Roman dates and their counterparts in the reformed Julian calendar and a table that allowed Roman times of day (in the same crucial year of 55 BC) to be converted to their modern equivalents, both of which appeared as appendices to the emperor's book, reflected a substantial investment of time.[106] The fact that Le Verrier's role in compiling the appendices went unacknowledged would not have concerned him: even though inquiries were communicated to Le Verrier by the head of the head of his personal library in the Tuileries palace, Alfred Maury, the sense of even that degree of closeness to the fount of imperial power was gratification enough.

Exchanges between the imperial household, Ministers of Public Instruction and their senior staff, and the designated elite of the nation's science and scholarship remained frequent throughout the empire. And they played their part in the fashioning of a close-knit administrative inner circle which, as even disgruntled liberal intellectuals had to admit, took its responsibilities seriously. The point was not lost on Duruy, who tempered his commitment to reform by leaving intact the privileged lines of communication that he found in place in 1863. Judiciously, he continued to take advice from the same men who had proffered it to Fortoul and Rouland: Nisard and Thierry in the humanities, and the trio of Le Verrier, Dumas, and Milne Edwards in the sciences. Yet Duruy's policies were seen to be of a different stamp from those of his predecessors, and his dealings with a group of advisers who had enjoyed the gratifications of power for so long showed frequent strains. Dumas did not conceal his indignation at seeing Duruy rather than himself appointed Rouland's successor, and it was some time before he and Duruy resumed their earlier cordial relations. Nisard, who had also hoped to succeed Rouland, was a much tougher nut to crack. His unyielding conservatism and opposition to the opening of French intellectual life to foreign (especially German) influences set him against all manifestations of reform. He remained a thorn in Duruy's flesh until student unrest in support of Sainte-Beuve's defense of the "advanced" literature that had found its way into the public library of Saint-Etienne precipitated the temporary closure of the Ecole normale supérieure in July 1867 and his humiliating dismissal as the school's director.[107] Duruy softened the blow by retaining Nisard as one of the three general inspectors of public instruction and as a member of the Council of Public Instruction; he also arranged for Nisard to be offered a seat in the Senate. But a humiliation it remained.

It was Le Verrier, however, who tried Duruy's patience most sorely and whose refusal to bow to the ministerial will eventually dragged astronomy into the mire of a prolonged and very public scandal.[108] Trouble had been mounting ever since Le Verrier took up his post in 1854. His dictatorial manner had made the observatory an unhappy place from the start. But the frictions had remained a largely internal matter until a simmering dislike between Le Verrier and his slightly younger colleague, Charles Delaunay, erupted in a series of intemperate exchanges in the Académie des sciences in February and March 1860. The conflict turned on an error that Le Verrier claimed to have found in Delaunay's recently published theory of the motion of the moon.[109] By November, it was Delaunay who took the offensive, again in the Académie and again on the grounds of errors, which he had found in a contribution by Le Verrier in the *Annales de l'Observatoire*.[110] The acrimony of what had begun as a discussion of issues of great technicality drew the inevitable attention of the press and of scientific writers ever eager for gossip. For the next decade, wherever Le Verrier was concerned, both gossip and overt dislike of the man flowed in abundance.[111]

Almost no one, it seemed, could retain Le Verrier's favor for long. Foucault had begun well enough; he was too good a catch as the observatory's senior physicist in 1855 for Le Verrier to turn on him immediately. Within two years, however, Foucault's preference for working in his private laboratory in the rue d'Assas rather than in the observatory was proving a source of irritation to Le Verrier, as was his interest in aspects of physics that did not bear centrally on the priority areas of meteorology, terrestrial magnetism, and weights and measures.[112] Le Verrier's disquiet on these counts was not unreasonable, but his attempts to bring Foucault to heel were, at best, gauche. His refusal, at one stage, to pay Foucault's salary was just the tip of an iceberg of destructively hostile acts and tart letters of the kind that contributed to the departure of sixty-four members of the observatory's astronomical staff between 1854 and 1867.[113] Foucault's hardship was alleviated by an exceptional direct payment authorized by Duruy. But other employees who suffered from Le Verrier's use of the weapon of a withdrawal or reduction of salary were almost certainly less fortunate.[114]

Especially sad was the case of Hippolyte Marié-Davy, the head of the meteorological service from 1862 and initially one of Le Verrier's most loyal collaborators. Le Verrier's charge that Marié-Davy had passed off as his own results that had been obtained as part of the work of the observatory seems to have been grotesquely unfair,[115] and no amount of tension between the two men could have justified Le Verrier's capricious treatment of a colleague who had compromised both health and family life in the service of the institution. Matters came to a head in the winter of 1865–66, when Le Verrier turned off the heat in the living quarters of the observatory that Marié-Davy occupied in order to be available for taking readings at odd hours.[116] As late as April 1866, Marié-Davy expressed his apparently genuine respect for Le Verrier in the preface to his *Météorologie*.[117] But before the end of that month he had composed a letter summarizing his grievances, and in the autumn he sent this to the emperor.[118] Relations between Marié-Davy and Le Verrier had soured irreversibly, and the letter, on which the emperor asked Duruy to take action, became an important document as internal tensions rapidly assumed the dimensions of a public scandal. On through 1867, few aspects of Le Verrier's administration escaped scrutiny. For this, Camille Flammarion, who was the best known of all of Le Verrier's victims (having been dismissed from the observatory in 1862), had a large measure of responsibility. Already in 1866, Flammarion had begun publishing a regular column in the liberal newspaper *Le Siècle* in which for the next four years, side by side with replies from Le Verrier, he regaled readers with tales of mismanagement.[119] In these exchanges, no punches were pulled, on either side.

The barrage of complaint and recrimination, both behind the scenes and in the daily and periodical press and other public documents, not only had a corrosive effect on Le Verrier's reputation; they also, and more seriously in Duruy's eyes, undermined the Ministry of Public Instruction's reputation for exercising responsible control over its employees. After turning a blind eye (at least in public) to Le Verrier's behavior during his first four years as minister, Duruy found his patience exhausted. In August 1867, a hectoring letter from Le Verrier accusing Marié-Davy of hypocrisy and a refusal to obey orders can only have strengthened Duruy's resolve.[120] At all events, on 4 October 1867, in a letter to the emperor, accompanying a dossier of evidence, Duruy wrote:

Sire,

For four years now I have not dared to look inside the Observatory. It is no longer possible to refrain from doing so. The Emperor will be persuaded of this if he will agree to cast an eye over the enclosed file.

The Emperor will see that if the Ecole normale was not governed, the Observatory is governed to excess.

Eleven astronomers or assistant astronomers are very nearly inactive.

All those holding permanent appointments at the Observatory in 1854 have been dismissed, except for one, who is left without work.

Of 68 calculators whom M. le Verrier has appointed at different times, 48 have left. The auxiliary staff, for their part, have reached breaking point. 33 of them have gone.

Salaries are arbitrarily suspended, reduced, or withdrawn.

Science is suffering as a result of constant changes of personnel and the ill feeling that these cause. Work has ceased at the Observatory, and the reputation of this great institution is in decline. It is said that astronomy is no longer being done in France.[121]

Duruy's closeness to the emperor and the notoriety of the case ensured that the scandal had at last to be faced. The institution of an eight-man committee of inquiry that followed was a gauge of Duruy's, and the emperor's, resolve.

From the start, Le Verrier's appointment to the committee risked compromising its capacity for decisive action, and he did nothing to conceal his rage. The committee duly received his withering denunciation in two published broadsides.[122] Nevertheless, it met. It did so thirteen times between November 1867 and January 1868, invariably in Le Verrier's absence (since he refused to attend) and despite his objection to two of its members, almost certainly Delaunay and the mathematician Alfred Serret (who had worked at the observatory, first as a calculator and finally as an *astronome adjoint* from 1856 to 1862).[123] The report that followed was uncompromising. It recommended a severe limitation of the director's power, which was henceforth to be subject to the decisions of a supervisory committee. The result was more than two years of spectacularly turbulent exchanges that brought science to public attention for all the wrong reasons.[124] Le Verrier, for his part, was unshakeable in his refusal to acknowledge not only the legitimacy of the committee of inquiry of 1867–68 but also the authority of the new supervisory committee; his critics, though, found abundant ammunition in the inquiry's report. It was a recipe for face-to-face squabbles in the Académie des sciences and printed statements and press comment of breathtaking vehemence. Duruy's *cri de coeur* in a personal letter of March 1868 to the empress about a "professor who does not profess, a general inspector who does not inspect, a director who directs too much"[125] was fired by frustration. But it was also pithily accurate.

As events transpired, Duruy was no longer in office by the time the affair reached its culmination. But six months after his departure from the ministry in July 1869, the turbulence at the observatory was undiminished. Delaunay's condemnation of Le Verrier in a letter of 9 December 1869 to the new Minister of Public Instruction, Olivier Bourbeau, was blistering: "The main facets of his character are an effrontery and charlatanism such as I have encountered nowhere else. Add then that falsehoods pour from his mouth as readily as truths: he lies with an impudence and shamelessness that beggar belief."[126] And so on, in a litany in which Delaunay elaborated on Le Verrier's exaggerated reputation and condemned his "capricious despotism," his "infernal character," and an indifference to the interests of science that had "killed" observational astronomy in France.[127] By late January 1870, the affair was set on a helter-skelter of events that could have only one end.[128] Criticism of Duruy in Le Verrier's recent annual report enraged the council of the observatory. And after all thirteen of the institution's senior astronomical staff—four *astronomes titulaires* and nine *astronomes adjoints*—had resigned and elaborated their grievances in a memorandum to the Ministry of Public Instruction, Emile Segris (now at the head of the ministry, following Bourbeau's brief tenure) set up a new committee to investigate the conflict.[129] At this point Le Verrier himself added the last straw. His demand in the Senate that Segris should be called to account displayed an impertinence that neither his fellow senators nor the emperor's inner circle of advisers could accept. The imperial decree removing him from the directorship of the observatory followed on 6 February. No manner of going could have inflicted greater humiliation, and the appointment of Delaunay as Le Verrier's successor a month later only added salt to the wound.

The episode can be interpreted at one level simply as the inevitable culmination of Le Verrier's record of intolerable behavior, which since his appointment in 1854 had turned the Observatoire de Paris into what Emile Ollivier, in his history of the later, liberal phase of the Second Empire, was to describe as nothing short of a battlefield.[130] But behavior that had been allowed to go unchecked in the empire's authoritarian phase was judged differently in the 1860s. The growing acrimony and increasingly public nature of political debates between conservatives and liberals now added a new dimension to what might otherwise have remained a tricky but resolvable matter of internal discipline. More immediately, the fact that by the late 1860s Duruy himself was coming under unprecedented fire from opponents of his attempts to liberalize French academic life can only have emboldened Le Verrier (as it emboldened other critics). It is not hard to imagine the encouragement as well as the satisfaction that Le Verrier must have drawn from the tide of hostile opinion, much of it clerically driven, that culminated in Duruy's dismissal by a reluctant emperor in July 1869.

At the time when Duruy left the Ministry of Public Instruction, in fact, the triumph of his liberalizing influence on the administration of the French world of learning was far from assured. The debate surrounding the chair of Hebrew, Chaldaean, and Syriac at the Collège de France, vacant again since 1867, following the death of the Jewish biblical scholar Salomon Munk, brought into stark focus the precariousness of the liberal position. When the question of filling the chair came up in 1869, the Collège and the Académie des inscriptions et belles-lettres urged the appointment of Ernest Renan, and it seems that this choice had powerful support: Ollivier, as head of the liberally inclined center-right administration that led France from January 1870 until the early days of the Franco-Prussian war, and Segris, his Minister of Public Instruction, were both in favor. But a decision to bring back Renan barely five years after his dismissal from the chair in 1864 would have been seen by conservatives, clerical and nonclerical alike, as a blatant provocation. It was a sign of the singular sensitivity of the issue in the months following Duruy's downfall that Ollivier and Segris bowed to what they saw as the inevitable and decided not to endorse Renan's appointment.[131]

When Renan had been suspended and then dismissed from his chair, the emperor had written to him to express his regret.[132] The contact is significant as an indication of where his sympathies lay. As he believed, Renan's travails during the 1860s gave unwelcome satisfaction to those who had most vehemently resisted the liberalization of the empire, and he deplored the criticism of Renan as much as he deplored the behavior of Le Verrier and sympathized with his victims, Marié-Davy in particular. The tragedy was that his rallying to the liberal cause in French intellectual life, which might have saved his reputation in the eyes of the contemporary cultural avant-garde, if not his regime, came too late. Nevertheless, liberalization left its legacy. It created an openness to at least some measure of academic freedom, and a context in which the personal rapport between the emperor and Duruy could have an effect. The reforms required the confidence and assertiveness of the Third Republic for their full realization. But, as the roots of what was to come, they form an essential part of any account of intellectual life under Napoleon III. For science, there was far more to the empire's achievements in its last years than taming Le Verrier, significant and revealing though this was.

The Roots of Academic Reform

Duruy came to the Ministry of Public Instruction in 1863 as a respected scholar in his own right and, more controversially, as someone who encapsulated the ideals of the liberal Empire. Under the July Monarchy, as an admirer of Michelet and while teaching history at the Collège Henri IV in Paris, he had already published two substantial volumes of Roman history[133] and so laid the foundations of the seven-volume *Histoire des Romains* that was to become his most important scholarly legacy.[134] Through the 1850s and early 1860s, his undimmed productivity, extending to frequently reprinted histories of ancient Greece and of France, further enhanced his reputation. He could also boast the signal battle honor, for a liberal, of a widely publicized confrontation with Désiré Nisard, one of the most hated incarnations of imperial intellectual orthodoxy. The occasion had been the defense of Duruy's thesis on the Roman world at the time of the first emperor, Augustus. During the defense at the Sorbonne in 1853, Nisard had abused his position as a member of the jury by drawing flattering parallels between the beneficial rule of Caesar and Augustus and that of Napoleon I and the French imperial dynasty, now represented by Napoleon III. The audience's hostile reaction and Duruy's personal indignation at the misuse of his own interpretation were predictable and set the tone for the contempt with which the older generation of the regime's favorites were viewed in liberal circles until the end of the empire.[135]

Universitaires who shared Duruy's tolerant but essentially anticlerical instincts were understandably heartened by his appointment, and they were not to be disappointed. While some hearts and minds remained obdurately opposed to him, most responded favorably to his manifest commitment to improve the lot of his corps of teachers at both the secondary and higher levels of education. One early measure, significantly reversing the practice introduced by Fortoul, was his decision that any proposal for the dismissal or enforced retirement of a professor should be referred to a committee of the Imperial Council of Public Instruction before it could be implemented.[136] Of more general consequence was his sensitivity to the long-standing grievances about the profile and gen-

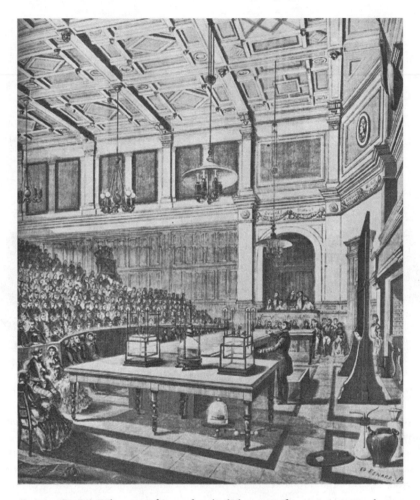

Figure 1. Eugène Péligot, professor of applied chemistry from 1841 to 1889, lecturing at the Conservatoire des arts et métiers, in the amphitheatre inaugurated in 1822. In this image, dating from about 1850, Péligot is performing an experiment on the synthesis of water. Courtesy of the Département médiathèque, Musée des arts et métiers, Paris.

Figure 2. Cartoon of a banquet during the Congrès scientifique in Strasbourg in 1842. Conviviality was essential to Arcisse de Caumont's conception of his annual congresses, and
outings and banquets became something of a public spectacle. On occasions, as at the
Douai congress of 1835, members of the public were even invited to walk about the room
and observe the diners. The cartoon appeared in the satirical magazine *Le Charivari* soon
after an article ridiculing the congress's intellectual pretensions (see chapter 2). The mention, in the caption, of toasts to such causes as the elimination of weevils, independence for
beetroot, and the promotion of candles without wicks and the diners' concluding cries of
"Long live Strasbourg pâté" conveyed the air of absurdity. From *Le Charivari*, 11ᵉannée, 22
October 1842. © British Library Board (BL F.15)

Figure 3. Menu card for a banquet of the Société zoologique de France, 1894. The society held the first of its annual banquets in this year, when the event honored Alphonse Milne-Edwards, director of the Muséum d'histoire naturelle. The banquets were normally held on the occasion of the society's annual general meeting and were seen as occasions for marking the achievement of a major contributor to the discipline and the work of the society. They also served the important function of maintaining the bonds between the society's senior academic members and its substantial membership among amateur enthusiasts. Courtesy of the Société zoologique de France.

Figure 4. The chemist Michel-Eugène Chevreul in the study of his house in the grounds of the Muséum d'histoire naturelle, where he continued to work regularly until the early 1880s, when he was 96. A pupil of Vauquelin, he made important contributions to dyeing and color chemistry at the Gobelins textile factory and was professor of organic chemistry at the museum from 1830 until his death. As the museum's director from 1863 to 1879, Chevreul consistently championed its autonomy. During the Second Empire he resisted attempts to reform the institution's administrative structure, based since its foundation in 1793 on the principle of an independent community administered by its professors. This image was published, from a photograph by Nadar, on the occasion of Chevreul's death at the age of 102. Published in *La Nature* 17, no. 1 (1889): 321.

TREIZIÈME MOIS.

BICHAT.

LA SCIENCE MODERNE.

Copernic.	*Tycho-Brahé.*	
Kepler.	*Halley.*	
Huyghens.	*Varignon.*	
Jacques Bernoulli.	*Jean Bernoulli.*	
Bradley.	*Roëmer.*	
Volta.	*Sauveur.*	
GALILÉE.		
Viète.	*Harriott.*	
Wallis.	*Fermat.*	
Clairaut.	*Poinsot.*	
Euler.	*Monge.*	
D'Alembert.	*Daniel Bernoulli.*	
Lagrange.	*Joseph Fourier.*	
NEWTON.		
Bergmann.	*Scheele.*	
Priestley.	*Davy.*	
Cavendish.		
Guyton-Morveau.	*Geoffroy.*	
Berthollet.		
Berzélius.	*Ritter.*	
LAVOISIER.		
Harvey.	*Ch. Bell.*	
Boërhaave.	*Stahl.*	
Linné.	*Bernard de Jussieu.*	
Haller.	*Vicq-d'Azyr.*	
Lamarck.	*Blainville.*	
Broussais	*Morgagni.*	
GALL.		

Fête universelle des MORTS.
Fête générale des SAINTES FEMMES.

Jour complémentaire.
Jour bissextile.

Figure 5. The thirteenth month of Auguste Comte's positivist calendar. This final month in the Comtean year, the month of "modern science," was named for Xavier Bichat, whose work in anatomy and physiology Comte admired for its positivist approach. Each of the month's twenty-eight days was associated with the name of an historical figure commemorated on that day, often with a supplementary name to be used as an alternative once every three years. Each week ended on a Sunday, when a figure of special importance for the week's commemorations, was designated. For the month of Bichat, Comte singled out Galileo, Newton, Lavoisier, and Gall in this way. From Comte, *Système de politique positive ou Traité de sociologie, instituant la Religion de l'humanité,* 4 vols. (Paris, 1851–54), vol. 4, facing page 402.

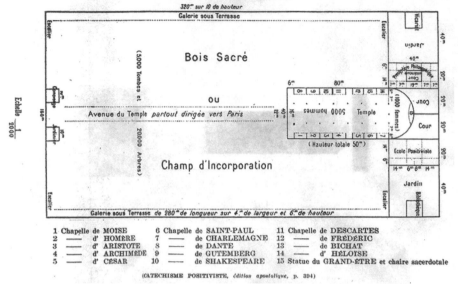

PLAN GÉNÉRAL
d'un grand temple de l'Humanité
PAR
AUGUSTE COMTE

320ᵐ sur 10 de hauteur

Galerie sous Terrasse

Bois Sacré

ou

Avenue du Temple *partout dirigée vers Paris*

Champ d'Incorporation

(5000 Tombes et   20000 Arbres.)

5000 hommes   Temple

(Hauteur totale 50ᵐ)

Galerie sous Terrasse *de 280ᵐ de longueur sur 4ᵐ de largeur et 6ᵐ de hauteur*

Echelle 1/2000

1	Chapelle de MOISE	6	Chapelle de SAINT-PAUL	11	Chapelle de DESCARTES
2	— d' HOMÈRE	7	— de CHARLEMAGNE	12	— de FRÉDÉRIC
3	— d' ARISTOTE	8	— de DANTE	13	— de BICHAT
4	— d' ARCHIMÈDE	9	— de GUTEMBERG	14	— d' HÉLOISE
5	— d' CÉSAR	10	— de SHAKESPEARE	15	Statue du GRAND-ÊTRE et chaire sacerdotale

(CATECHISME POSITIVISTE, *édition apostolique*, p. 394)

Figure 6. Floor plan of an ideal temple of humanity, drawn by Georges Audiffrent from a sketch by Comte now in the archives of the Maison d'Auguste Comte. Most of the area of 320 meters by 160 meters was to be occupied by a wooded cemetery with space for five thousand tombs. The temple itself, aligned in the direction of Paris, would accommodate five thousand men and (separately, in the sanctuary, with its all-important female figure representing humanity) a thousand women. The side chapels were destined for representations of the historical figures associated with the thirteen months of the year, the fourteenth chapel being dedicated to Héloise, representing the Comtean category of "holy women." The positivist school and library occupied a characteristically prominent place, close to the priest's "philosophical presbytery." The plan was first published in the *Revue occidentale philosophique sociale et politique* in January 1880, facing page 176. This version is from the 1891 "apostolic edition" of Comte's *Catéchisme positiviste*, facing page 394. Courtesy of the Maison d'Auguste Comte, Paris (Fond Gauge, BP).

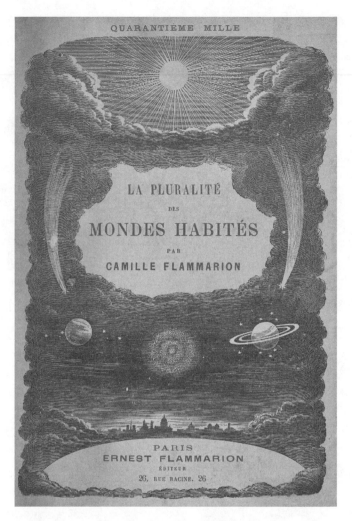

Figure 7. Cover of Camille Flammarion's *La Pluralité des mondes habités*. First published in 1862, the book went through many editions (thirty by 1883) and had sold forty thousand copies by the time of this popular edition of 1909. It offered a lucid introduction to planetary and stellar astronomy, with the added spice of Flammarion's speculations on the conditions of life on the planets of the solar system, which he believed to be (or to have been) inhabited. Despite its popularity, sales did not match those of Flammarion's *Astronomie populaire*, which sold more than 125,000 copies between its initial publication in 1880 and the First World War.

Figure 8. Raising of a table during a séance conducted by the Italian medium Eusapia Palladino. The image (drawn from a photograph taken with the aid of a magnesium lamp) depicts a trial conducted in Milan in 1892. The seated observers, in positions that allowed them to control the hands and feet of the medium, were the Italian criminologist and positivist Cesare Lombroso (*left*) and the physiologist Charles Richet (*right*). From Flammarion, *Les Forces naturelles inconnues* (1907 and subsequent reprints), plate 8, facing p. 210 in the 1907 edition.

Figure 9. Advertisement for public lectures in the rue de la Paix, Paris, March 1865. Emile Deschanel organized the lectures as a commercial venture, for predominantly society audiences willing to pay 2.50 francs for a reserved seat. The lectures often broached cutting-edge topics such as spontaneous generation and the implications of fossilized human remains, both treated here by Nicolas Joly, a prominent champion of Félix-Archimède Pouchet. The advertisement, on a single sheet, is in the Archives départementales de la Haute Garonne, Toulouse.

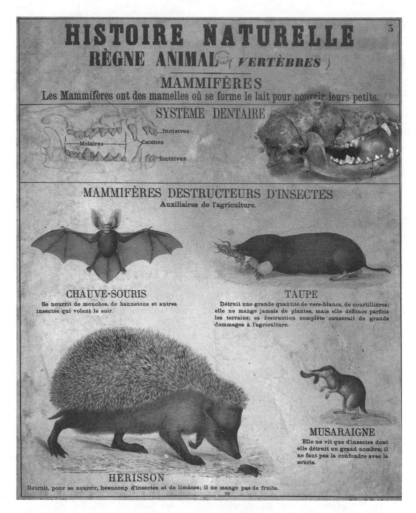

Figure 10. One of twenty wall charts illustrating all areas of natural history, published by the firm of E. Deyrolle, fils in the 1880s. Many of the charts, like the one shown, were supplied with real specimens, in this case a mammalian skull, in accordance with Deyrolle's principle that observation was a better route to understanding than abstraction. Founded in 1831, the Deyrolle firm catered for amateur as well as specialist collectors. In the later nineteenth century, it benefited from the demand for specimens and charts stimulated by the growth of natural history as a subject in primary education. © British Library Board (BL Tab.1800.c.1).

Figure 11. Foucault pendulum in the Panthéon. Originally installed in 1851, the pendulum was removed after Louis-Napoléon's coup d'état in December of that year, when the building reverted to its original function as a church. Following the Panthéon's restoration to public use in 1885, on the occasion of the funeral of Victor Hugo, the pendulum was reinstalled there in 1902, at the initiative of the Société astronomique de France and under the supervision of the architect of the new Sorbonne, Henri-Paul Nénot. From Camille Flammarion, *Notice scientifique sur le pendule du Panthéon. Expérience reprise en 1902 au nom de la Société astronomique de France* (Paris: Société astronomique de France, 1902).

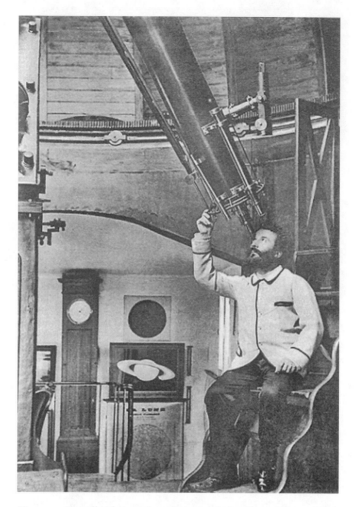

Figure 12. Camille Flammarion observing in his observatory at Juvisy. He is shown using the 240 millimeter equatorial telescope, constructed by the Parisian instrument maker Bardou. The house there had been given to Flammarion in 1882 by Eugène Méret, an elderly bachelor who had himself pursued amateur astronomy at Juvisy from 1856 until advancing blindness and the ransacking of the property by Prussian troops led him to abandon it in 1870. Following Méret's departure, the house and fine gardens had fallen into disrepair. But restoration by Flammarion, including the construction of the dome shown in this postcard image, turned the house into an important center for observational astronomy by amateurs of high standing, such as Eugène Antoniadi and Ferdinand Quénisset. Courtesy of Françoise Launay, private collection.

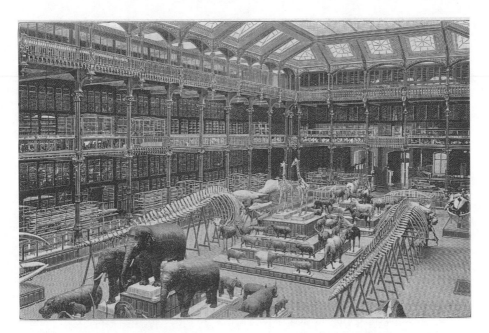

Figure 13. The main display in the new zoological galleries at the Muséum d'histoire naturelle, a popular public attraction inaugurated in 1889. The gallery retained its original style of presentation until its refurbishment and reopening as the Grande galérie de l'évolution in 1994. The large whale skeletons were always a particularly attractive exhibit. From Gaston Tissandier, "Nouvelles galeries de zoologie au Muséum d'histoire naturelle de Paris," *La Nature* 17, 2ᵉ semestre (1889): 313.

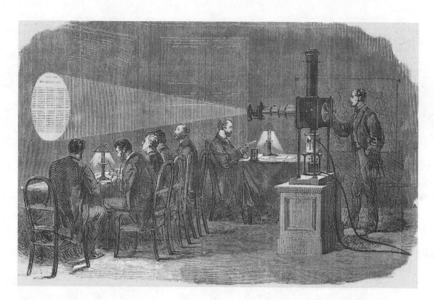

Figure 14. The projection and copying of photographed messages brought to Paris by carrier pigeon during the siege of 1871. Thanks to what *L'Illustration* hailed as "Another miracle of science," it was claimed that a whole page of the official government gazette, the *Journal officiel,* could be reduced to a square millimeter, so allowing one pigeon to transport as many as thirteen thousand messages, which were enlarged, copied, and delivered within Paris as normal post. In this way, the main organizer of the service, the photographer René Dagron, assumed the status of a national hero. From *L'Illustration*, 29, no. 57 (21 January 1871): 25. © British Library Board (BL F.63).

Figure 15. The electricity fairy, as depicted in the 1880s on the title page of bound copies of *La Lumière électrique*, a weekly journal of electricity launched in 1879. Such images of the "fée électricité" as a benign female figure bestowing light and material blessings on the world were common in the late nineteenth and early twentieth centuries.

Figure 16. An alternative depiction of the electricity fairy, by Albert Robida. The image appeared as the frontispiece to Robida's *La Vie électrique* (Paris, 1892), a futuristic novel describing life in the 1950s. In representing the fairy as a slave with a distinctly menacing aspect, Robida was conveying the anxiety that tempered his fascination with modern technology, which he believed would have transformed lives by the mid-twentieth century for ill as well as for good.

eral level of salaries in the faculties of science and letters. An income of 4,000 francs a year, plus the supplementary *traitement éventuel* of between 15 and 60 percent drawn from the fees for examining, set even the most eminent professor in a provincial faculty below a senior teacher in a Parisian *lycée*, for example, and far below a prefect or other highly placed employees of the state, who could expect to earn 20,000 francs or more.[137] Even the eighteen chairholders in science at the Sorbonne had professorial salaries (including the *traitement éventuel*) of no more than 11,500 francs (although the tradition of *cumul* did allow at least some of them to augment their incomes by taking other, additional posts). In the event, it was some years before Duruy's persistent pleas secured funds for a modest improvement, but at least his good will was recognized and appreciated.[138]

The innovation that most forcefully signaled Duruy's intent, however, was his dismantling of what remained of Fortoul's *bifurcation*. In fact, the system had already been somewhat eroded under Rouland. But within four months of his appointment, Duruy had further undermined it by putting back the separation between the streams in science and letters, and so postponing specialization, by one year.[139] By December 1864, a complete restructuring had brought even this emasculated form of *bifurcation* to an end by introducing an undifferentiated core program of study, including an enhanced dose of science, that took pupils through to the *baccalauréat-ès-lettres*.[140] Candidates wishing to go on to the *baccalauréat-ès-sciences* would then be expected to embark on a final one or two years of preparation in the class designated *mathématiques élémentaires*. Although the option of bypassing the *baccalauréat-ès-lettres* and moving straight into *mathématiques élémentaires* existed, it did so as a last vestige of the old parity between the scientific and literary streams, and was discouraged as undesirable corner cutting. The element of literary study that was maintained in the program of *mathématiques élémentaires* did not provide the cultural leavening in the humanities that Duruy saw as an essential counterweight to scientific and mathematical study. The report he submitted to the emperor as a preamble to the legislation of December 1864 pressed the case for balance:

> Through literature, we foster affective sentiments, moral ideas, lucid argument, imagination, the taste for the good and the beautiful, and the experience of life. Through the sciences, we create a healthy counterweight to the faculties of imagination and feeling, whose flights have to be tempered and restrained ; we subject the mind to the strict discipline of methods of reasoning and point the way on the hard, austere path that has to be followed in the pursuit of truth.[141]

Duruy's belief in the complementarity of the humanities and the sciences and the importance of both (though especially the humanities) in a rounded education reflected his commitment to the highest ideals of intellectual life, at least for the elite of French society. It also signaled his rejection of the blanket utilitarianism that had so profoundly marked the policies of Fortoul and, in a lesser degree, Rouland. For Duruy, the goal of any curriculum at the secondary level was and should remain the disinterested cultivation of the mind, the fount of all human endeavor.[142] To a minister with that lofty vision, freedom of thought, tempered by discipline and the rigors of academic apprenticeship, was an objective that nothing should be allowed to jeopardize. It was entirely consistent with this view that on 10 July 1863, after only days in office, Duruy reinstated the most contentious of all the *agrégations* that had fallen foul of Fortoul's axe, that in philosophy. No single act could have trumpeted more loudly the change in the educational climate, and (along with the creation of a new chair of philosophy at the Ecole normale supérieure later in the year) it had a profound symbolic impact. Reforms in the scientific *agrégation* too signaled Duruy's intent, albeit less flamboyantly. Here, Duruy's aim was to bring candidates closer to the cutting edge of their various disciplines rather than subjecting them to an exercise in the uncritical accumulation of a vast range of knowledge. To that end he retained the two specialized *agrégations*—one in mathematics, the other in "sciences physiques et naturelles"—that Rouland had introduced in 1858 to replace Fortoul's unified *agrégation* in science. In the same spirit, in February 1869 he even added a third *agrégation*, in "sciences naturelles," albeit one that did not function until 1881 when the stock of natural history in both secondary and higher education was quite suddenly beginning to rise, under the influence of the botanist Gaston Bonnier.[143]

Throughout Duruy's six years in the ministry the scent of reform remained in the air. The balance that he struck was perforce a delicate one. The publication, under ministerial auspices, of twenty-nine volumes of the Recueil de rapports sur les progrès des lettres et des sciences en France, each devoted to a review of the state of a particular scientific or scholarly discipline, was an exercise in patriotism stimulated by the Exposition universelle of 1867;[144] the emphasis of the reports was firmly on "progress," with French scientists and scholars cast as the key instigators. More valuable as a springboard for reform, though still domestic in its focus, was the major ministerial report on the state of French higher education that appeared in 1868.[145] It was behind the façade of the carefully phrased and self-congratulatory language of this and other official documents that Duruy showed his mettle as a minister committed to a portfolio of objectives balancing modernization, efficiency, and discipline. Despite the diversity of objectives, Duruy intended that French higher education and research should become a seat of seriousness and clear thinking about objectives. Hence while he understood the temptation for professors to respond to the shortage of enrolled students by moving their lectures to the evening and making them more palatable to lay audiences, he believed that such rescheduling should be discouraged. The specter, as he expressed it in a circular message to all rectors in 1864, was superficiality. Evening lectures risked reducing science to the level of an "agreeable conversation," thereby undermining the role of the faculties as the guarantors of science's dignity and true purpose.[146] The faculties certainly had a role in fashioning the public face of science and scholarship. But Duruy never wavered in his insistence that that role should not be allowed to obscure the professoriate's primary duty to practice their disciplines at the highest level.

An essential element in all this was the opening of French eyes to what was afoot in education and intellectual life elsewhere in Europe. One of Duruy's favored practices, of commissioning employees of the University to study institutions of higher education abroad, was a touchstone of his priorities. The practice, it is true, was not new; a report by the Swiss-born professor Sigismond Jaccoud on medical education in Germany, for example, had been requested by Rouland, even though it only appeared in 1864.[147] Duruy, however, took the exploitation of foreign models as a way of raising concern in France to new levels. Accounts of practices in Britain, Italy, and Belgium all contributed to his thinking.[148] So too did the judgments of the widely traveled French-born geologist Jules Marcou, whom Duruy encouraged to write a series of vignettes of France's scientific institutions in the light of his experiences abroad, especially in the United States, where he had worked with Louis Agassiz. The result was *De la science en France*, a devastating indictment of the state of the country's science. Marcou described an Académie des sciences in thrall to the preferences and prejudices of its immovable permanent secretaries and a Muséum d'histoire naturelle inveterately resistant to change.[149] Duruy may well have found Marcou's virulence excessive. But much of *De la science en France* would have rung true and served to strengthen his will for reform.

Although Marcou knew the United States better than any other country, he could also draw on firsthand experience of working in Germany and German-speaking Switzerland. This gave his observations special authority. For it was the German model that most profoundly marked discussions about possible ways forward, as it also marked Duruy's thinking, even if on occasions he had to be careful to adjust to patriotic sensibilities by qualifying his respect for Germany in public.[150] Paul Lorain, a doctor in the Saint Antoine Hospital and professor in the faculty of medicine in Paris, had no such inhibition; his statement of admiration for German medical studies and the "scientific" character of medical practice in Germany led on to the categorical statement that "in the realm of science Germany has taken the lead over France."[151] Whatever he believed, Duruy could not quite say as much. Nevertheless, he showed his colors through his acts, for example by despatching Adolphe Wurtz on a fact-finding mission to Germany in 1868. By the time Wurtz left on his two-month journey in June, armed with the inestimable advantages of bilingualism in French and German and the experience and contacts he had gained as a young chemist in Liebig's laboratory at Giessen in the early 1840s, he was well prepared to give advice on the sciences. He was at once more trustworthy than Le Verrier and, as a man of characteristic Alsatian openness to the world beyond the borders of France, far more in tune with Duruy's thinking. He had already undertaken inquiries into the conditions for science in the German and Italian states, Russia, and Norway, and the observations of those countries that he had gathered since 1864 informed the stream of reports from across the Rhine that he now relayed to

Duruy.[152] Within the ministry, Duruy's inner circle had no need to be persuaded of the superiority of German provision for scientific education and laboratories. But the task of marshalling wider support for such a view remained. For this job, there was no better man than Wurtz. He spoke with the authority of a dean of the Paris faculty of medicine (a position to which Duruy appointed him in January 1866) and from a central position in an alliance that Duruy and he had forged during the 1867 Exposition universelle, when Liebig and A. W. Hofmann had visited Paris and spoken not only with the two of them but also with the emperor.[153]

The year 1867 marked a turning point in Duruy's campaign. In the introduction to the official French report on the exhibition, the influential senator and champion of free trade Michel Chevalier added his voice to the case for enhanced investment in science; the contrast he drew between the present levels of expenditure (not least on the Ecole polytechnique, where he had studied) and the amount lavished, and still to be lavished, on the new Opéra was a pointed one.[154] With the matter having been aired at such high levels, the information that Wurtz brought back from Germany in 1868 could scarcely have carried greater weight. In his final report, which he presented to Duruy in March 1869 and published shortly before the war of 1870, Wurtz gave detailed descriptions of a dozen recently constructed laboratories for chemistry, physiology, and anatomy and pathology and briefer accounts of several others.[155] In chemistry, he singled out for special attention the great university laboratories of Berlin, Bonn, Leipzig, and Vienna, all of which he presented as showplaces (although the last of them was still under construction). Those of Berlin and Bonn, for example, had each cost 3 million francs, as Paul Lorain (a relentless campaigner for greater expenditure on facilities in French medical faculties) also observed, and even the tiny German university of Greifswald in Pomerania had spent more than 560,000 francs on its chemical laboratory.[156] The contrast between France's leading scientific institutions and the very least of those in Germany was damning. Expenditure in the German-speaking world was on a scale unimaginable in a French faculty and had been so for almost two decades. How different it all was from the plan for the reconstruction of the faculty of science at the Sorbonne, which had gone no further than the foundation stone, laid in 1855 but now, poignantly, lost.

If science was advancing more rapidly in Germany, the fault, as Wurtz insisted, did not lie with France's men of science but with the conditions in which they had to work. In a withering résumé of his case, he wrote:

> It is not genius, it is not men that we have lacked. It is the material resources and tools for work. On this count, Germany has been better-served. In those many seats of intellectual life, each with its own creative force, in those centers of serious study and high civilization that we know as the German universities, the number, scope, and richness of laboratories have markedly increased in recent years. This trend, which began fifteen or twenty years ago, is rapidly gathering

strength. Public opinion supports it, and governments make commendable efforts to give it succor. It has already borne fruit and it will bear fruit in the future. With us, it has scarcely begun.[157]

Wurtz's point about the importance of mobilizing both public and governmental opinion was one on which Duruy required no persuasion. That had been the foundation of his strategy since 1863.

By the time Wurtz's report appeared, the campaign for a greatly enhanced provision for science had assumed a momentum that would have been unimaginable a decade earlier. Its strength lay in the breadth of the support it won. This came not only from those (such as disaffected provincial professors) who in other contexts might have been dismissed as marginal malcontents but also from savants of the highest standing. It was as dean of the faculty of science at the Sorbonne, for example, that Henri Milne Edwards had bombarded Duruy and other senior administrators for years with tales of woe and pleas for greater funding: "I know of no comparable body of teachers in Europe whose facilities are as wretched" was his damning judgment in 1864.[158] Claude Bernard had added another powerful voice in 1867 in his contribution, on physiology, to the series of "Recueil de rapports sur les progrès des sciences et des lettres." In a work commissioned to glorify French intellectual achievements, Bernard peppered his text with envious comments on the superiority of foreign, especially German, laboratories in his discipline and with accounts of the material impediments that had hindered both his own experimental research and that of his predecessor, François Magendie, as head of the laboratory of experimental physiology at the Collège de France.[159] The need, as Bernard argued, was not for chairs or lecture theaters for the dispensing of the accepted truths of science ("la science faite");[160] it was rather for laboratories in which specialists with particular skills in the various divisions of physiology—anatomy, histology, chemistry, and so on—would be able to work together and in the best possible conditions.

An equally insistent voice was Pasteur's. Frustrated not only by the general pattern of neglect but also by the exiguity (by international standards) of his own facilities at the Ecole normale supérieure, Pasteur followed the example of his colleague Henri Sainte-Claire Deville, who used a personal approach to the emperor to secure an abnormally high level of support for his laboratory.[161] His letter to the emperor, in September 1867, yielded sympathy in return but no action, and Pasteur was soon on the warpath again. This time he spoke publicly, in a blistering indictment of French neglect of science that he published in 1868 as *Le Budget de la science*.[162] In fewer than two thousand words, the document struck home on all the familiar fronts. Laboratories lay at the heart of the case. They were "temples of the future, wealth, and well-being"; without them, physicists and chemists were reduced to the condition of "soldiers without arms." But whereas for thirty years Germany had been constructing large, well-endowed laboratories, France had slumbered, content to see its savants risk their health in the dark, damp, unventilated rooms in which many were obliged to work. Claude Bernard's recent re-covery from an illness contracted in the unwholesome premises where he worked at the Collège de France had to be accounted a miracle, but the deceased professor of chemis-try and dean of the faculty of science in Lyon, Amand Bineau, had been less fortunate; according to Pasteur, Bineau had succumbed to the consequences of working in the cel-lar that passed for a laboratory. With the exception of Sainte-Claire Deville at the Ecole normale supérieure, only those with the means to finance their own facilities—Dumas, Fizeau, Foucault, and Boussingault among them—could buy their way out of the gen-eral misery. The rest had no choice but to sustain their research by diverting such paltry resources as they could from the funds that were allocated for teaching. "Will anyone believe me when I say that in the budget of the Ministry of Public Instruction not a *sou* is allocated to laboratories for the advancement of physical science?" Pasteur wrote. The question was rhetorical. But, for French scientists across the board in the late 1860s, its implication and every word of *Le Budget de la science* rang true.

By January 1868, the emperor, who had seen Pasteur's as yet unpublished sally, was convinced that his earlier sympathy should be transformed into concrete measures. A visit to the chemical laboratories of the Ecole normale supérieure with the empress late in that month[163] and a meeting at the Tuileries palace on 16 March with Pasteur, Ber-nard, Sainte-Claire Deville, Milne Edwards, Duruy, Edouard Vaillant (a former *cen-tralien* with socialist leanings and a firsthand knowledge of Germany as a student), and Eugène Rouher (who held the nation's purse strings in his capacity as head of the Com-mission du budget) kept the momentum going. Pasteur followed with an attack on the administrative weaknesses of the French system, including most notably *cumul*, a prac-tice that ate into senior scientists' time for research and impeded the careers of their younger colleagues.[164] And soon, with Rouher won over as an essential intermediary, Duruy had the emperor's personal assurance that the 50,000 francs he needed as a con-tribution to the provision of new laboratories for Berthelot, Pasteur, Bernard, and Wurtz would be made available.[165] Duruy's landmark address to the annual meeting of representatives of the nation's *sociétés savantes* at the Sorbonne on 18 April reinforced the sense of optimism. Skipping quickly over the conventional pleasantries of congratu-lation for the individuals and societies that were to receive prizes, Duruy made the mo-mentous promise of ministerial action to encourage the best professors (through selec-tive increases in salary) and to provide funds not only for teaching laboratories but also for the creation of laboratories dedicated to research.[166]

Duruy's statement convinced modern-minded academic scientists that their voices had at last been heard. Scholars in the humanities too were promised enhanced support, and their optimism merged with that of their scientific colleagues in a shared sense that a powerful alliance in favor of learned culture had suddenly been consolidated after years of neglect. In their advocacy of a German model of university education and criti-cal inquiry, Gabriel Monod (in history), Gaston Paris and Michel Bréal (in philology and linguistics), and Karl Hillebrand (in modern European literature), as well as the towering figure of Renan, had shown themselves to be quite the equals of Pasteur and Wurtz in persuasiveness and zeal, and they now looked forward to a new world in which, as they hoped, their particular aspirations would be realized. With minds open to rigorous scholarship in the German style, their goal was not just greater funding for libraries and travel (much-needed though this was) but rather the implantation in France of the kind of seriousness that would make the faculties (of science as of letters) centers of true intellectual engagement rather than, in Hillebrand's words, "a cog in the workings of the great centralized machine that we call the French state."[167] Essential to this goal would be closed, specialized research seminars such as Monod had experienced as a young medievalist in Berlin (under Philipp Jaffé) and Göttingen (under Georg Waitz).[168] The humanities, as Monod insisted, needed nothing less than a transformation in the whole spirit of intellectual life.

These were heady days for the champions of educational reform; it seemed that what had always been recognized as an efficient, if not always benign, administration was about to promote root-and-branch change. By July, Duruy's promises had begun to take concrete form in legislation that put in place his most enduring institutional achievement, the Ecole pratique des hautes études. The new organization was a masterpiece of administrative improvisation, creative and far-sighted in its objectives, though cruelly constrained by budgetary limitations. A mere 50,000 francs could be allocated in the first year, and Duruy had no choice but to pursue his aims by directing the modest funds at his disposal to the improvement of facilities already under his control. Ingenuity, though, paid off, with immediate effect. The Ecole's four sections—mathematics, physics and chemistry, natural history and physiology, and historical and philological sciences[169]—were soon providing for research and advanced teaching at the cutting edge across the board of the academic disciplines.[170] Within five months, twenty-seven laboratories (all but five of them in Paris) and eight centers for research led instruction in history and philology were functioning under its aegis, and 395 students (including 85 in the provinces) were formally enrolled by March 1870, the ablest of them working for the doctorate.[171]

The Ecole pratique des hautes études was an oddity in many ways, not least for the absence of any requirement of formal qualifications for entry. This stood in sharp contrast with the progression through the hierarchy of the *baccalauréat*, *licence*, and doctorate and the various *concours* of the *agrégation* that was traditionally required of aspirants to an academic career. Faced with such a marked departure from French norms, carping voices were inevitable. For the liberal writer and historian, Charles Louandre, Duruy had fallen into the trap of undervaluing the existing intellectual achievements of his professors and, through an exaggerated admiration for conditions in Germany, of trying to turn French science into a mere "pastiche" of the German reality.[172] For the author of an unsigned editorial in the reforming *Revue des cours scientifiques*, the Ecole's greatest weakness lay in the all-embracing nature of its activities ("amounting to nothing since it seeks to embrace all areas of knowledge") and an administrative complexity that would have been avoided if the funds had been channelled directly to the chosen professors.[173] But the critical voices represented a minority, and, given the financial constraints, the level of activity in the first year of the Ecole has to be deemed a triumph. It was a triumph of which Duruy was justly proud.[174]

Although publications in history and philology were plentiful (especially in the 1870s[175]), the most substantial successes were in the sciences. This reflected the priorities that Duruy harbored for his new institution, despite his own main disciplinary activity as an historian. Science, as he insisted in the meeting of 18 April 1868 in the Sorbonne, had a centrality in the world of learning that no other discipline could match. To the question, "Is science not the force that today sustains all others?" he delivered a resounding "yes": "To commerce it furnishes steam and the electric telegraph, to industry the machine-tools that master nature, and chemical analyses that transform it. In agriculture it will give new life to the work of the farm, as it has done to that of the factory."[176]

Among the sciences represented in the Ecole, the flagship discipline was physics, pursued in two laboratories at the Sorbonne—one led by Jules Jamin, the other by his professorial colleague Paul Desains. While Desains's Laboratoire d'enseignement de physique concentrated primarily on advanced teaching to a relatively large number of students (more than fifty by the mid-1870s, most of them working for the *licence* and the *agrégation*), Jamin's Laboratoire des recherches physiques offered a setting for research pursued by a small number of students (still fewer than a dozen in the 1870s) preparing the doctoral theses that were by now the essential preliminary to an academic career.[177] For both Desains and Jamin, a clearly defined, if broad intellectual focus, with an emphasis on precise measurement, was essential: most of Desains's students were set to radiant heat, latent heat, and optics, especially spectroscopy, while Jamin's emphasis was on optics and electricity and magnetism. Although Jamin's laboratory had been refurbished in 1867 and was regarded as a rather fine one for the period,[178] the conditions were far from opulent and the funding was minimal. But the modesty of the provision became almost a matter of pride. Much later, the long-serving vice-rector of the academy of Paris, Octave Gréard, recalled a visit he had made to the laboratories of the "old" Sorbonne in the 1870s, with Desains as his guide:

> One day . . . during a visit, with the Professor of Physics Paul Desains, to the rooms, with their low, sloping roofs, that had been allocated to him—as he showed me the windows that would not close properly, the ill-fitting doors, the sweating tiles on the floor, the worm-eaten stairs on which at almost every step he had to interrupt his explanations with a watchful "Be careful!"—he stopped in a dark corner and said: "Despite all this, how good it is to be here."[179]

As both Desains and Jamin knew to their cost, dilapidation could be tiresome. But it served, if anything, to promote something of the sense of community that French observers saw as an essential quality of laboratories in Germany. The spirit of shared endeavor transcending differences of age and seniority also owed much to Jamin's affability, which struck the young Arthur Schuster during a brief visit in the mid-1870s.[180] But the key to the atmosphere and a mark of the revolution that the Ecole pratique des hautes études was intended to effect in French academic life was the creation of short-term junior posts that allowed a select band of young men to devote an early period in their careers to research rather than (in the manner of Jamin himself) to the traditional years of humdrum service in a *lycée*. Edmond Bouty and Gabriel Lippmann, under Jamin, and Edouard Branly, as assistant to Desains, were among the future physicists of distinction who benefited from these new opportunities, garnered from the meager resources that Duruy was able to direct to the Ecole.

The imaginativeness of Duruy's conception of the Ecole went hand in hand with a vulnerability that soon crystallized within a crescendo of threats to the whole program of academic modernization. By the summer of 1868, Haussmann's plans for the renewal of Paris were draining precious public resources, and thoughts were turning to the gathering prospect of war with Prussia. More immediately for the Ministry of Public Instruction, conservative elements in the upper reaches of the Catholic clergy were renewing their campaign against the state's monopoly in higher education by sniping with new vigor at what they saw as the laxity of Duruy's handling of the intellectually and morally suspect employees of his ministry.[181] Such attacks unfairly targeted a minister whose vigilance with regard to ideological deviance was of necessity and by any but the most repressive standards exemplary. But the charges were hard to rebut at a time when clerical and conservative opinion could all too easily portray any form of liberal opposition to the empire as a threat to the entire fabric of society. Once the elections of May 1869 had confirmed the intensity of discontent with the imperial regime, the voices of anti-imperial protest from the left and center left grew in strength, as did Catholic concern about Duruy's perceived laxity on intellectual matters and the disruptive potential of his educational reforms, including his support for the extension of primary education under state control and the education of girls. Duruy found himself caught between left and right in a way that fatally undermined his position. By July he was gone. The seat in the Senate that he received on his dismissal by a reluctant emperor was no more than a conventional gesture, and although he remained a real presence in debates on educational matters, his program of reform appeared to have come to a halt. But providentially, with other matters pressing more urgently in the year leading to war, the innovations already in place were left untouched. Among those that were beyond dismantling, the Ecole pratique des hautes études was, in Duruy's eyes, the most important. It was still "everywhere and nowhere," as he put it,[182] and it still ran on a shoestring. Nevertheless, the laboratories and seminars that were created in the Ecole's first six months survived and were even (very cautiously) added to.

Another notable survivor was the municipal meteorological observatory of the parc de Montsouris at the southern extremity of Paris. The observatory was a typical product of Duruy's pragmatic ingenuity. Seizing the opportunity of the relocation of the Palace of the Bey of Tunis from the site of the 1867 exhibition in the Champ de Mars to the park, Duruy persuaded Haussmann to back the use of the building for meteorology rather than as a refreshment room.[183] The measure had the considerable attraction of encroaching on intellectual territory that Duruy's old adversary Le Verrier saw this as his preserve. But above all it reflected a determination to get things done, within a political and administrative structure that was often unhelpful, even (by the late 1860s) positively obstructive. Faced with the prospect of what would almost certainly have been a fruitless quest for designated funding, Duruy was resolved to push ahead undeterred and to build on improvised beginnings. By 1869, small contributions from the relevant ministries (war, the navy, and public instruction) offered a modest assurance of permanence; still more important were a thirty-year lease and a guarantee of the cost of material upkeep from the city of Paris. But the benefits of the service, as these materialized in the 1870s, were out of all proportion to the investment. Once the observatory began functioning in 1872, it became a pioneering element in the relaunched national provision for meteorology under the aegis of the Ministry of Public Instruction. And its importance grew still further in the mid-1870s, when the establishment of an astronomical observatory in the parc de Montsouris turned the area into something of a Parisian public space for science.[184] Its eventual incorporation as part of the new national meteorological service administered by the Bureau central météorologique in 1878 confirmed its passage from an initiative of modest origins to a significant contributor to a major national institution.

The positive balance sheet of achievement against expenditure was a recurring theme of Duruy's accounts of his administration. The results, on this score, were astonishing. And they have to be counted in terms that go far beyond new institutions and a modest, though painfully won, increase in the budget for public instruction.[185] More profoundly important was a sense, new in the University, that professors were working for a minister who understood their aspirations. Another result was the affirmation that academic life should embrace disinterested research and criticism and not just the elegant rehash of safe, familiar material or the pervasive utilitarianism of Fortoul's time at the Ministry. With Duruy gone, however, there was soon a reminder of the continuing menace of reaction. In the spring of 1870 his successor Segris had no choice but to use his ministerial authority to close the faculty of medicine in Paris; he did so as a mark of firmness following a politically motivated student disturbance at a lecture by the professor of forensic medicine Auguste-Ambroise Tardieu, who had inflamed liberal opinion by acting as an expert witness in the acquittal of Pierre Napoléon Bonaparte on a charge of murdering the radical journalist Victor Noir.[186] Thereafter, the abnormal period of the war with Prussia and then the Commune served only to fan the tensions between those who sought to put the clock back to an age of conformity, control, and order and those who regarded the educational achievements of the liberal empire as a springboard for further, more adventurous reform. In such an atmosphere, it was initially difficult to sustain the momentum that Duruy had created. But if postwar uncertainty about the political destiny of France inevitably trimmed the sails of reform, by 1880 much of the vision for which Duruy (in public) and the emperor (in private, though no less resolutely) had battled reasserted itself.

Writing in 1874, Charles Laboulaye, a moderate republican and no friend of the imperial regime, asserted in the National Assembly that the empire had done little for higher education. This was a preposterous misrepresentation to which Duruy responded with just indignation.[187] Arguably, he had tried to do too much and had spread his reforming zeal too widely.[188] But his achievements were formidable. He had not only carried through major changes but also laid the foundations for key measures that came to fruition later, in the more clement atmosphere of the Third Republic. In particular, the fundamental reorganization of the disparate network of faculties and their restructuring as universities, though delayed until the 1890s, was already on the agenda when enemies and events conspired to force him from his post. The truth was that, in difficult circumstances, an unfailingly responsive minister had scarcely put a foot wrong in his struggle to temper discipline with a significant measure of intellectual freedom. He had irreversibly opened French science and scholarship to the currents that were transforming the life of the mind on the wider, international stage.

CHAPTER FOUR

Science, Philosophy, and the Culture of Secularism

• • •

Recent scholarship has made it difficult to see science as the main cause of the tide of secularization that traversed western Europe during the mid and later nineteenth century. Notions bred of the works of J. W. Draper and A. D. White, who interpreted the relations between science and religion in terms of relentless warfare or conflict,[1] have given way to analyses that emphasize diversity and the importance of local context. Among modern authors, John Hedley Brooke has made a particularly important contribution with his nuanced demonstration of the difficulty of establishing any general thesis about the ways in which science and religion have interacted and still interact today. As he has said, the real lesson arising from his classic study of key episodes at the shifting interface between scientific thought and religious belief since the sixteenth century is one of complexity.[2] For Brooke, there is "no such thing as *the* relationship between science and religion"; historical inquiry about such a relationship, as he argues, bears its richest fruits when it focuses on "what individuals and communities have made of it in a plethora of different contexts."[3]

This statement rings resoundingly true in the case of nineteenth-century France, which shows a predictable mixture of general pan-European trends and the particular characteristics of a Catholic country in which confrontation between, on the one hand, the forces of social as well as religious conservatism, broadly represented by the Church, and, on the other, those of secularism, drawing on a deep well of predominantly republican sentiment, was endemic. While science was not the cause of the confrontation, it became inextricably involved. Supporters of secularism made consistent appeals to scientific modes of thought as the symbol of a progressive, modern body of knowledge that was to be contrasted with the static teachings of Catholicism. In response, some modernizing voices within the Christian tradition struggled to show that faith had nothing to fear from science, provided that science was correctly pursued. But generally the exchanges served only to heighten the divisions, which became more clearly marked than ever under the resolutely anticlerical Third Republic and culminated in the separation between church and state in 1905.

With respect to the period of the Second Empire and early Third Republic, with which this chapter is mainly concerned, it is striking how many scientific issues became involved in exchanges that were essentially political in nature. From the 1850s, French opinion was racked by profound cleavages on the moral issues raised by evolution, as happened in many other countries. But passions were aroused no less by the topics of spontaneous generation, polygenetic beliefs about the diverse ancestry of the human race, and (perhaps most profoundly) the challenge of philosophical materialism of the kind that Karl Vogt, Jakob Moleschott, and Ludwig Büchner unleashed from German-speaking Europe. Individually, these were not beliefs that necessarily set scientific against faith-based modes of thought. When bundled together in what can properly be described as a "radical synthesis," however, they became a powerful weapon in the hands of those whose aim was victory not just in a scientific battle but also in higher struggles, either to legitimate their own secular worldview or to rid France of the incubus of religious or political conservatism, or in many cases both of these. Hence, in the way Brooke has suggested, their historical significance lay less in their logical incompatibility or otherwise with traditional beliefs than in the way in which they were deployed on a far broader stage of public debate—in this case the profoundly divisive debate that accompanied France's passage from monarchy to empire and on to republic—with the traumas of the revolution of 1848 and the defeat of 1870 along the way.

The Midcentury: Conformity and Dissent in French Philosophy

During the Bourbon Restoration, the leaders of the scientific community in France adopted a predominantly conformist stance in religion as well as in politics. The Protestant Georges Cuvier worked hard to portray his strong personal faith as perfectly compatible with science, maintaining that faith and science belonged in separate realms, only tenuously connected in his case by a belief in nature as a very general manifestation of God's providence.[4] Among Catholics, Pierre-Simon Laplace's famous statement that his system had no need of a god carried a faint threat, but it never diverted him from a religious journey that took him from a public stance of agnosticism to a conventional private piety toward the end of his life in 1827.[5] Among other leading savants with declared Christian commitments, André-Marie Ampère, Jean-Baptiste Biot, and Augustin-Louis Cauchy stood out and never wavered.[6] The conciliatory effect of these demonstrations of orthodoxy is beyond question. Within a decade of the return of the Bourbons, reactionary attempts to portray science as subversive of both the moral and the political order had lost most of their plausibility. Hence when the long-serving professor of mineralogy and geology at Montpellier, Marcel de Serres, published a scholarly two-volume work arguing for the compatibility between modern geology and the Mosaic account of creation in 1838, he was not contributing to a particularly heated debate.[7] Most readers with the religious predisposition to belief would have found reassurance in Serres's evidence. It took no great adjustment to accept, with Serres, that biblical references to the days of creation could be read as an account of successive epochs of indeterminate length going back into the deep mists of time; and any unease that there might have been was assuaged by Serres's accompanying demonstration that the history of mankind, as recounted both in Genesis and in the annals of secular historians, began no more than 7,600 years ago (the Noachian deluge and subsequent renewal of the human race following two thousand years later).[8] Those who had no such religious disposition, in contrast, were unlikely to give the book more than a cursory glance; they would not have seen it as a threat to their vision of a law-bound universe. The paucity of comment from within the scientific community, both then and in response to the later editions that appeared in 1841 and 1859, makes the point. The only serious critical engagement with Serres, in fact, seems to have come from the extreme conservative Louis de Bonald, one of the rare Catholic commentators to be disturbed by Serres's readiness to reinterpret scripture in the light of scientific knowledge.[9]

All that remained under the July Monarchy, in fact, was a rapidly fading suspicion that scientific and, more particularly, mathematical studies desiccated the mind and diminished the spiritual dimension of those who pursued them. It was this vestige of the legacy of hostility toward the Age of Reason that provoked the Romantic poet Alphonse de Lamartine to look back on the Napoleonic Empire as a period in which mathematics had fettered his young generation, like "chains of human thought."[10] Such anxiety about the scientific worldview and its dehumanizing, materialist connotations was never widespread, however, and by the 1830s it would have found even fewer echoes in public opinion. Now, most contemporaries would have been reassured by the attempts of Cauchy and a circle of Catholic scientists close to him to present the search for truth in science as analogous to the quest for an understanding of the eternal truths of the Catholic Church: their contributions to the programs of the Institut catholique, a loosely organized structure founded in 1842 to provide lectures on both scientific and nonscientific subjects for the general public, had a calming effect and placed yet another brick in the wall of reconciliation.[11] Those who feared for either faith or civic order saw far greater threats in the political radicalism and the working-class unrest that surfaced spasmodically in the 1830s and 1840s. By comparison, science seemed remote from the great challenges of the age.

The turbulent events of 1848–52 changed much in this as in many aspects of French intellectual and cultural life. Among the casualties was the cosy image of science in lay consciousness. In the aftermath of France's passage from Louis Philippe's bourgeois monarchy to the authoritarian regime of Napoleon III, via the revolutionary upheaval of 1848, the Napoleonic coup d'état of December 1851, and the declaration of the Second Empire a year later, philosophies modeled on or informed by science once again became a focus for conflict. The context was a tide of scientism that forced itself on the attention of educational administrators and educated laymen as well as on that of academic philosophers. What happened was not that scientifically inspired ways of thinking succeeded on their own in toppling the reigning philosophy of the university world, with its roots in the teachings of the philosopher Victor Cousin and what was variously described as the eclectic or the spiritualist tradition. It was rather that the new scientism gathered force just at a time when eclecticism, with its very different, literary cast, was beginning to falter. As scientism rose, it found its readiest audience among opponents of the Empire, and from that base it fed insidiously into a reaction that by 1870 had lowered Cousin and his followers from the rank of unassailable authorities to that of exponents of an interesting but essentially outmoded system of thought.

Eclecticism was no easy target: there was something attractive about its rejection of dogmatism, whether the dogmatism stemmed from church authority, German idealism, or the more secular traditions of sensationalist psychology. But the very fact that eclecticism had acquired an almost official character in the University during the July Monarchy made it immediately vulnerable to the change of regime, and that danger duly turned into reality. The greatest single blow was delivered early by the legislation establishing the *bifurcation*, Hippolyte Fortoul's rigid separation between the scientific and nonscientific tracks in education, with its attendant attack on the *agrégation* in philosophy and the reorganization of the six existing *agrégations* into just two, one in letters, one in science. As we have seen in chapter 3, Fortoul conceived the reform as an attempt both to modernize French secondary and higher education, in particular by raising the status of science, and to trim the waywardness of those teaching under his ministerial aegis. Although the reform was primarily administrative, it was in accordance with Fortoul's broader intellectual objectives that he did away with one of the two philosophical chairs at the Sorbonne, where the heady stuff of eclecticism held sway, under Cousin's still powerful influence. The target of Fortoul's measure, as of his decision to reduce the teaching of philosophy to pupils at the Ecole normale supérieure to the bare needs of the future pedagogues of the bourgeois young,[12] was the indiscipline that he believed had invaded philosophy in the eclectic tradition, and there can be no doubt that it had the desired effect: deprived of its main sources of power in the academic system, the eclectic school faltered and never fully recovered its old primacy in French philosophy.

In another and less obvious sense, however, Fortoul's intervention failed. For already, by the mid-1850s, interest in precisely the controversial, open-ended issues from which he had sought to divert attention had actually gathered strength. Paul Janet, speaking from the chair of the history of philosophy at the Sorbonne in 1865, identified what had occurred over the previous decade with a clarity born of a mixture of objectivity and anxiety.[13] Not only eclecticism, he insisted, but the whole spiritualist tradition was in "crisis." This was not to say that eclecticism was dead: Janet himself had retained some sympathy for it from his days as a *normalien* in the early 1840s, and there were still a number of contemporaries in philosophy—including such significant figures as Emile Saisset, Jules Simon, and Elme-Marie Caro—who would have regarded themselves as belonging to the eclectic school in a more formal sense. But residual eclectic resistance could not conceal the fact that since the beginning of the Second Empire, the creative vanguard of French philosophy had increasingly been occupied by the newer, "scientific" philosophies. One result had been the revival of interest in the Enlightenment.

This was reflected in the turn of a true "son of the eighteenth century," Jules Michelet, toward nature in a series of books—*L'Oiseau* (1856), *L'Insecte* (1857), *La Mer* (1861), and *La Montagne* (1868)—that took the material world, living and nonliving, as a source of literary inspiration.[14] While Michelet's gentle romanticized prose poems had no pretensions to a place at the cutting edge of midcentury thought, both their tone and their success were the touchstone of a new openness to science as a complement or even an alternative to conventional sources of creativity and philosophical reflection.

As Janet saw it, and he saw rightly, the real challenge to the traditions with which he more or less loosely allied himself had lain, and still lay, in a resurgent positivism. Once the "foolish utopias" of the most extreme adepts of its political and religious teaching were set aside, positivism, for Janet, now stood stronger than ever before: it was no longer possible to ignore the claims of those ambitious but sober positivists who wanted to advance their ideas not just as a philosophy of science but as a complete philosophy extending scientific analysis to the study of man, the human mind, and society. Positivism, in Janet's view, had become the cutting edge of a wider modernist movement that sought to advance what he variously called the "empire" or "tyranny" of science at the expense of the older "literary" traditions that had traditionally been at the heart of philosophy.[15] Janet's analysis, with its clear delineation of a conflict between scientific and nonscientific worldviews, had a clear polemical purpose. It provided an essential foundation for the presentation of his own goal as a philosopher, which was reconciliation. As a man of the middle ground, Janet sought to open philosophy to the new scientific spirit while defining the limits of the relevance and efficacy of science and retaining a role for philosophical reflection on the immaterial entities that positivists regarded as beyond the domain of rational inquiry. Thirty years earlier, "enthusiasm, passion, and sensibility" had been the hallmarks of philosophical debate. To Janet's regret, contemporary philosophers affected "coldness, desiccation, and contempt for the feelings of the heart."[16] Between these two extremes, there had surely to be a path of compromise, if the resurrected (and deformed) philosophy of the Enlightenment, with its doctrinaire allegiance to sensationalism in psychology and its disdain for metaphysics, was not to suppress "the freedom and dignity of the human spirit."[17] The threat lay in the "inordinate pretension" of science;[18] the defense in a resolute quest for a philosophy that embraced but also transcended sense experience and reason.

Despite the personal goals of Janet's scene setting, his description of the polarization of French philosophy was essentially correct. It was endorsed, for example, by the semiofficial report, *La Philosophie en France au XIXe siècle*, that Félix Ravaisson prepared at the time of the Exposition universelle of 1867 for the Ministry of Public Instruction's series Recueil de rapports sur les progrès des sciences et des lettres.[19] In this work, Ravaisson confirmed the new prominence of scientistic trends in philosophy by devoting more space to positivism and the spins that John Stuart Mill, Littré, and others had given it than to any other single doctrine, even though he himself was no positivist.[20]

Janet's description was also given substance by Hippolyte Taine's recollections, previous to his own, of the young, scientifically minded circle that he frequented as a student in the Latin Quarter in 1852. Here at least, a familiarity with science had bred contempt for the older "official" philosophy, which Taine and his friends saw as nothing more than "elegant rhetoric," good for testing by examination but unfit as a foundation for serious thought.[21] The experience bred Taine's withering review of the rise and influence of eclecticism that formed the core of his much reprinted work, *Les Philosophes français du XIXe siècle* in 1857. By the time Taine published the second edition of this work in 1860, the reaction against "literary" philosophy that he helped to provoke had become far more widespread. It was now at the heart of a polarization of precisely the kind described by Janet. As Taine put it in his preface to the second edition: "Today two main philosophies exist in France and are to be found, with minor differences, in Germany and England as well. One of them finds its exponents in the humanities; the other among savants. In France, one bears the name spiritualism; the other we call positivism."[22]

It is no easy matter to unravel the causes of the revival, after 1848, in the fortunes of a range of scientifically oriented philosophies of which positivism was just one expression. An attractively simple explanation would be the immediate personal influence of Comte. But, in reality, Comte's contribution to the resurrection of his ideas was of questionable value. He certainly published a great deal in these years, including the four volumes of his social and political philosophy, the *Système de politique positive* (1851–54). However, the *Système* seems to have been read by most as the eccentric work of a man whose instability and failure to achieve academic advancement through the 1830s and 1840s had left him on the fringes of the intellectual elite. Positivism, as a significant philosophical force, stood (at the very least) in need of trimming and refashioning. The man for that task was not Comte himself but rather the most prominent of his "philosophical" disciples, Emile Littré, and it was Littré, more than anyone, who drew positivism into the mainstream of public debate. In a manner typical of even well-read scholars, Littré knew nothing of Comte until 1840, when a friend lent him some of Comte's work, presumably the early volumes of the *Cours de philosophie positive*. But he became an instant convert. "His book captivated me," Littré recalled of Comte over twenty years later: "I immediately became a disciple of the positive philosophy, and a disciple I have remained."[23] Like the great majority of his contemporaries, however, Littré was unable to follow when Comte's interests passed, from the mid-1840s, to the Religion of Humanity and the development of a system of "positive polity" (*politique positive*) that, unlike Comte, he saw as having no logical connection with the philosophy that preceded it.[24] Consequently he conceived his main task as that of detaching the rigorous essentials of positivism from the "aberrations" of Comte's "second life," in which a "subjective" method had replaced the objective method of the *Cours*.

The consistency of Littré's commitment to his task is remarkable both in itself and for our understanding of the resurgence of interest in Comte's ideas during the Second

Empire. In 1844, he began with the first of three lengthy articles on positivism in its widest philosophical, moral, and political dimensions in the liberal-leaning newspaper *Le National*,[25] and his campaign continued on into the Third Republic. Throughout this period, Littré strove to preserve the Comtean legacy from two of the main weaknesses that beset it in contemporary eyes: the extravagance of its analysis of human behavior (based on a simplistic psychology) and the tendency to materialism that Janet portrayed as one of its leading characteristics.[26] As a champion, Littré was astute and never uncritical. He presented positivism as a doctrine consistent with the ideals (his own) of republicanism, modernity, and the ending of the old philosophical order. Yet positivism, as he conceived it, was essentially a force for stability, one that bore none of the threat of upheaval and destruction that had reared its head briefly in 1848.[27] The enemies of modernity, however, especially those who wrote (as most of them did) from a Christian perspective, remained unconvinced. They portrayed Littré as a dangerous thinker whose writings conveyed a congeries of doctrines—vaguely identified as materialism, atheism, and socialism, or various combinations of them—that were abhorrent to the conservative mind. Such perceptions were hard to eradicate. In 1859, after well over a decade of careful presentation and subdued polemic, Littré could still write that positivist philosophy was (unfairly) "buried in obscurity," being either ignored or misunderstood.[28] But there was something disingenuous about this assessment. For while Littré's comment accurately conveyed the low level of formal adherence to positivism in intellectual circles, it underplayed the attention that he had drawn to the philosophy and the scientism that positivism enshrined during the 1850s.

Four years later, any suggestion that positivism was neglected was made totally untenable by a confrontation that no one could ignore. The confrontation centered on Littré's candidature for election to the Académie française in 1863. His academic credentials were impeccable: he had been a member of the Académie des inscriptions et belles-lettres since 1839, his ten-volume translation of the works of Hippocrates had recently been completed,[29] and the first fascicules of his great *Dictionnaire de la langue française* had appeared more than two decades after its initial commissioning by its publisher, Hachette.[30] But the Bishop of Orléans, Félix Dupanloup, saw Littré as more than just a distinguished scholar; he was also the incarnation of the whole anticlerical movement of which Renan and Taine were the equally detested representatives. So long as Cousin's influence had prevailed in philosophy, the enemy had been kept at bay; but, to Dupanloup's dismay, the floodgates of secularism were now open. In the resulting tempestuous exchanges, the representatives of moderate opinion, such as Adolphe Thiers, tended to align themselves with Littré.[31] As the election loomed, Dupanloup did not see the battle as lost. The only question for him was whether he should speak out before the Académie voted, as Cousin advised him to do, or afterward, as some of those closest to him believed he should if the church was not to appear to be conducting a self-destructive struggle against good scholarship as well as against modernity. As far as the election was concerned, Dupanloup's decision to opt for immediate action proved the right one. A combination of personal lobbying and a vitriolic attack of some 120 pages on materialism and positivism, distributed on the eve of the election to all the academicians whom he thought were not committed to Littré, was decisive in carrying the day.[32] Littré was defeated by the Catholic historian and man of letters, the comte de Carné, by eighteen votes to twelve on a third ballot.

Neither party saw the matter as closed, and hostilities resumed when Littré reappeared as a candidate in December 1871, with François Guizot now espousing his cause. Dupanloup spoke as vehemently as he had done eight years earlier against "the apostle of the most destructive doctrines that exist in the realms of religion, morality, and society."[33] But this time he spoke in vain, and Littré won comfortably, with seventeen votes to nine for Saint-René Taillandier and three for Louis de Viel-Castel. The failure of Dupanloup's campaign spoke volumes on the changed philosophical climate that prevailed by 1871. The point was not lost on him, and within hours of the election he took the profoundly symbolic step of resigning from the Académie.[34] The point was not lost either on the press, notably the *Journal des débats*, which articulated a moderate liberal position in criticizing Dupanloup's resignation as a mark of the impotence of religion in the modern world.[35]

It is attractive and not wholly misleading to analyze the philosophical divide during the empire in terms of the confrontation between Dupanloup and Littré as advocates of the main opposing strains in mid-nineteenth-century philosophy, much as Janet and Taine perceived them. On one side, Dupanloup represented a strong current of bourgeois opinion in portraying positivism as preeminently the philosophy of the new age of science and as the embodiment of the spirit of opposition not only to traditional religion but also to public morality and good government, whether of the empire (which he supported, albeit with significant reservations) or of the republic (which, as a monarchist, he abhorred). On the other, Littré spoke as the champion of a scientifically grounded liberalism that rejected the prevailing values of contemporary intellectual life and set reason and modernity above tradition. Elaboration of this dichotomy is required, however, if we are to understand the intensity both of the opposition between the two men and of the public interest it aroused.

Although essential elements in the background to the events of 1863 and their aftermath in the early 1870s lie, as I have argued, in the faltering of eclecticism and the gathering interest in positivism, other more specifically scientific issues also watered the seeds of conflict. The most corrosive of these in conservative eyes turned on materialism, a philosophy that left virtually no ideological debate of the 1850s and, more particularly, the 1860s untouched. Its impact in France, however, only slowly became apparent. When materialist doctrines arrived from Germany in the early 1850s, the overwhelming reaction among the French reading public was one of surprise and a slowness to recognize their full significance. The first wave of the doctrines came in the form of three books that appeared in Germany between 1852, the year of Moleschott's *Der Kreislauf des Lebens*, and 1855, the year of both Büchner's *Kraft und Stoff* and Vogt's *Koehlerglaube und Wissenschaft*.[36] Translations were slow in coming: Büchner's *Kraft und Stoff* and Moleschott's *Der Kreislauf des Lebens* were unavailable in French until 1863 and 1865 respectively,[37] and Vogt's *Koehrlerglaube und Wissenschaft* was never translated. Nevertheless, the books were not ignored, and reviews and commentaries began to appear in the periodical press by the mid-1850s, though with nothing like the intensity that became apparent in the 1860s, when Büchner's *Force et matière* went into a second and, by 1869, third French edition[38] and when Vogt's Swiss pupil Jean-Jacques Moulinié promptly translated his *Vorlesungen über den Menschen* of 1863.[39] By now, too, the close personal bonds between "Vogt and company" (to use the language of the critics) and a resurgent French tradition of free thought were adding a new degree of menace.[40] The free-thinkers' leading organ, the weekly *Pensée nouvelle*, carried contributions by and about Büchner in 1867 and 1868,[41] and in September 1867 Vogt (along with Rudolf Virchow, whom the French always saw as a leading inspiration of the materialist movement) took the symbolically significant step of attending a much-publicized dinner of free-thinkers in Paris.

The collective effect of the writings of Vogt, Moleschott, and Büchner was to place physiological evidence at the heart of renewed discussion about the relations between the mind and the body. Eclecticism's approach to this recurring philosophical question was embedded in a long tradition. From about 1840, Théodore Jouffroy and Félix Ravaisson, writing in a broadly eclectic mold, resurrected the notions of the duality of the material body and the immaterial mind that Maine de Biran had developed a quarter of a century before. In doing so, they were generally regarded as having done little to move the debate forward, and their analysis entered the orthodox canon of July Monarchy philosophy without arousing excitement or any clear sense either that dualism should be challenged or, for that matter, that it signalled a constructive way forward in psychology. From the early 1850s, however, with eclecticism losing its grip in more progressive philosophical circles and with the Church as uneasy as ever about the implication of secular theories of the mind for traditional Christian doctrines of the soul, the intellectual climate was very different. It was a climate of shifting opinion that gave the alternative, materialist approach a rare opportunity of asserting itself.

Despite a broad common thrust and a tendency for contemporaries to conflate their teachings, the emphases of the three German pioneers of materialism were different. Büchner's materialism was essentially a philosophical riposte to German idealism and theological interpretations of nature and natural phenomena; the main focus of Moleschott's work, in contrast, was the physiology of the brain and nutrition, analyzed in exclusively material terms that set him explicitly at odds with Liebig, in particular with the latter's defense of the notion of vital force; and Vogt's interests, while rooted in physiology and given vehemence by his opposition to the ideas of the soul espoused by the Göttingen physiologist Rudolph Wagner, quickly moved into the realm of paleontology and anthropology. But all three interpreted the processes of life in ways that took no account of any immaterial spirit or force distinguishing living from nonliving matter. Their argument was encapsulated most memorably in 1855 in the ringing assertion of the preface to the third German edition of *Kraft und Stoff,* where Büchner wrote that "there are no other forces in nature beside the physical, chemical, and mechanical."[42] It followed not only that the processes of living organisms were carried along entirely by physical, chemical, and mechanical forces but also that those same forces could, in principle, explain the origin of vegetable and animal organisms. The materialism of Büchner's dictum "no force without matter—no matter without force" brooked no qualification.[43]

Such notions spread alarm in the Catholic Church and among those who were most receptive to the church's teaching and to the empire's promise of social order. Once life was explained solely in terms of matter, the concept of an immaterial, immortal soul was relegated firmly to the realm of untestable metaphysical speculation. And along with that concept there was also relegated (especially in Büchner's analysis) a range of associated doctrines, including the idea of a moment of divine creation, which was inconceivable in a universe of eternal, indestructible matter.[44] For Büchner himself, jettisoning traditional ideas of creation, providence, morality, and free will had a liberating effect, in much the same way as it had already had for Harriet Martineau in Britain, under different circumstances at about the same time.[45] Once belief in a personal god was rejected, a human being ceased to be the plaything of an otherworldly despot and became instead, in the words of Büchner's "introductory letter" to the English translation of *Kraft und Stoff,* "nature's noblest and best son," released from "childish fear of spirits or supernatural influences" and free to pursue objective truth and progress.[46] No Christian, though, could accept such freedom as anything but an illusion: for the Christian, materialism was tantamount to our being the slaves not the lords of Nature.

There was more than enough in the writings of the early German materialists to provoke a public perception of science as the enemy of religion and morality and for that perception to carry a conviction, in conservative eyes, that it had not had since the days much earlier in the century when Chateaubriand (in *Le Génie du Christianisme,* 1802) and Maistre (in *Les Soirées de Saint-Pétersbourg,* 1821) had inveighed against the excesses of Enlightenment rationalism. If further reinforcement of the perception was needed, it was provided in abundance by the political overtones that the materialist doctrines brought in their wake. In Germany Vogt, in particular, had been active in the revolution of 1848 and afterward had fled from Giessen, where he had taught zoology since the early 1840s, to Geneva, where he taught geology and became a naturalized citizen. Büchner too had been identified as a student radical in the 1840s, and after the mid-1850s both he and even the temperamentally milder Moleschott were obliged to resign their university posts (respectively at Tübingen and Heidelberg) in response to the outrage that their books caused. Büchner remained in Germany, practicing medicine in

Darmstadt and writing and lecturing unrepentantly on materialism, while Moleschott moved on to academic posts in Zurich, Turin, and Rome and took Italian citizenship.

In view of these ideological and political associations, it is hardly surprising that the corpus of German materialist teaching first took root in France in the 1850s and 1860s among those who opposed the empire either for its policies or for the pious, unthinking conformity that critics saw as pervading bourgeois society. This is not to say that political opponents of the established order were necessarily materialists. Nor does it imply that materialism commended itself across the board of the philosophical avant-garde. On this point, Littré in particular argued strenuously for a distinction between the positivist and the materialist positions. As he maintained in his "Préface d'un disciple" in 1864 and in the following year in the preface to a book by the long-standing positivist Alphonse Leblais, positivism (at least as he conceived it) was far more circumspect than materialism, since it relegated to the realm of the unknowable such questions as spontaneous generation, the physiological basis of thought, and the nature of matter.[47] For the positivist, unlike the materialist, the existence of a nonmaterial world of the spirit was not an impossibility; it was simply one that lay beyond the reach of investigation.

Littré's distinction was perfectly just and had ample backing in the teachings of Comte. Nevertheless, it did little to allay the charged polarization between new and old systems of thought that followed the arrival of German materialism in France. The polarization made its mark repeatedly in a series of intricately related and hotly contested scientific debates—on the origin and nature of life, Darwinian evolution, and anthropology—that raged from 1858 until the end of the empire. In these debates, personalities loomed large, and philosophical niceties were lost from view as opinion divided between those who systematically championed traditional doctrines in science and religion and those who no less systematically espoused free thought in opposition to orthodox teaching. As I now argue, the cumulative vehemence of this confrontation reached a fever pitch by the late 1860s that was unmatched in any other public discussion of science in the modern period in France.

The Nature of Life: Pasteur–Pouchet Revisited

The first of the scientific debates to capture the attention of the wider reading public concerned spontaneous generation or, more correctly, heterogenesis: that is, the view that new living microorganisms could be created from dead organic matter. Such a view had already received a measure of somewhat equivocal support in the writings of Cabanis and Lamarck about the turn of the nineteenth century, in particular in Cabanis's *Rapports du physique et du moral de l'homme* and Lamarck's *Recherches sur l'organisation des corps vivans*, both published in 1802. And thereafter it had survived in the tradition of transformist speculation associated chiefly with the work of Etienne Geoffroy Saint-Hilaire.[48] Since the 1830s, however, spontaneous generation had been eclipsed through the dominating orthodoxy of Cuvier, whose victory in his debate with Saint-Hilaire in 1830 had served to remove it—to all appearances definitively—from the realm of respectable science. For Cuvier, the origin of living beings was one of the greatest mysteries of science. While stopping short of a definitive condemnation of spontaneous generation, he made his skepticism clear in 1808:

> We have always seen life as being born of life; we observe life being transmitted, never created, and although the impossibility of spontaneous generation cannot be demonstrated with complete certainty, the efforts of those physiologists who believe such generation to be possible have yet to yield a single example. Consequently, the human mind has to choose from, on the one hand, a variety of hypotheses concerning the development of germs or, on the other, occult qualities advanced under such terms as "inner mold," "formative instinct," "plastic virtue," "polarity," or "differentiation." Nowhere does it encounter anything but mist and obscurity.[49]

And he remained loyal to that statement until his death in 1832.

The manner and the timing of the resurrection of spontaneous generation as a focus for controversy owed much to the changing climate of intellectual life that in the 1850s was opening radical minds in France to materialism. As with many other aspects of the materialist doctrines, Büchner's advocacy carried special weight. In *Kraft und Stoff*, Büchner vehemently declared his sympathy not merely for spontaneous generation but more specifically for abiogenesis.[50] This was a particularly strong version of the doctrine that allowed for the generation of life from inorganic and not just organic matter. Not all materialists shared Büchner's extreme position. Vogt, in particular, was far more cautious about spontaneous generation in any form. In 1851 and 1858, he laced the two editions of his German translation of Robert Chambers's *Vestiges of the Natural History of Creation* with criticism of the as yet unidentified author's endorsement of the eye-catching experiments of Andrew Crosse and William Henry Weekes, who in conditions that supposedly made contamination impossible appeared to have produced tiny living creatures by applying electricity to a prepared silica solution.[51] While Vogt did not rule out the possibility that life might one day be produced in the laboratory, he consistently maintained that the whole question of spontaneous generation belonged to a domain that experiment and observation could not penetrate. Even agnostic declarations of this kind, however, served only to draw further attention to the subject; they certainly did nothing to arrest the tendency (no less apparent in Britain) for the nonscientific public to assimilate spontaneous generation to the broader materialist threat.[52]

We cannot know how far the science of the main champion of spontaneous generation in France, the long-serving director of the municipal museum of natural history in Rouen, Félix-Archimède Pouchet, was affected by the philosophical materialism of the 1850s. Certainly there were elements in Pouchet's earlier life that would at least have made him familiar with its tenets. One of these elements was his medical training, first under Gustave Flaubert's father, Achille, the head surgeon at the Hôtel-Dieu in Rouen, and then, in 1827–28, in Paris, where he established a close attachment to Henri-Marie Ducrotay de Blainville (at the faculty of science) and François Broussais (at the faculty of medicine) and at least met Geoffroy Saint-Hilaire:[53] belief in spontaneous generation had deep roots in the medical tradition and in all countries it found its main support there. Equally influential, however, was Pouchet's family background, which provides the most important single key to his worldview. The son of a prominent textile industrialist in Rouen, Pouchet had been raised in an atmosphere of committed Protestantism. It was in keeping with his upbringing that he was able to deliver a strongly providentialist homily on the wonders of the creation at the annual prize-giving of the municipal schools of Rouen in 1854,[54] and throughout his life references to divine providence and the limitations of human understanding of God's creation surfaced repeatedly in his writings.[55]

There is no reason to think that Pouchet's declarations of faith and intellectual humility were anything but sincere, and I certainly do not see him as a closet materialist. This conclusion stands, I believe, despite the description of him as a "free-thinker and rationalist" that appeared in 1876, four years after his death. The source of the description was the warm appreciation of him by Georges Pennetier, his devoted assistant and eventual successor in the directorship of the Rouen museum of natural history.[56] The fact would appear to be that Pouchet insisted on maintaining absolute freedom in his science and that he strove consistently to eliminate from scientific inquiry any considerations, including those of religion, that did not lend themselves to rational and empirical scrutiny: in Pennetier's words, he was a "free inquirer on the field of science."[57] In religion, however, all that can be said is that he was no atheist and that his providentialism was compatible with anything from a rather old-fashioned Enlightenment deism to rational protestant piety of a liberal kind. Within that broad range of opinion, it would be hard to place Pouchet's beliefs—beliefs that in any case may have changed, though probably rather little, in the course of his long life.

The distinction I draw between Pouchet's scientific and religious views was not a matter of concern to his contemporaries, and it has to be said that Pouchet himself sometimes offered a hostage to fortune. For example, those who read his *Théorie positive de la fécondation des mammifères*, a book-length study of mammalian ovulation and conception that he published in 1842, not unreasonably perceived anticlerical and, more specifically, positivist tendencies in a passage about the importance of observation and sense experience that would have done credit to a Comtean harangue. Pouchet wrote: "We are presently at the stage of observation and experiment. We are embarked, in other words, on the only sure path to positive results. Let us enrich our knowledge with the power that comes with the application of logic. In that way, we shall then have all the elements that go to make up truth."[58] I read that statement as an endorsement of a rigorous experimental method, and nothing more. Although Pouchet's son, Georges, did become a leading advocate of Comte's teachings from the 1860s, Pouchet himself appears never to have engaged in the positivist movement in anything approaching a formal sense.

In my judgment, in fact, Pouchet was not a philosophical rebel but simply a man whose detachment from Paris in the forty-four years that he spent at the museum in Rouen dated him and placed him, to his regret, out of the mainstream. Until he suddenly became involved in the debate over spontaneous generation in the late 1850s, he had never sought to demarcate himself in any way from the "establishment" of Parisian science. Indeed, from his earliest days in the museum, he had maintained a friendly though deferential correspondence with the leaders of the life sciences in the Académie des sciences and the Muséum d'histoire naturelle, as he had also done with disciplinary peers abroad, especially in England.[59] Among those leaders and peers, his work was evidently regarded as worthy, if not particularly exciting, and that respect was duly expressed in 1845 by the decision of the Académie des sciences to award him its prize for experimental physiology for his *Théorie positive de la fécondation des mammifères* and some related manuscript material.[60]

In retrospect, it is possible to see in the *Théorie positive* and the enlarged version that appeared under a somewhat different title in 1847[61] the origins of the research that transformed Pouchet into an object of notoriety scarcely a decade later. But in the late 1840s, Pouchet's views aroused no excitement, and it was still as an outsider, albeit (as a corresponding member of the Académie since 1849[62]) not an unknown one, that he was propelled to public attention when the spontaneous generation controversy broke in December 1858. The catalyst was a brief communication by him to the Académie.[63] In less than three pages, Pouchet summarized in rather flat detail experiments in which he had seen microorganisms emerge from infusions of hay immersed in mercury and supposedly uncontaminated by living matter of any kind. In an atmosphere made more sensitive by the incipient materialist debate, the stir was immediate. Two weeks later, Henri Milne Edwards replied with a denial of the evidence, which he believed to have been flawed by the inadequacy of Pouchet's procedures for the elimination of living organisms from his hay.[64] The confrontation, launched in these two low-key academic papers, was to run for almost a decade, raising Pouchet, now in his late fifties, to the position of a distinctly improbable controversialist.

The debate that followed was conducted essentially as a series of personal exchanges between Pouchet and Pasteur, who quickly emerged as Pouchet's main adversary. It has come to be conventionally interpreted as one in which conservative political and religious prejudices explain Pasteur's position. Pennetier, a militant free-thinker, presented it in these terms, and Pouchet himself, though always less vehement than Pennetier in his comments on the ideological dimensions of his defeat, certainly felt that his experiments had not received the unprejudiced hearing they deserved.[65] This perception of the debate as a struggle in which the blinkered conformity of Pasteur and his supporters ensured the defeat of compelling evidence in favor of spontaneous generation has entered modern historiography largely through the influential article that John Farley and Gerald Geison published in 1974.[66] The interpretation by Farley and Geison has had its critics, and I certainly find myself more persuaded by Geison's later and somewhat different account, which sees Pouchet's stance as colored no less than Pasteur's by elements in his background that lay beyond the strictly defined realm of science.[67] Even when that essential symmetry is conceded, however, it remains to be determined just how far non-scientific considerations on either side really did fashion the outcome. Despite the intense public interest that the debate engendered, my own inclination is to give such considerations less significance than has been customary in recent writing.

One point on which all modern commentators have agreed concerns the different predispositions that Pouchet and Pasteur would have had with regard to spontaneous generation and to the importance of engaging in public debate about it. Pouchet's long-standing interest in reproduction and characteristically Protestant familiarity with German thought (including the contemporary attacks on vitalism by Johannes Müller, Theodor Schwann, and Hermann von Helmholtz[68]) would have drawn him naturally enough to spontaneous generation, and this appears to have happened after about 1855.[69] With regard to Pasteur, tracing the roots can be more specific. There seems no reason to doubt his own assertion that his interest grew from his earlier work on fermentation, on which he began to publish in 1857.[70] From the time of his first paper on the subject, the classic study of lactic fermentation that he read to the Société des sciences, de l'agriculture et des arts de Lille in August 1857, Pasteur took the view not only that lactic yeast and other ferments were made up of living microorganisms, as many (though not all) agreed by that time, but also that they were the cause of the process of fermentation and not (as Liebig's purely chemical interpretation suggested) a product of it.[71] It is important to realize that Pasteur did not raise the question of spontaneous generation in the paper of 1857. Nevertheless, his argument dealt an oblique blow at heterogeny by suggesting that the character of a fermentation was determined solely by the nature of the microorganisms that were added to the fermentable medium: as Pasteur saw it, lactic yeast and brewer's yeast had quite different effects, producing lactic and alcoholic fermentation, respectively, simply because the organisms composing them were different.

This interpretation meant that when Pouchet's brief communication came before the Académie des sciences in December 1858, Pasteur's mind was primed, and eight weeks later he entered the fray. He did so in a contribution to what had now established itself as a matter of keen interest within the Académie, with Pouchet as a resolute contributor, though one who was to find himself consistently on the defensive.[72] Pasteur's evidence that, in the absence of air contaminated by living organisms, there were no signs of life, fermentation, or any other kind of activity in a sealed vessel containing a solution of fermentable liquid evidently drew a rapid response in the form of a letter from Pouchet. But only Pasteur's reply to this letter has survived. The reply conveyed, first, Pasteur's reluctance to come out categorically against the possibility of spontaneous generation (a residual caution that he carried over from his earlier research) and, second, his critical attitude to the particular observations that Pouchet had advanced as evidence.[73] In Pasteur's view, Pouchet had failed to eliminate the possibility that ordinary air, carrying an impurity or impurities essential to life, had contaminated the fermentable medium. At this stage Pasteur did not use the term "germs," in describing the impurities. But "germs" were clearly what he had in mind.

Thereafter, the central point at issue remained essentially unchanged. On the one hand, Pouchet tried repeatedly to demonstrate the sterility of his vessels and the other materials he used; on the other, Pasteur was concerned to show that microorganisms (usually referred to as "organized corpuscles") had unwittingly been introduced from the environment, usually in imperfectly calcined air. The debate spawned some classic experiments, among them Pasteur's studies of fermentation in a series of swan-necked vessels.[74] Each of the vessels contained a fermentable liquid, such as sugared yeast water, milk, or urine, that was boiled vigorously before the vessels were sealed, leaving a partial vacuum. The necks of the vessels were then opened at three different altitudes: twenty of them in the foothills of the Jura Mountains, twenty at the summit (850 m above sea level), and twenty on a windswept glacier at 2,000 meters, close to the Mer de Glace near Chamonix. Of the twenty vessels opened at the lowest level, eight in due course showed microbial growth; the corresponding figures for the two other locations were five and one, respectively. It was hard to avoid the conclusion that the difference in the degrees of contamination of the air entering the flasks at the three altitudes was the decisive parameter. In the face of this and related evidence, however, Pouchet stuck resolutely to his position. He defended himself with the claim that the species observed in his flasks were of a different kind from those in the surrounding air and that, if the emergence of life on the scale he had observed were to be ascribed to an external agent, the air must be teeming with the all-important plant spores and eggs of infusoria in numbers that even Pasteur himself had never envisaged. Pouchet's own view was that, in reality, such organisms were infinitely rare in the atmosphere. The stage was set for deadlock, and amid a rising tide of claims, denials, and counterdenials, both men proceeded to bombard the Académie with reports on their work until the steam went out of the debate in 1864.[75]

At an early stage, the sense of engagement, on both sides, was heightened by the decision of the Académie, in March 1859, to offer its Alhumbert Prize of 2,500 francs for experiments that would throw new light on spontaneous generation.[76] The outcome of the prize competition has come to be interpreted as evidence of the closed minds that from the start made Pouchet's struggle difficult. Certainly Pouchet perceived it in this way, and throughout the competition he did what he could to fight against prejudice and present himself in the most favorable light. One function of his large and learned book *Hétérogénie* was to achieve precisely this. In the book, which remained, from its publication in 1859 until his death in 1872, the most detailed exposition of his theories, Pouchet disarmingly insisted on the moderate nature of his views. His was the middle ground: "If we follow philosophers, we often find ourselves caught up in the excesses of spiritualism; if we go with physiologists, we find ourselves as materialists. The truth hovers somewhere between these two opposed positions."[77] Hence while he endorsed heterogenesis without reservation, he firmly distanced himself from abiogenesis and thereby from the more extreme (though admittedly rather confusing and shifting) ideas of Lamarck, an author for whose writings, along with those of Geoffroy Saint-Hilaire, he otherwise expressed profound respect.[78]

For Pouchet, in fact, life could only develop under the influence of an immaterial entity, which he variously called a "plastic force" or "vital power." This was the indispensable organizing agent that alone could turn dead into living organic matter. Such a view, with its insistence on the inability of physical forces alone to effect the change, could be made to square elegantly with the pattern of the history of life that Pouchet, along with most Christians, favored: that of progressive, stepwise development through periods of relative quiescence punctuated by discrete violent revolutions, such as Elie de Beaumont's *soulèvements*, or (as the British preferred to call them) catastrophes.[79] In Pouchet's view, it was in the aftermath of these episodes that heterogenesis came into play, as the organic debris from the preceding epoch produced the eggs of the new species that would characterize the next: "the particles composing our own corpses constitute new materials for the beings that are to come after us," as he put it in 1862.[80] With the "plastic force" an essential element in his interpretation, Pouchet presented himself not as a materialist but as a champion of true religious belief. "The heterogenists' theories," he wrote, "far from detracting from the attributes of the Creator, merely enhance their divine majesty."[81] How God acted in the successive phases of creation was left vague. But on one point Pouchet had no doubt: when heterogeny came into play, it did so both as a natural event and as a manifestation of God's infinite wisdom.[82]

Pouchet's insistence on the religious dimensions of his opposition to materialism, reinforced as it was by statements of his going back to the 1840s, made it hard for him to be attacked on moral grounds. Where the odds were loaded against him was rather on the terrain of science, on which he had to trade experimental evidence with Pasteur. Although Pouchet expressed gratitude to Charles-Amédée Verdrel, the mayor of Rouen, and Baron Ernest-Hilaire Le Roy, prefect of the department of the Seine-Inférieure for the facilities available to him in Rouen,[83] he was under no illusion: the conditions in which he worked were far inferior to those to which Pasteur, a master of meticulous research, had access at the Ecole normale supérieure. However, the extent and influence of any supposed prejudice on the part of the judges for the Alhumbert prize are not easy to determine. Pouchet's opinion, though, was clear: it was that a committee composed of known opponents of spontaneous generation—Henri Milne Edwards, Adolphe Brongniart, Pierre Flourens, Victor Coste, and Claude Bernard—deprived him of any hope of victory. It is easy to see why Pouchet felt as he did, and fortune certainly did not favor him. The committee that had set the subject for the prize had included two potential sympathizers, Etienne Geoffroy Saint-Hilaire's son Isidore and Augustin Serres, the professor of comparative anatomy at the Muséum d'histoire naturelle, who had allowed Pouchet to spend two weeks working in his laboratory. But Saint-Hilaire's death in November 1861, the aging Serres's virtual cessation of scientific activity after 1860, and their replacement on the committee by Coste and Bernard tipped the balance decisively. Nevertheless, Pouchet persisted, and it was only late in the day, in November 1862, that he withdrew,[84] leaving the prize to be awarded in the following month to Pasteur for his definitive account of his experiments with the swan-necked flasks.[85]

With Pouchet disenchanted and convinced that his experiments had been denied the serious examination they deserved,[86] the matter might well have rested there had it not been for one important academic peer who bolstered his resolve. This was Nicolas Joly, a long-established professor of zoology in the faculty of science in Toulouse (where he had been appointed in 1840) and of anatomy and of human physiology in the city's Ecole préparatoire de médecine et de pharmacie. Joly was a noted public speaker, and his lectures on human reproduction had already on one occasion scandalized local opinion.[87] It was both as a professor and as someone with broad interests, extending to philosophy and languages, that he gave the high-profile endorsement from the official world of learning that Pouchet's cause had lacked. Joly's support (backed by a grant from Rouland as Minister of Public Instruction[88]) and that of his assistant, Charles Musset, persuaded Pouchet to resume the struggle in the summer of 1863. The pretext for reopening the question and requesting a reexamination of Pasteur's observations, were new experiments that Pouchet, Joly, and Musset had conducted high in the Pyrenees, in which microbial growth had consistently appeared in boiled infusions of hay exposed to the atmosphere at widely different altitudes.[89]

The challenge was directed once again at Pasteur's claim that in mountainous regions, where the air could be supposed to bear fewer germs, contamination from the atmosphere would be less likely to give the appearance of the generation of new life. While Pasteur remained as convinced as ever of the inadequacy of the precautions that the Pouchet camp had taken against contamination, the Académie responded with exemplary patience. In January 1864 it acceded, with Pasteur's support, to the request of Pouchet, Joly, and Musset that it should establish another committee to judge the rival claims.[90] It was evidently the committee's intention that a repetition of Pasteur's experiments with the swan-necked vessels should be the main focus of the reinvestigation. But Pouchet and his associates took a different view. The detailed program of investigations that they eventually proposed so palpably transcended the terms of their original request for reopening the case and was accompanied by such prevarication and generally uncooperative behavior on their part that the committee (composed entirely of known opponents of spontaneous generation) rejected it and insisted on proceeding with the inquiry as originally planned.[91]

This response, which marked the irreconcilable disagreement between the two parties on the criteria and procedures for judgment, in effect marked the end of the affair as a major scientific debate. With a capriciousness that did nothing to advance their cause, Pouchet, Joly, and Musset attended the meetings in the chemical laboratory of the Muséum d'histoire naturelle at which the decisive experiments were to have been performed, on 22 and 25 June 1864. Suddenly, however, they withdrew, leaving Pasteur and the Académie's far from neutral committee free to pursue the program of experiments that Pasteur had all along maintained would resolve the issue. Over the next few months, memebers of the committee continued the experiments with the sealed vessels until, in February 1865, Pasteur was predictably vindicated in a formal statement backed by the full authority of the Académie.[92] Three years later Pouchet received the consolation of promotion from the rank of *chevalier* to that of *officier* in the Legion of Honor, though in the same list in which Pasteur advanced from *officier* to the highest rank in the order, *commandeur*.[93]

After his final withdrawal, Pouchet devoted his main energies to advancing secular education and to popularization, his talent for which is conveyed in his dry, unpolemical, but immensely successful account of the wonders of the animal and vegetable kingdoms, the earth, and the heavens, published as *L'Univers* in 1865.[94] Although he made no significant contribution to further discussion of spontaneous generation himself, he never admitted defeat and remained obstinately, though discretely, loyal to his vision of a history of life punctuated by episodes of heterogenesis.[95] He remained equally loyal to his broader ideals, which he presented as being in the traditions of the Enlightenment. Writing in September 1865 to Eugène Noël, his friend and editor of the *Revue de Rouen*, he was defiant: "In our city—it is you who have said this—there will be a true school of free-thinkers, and we shall perpetuate the work of the Encyclopedists."[96] It was in the same spirit that he supported the educational ideals of the leftist Ligue de l'enseignement, from the Ligue's foundation in 1866, and so identified himself with the political opposition to the empire.[97] But the chances that a perpetuation of the scientific debate about spontaneous generation would bear fruit on this or any other front were slight, and further discussion, such as it was, soon became an affair of the periphery. It was conducted with undimmed conviction from the margins of the scientific community by Joly (from his chair at Toulouse) and Pennetier (in the museum of natural history in Rouen) but was now largely ignored at the center.

Despite his public silence on spontaneous generation in the mid-1860s, Pouchet himself became a symbol of vain but courageous opposition to the might of the Parisian establishment that, as Pennetier presented it, had crushed him.[98] The symbol gained potency across a broad spectrum of educated opinion. Among nonscientists Gustave Flaubert may be regarded as a less than objective witness: not only had his father taught

Pouchet at the Hôtel-Dieu in Rouen but Flaubert himself had been Pouchet's pupil during his time at the city's *lycée*. Nevertheless his declaration of having been "dazzled" by his reading of *Hétérogénie*[99] and his description of the opponents of Pouchet's ideas as either "impostors" or "cretins"[100] reflect the seriousness of his engagement with the issues. For scientists, the debate was of greater immediacy. With them, it was bound up with the wider campaign of liberally minded opponents of *la science officielle* who ascribed Pasteur's victory to the tyranny of the center rather than its scientific superiority. Preeminent among these opponents, both for his virulence and for his persistence, was the radical science writer Victor Meunier, whose campaign against the Académie des sciences was already well under way by the time Pouchet resurrected the question of spontaneous generation in 1858.[101] In Meunier's eyes, Pouchet stood (with his loyal supporters Joly and Musset) as both martyr and hero: a victim of the ideology of the empire and a disadvantaged crusader for the true experimental method, he had revealed for all to see "the decrepitude of our abuse-ridden edifice of science."[102]

The reaction of Louis Figuier, the most widely read of all popular writers on science at this time, lacked Meunier's vehemence. Indeed, the evenhandedness of his discussions of spontaneous generation in *La Presse* and successive issues of his *Année scientifique et industrielle* is striking. Nevertheless, his insistence that the experiments of Pouchet, Joly, and Musset had to be taken seriously and his judgment that the exchanges left the question open reflected a reluctance to accept Pasteur's evidence and a basic sympathy for Pouchet's arguments and more generally for the provincial savants whose impotence in the face of "official" science the case encapsulated.[103] A similar mixture of reactions (though fired with additional elements of pantheism and a politically motivated suspicion of any view that smacked of the imperial regime's authoritarianism) was probably at the root of Jules Michelet's sympathetic urging of Joly to pursue "the great question."[104] But, where Figuier and Michelet were arguing essentially for the question to remain open and hence for mutual tolerance in the debate, others followed Meunier in perceiving conspiracy or a distortion of the scientific issues. For them, the distortion resulted from extraneous factors that almost invariably weighed in Pasteur's favor. Even in one of the most reasoned lay defenses of Pouchet, published in 1864, Eugène Noël ended an otherwise soberly scientific pamphlet with a categorical assertion that the opposition to spontaneous generation had been fired, first and foremost, by a hatred of atheism.[105] There were unquestionably grounds for believing that the charge of atheism had been misdirected or exaggerated; sympathetic reactions to Pouchet's work in the writings of Cardinal Donnet, Archbishop of Bordeaux, and the rather eccentric explorations of human belief in immortality by Noël's close friend Alfred Dumesnil both point in that direction.[106] The fact remained, however, that the charge had been made. And that more than anything, in Noël's view, accounted for the dismissiveness with which Pouchet, Joly, and Musset had been treated.

It is easy to see how interpretations of the debate such as Noël's gained currency in the 1860s. Writers who, for political or other reasons, chose to speak for those on the margins of science were drawn to any cause that set a disadvantaged provincial figure against a leading savant of the capital or an unorthodox doctrine against received opinion. When the spokesman for orthodoxy was someone as firmly entrenched in the scientific elite as Pasteur, the temptation was irresistible, the more so as (at least in the later stages of the debate) Pasteur did not hesitate to parade his credentials as an unimpeachable champion of order and propriety. When he lectured on spontaneous generation at one of the first "soirées scientifiques et littéraires" of the Sorbonne in April 1864 (see chapter 5), he declared his position unequivocally. He reserved the full force of his attack for the materialistic and atheistic tendencies of the age that had helped to give improper credence to the doctrine: if spontaneous generation were vindicated, the cause of the materialists would be advanced, and the cherished Catholic belief in the mystery of God's role in creation would disappear. In what Pasteur presented (though without explicitly implicating Pouchet) as a facet of the broader conflict between the two great currents of contemporary thought—materialism and spiritualism—such an outcome would be demoralizing.

> What a triumph, gentlemen, what a triumph for materialism if it could claim to rest on the demonstrated fact of matter organizing itself, coming to life independently, endowed with its familiar forces . . . Ah ! If we could add to matter that other force that we call life—life in the diverse manifestations it displays in the varied conditions of our experiments—what could be more natural than then to deify it? What good then the idea of a primordial act of creation, before whose mystery we should bow? What good then the idea of a creator God?[107]

The *bien-pensant* members of Pasteur's audience and those who read the published lecture could only have felt anxiety as they reflected on these chilling questions. Unconvinced by Pouchet's consistent dissociation of his doctrines from any atheistic tendency, they would have been relieved to see spontaneous generation so persuasively rebutted. Such relief, however, and Pasteur's readiness to exploit bourgeois unease for rhetorical effect on the public stage do not imply that the religious implications of spontaneous generation colored deliberations in the Académie. In his lecture of April 1864, Pasteur insisted that he had come to the question of spontaneous generation "with no preconceived ideas" and that "neither religion, philosophy, atheism, materialism, nor spiritualism bore on the case."[108] That declaration has surely to be believed, just as Pouchet's providentialist statements too have to be taken at their face value. For while it is true that Pasteur was reluctant to accept the results of early experiments of his own that accommodated at least the possibility of spontaneous generation, such reluctance can be explained well enough by his exposure to existing evidence, notably Schwann's,[109] that had already suggested a possible role for yeast as a living fungus causing fermentation.

More immediately, though, Pasteur was persuaded by his just recognition of the many possible sources of contamination in Pouchet's experiments—a contamination which, by the end of the debate, he had succeeded triumphantly in demonstrating. What determined the scientific outcome of the Pasteur–Pouchet debate, therefore, was not the ideological divergence between the two protagonists. Pasteur's conventional piety and Pouchet's rather old-fashioned deistic providentialism may have had little in common, but they did not set an unbridgeable gulf between the two men and they certainly do not justify the Manichean terms in which their exchanges have commonly been analyzed. Pasteur and Pouchet were as one in assiduously restricting their exchanges to what they perceived as the strictly scientific issues. In this, their priorities stood in marked contrast with those of the vociferous contemporaries, most of them outside or on the margins of science, who interpreted the exchanges primarily in terms of a conflict between authority and liberalism. What occurred was that a scientific debate of some complexity had been drawn, largely despite itself, into the wider discussions about secularization and the dominant ideological values of the empire that characterized the 1860s. It is to those discussions that we now return.

The Radical Synthesis and Its Enemies

To speak of the consolidation of a radical synthesis in the later years of the Second Empire runs the risk of endowing the disparate criticisms of "official" science and its senior Parisian representatives with a measure of coherence that they did not possess. The fact remains, however, that by the mid-1860s critics had available to them a recognizable menu of beliefs that were in varying degrees at odds with science as it was taught and debated in most of the great national institutions. The intellectual profile that an individual fashioned from the menu was a matter of personal choice. But opponents of traditional academic authority and religiously founded bourgeois conformity could be expected to have a sympathetic interest in materialism or positivism (without necessarily being a formal adherent of either) and an openness to (if not acceptance of) the evidence for spontaneous generation and the polygenism of both Paul Broca's and Georges Pouchet's doctrines of human ancestry.[110]

From the early 1860s, a new element joined the existing heterodox beliefs. This was Charles Darwin's theory of evolution by natural selection. Although the theory had been a focus for public debate in Britain since the publication of the *Origin of Species* in 1859, in France it entered rather slowly even into discussion among specialists. The book was certainly read, and for three years it was reviewed, but without assuming the status of even a minor literary or philosophical sensation.[111] From this initial treatment it did not emerge unscathed: Alfred Sudre's review in the *Revue européenne*, for example, was determinedly, if rather superficially, dismissive.[112] No criticisms, though, did significant damage. The prevailing view was that the *Origin* was an informative but speculative work that offered no profound novelty, to be read as at most a gloss on Lamarck's long-familiar transformist doctrines.[113] As such, for the leaders of the life science community, the book did not merit serious attack. And it certainly gave no reason for questioning contemporary orthodoxy, as expressed in a two-volume work by Alexandre Godron, the professor natural history and dean of the faculty of science in Nancy, in the very year of the *Origin*. In a manner typical at the time, Godron's orthodoxy embraced a firm reassertion not only of the fixity of species but also of the unity of the human species, all argued unyieldingly.[114]

Among the few who engaged sympathetically with the Darwinian theory was the well-traveled engineer, writer, and naturalist Auguste Laugel. But even he was unconvinced: the objections, including ones such as the sterility of hybrids that Darwin himself had rehearsed in the *Origin*, were just too powerful.[115] Still more striking, and indicative of the uphill struggle that Darwin had to face, was the reticence of Isidore Geoffroy Saint-Hilaire, who might have been expected to be open to transformist ideas; disappointingly for the Darwinian cause, in his one brief reference to natural selection before his death in November 1861, Isidore rejected the crucial analogy between the agricultural breeder and nature as the selecting agent.[116] Yvette Conry has argued that a failure to recognize the importance of Darwin's theory was widespread and long-lasting, to the point that, in her view, full-blown Darwinism was never introduced into France.[117] While Conry's interpretation has been qualified by a number of historians since it was fully articulated in 1974, it is beyond question that the acceptance of Darwinian teachings by the scientific elite, whether complete or partial, was a slow process that only gained significant momentum from the mid-1870s. Among nonspecialist readers, in contrast, the publication of the first French translation of the *Origin* in 1862 marked the transformation of a hitherto largely ignored, or misunderstood, book by an eminent but far from revolutionary foreign naturalist into a work of incendiary potential.

The catalyst of this change in public perceptions was not so much the translation itself as the fifty nine-page preface by the translator, Clémence Royer.[118] In her early thirties at the time, Royer had begun to acquire a modest reputation as an independently minded member of the intellectual circles she frequented in Switzerland, where she lived in the late 1850s and early 1860s.[119] By then she had rejected the Catholic piety and legitimist nostalgia that had dominated her childhood and early education. Born in Nantes, the daughter of a former naval captain who was fired with an undying loyalty to the Bourbons and an intelligent, strong-willed but poorly educated Breton seamstress, Royer reacted to the 1848 revolution by becoming the free-thinking republican that she was to remain until her death in 1902. After training as a teacher, she perfected her English during a year spent at Haverfordwest in Wales and, after finding her way to Switzerland in 1856, pursued her scientific interests in the circle around Karl Vogt and the Natural History Museum of Geneva, while writing and lecturing, mainly on economic matters. It was in Switzerland, in Lausanne, that she became the life-long companion of a married man, a French publisher and writer with political sympathies close to hers, Pascal Duprat; and it was almost certainly in Geneva that she learned of the *Origin* through the paleontologist Jules Pictet.

How Royer then came to be chosen to translate Darwin's book is not clear. But her zeal for the task, fanned by her reading a well informed and overwhelmingly favorable review of the *Origin* by her friend the Genevan physiologist Edouard Claparède,[120] knew no bounds. With Claparède on hand to advise her on the scientific aspects of the book, what may well have begun as an essentially commercial engagement with the Parisian publishers, Guillaumin and Victor Masson, turned into an ideological crusade. For Royer, in fact, the *Origin* was part of a progressive secular revelation, one that evolved as science advanced. In such a worldview, as Royer insisted, belief in a god might be logically possible, but it was unnecessary and hence alien to the scientism she espoused in all her writings.[121] On the history of life, therefore, the choice was stark. It was between the Darwin's "rational revelation," on the one hand, and, on the other, the "irrational revelation" of Christian teaching (whether Catholic or Calvinist) with its hazy notions of the fall and redemption of the human race. Equivocation, for Royer, was inadmissible: "It is a case of either yes or no, and we have to choose. Whoever declares for the one is opposed to the other. For my part, the choice is made. I believe in progress."[122]

Royer's translation had its failings. In a letter to Quatrefages, Darwin (who had not been consulted about the choice of the translator) judged Royer to be "one of the cleverest & oddest women in Europe" and acknowledged that she had made "some curious & very good hits"; but he thought her extrapolation from natural selection to issues of human morality extravagant and wished she had known more natural history.[123] Claparède, for his part, wrote apologetically to Darwin; he found the translation "heavy, indigestible, and in places wrong" and feared that Royer's numerous footnotes, over which he had done his best to exert a friendly restraining influence, would not be to Darwin's liking.[124] A wholly unauthorized but revealing subtitle—"des lois du progrès chez les êtres organisés"[125]—served only to make the work even more contentious. But that was precisely what Royer intended, and her strategy succeeded. Following her translation, the *Origin* was wrenched from the realm of still superficial and largely inconclusive discussion and forced on public attention. Who could remain indifferent to a theory that explained the ineradicable physical and intellectual differences between men and women and invited debate about the inequalities between races? It was a theory that, in Royer's words, embraced "a complete philosophy of nature and a complete philosophy of humanity"; as such, it demanded serious consideration.[126]

The impact of the translation was heightened by Royer's explicit linking of Darwin's theory with the belief that life had initially emerged from nonliving matter and hence with the debate about spontaneous generation, then at its height. This association of Darwinian evolution with spontaneous generation was not new: Alfred Sudre, for instance, had conflated the two doctrines in his review of the *Origin* in 1860 and stated it as virtually axiomatic that transformist doctrines and spontaneous generation stood or fell together.[127] But the dogmatic tone of Royer's preface projected the conflation into the common currency of debate. With the *Origin* now at center stage, readers of a radical philosophical persuasion quickly added evolution to their portfolio of heterodoxy, so reinforcing the vision of two irreconcilable camps that Royer and, before her, Sudre had articulated. As Sudre had put it in a comment reflecting the starkness of the choice that now confronted thinking people,

> the realm of the human intellect is divided between two great metaphysical systems. Under different names, these lie at the heart of most of the profound disagreements that divide the devotees of the physical and natural sciences. One system conceives the universe as autonomous, eternal, and infinite, governed by blind immutable laws, and going its way and undergoing change as a result of forces within itself and the inexorable interplay between its various parts. The other theory states that beyond the world there exists a supremely intelligent, powerful, just, and conscious being and that this being alone is eternal, unchanging, and infinite.[128]

On the one side, for Sudre, lay atheism; on the other, theism.

Perceptions of the *Origin* as an element in the tide of radical thought struggling with the forces of conformity and conservatism were reflected in the acerbity that came quite suddenly to characterize judgments of the book by the Parisian scientific elite. Writing in 1863, with the authority of a permanent secretary of the Académie des sciences and professor at the Muséum d'histoire naturelle, Pierre Flourens gave no quarter. The woolly, figurative language of the *Origin* and a cavalier attitude to evidence set Darwin, in Flourens's judgment, far below Cuvier. In presenting nature as capable of "selection" ("élection" was the word that Flourens used, as Royer had done in the first though not the second edition of her translation[129]) and falling into the trap of extrapolating from the familiar variability of species to the unsubstantiated supposition that one species could be transformed into another, Darwin had created a "metaphysical gorgon": his argument was founded on language that was "pretentious and empty," and his ascriptions of intention and will to nature were "puerile and outdated."[130] Scientifically, Flourens's attack broke no new ground and simply ignored the persuasive responses that had already been advanced in Darwin's defense. Like so many opponents of Darwinism in France, he was exploiting well-worn criticisms (of the kind that made much of the lack of anatomical change between mummified remains from ancient Egypt and present-day species[131]) in fighting an old battle against the whole lineage of transformist doctrines going back to Lamarck and Geoffroy Saint-Hilaire.

Flourens's unsubtle and unoriginal arguments served to harden entrenched convictions rather than subject them to fresh scrutiny. No one who read his evaluation of Darwin, followed with the by now familiar dismissal of spontaneous generation as well,[132] could doubt that the core issue was one of confrontation between irreconcilable world-views, just as Sudre had stated and as Paul Janet was to do in 1864, when he incorporated an attack on Darwin (as well as on spontaneous generation) in his more general protest at the infiltration of German materialism into French thought.[133] From such a confrontation, the Catholic Church could not stand aloof, and Catholic writers drifted almost without exception into the anti-Darwinian camp. They did so slowly. In fact, a clearly stated Catholic opinion did not form until the later 1860s, when it was best-expressed in Guillaume Meignan's *Le Monde et l'homme primitif selon la Bible*. Although published in 1869, the book was based (how fully is not known) on lectures that Meignan had given in the faculty of theology at the Sorbonne in 1861–62, in his capacity as professor of holy scripture.[134] Meignan, who had been appointed Bishop of Châlons-sur-Marne in 1864 and was to go on to ever-higher offices in the church, as Archbishop of Arras, then Tours, and, from 1893, a cardinal, set his discussion of Darwin's theory in the context of an elaborate affirmation of the conformity of geological evidence with the Genesis story of creation. His discussion, incorporating an interpretation of the days of creation as periods of indefinite length, was far from ignorant: in geology, he had been prepared by the abbé Edmond Lambert, the keeper of the archaeological and geological collections of the seminary of Châlons-sur-Marne. But the end point was predictable. Darwinism emerged as just one of the subversive doctrines that had emanated from and been promoted by a "school" he despised as "positivist, materialist, atheist."[135]

Preeminent among the subversive doctrines was polygenism. For Meignan as for many Catholics, the fault of the polygenist account of human history lay in its making light not only of the similarities between European, black, and Australasian peoples but also, and more crucially in Meignan's view, of the endemic immorality that over countless generations had brought the native populations of Africa and Australia, untouched by Christianity, to their present degraded state.[136] Meignan, in fact, did not regard the ancestry that Darwin proposed, with its reduction of man to a superior ape, as the most corrosive weapon in the atheists' anti-Christian armory. Nevertheless, it was of a piece with the pernicious teachings that in recent years had undermined the Mosaic narrative and hence faith itself. Like spontaneous generation (seen by Meignan as a throwback to the dark doctrines of Epicurus[137]), it had to be combated, in this case by points that Meignan saw less as arguments than as facts. One such "fact" was the intellectual and moral gulf that separated the brute from man (with his unique capacity for reflection and a knowledge of God). Another was the philosophical absurdity of supposing chance to be a sufficient cause of the process that, on Darwin's analysis, had led from the "primordial cell" (its origins fatally unexplained, according to Meignan) to the wonders of the evolved world.[138]

By addressing his lectures and book to a broad audience of believers concerned by the accumulation of challenges to their faith, Meignan set the Darwinian debate in the broader ideological struggle that Sudre and Royer, from their respectively conservative and radical perspectives, had identified. In a climate of heightened sensibilities, encouraged by the more liberal tone of the middle and later years of the Second Empire, contemporaries needed little encouragement to move to the barricades. In the name of religion, the unsigned manifesto for a new weekly journal, *La Science et la foi*, stated the position unequivocally in the first issue, of December 1864:

> Today no serious observer can fail to be struck by the strange spectacle of a relentless struggle in the world of ideas between two adversaries fighting for the control of human minds. On the one side is Catholicism, with its union of faith and the teaching on which faith depends. On the other, rationalism with its retinue of philosophical and religious systems.[139]

For *La Science et la foi*, the outcome of the struggle was not in doubt. Rationalism would have its "partial successes," but in due course the leaders of the church, sustained by the "countless battalions" of the faithful, would triumph.

On both sides, passionate reactions were inevitable, and among the opponents of religion no one did more to arouse passion than the positivist, political radical, and pioneer of histology, Charles Robin. Although Robin did not subscribe to the full panoply of radical opinion—he opposed Darwin's theory, for example, on the grounds that it did not rest on empirical evidence[140]—he became, in conservative eyes, the incarnation of all that was most hateful about the new doctrines. In fashioning his militant brand of positivism, Robin was speaking from the scientific heart of the movement. He had imbibed positivist philosophy from the purest sources, first from Littré and later, in 1849, at the lectures on the history of humanity that Comte gave at the Palais national (as the Palais royal was called in the heady days of the Second Republic).[141] As a young *agrégé* in the faculty, newly infatuated with Comtean teachings, he took a leading part in the foundation of the Société de biologie in 1848, associating closely with the eminent physician and first president, Pierre Rayer, and with Claude Bernard, for many years Robin's fellow vice president. Although neither Rayer nor Bernard formally subscribed to positivism, their sympathy for many of its principles is beyond question and it helped to give the Société de biologie a radical edge.[142] They would certainly have found nothing to fault in Robin's statement of the aims of the society in the first issue of its *Comptes rendus*.[143] The statement, which began with an explicitly Comtean sixfold classification of the sciences, expressed Robin's ideal of a biology that would be "positive, abstract, and concrete" and in which (echoing the words of Comte and Blainville[144]) "the static state" (essentially anatomy and the work of classification that Comte termed *biotaxie*) and "the dynamic state" (physiology and the all-important relations between an organism and its environment) would be treated together.[145]

The air of dangerous nonconformity that surrounded Robin was reinforced by his known republican sympathies, which survived undimmed from the Second Republic and made him a natural ally of the intellectual critics of the empire who gathered for the weekly *dîners Magny*.[146] At these convivial gatherings of like-minded liberals, Robin was a leading spokesman for the materialist philosophy, as well as an authority on diet equally at home in the physiological and the gastronomic aspects of the subject.[147] And both at the Magny dinners and, from 1869, at the *dîners Brébant* that succeeded them[148] his anticlericalism and republican politics found a sympathetic audience, including Sainte-Beuve (by this time disenchanted with the regime he had once supported), Taine, Broca, Renan, Berthelot, and the great literary gossips, the Goncourt brothers. For a man of such unyielding conviction, career making was not easy, and Robin might well have remained in the lower reaches of the medical teaching profession had it not been for the support he consistently received from Rayer. It was Rayer who persuaded Rouland to create the new chair of histology in the Paris faculty of medicine and to appoint Robin to it in 1862. Rouland would never have chosen such a controversial figure as Robin of his own accord. But, as physician to the emperor for the previous ten years, Rayer was in a sufficiently strong position to make Robin's appointment a condition for his own acceptance of the post of dean of the faculty—a post that the emperor himself was anxious he should take.

Robin's appointment, though a shot in the arm for French physiology (notably in helping to open it to foreign influences),[149] incensed the clerical party, and he soon found himself the target for abuse in the press and for the organized interruption of his lectures.[150] By the mid-1860s he was a marked man whose public acts came to be ever more heavily charged with ideological significance as public debate on scientific issues became polarized. His candidature for membership of the Académie des sciences, in the section of anatomy and zoology, following the death of Achille Valenciennes in April 1865, brought predictable prejudices to the surface. By August 1865, it was known that Henri Milne Edwards and Quatrefages were opposed to Robin and that they intended to nominate Henri de Lacaze-Duthiers, "a man from their camp," as Taine disparagingly described him.[151] Supporting Robin were Bernard and the mathematician Joseph Bertrand. And, at Taine's suggestion, Sainte-Beuve even took up Robin's case with the emperor's cousin, Princess Mathilde, whose salon was a favored resort of liberal intellectuals.[152] By the time the election took place, in January 1866, politics and religion had become inextricably entwined with science. Minds, conservative and liberal, went back three years, to when Littré had been kept out of the Académie française through the intervention of Félix Dupanloup, and much the same battle lines were drawn.

As expected, the five members of the section of anatomy and zoology (Milne Edwards, Quatrefages, Emile Blanchard, Victor Coste, and Achille Longet) put forward Lacaze-Duthiers as their first choice, with Robin in second place, whereupon the Académie as a whole (apparently with the support of Pasteur) exercised its right by rejecting the recommendation and electing Robin.[153] The fact that Pasteur supported Robin, apparently in response to Saint-Beuve's lobbying, indicates that the pattern of voting did not divide on liberal and nonliberal lines.[154] Nevertheless, the liberal press led sympathizers in applauding Robin's victory as one for the broader political causes he represented. Entrenched conservatives, for their part, saw their worst fears about the insidious advance of free thought as confirmed, and by the end of the year they were on the offensive again. Their charge was the specific one that Robin had preached materialism in the faculty of medicine in the opening lecture of the academic session in November 1866. A disciplinary interview followed, as did a predictable compromise. Duruy displayed his known liberal leanings in stating that he believed the lecture had been in no way subversive and doing no more than remind Robin that he had been appointed to teach anatomy, not metaphysics.[155]

To all appearances, the confrontation had been defused. But worse was to follow, in the form of a petition that Léopold Giraud, a conservative journalist and editor of the popular Catholic periodical *Journal des villes et des campagnes*, submitted to the Senate in June 1867.[156] In the words of the petition, an unnamed professor in the Paris faculty of medicine (probably Robin) had declared: "The nervous substance possesses the property of thought. When it dies, it does not go on to a second life in a better world." In the same faculty, it had also been stated (again, almost certainly by Robin) that "matter is the god of savants.... If the monkey has a soul, so does man; if not, neither does man." And a doctor in the Salpêtrière hospital had scoffed at a woman who was wearing a medallion of the Virgin. All this, according to the petition, before a student body that openly subscribed to materialist doctrines without any restraint being exerted by the professoriate. Such behavior, the petition declared—and this was clearly the worst of it—undermined "the very foundations of the social order."

Despite its strong wording, the petition sought ostensibly to do no more than alert the government to the doctrines that certain "official professors" were peddling in the faculty and to point a familiar accusing finger at ministerial negligence as the guardian of morality. But carefully engineered publicity over the next year swelled the number of signatories from an initial 719 to nearly three times that number, including three hundred priests. This ensured that the matter did not rest, as it might have done, with the Senate's formal noting of the complaint and of the report of a senatorial committee of inquiry chaired by the distinguished lawyer and moderate senator Gustave Chaix d'Est-Ange.[157] The committee did what it could to defuse the confrontation by concluding that the excesses of youth and occasional exaggerated or badly phrased statements by professors did not constitute proof of widespread immorality; still less were they a reason for allowing the creation of universities free of state control or for the condemnation of Duruy for failing to enforce discipline. The whitewash, though, was in vain. Conservative and clerical interests were not to be fobbed off with such prevarication, and in May 1868, in an atmosphere of renewed Catholic hostility to the state's monopoly of higher education, the Senate felt obliged to devote four days to a debate on the petition.[158]

In the debate, the ultramontane Cardinal Archbishop of Rouen, Henri Marie Gaston de Bonnechose, led the attack on those he saw as the ringleaders in the faculty: Broca, Vulpian (recently appointed to the chair of pathological anatomy in the face of vigorous conservative opposition), and Germain Sée (whose Jewish religion in itself made him vulnerable). His special venom, however, was reserved for two recent editions of Pierre-Hubert Nysten's *Dictionnaire de médecine*, for which Littré and Robin had been responsible. Ever since Littré and Robin had been engaged by the publisher J.-B. Baillière et fils to undertake a major revision (in effect a rewriting) of the *Dictionnaire* in the mid-1850s, their work had been the subject of controversy, and Bonnechose was neither alone nor off the mark in reading a number of articles as oblique attacks on the foundations of Christian belief. How else was he to interpret the entry "Ame" ("Soul") in the eleventh edition (1858)? The article began provocatively: "A term in biology that conveys, anatomically, the various functions of the brain and the spinal cord, and, physiologically, all those functions that we term encephalic sensibility, that is to say perception, whether of external objects or inner sensations."[159] And so it continued, with will, emotion, and imagination all being attributed to the physical operations of the brain and the nervous system. Likewise under "Esprit" ("Spirit") the concept of immaterial spirits to explain the phenomena of life was dismissed as an outdated hypothesis: "one that suggested itself naturally to the human mind in earlier times but whose function is beginning to be entirely fulfilled by the positive conception of the world and man."[160]

The articles "Idée," "Pensée," "Métaphysique," and "Conscience" were in the same vein, all infused with a mixture of materialism and positivism, which Bonnechose insisted were essentially (if not on every point) "one and the same thing."[161] With only moderate exaggeration, Bonnechose claimed that the malign influence of Comte, "the dismal author of positivism,"[162] pervaded the whole work. Anyone who doubted the charge had only to read the article "Positive (Philosophic)," which with true Comtean zeal roundly disparaged the theological and metaphysical phases in the history of thought.[163]

Duruy's position at the head of the University, as well as his known antipathy to the clerical interest, made him no less a butt of Bonnechose's fury than Robin and his colleagues were. The defense that the texts that students used in higher education were beyond Duruy's control was interpreted by Bonnechose as neglect of ministerial responsibility: if Duruy's power was really so limited, what comfort could citizens derive from his assurance that he would not tolerate the teaching of impious doctrines in any institution under his control?[164] Bonnechose conceded that Duruy's recent intervention to annul the award of a medical doctorate for a thesis that was said to deny free will, the distinction between man and beast, and moral responsibility had shown good intentions. But he could not accept that either the intervention itself or the reprimand that Duruy delivered to the author had done anything to stem the tide of materialism that he saw emanating from the Paris faculty.[165] Moreover, it did nothing to reassure those who were concerned to preserve a stable social order. For, as Bonnechose argued, materialism fostered radical, left-wing opinions in politics, an assertion to which the militant revolutionary activities of Alfred Naquet, who had taught for some years in the faculty, added substance.[166]

Defense against these charges was difficult. Jean-Baptiste Dumas may have been correct in asserting that it was an illusion to suppose that the majority of the faculty's professors favored materialism, and his position as a trusted friend of the empire and a respected savant lent authority to his statement.[167] But the public perceptions that Dumas was trying to counter would not go away. In the end, it was the aggressively intolerant stance of the critics and the excessive retribution they sought that undermined their case. In the late 1860s, with liberalism as well as conservatism on the march, the idea that clerical outrage had to be appeased at all costs was simply unacceptable. By majorities of roughly two to one, the Senate followed the lead of Sainte-Beuve, Prosper Mérimée, and Duruy himself in rejecting the charges against the faculty. In a vote with even more far-reaching implications, it then also rejected the proposal for ending the state's monopoly in higher education, so denying the conservative Catholic interest the prize it had most coveted throughout the affair.[168]

The votes in the Senate conveyed an unequivocal message. They were votes for controlled modernity and at least a measure of free speech on unorthodox philosophical issues; and they were, no less importantly, votes of confidence in Duruy. Without that support, it would have been far more difficult, even impossible, for Duruy to implement the important reforms that marked the last year or two of his term of office at the ministry. Outstanding among these reforms was the founding of the Ecole pratique des hautes etudes in 1868, with its commitment to the untrammelled pursuit of new ideas (see chapter 3). By appointing Robin as director of the Ecole's research laboratory for histology and Broca to direct the anthropological laboratory, Duruy left no doubt as to where his sympathies lay. With respect to research (though not to teaching, over which he continued to exercise such control as he could), he stood unequivocally for freedom of inquiry of the kind that the savants of Germany enjoyed. Despite a Parliament that remained stubbornly parsimonious when it came to granting funds for research and a clergy that opposed him as bitterly as ever, his stand was successful, to the extent that many of the aspirations of the reform movement had become part of ministerial policy. But by now Duruy's days as a minister were numbered. When the end came, with his dismissal in July 1869, the emperor felt genuine regret at the departure of an old friend and a trusted ally in the quest for the cautious liberalization of the empire. As he knew only too well, the sacrifice of Duruy was one of the few cards he still had to play if a decade of strained relations with the church was to be eased.

What more might have been achieved, even with Duruy gone, remains a matter for speculation. For only a year later, the beginning of the war with Prussia brutally diverted attention from domestic reform. It also interrupted the debate on scientific issues that had riven public opinion for much of the 1860s. Once the war was over, the debate resumed, but with a distinctly different tone. Among those who, in the name of science, set their faces against a religiously inspired worldview, opinions after 1870 tended, if anything, to harden. The writer and teacher Emile Ferrière was just one champion of scientific naturalism whose advocacy became more aggressive. He had already signaled his philosophical radicalism under the empire with an essay on free will that attacked what he saw as the illusion of a human freedom of choice sanctioned by a belief in an immaterial soul and a divine presence.[169] But by 1872 he had moved to the heart of debate on the moral implications of science by going beyond mere support for Darwinian theory; provocatively, he now set two of the most controversial figures of the day— Thomas Huxley and John Tyndall—alongside Darwin and Lyell in a four-man "elite of savants who stand today as the glory of England."[170] Thereafter, no aspect of the Christian tradition escaped Ferrière's sustained reexamination of the foundations of faith. In 1879, he flaunted the power of scientific method in a study of the context and influence of the apostles that laid bare the recurring corruption of the church and the frail foundations of such leading Catholic principles as priestly celibacy and opposition to divorce.[171] And, in two books published in the early 1890s, he rehearsed what he saw as the manifest absurdities of most scriptural references to the natural world in a revisiting of the familiar but still sensitive argument that the Bible was the work of human hands and not of an omniscient, infallible God.[172] It was unrelenting stuff, and no one could have been taken in by Ferrière's air of detached objectivity and his acknowledgement that faith did not stand or fall by the truth or falsehood of the Bible: the controled antireligious thrust of Ferrière's writing was as corrosive as the head-on attacks of any declared free-thinker.

Among religious believers, the evolution of opinion tended to take a more compromising form. As Harry Paul has observed, Catholic intellectuals became conspicuously more open to views on the history of the human race that had once been anathema within the church. One sign of this was the openness with which such views were discussed at the five international congresses of Catholic scientists, philosophers, and theologians that took place between 1888 and 1900.[173] The congresses were anything but marginal events without scientific or social significance, despite the efforts of the left-wing press either to ignore them or to treat them with the disdain judged appropriate for an initiative of pathetic irrelevance. The first congress got off to a bad start; its postponement from 1887 to 1888 was easily interpreted, in *Le Temps, Le Siècle, La République française*, and other liberal quarters, as a mark of failure.[174] But when the congress, a four-day affair in Paris, got off the ground, it attracted 1,605 subscribers, including 586 from outside France (almost half of them from the Austro-Hungarian Empire).[175] And there and in the subsequent gatherings in Paris (a second congress, in 1891), Brussels (1894), Fribourg (1897), and Munich (1900) reflective Catholics exploited a forum in which they could engage in unfettered debate on new departures in science and the social sciences (especially anthropology), as well as philosophy and religion. In this privileged setting, conservative opposition to evolution began to be tempered by a growing readiness at least to discuss the issues. Paul Maisonneuve, professor of zoology at the Catholic faculty of science in Angers, set the tone of compromise at the 1888 congress by pleading for evolution, of which he was unconvinced himself, to be given a hearing. It, as a purely scientific theory, held no threat to the uniqueness of the human race or the concept of a divinely created, even a divinely controlled universe.[176]

While Catholic opposition to evolutionary and other modern doctrines continued to be voiced until well after the turn of the century, criticism became perceptibly more nuanced, as the doctrines came to be distinguished more carefully from the broader specter of materialism that had caused alarm during the empire: at the Paris congress of Catholic scientists in 1888 debate was so open that occasional conversions, such as that of the Belgian abbé Gerard Smets, were owned up to without any sense of betrayal.[177] In advancing this process, the opinions of Albert Gaudry, professor of paleontology at the Muséum d'histoire naturelle and a committed Catholic, carried special weight. As early as 1878, Gaudry declared himself a believer in evolution and wrote convincingly in that vein. He wrote with caution, however, and his advocacy was all the more persuasive for its moderation.[178] In particular, on the causes of evolutionary change, he maintained an agnosticism, now and in his later writings, that helped to calm suspicious minds.[179] While he briefly noted Darwin's ideas on natural selection, he never endorsed them; instead, he presented natural selection as at best one among many ways in which the Creator had advanced the evolutionary process.

Although such moderation was never formally endorsed as Catholic teaching, it gathered support, in France as elsewhere, as part of the church's wider Modernist movement. During the papacy of Leo XIII (1878–1903) the Modernists' exploration of new ideas was condoned, even mildly encouraged. The challenge by the late 1870s was clear, and it could not be avoided. Was the church to retreat to a traditionalism that took no account of modern thought? Or was it to seek a new orthodoxy fitted for a world irreversibly marked by scientific advance? One way forward, commended in the papal encyclical of August 1879, *Aeterni patris*, was Thomism, a philosophy founded on a constructive adaptation of the thirteenth-century teachings of Thomas Aquinas that had won significant support among Catholic philosophers in France since the 1850s.[180] As the abbé Paul de Broglie presented it in two large volumes in 1880, the Thomist approach was capable of embracing the results of experimental science while rejecting any dichotomy between a world accessible to sense experience (the world of the positivists and their fellow travelers who were de Broglie's main target) and one beyond our capacity, at least our present capacity, to observe.[181] When applied to specific theories, the effect of de Broglie's discussion was to stress the uncertainty that was part and parcel of science. Atoms, for example, lay beyond the realm of our senses, and Darwinian evolution was mute on the origins of life. But that did not imply that causation should be of no concern to the scientist. Guided by a rigorous system of philosophy, it was as much a part of science as observation and experiment.[182]

The general principles of Thomism had undeniable force, especially when expounded by someone like de Broglie, who spoke with the authority of both a priest and a former *polytechnicien* and naval engineer. But they provided less than complete guidance to Catholics facing the ever-quickening pace of contemporary science and scholarship. One who found Thomism an inadequate vehicle for engagement with modern thought was Alfred Loisy. Like de Broglie, Loisy responded quickly to the tide of moderate liberalism that began with the accession of Leo XIII. As a young priest and lecturer in holy scripture at the Institut catholique in Paris in the early 1880s, Loisy attended Renan's lectures on Hebrew at the Collège de France; and thereafter, in the course of a clerical and academic career in which he held chairs not only at the Institut catholique from 1890 to 1893 but also at the Collège de France (in the history of religions) from 1909 until his retirement in 1932, he did much to open Catholic minds to modernity.[183] Tragically for Loisy and for the church, however, the favorable conditions that prevailed until Leo XIII's death in July 1903 were abruptly dissipated under the more authoritarian Pius X. By October, the new pope's first encyclical, *E supremi apostolatus*, had warned the clergy against the "insidious maneuvers of a certain new form of science that bedecks itself in the mask of truth . . . a deceitful science that seeks, through false, perfidious arguments, to beat a path to the errors of rationalism and semi-rationalism."[184] Although the encyclical was directed more forcefully at trends in biblical criticism than at science, such a statement created a chill atmosphere for Catholic discussion of modern thought in all its forms. This is not to say that discussion ceased: Loisy's campaign to achieve an integration of modernity with the essence of Catholicism survived as a beacon for liberally inclined believers despite the appearance of several of his books on the *Index librorum prohibitorum* and his excommunication, after years of tension, in 1908. But, in circumstances made more acrimonious in France by the gathering storms of conflict between church and state, such discussion moved to the margins, where "Modernism" became the preserve of Catholics whose beliefs set them outside the main stream of orthodoxy, sometimes (as in Loisy's case) far outside.

In circles less anxious about the religious and social implications of modern thought, opinions changed more quickly, although here too in the face of pockets of tenacious conservatism. Of the various ingredients of the radical synthesis of the 1860s, spontaneous generation remained a discrete presence on the fringes of serious debate after 1870, still advocated by Pouchet's loyal collaborator Georges Pennetier but by few others. Polygenist teachings in anthropology had greater success, under the influence of Paul Broca and the Ecole d'anthropologie, an independent institution for the teaching of physical anthropology that Broca was instrumental in founding in Paris in 1875. But despite the strength that polygenism drew from its association with the improving fortunes of evolution, the success was never total. Monogenist resistance, led by Armand de Quatrefages (to the end of his life in 1892[185]) and the elderly and respected Catholic anthropologist, the marquis de Nadaillac, was resolute, the more so as it could feed on the menacing public image of the chief opponent of monogenism, Gabriel de Mortillet.[186] As an outspoken partisan of the most anticlerical tendencies in republican opinion, Mortillet gave his advocacy of polygenism and an evolutionary history of hominoid ancestry going back to the Tertiary an ideological edge that he was always ready to flaunt. In the process, he came to embody, in conservative Catholic eyes, all that was most hateful about scientific modernity. From 1868, when Duruy appointed him to a curatorial position at the imperial museum of Celtic and Gallo-Roman antiquities at Saint-Germain-en-Laye, until his death in 1898, he never wavered in his vision of the damage that had been done to science "in the name of religion and supposedly whole-some doctrines."[187] Such language, entirely consistent with Mortillet's known hostility to religion, did far more to provoke than to placate.

It is hard to separate discussions of biological evolution from the other, adjacent de-bates in which they had become inextricably involved during the empire. Broadly, how-ever, evolution after 1870 benefited from republican tolerance, even encouragement, of doctrines that were at odds with traditional teachings, many of them seen as vestiges of a clergy-ridden old order. But there was no immediate wholesale conversion. Henri Milne Edwards had been respectful but essentially hostile to Darwin's theory when he treated the *Origin* with disparaging brevity in 1867: although, like Flourens, he accepted a variability capable of producing changes within a given species, he saw no reason to believe that this was sufficient to yield new species.[188] Darwin's book emerged from Milne Edwards's handling as a work that had contributed little to the advancement of science, certainly less than the contributions, from Lamarck to Isidore Geoffroy Saint-Hilaire, originating in France. Thereafter, until his death in 1885, Milne Edwards (like the influential journal he edited, the *Annales des sciences naturelles. Zoologie*) made no significant contribution to the evolutionary debate.[189]

The views of Quatrefages, the best-informed and most open-minded of Darwin's French critics and the one Darwin would have most wished to win over, similarly re-mained unchanged. His position, articulated in a series of articles in the *Revue des deux mondes* in 1868–69 and then reprinted in revised form as a book, was firmly though not dogmatically anti-Darwinian.[190] While (in the characteristic French fashion) he assimi-lated Darwin's contribution to the long tradition of French transformist speculation, he appreciated more clearly than most life scientists in France the originality and impor-tance of the mechanism of natural selection. For Quatrefages, the struggle for survival was an observational fact and a force constantly at work in nature. Darwin, however, had advanced no evidence of its capacity to effect the alteration of species. As in so many French criticisms of Darwin, it was the impossibility of observing the evolutionary pro-cess in action that was seen to condemn his theory. Quatrefages expressed the point in the late 1860s in terms that appeared to set the evolutionary history of living organisms beyond the realm of further inquiry: "That is where we stand as far as living organisms are concerned. We study them in their finished state. We have been unable as yet to enter the workshop from which they emerged. We can therefore say nothing about the processes by which they were fashioned."[191]

More than twenty years later, shortly before his death, Quatrefages reiterated his agnosticism. Although he conceded that Darwinism had gained ground, he insisted yet again that the doctrines lacked empirical evidence. While future inquiry might clarify the issue, there could, for the moment, be only one conclusion:

> The origins of life on earth remain for all of us an impenetrable mystery. We can advance no plausible explanation for the changes that have come about in our flora and fauna. The modifications that today's living forms have undergone may well be sufficient to yield new varieties or races. None has been capable of producing a new species. The species remains an immutable entity, similar to those composing simple chemical substances.[192]

It is important to stress that Quatrefages's position was not that of a die-hard extremist. Also he was not alone: one of his most influential professorial colleagues at the Muséum, Emile Blanchard, was (though almost a decade younger than Quatrefages) equally resolute in his opposition to Darwin and for similar reasons.[193] Against the general skepticism, however, Gaudry's openness to evolutionary thinking has to be accounted as an indication of the broader change of opinion that began to gather strength in the mid-1870s. Two new French translations of the *Origin*, in 1873 (by Jean-Jacques Moulinié) and 1876 (by Edmond Barbier), helped by avoiding the extraneous polemical trappings that Clémence Royer had injected in her edition of 1862; this had the effect of detaching the Darwinian debates from the ideological confrontation of the previous decade.[194] Also there now emerged just a few younger voices ready to speak out unequivocally in favor of Darwin.

Of the new wave of Darwinians, the most influential was Alfred Giard. Appointed as assistant to Camille Dareste in the faculty of science in Lille in 1872, when he was still in his mid-twenties, Giard broke with the insistence of his senior colleagues on the need for empirical proof and declared himself a champion not just of transformist ideas but also, and quite specifically, of Darwin's mechanism of natural selection. In an important address to the meeting of the Association française pour l'avancement des sciences in Lille in 1874, he declared his aspiration to see the faculty in Lille, with its associated laboratory of marine zoology at Wimereux (initially financed from Giard's own resources), become a center for the diffusion of the teachings of Darwin and other foreign leaders in the life sciences.[195] In naming Vogt, Edouard Claparède, Vladimir Kovalevsky, and Ernst Haeckel among those leaders, Giard was displaying his commitment to the new departures in his field and he soon became Darwin's leading champion in France. For his advocacy to have its full effect, however, he needed a more prominent stage within his discipline. A move from Lille to a teaching post at the Ecole normale supérieure in 1887 was decisive in this respect, as was his appointment, in 1888, to lecture on evolution at the Sorbonne, a position financed by the city of Paris that was raised to the rank of a chair four years later.[196]

With the benefit of official approval, good facilities, and a Parisian chair, Giard proceeded to advance the Darwinian cause through the pupils he was now able to fashion in his own image. He also reaped the benefit of the sheer passage of time, which by 1900 had taken its toll on the old guard and dulled the radical ideological dimension of evolutionary doctrines. By the time of his death in 1908, the victory of Darwinism (or, for the more cautious, Lamarckism) was largely assured. In the scientific world, evolutionary thought in one form or another now ranked as the prevailing orthodoxy, while among the religiously minded general public many found that they could reconcile themselves to a Darwinian or Lamarckian account of the physical aspects of the history of life by insisting on the spiritual dimension that lay, for them, at the heart of the distinction between the human race and the brute creation. In this way, the sway of science was acknowledged across a broad spectrum of informed opinion, but with a caveat, for those who needed it, that left room for divine action.

A Faith for the Age: The Religion of Humanity

Although, in their general thrust, the radical doctrines that fired public debate in the 1850s and 1860s raised substantial difficulties for Christian belief, it was rare for even the most aggressive opponent of established religion categorically to deny the existence of a god or of some immaterial entity corresponding more or less distantly to the conventional notion of the soul. The case was rather that such entities were unnecessary for the explanation of the working of the natural world and irrelevant to either morality or human happiness: hence for a church to preach their existence as received doctrine was inadmissible. It was in this polemical spirit that the majority of those who criticized Christian teachings and practices saw institutional religion of any kind as the vestige of a phase of human history that had no place in the modern world they were seeking to build. However, there were also those who, while rejecting Christianity, remained open to some form of organized religious observance. The disparate contributors to this tradition were united by a recognition that the Christian heritage bore with it something of value—moral, psychological, or social—for human life and that, as it faltered, a secular substitute needed to be constructed to take its place.[197]

The new religions that grew from this belief took many forms. An important group has been classed by Donald Charlton as "metaphysical": Victor Cousin, Ernest Renan, the anti-positivist philosopher Etienne Vacherot, and a number of thinkers with pantheistic tendencies, including Hippolyte Taine, all subscribed to positions that fell broadly into this category.[198] Others in Charlton's classification explored occult, Eastern, and neopagan religions or, like Edgar Quinet and Jules Michelet, took the form of ideas based on what he describes as the "cult of history and progress."[199] And others again developed "social religions" of the kind espoused by Charles Fourier, Pierre Leroux, and Saint-Simon.[200] While Saint-Simon's "New Christianity" can indeed be seen as "social" in Charlton's sense (as a religion whose main goal was one of "social utility"), it had much in common with the religions in another of Charlton's categories, those driven by the "cult of science."[201] In this respect the religion that Saint-Simon's dissident disciple Auguste Comte began to elaborate from the 1840s can be seen as "New Christianity's" intellectual heir.

To most contemporaries, Comte's conception of his Religion of Humanity (a term first used by Leroux, in a quite different context, in 1838[202]) appeared as an aberration. It came after two decades that Comte had devoted to fashioning positivism as a cerebral philosophy in which the irrational and the emotional had no place. The change shocked the many positivists who, with Littré and John Stuart Mill, could not accept the sacerdotal turn in Comtean teaching, even though they regarded themselves as being among the truest adepts of the positivist philosophy. Modern scholars are divided in their treatment of the relation between the two broad phases in Comte's intellectual life. Some have rejected the perception of Littré and Mill and argued for a strong element of continuity or at least an unbroken evolution between the earlier and the later Comte.[203] More commonly, however, the author of the *Système de politique positive* (1851–54) and the religious writings has been interpreted as someone very different from the author of the *Cours de philosophie positive*.

Comte's own account would certainly point to a logical connection between the two facets of his philosophy: as he put it, having rendered to humanity the services of an Aristotle, he felt that the time had come for him to pursue the career of a Saint Paul,[204] and this image of an elegantly smooth, almost planned transition has substance. Yet it has at least to be tempered in the light of the fundamental changes in Comte's personal circumstances that left him a very different person just at the time when the transition occurred. The changes took the form of emotional and material crises that came to a head in the early 1840s. The strain of completing the last three volumes of the *Cours*, the lawsuit he brought against Blanchard, the departure of his wife in 1842, and the loss of his post as entrance examiner at Polytechnique two years later left Comte more vulnerable than ever to misanthropy and prone to the corrosive introspection that was to plague him until his death (see chapter 1). Hence he decided on the new departure in his philosophy in a context of mounting stress and the prospect of a future of irremediable financial hardship. There was also the unexpected catalyst of Clotilde de Vaux, a thirty-year-old would-be *femme de lettres* and the wife of an absent husband, whom Comte met in 1844 and with whom he exchanged letters of profound, though chaste, tenderness from the spring of 1845 until her death a year later. As Beatrice had been the personification of philosophy for Dante, so Clotilde was Comte's personification of humanity. To love his "angelic interlocutor" was therefore to love humanity,[205] and it was humanity, of which Clotilde was the purest example, that now became the object of Comte's thinking and the focus of the new religion.

Comte made the first tentative references to the religion in his public lectures in 1847 and then in the "Conclusion générale" of his *Discours sur l'ensemble du positivisme* (1848).[206] By 1849, a full-blown Religion of Humanity had assumed its definitive form in his mind, and by October 1851 he was able to end a five-hour conclusion to his annual course of philosophical lectures on the "general history of humanity" at the Palais royal (or Palais cardinal, as it had briefly been renamed since the end of the July Monarchy) with a ringing assertion of the transition that would now mark the end of the old religious order:

> In the name of the Past and of the Future, the servants of Humanity—theoricians and practicians [sic]—come forward to claim as their due the general direction of this world, in order to construct at length the true Providence, moral, intellectual, and material ; excluding once for all from political supremacy all the different servants of God—Catholic, Protestant, or Deist—as at once belated and a source of trouble.[207]

The statement prepared the way for the full expositions of the religion that followed in the *Catéchisme positiviste* (1852), cast as a series of eleven dialogues between a female aspirant (unmistakably modelled on Clotilde) and a priest (no less obviously Comte himself), and soon afterward in the fourth volume of the *Système de politique positive*.[208]

In the religion, the cult of science and reason was formalized as a scientifically inspired recasting of the outward forms of traditional worship. At the heart of it was a new calendar that began retroactively on 1 January 1789, the year in which the French prepared for and executed their revolution. The positivist year was divided into thirteen months, each of four weeks of seven days and each named after a major figure in the intellectual or material development of humanity toward its future "final regeneration." The sequence of the months reflected humanity's gradual passage through its theological and metaphysical phases on its way to the end point of full-blown positivism. It began with the month of Moses, the month of "the initial theocracy" (*théocratie initiale*) and passed through months marking such phases as "military civilization" (César) and "Catholicism" (Saint Paul), before reaching the last six months of the year, which bore the collective epithet "modern." The culmination of the year came perforce with a month, bearing the name of the physiologist Xavier Bichat, that evoked the conquests of "modern science" (see figure 5). Everything about the calendar, in fact, conveyed the Comtean notion of progress achieved through human endeavor and the gradual triumph of rational inquiry. The succession of secular saints whose names were attached to the days and months flagged a believer's year, as reminders of advancing enlightenment. The days of the third week of Bichat, for example, were named after chemists who had led their discipline toward (though not yet *to*) the positivist state: Bergman, Priestley, Cavendish, Guyton de Morveau, Berthollet, Berzelius, and (for the day of rest, always distinguished as a special day and by a special name) Lavoisier.

Behind the aggressive secularism, however, the lingering thumbprint of Catholicism was unmissable, despite Comte's insistence on the corruption and decline that had diminished the Catholic tradition since the time of Thomas à Kempis, an author he admired as a model of ascetic spirituality. A generous provision of feasts and other special days, including in due course the anniversary of Comte's death, observed on 5 September, underlined the parallels with the Christian year, as did the "universal feast of the dead," a secular imitation of All Souls Day that provided the extra day needed to complete the 365 days of the year. Mariolatry too made its mark in a "universal feast of holy women," a day added every four years to accommodate leap years. As indications of the importance accorded to women, evocations of Clotilde were even more pointed.[209] In the preface to the *Catéchisme positiviste*, Comte explicitly invoked Clotilde's name. It was through her, his "ange méconnu" ("angel who never received her due recognition" in the authorized English translation), that he had experienced "the presiding female influence" and so been prepared for his new existence. More broadly, for humanity as a whole, the gathering revolution in the position of women was yet another mark of human progress. It was preparing women for the final regeneration, in which they would at last play a full role in society and join with the proletariat, whose members had become integrated in the "modern order" through the events of 1789.[210] Once women had assumed their proper place, a series of complementarities would exercise their beneficent influence. "The reason of man" (*raison masculine*) would be subordinated to "the feeling of woman" (*sentiment féminin*), male aggression would be tempered by female affection, and the excesses of proletarian disorder, recently manifested in the anarchy of 1848, would be calmed. This, and only this, would allow the key Comtean trio of love, order, and progress to establish themselves as the foundations of society.

The Catholic past that contributed so much to such beliefs was nowhere more evident than in the nine sacraments that were to be administered at intervals throughout life, most of them identified by the passage of years in multiples of seven.[211] In the sacrament of "presentation," a baby would be dedicated to the service of humanity, and parents and the secular equivalent of godparents would commit themselves to guiding the child accordingly. "Initiation" came at the age of 14, when instruction in accordance with the positivist catechism would begin, followed seven years later by "admission," the sacrament of full entry into the service of the Great Being (*Grand-Etre*), made up of the whole of humanity. After another seven years (at the age of 28), for men only, came "destination," which marked entry on a chosen career. And so on, through the sacraments of "maturity," received only at the age of 42, "retirement" (*retraite*) at 63, marking the end of economically active life, "transformation" (on death), and "incorporation," a sacrament of commemoration celebrated seven years after death. In contrast to these temporally fixed sacraments, that of marriage could be received at any time in the period of seven years (longer, in exceptional circumstances) that were thought necessary for finding a partner; for women, the period began at the age of 21, while for a man the seven years could not begin until he was 28 and set on the course in life to which he would by now have dedicated himself in the sacrament of destination.

Putting the rites of the Religion of Humanity into practice proved more difficult than their exposition within the body of abstract positivist principles. In Comte's lifetime, sacraments appear to have been administered rather infrequently, although in 1853 he did remark favorably on the religious observances that Georges Audiffrent, a doctor, former *polytechnicien*, and close disciple of positivism in its fullest sense, had organized in Pertuis and Aix-en-Provence.[212] Thereafter, the development of the religion was inhibited not only by the critical stance of some of those who opposed the sacerdotal aspects of positivism—Littré and Mill in particular—but also by the legal constraints on the promotion of public religious observances, other than traditional Catholic rites, during the empire.[213] Comte's death on 5 September 1857 dealt a further blow, compounded by the conspicuously lukewarm commitment to the priestly role on the part of his long-standing disciple, chief executor, and successor as the director of positivism, Pierre Laffitte.

The transition to a positivist movement without Comte was not easy.[214] When, in late September 1857, Laffitte acceded to the wishes of those closest to Comte that he should become head of the movement, he seemed initially to offer a safe pair of hands. A mathematician by training and now in his mid-thirties, Laffitte had shown no signs of deviation from Comtean orthodoxy. But, under him, the momentum of Comte's last years was soon lost. While Laffitte never spoke out categorically against the religion and its social underpinnings in the way that Littré did, he undermined it by giving the administration of the sacraments a low priority. We know of a handful of sacraments that he conducted in the late 1850s and 1860s, including those of destination for the prominent agriculturalist, engineer, and patrician of the positivist cause, Auguste Hadery, presentation for the son of the positivist doctor Eugène Robinet, and commemoration for the deceased Dutch positivist Willem de Constant Rebecque.[215] But it seems unlikely that more than a dozen or so sacraments were administered over the decade, the majority of them taking place with a minimum of publicity or ostentation in Comte's former apartment at 10 rue Monsieur le Prince in the Latin Quarter.

Laffitte's lack of serious engagement with the Religion of Humanity served only to hasten the weakening of that aspect of the positivist movement, and during the Second Empire and on into the 1870s, his idiosyncratic conception of positivism won few friends among what in any case had never been more than a limited number of committed adepts. In 1877, there were only ninety-six French contributors to the maintenance of Laffitte and his work, with a further fifty-eight abroad.[216] The sums raised in the late 1870s, just over 10,000 francs a year, were enough to provide an income of 6,000 francs for Laffitte himself. Even if they far exceeded the total contributions of 3,991 francs for 1863 (when financial support had reached its lowest ebb), they left little to spare for Laffitte's various projects. And they left even less for the promotion of the broader, more elevated body of doctrine, including elements of both the philosophy and the religion, to which the more orthodox positivists remained committed. In these circles, admiring eyes were cast across the Channel to London, where Rev. Richard Congreve, a former fellow of Wadham College, Oxford, whom Comte had designated as the head of his British disciples shortly before his death, was pursuing a form of positivism that fully incorporated Comte's religious teachings.[217] From 1870, Congreve's position was strong enough for him to lead a small group of English positivists—among them Frederic Harrison, J. H. Bridges, and Edward Spencer Beesly—in establishing a hall in Chapel Street, London, for holding ceremonies. While the number of English positivists in Congreve's circle was not large, even the modest Chapel Street initiative was one that Comte and his followers in France never managed to emulate.

With respect to places of worship, in fact, little remains on the French side. The principles to be observed in an ideal temple of humanity, though, were clear enough: set out in brief descriptions in the *Catéchisme positiviste* and the *Système de politique positive*[218] and a floor plan that was found in Comte's papers after his death (see figure 6), they conveyed the Religion of Humanity's debt to the outward forms of Christian ritual. For almost half a century after Comte's death, however, no action was taken. And even when a dedicated place of worship was finally established, in the early twentieth century, the lead was taken not by the French but by a group of Brazilian positivists who, in 1903, bought a house that they thought to have been Clotilde de Vaux's in the rue Payenne in the Marais area of Paris. Unfortunately, the purchasers did not consult their Parisian coreligionists or any other positivists on Clotilde's precise address, with the result that in error they bought number 5 rue Payenne rather than number 7. Nevertheless, they fitted out their chapel on the first floor, in accordance with Comte's prescriptions, and inaugurated it in 1905.[219]

It was a mark of the weakness of the religious side of positivism that the chapel was never put to significant use. By the time it opened, crippling disagreements had long divided even those who adhered to positivism in its sacerdotal as well as its philosophical sense. Essentially, there were two groups. One was led by Audiffrent, who had already emerged under the empire as a discrete rallying point for the purists, those he described as "the few disciples who have stayed faithful to the true traditions of the master."[220] The other group remained loyal to Laffitte, despite a record of eccentric administration that made him the target of accumulated criticism on a number of fronts. As early as December 1855, almost two years before his death, even Comte himself had qualified his wish that Laffitte should in due course preside over the thirteen executors of his will with a criticism of Laffitte's "deficiency in energy of character" inappropriate in someone he expected to succeed him as head of the Religion of Humanity.[221] That rather nebulous reservation soon proved to have substance, and by the later 1870s the accusations against Laffitte had hardened. They included both persistent negligence in the day-to-day tasks of the movement and the fundamental impropriety of his assuming the formal role of the High Priest of Humanity when, as his critics argued, he had merely been designated as the interim director of the positivist movement following Comte's death.

For Audiffrent, Laffitte's greatest failing lay unequivocally in his limited vision of the sacerdotal function, which led him to a preoccupation with public lecturing, often on science and mathematics, and a neglect not just of the religion but even of positivism as a philosophy. A typical course, of sixteen lectures delivered at 10 rue Monsieur le Prince, was the one that Laffitte offered on differential geometry in January 1879; the advertised program suggests that the lectures had no bearing on positivism in its sacerdotal or any other form, except for a brief and presumably perfunctory "Conclusion religieuse" at the end of the sixteenth lecture.[222] Such a course, of which Laffitte intended to deliver two in each year, can only have enraged someone with Audiffrent's priorities. Inevitably, between 1876 and 1879 division degenerated into a schism culminating in the expulsion of Audiffrent and his fellow doctor Eugène Sémérie from the Société positiviste that Comte had founded in 1848.[223] Thereafter, Laffitte remained in possession of Comte's apartment at 10 rue Monsieur le Prince, where he launched and directed a new and rather successful journal, the *Revue occidentale philosophique sociale et politique*, while Audiffrent's circle established itself, with Congreve's help, at 30 rue Jacob, nearby in the Latin Quarter. The Audiffrent circle's tactic of fostering its alliance with Congreve and the more orthodox British disciples only aggravated the divergence between the two tendencies within the French community, and schism soon widened into an unbridgeable chasm.

The divisions among the leaders of French positivism left the Religion of Humanity gravely weakened. Laffitte's already limited commitment to the religion waned still further, as he continued to give priority to his pedagogical interests, chiefly through the Société positiviste d'enseignement populaire supérieur, which he founded as a vehicle for popular lecturing in 1876. In the late 1870s, other distractions came with the political consolidation of the Third Republic. Laffitte now seized the opportunity of moving close to the aggressively secular republican seats of power, though at the cost of further hostility from Audiffrent and his allies, who interpreted Laffitte's political manoeuvring as self-interested opportunism incompatible with Comte's insistence on the need for positivism to retain its independence. By Laffitte's own lights, however, the strategy reaped rewards. Through his friendship with Jules Ferry, positivism acquired something of the character of an approved philosophy of the early republic, given physical expression in the elaborate statue to Comte that was erected in front of the Sorbonne in 1902. At a personal level, it was approval by some key figures of the republican intellectual left that eased Laffitte's appointment to a new chair of the "general history of science" at the

Collège de France in 1892, sixty years after Comte had first proposed the creation of such a post in a letter to Guizot, then Minister of Public Instruction.[224] Once in the chair, Laffitte was able to promote his vision of the positivist cause, until illness overtook him in 1900. He did so by presenting a progressivist interpretation of the history of science, focussed mainly on mathematics but with excursions into theories of society, that was well calculated to please his secular political patrons. It was an interpretation, however, that had no bearing on the religion of which he was at least nominally the head, and the divisions within positivism went unhealed.

Even Laffitte's death in 1903 changed little. His chair at the Sorbonne went to Grégoire Wyrouboff, a positivist of the Littré school with no evident sympathy for the Religion of Humanity. At the Comtean shrine in the rue Monsieur le Prince, Laffitte was succeeded by Charles Jeannolle, whose difficult personality, administrative shortcomings, and indifference to the religion made bridge building between the positivist factions impossible. With Audiffrent now over eighty, it was left for the journalist and republican politician Emile Corra to maintain some of the traditional ceremonies that marked the feast days of the positivist year. Under him, the Feast of Humanity (1 January) and the Universal Feast of the Dead (31 December) were both celebrated, the latter attracting a public from beyond that of committed positivists.[225] So too were the feasts for the birth and death of Comte and those that marked the great phases in the history of humanity: the Feast of Theocracy, the Feast of Ancient Science, the Feast of Catholicism, the Feast of Modern Industry, and so on, each feast associated with a particular month of the positivist year. But all this was done from the relative obscurity of Corra's home in Neuilly-sur-Seine, in an atmosphere of intense formality and seriousness though with no evidence of the administration of any of the Comtean sacraments. And it was done against the background of irreconcilable animosity between Jeannolle (who clung tenaciously to his apartment, next door to Comte's until his death in 1914) and Corra, with both men simultaneously assuming the all-important tile of "Director of Positivism."

We shall never know what might have come of a last flickering of Corra's more religiously oriented conception of positivism on the eve of the war. In May 1914, the efforts of over a decade of relentless proselytizing were rewarded when Corra was able to preside at the inauguration of premises, at 54 rue de Seine, in the Latin Quarter close to the rue Monsieur le Prince. At last, in Corra's words, positivism had "a civic temple and home of its own," a setting (though hardly a "temple") fit for the religious observances that would allow positivism to realize its goal of ministering to the heart and human character as well as to the mind.[226] But within weeks the war had extinguished Corra's fleeting optimism. The already modest income that he garnered from the subscriptions of the faithful dropped by half, and a reduced level of activity never picked up. After 1918, postwar inflation provoked a further downward turn, before Corra's withdrawal from his leadership role in 1925, at the age of 79, inaugurated a long phase of terminal

decline that ended, for the tiny French community, with an acknowledgment of the religion's demise at the end of the Second World War.[227]

Despite the protracted death throes to come, all but a dwindling band of obdurately
loyal sympathizers knew by 1914 that the most elaborate and formalized of all the secular religions of the nineteenth century had failed. As debate on the relevance, or even
the possibility, of religious belief in the modern world had gathered intensity, as it had
done in France from the 1880s, the wider confrontation had become essentially one
between the champions of one form or another of the Christian tradition and those
who rejected religion of any kind. In the process, the middle ground that the Religion
of Humanity had once promised to occupy as a system of belief appropriate to the age
of science was left thinly populated. The niceties of the *Catéchisme positiviste* had come
to appear irrelevant amid the momentous ideological debates between clerical and secular worldviews that racked France before 1914. And both before 1914 and after the peace
of 1918 even the purely social ideals that leading adepts of the religion had promoted,
such as international peace and the welfare of the proletariat, had become the province
of other more powerful movements with which the impoverished dwindling band of
positivists could not compete.

By 1914, the mounting polarization between secular and religious worldviews, especially
over the previous decade, had done more than condemn the niceties of the Religion of
Humanity to irrelevance. It had also had the effect of immersing science in a broader
confrontation with stakes that far surpassed the fate of any individual scientific theory
or way of doing science. Catholic protests following the legislation on the separation of
church and state in 1905 were engendered by the momentous loss of status and material
possessions that went with the passage of churches and church lands into the hands of
the state or local authorities. More profoundly still, they fed on concerns about the
course to be followed by a nation at the crossroads of history. As industrialization and
urbanization accelerated and war with Germany moved inexorably closer, was France to
continue on the path of secularism and modernization that it had taken since 1870 in an
attempt to compete internationally, not least with the economic and military power of
its neighbor across the Rhine? Or was it to reaffirm its national traditions by building
on the signs of a late resurgence of Catholic values and a faltering of positivism, as Henri
Massis and Alfred de Tarde urged, from the political right, in their address to the "young
people of today" in 1913?[228] Such questions transcended differences with regard to the
relations between science and religion. But they never eradicated them. Militant spokesmen for the ideals of free thought or liberal values, many of them in the republic's resolutely secular system of state education, still contrasted science as the touchstone of
modernity and progress with the obfuscating prejudices of faith, just as Charles Robin,
Clémence Royer, and other free thinkers had done in championing materialism during
the Second Empire. Conservative elements within the Catholic Church, for their part,

lent substance to that Manichean view by a continuing suspicion of modern thought, displayed in the Senate debates of 1868 and promoted with renewed vigor by Pope Pius X after his accession in 1903. Between these rhetorical extremes, attempts to chart a middle course that would allow science and faith to live in harmony bore some fruit, for example in Albert Gaudry's decision to adopt Darwinian evolution while keeping an open mind on natural selection. But attempts at accommodation were easily lost from view in the heat of more eye-catching debate. To insist again on the point that I made at the beginning of this chapter, where science and religion came into conflict, the conflict had little to do with intrinsic incompatibility. What made science contentious was not science itself but the way in which it was deployed. That point emerges strongly from the exchanges about materialism, spontaneous generation, and other new scientific ideas that inflamed public debate during the Second Empire. And it will emerge no less strongly from the further conflicts to which I shall return in my discussion of republican ideology and its critics in chapter 6.

CHAPTER FIVE

Science for All

• • •

During the nineteenth century, the creation of forms of science for consumption by the general public became a matter of escalating concern, sophistication, and, for some, economic profitability. Initially, the dominant style was deferential, encapsulated in a tradition of didactic poetry with roots going back to the eighteenth century and in a periodical press that began, under the Bourbon Restoration, to offer a respectful, rather cloudy window on the activities of the great savants. From the midcentury, however, the tone of scientific books and articles for the reading public began to be transformed. This transformation was the work of a new generation of specialist writers who saw the task of communicating science as something more than a straightforward process of transmission from a remote world of arcane knowledge to a passive untutored consumer. The new style of communication was an activity that gave scope for adventurous, engaged forms of exposition and interaction. Especially in the freer atmosphere that prevailed from the liberal phase of the Second Empire of the 1860s, popular writers and lecturers purveyed science and other forms of learned culture to audiences that were not only interested and well informed but often responsive and critical as well. And they prospered, both financially and as major figures in the nation's cultural life. The result was a golden age of popularization that lasted into the early years of the new century.

In this golden age, when some of the greatest popularizers in the history of scientific communication were at the height of their powers, "science for all" assumed a profile that distinguished it quite clearly from the science of the Académie des sciences and the other institutions of research and higher education. The abbé François Moigno, Victor Meunier, Louis Figuier, and Camille Flammarion, all of whom came to prominence during the Second Empire, were masters of the written or spoken word (or both) whose work embraced the science of the learned journals and national educational curricula but also transcended it. Their science was perforce simpler than "official" science, both conceptually and, above all, mathematically. Yet it managed to offer a vision of the cutting edge of discovery and controversy that condoned and even invited participation by educated nonscientists. Participation might take the form of a critical stance on the views of the leaders of science or an engagement in collecting or simple observational astronomy pursued in the field or at home. As a result, by the beginning of the twentieth century, popular science was not just a commodity to be consumed through the printed word, although many still regarded it in this light. For the more adventurous, it had also become something to be thought about critically and in certain specialities practiced.

Fashioning the Audience

Throughout the eighteenth and nineteenth centuries, science made an indelible mark on French literate culture. Although it would be extravagant to claim that it provided a primary source of inspiration, it was enduringly popular as a literary resource. At the turn of the nineteenth century, the fruits were predominantly didactic, as they had been from the time of Fontenelle's *Discours sur la pluralité des mondes* (1683) and still were in Buffon's much-read works on natural history. Although the dominant form was prose, a popular alternative vehicle by the end of the eighteenth century was the classically structured philosophical or historical poem. Work in this form was seen in the late Enlightenment as elegantly adapted to the purveying of scientific principles and the wonders of nature, as examples published in English as well as French demonstrated: Erasmus Darwin's *Botanic Garden* (1790) was part of the same tradition to which the most distinguished of France's "scientific" poets also belonged.[1]

In the French style, the tone of such poetry was set in the 1790s by *La Sphère*, by Dominique Ricard, a devotee of astronomy and former honorary canon of the cathedral of Auxerre, best known for his translation of Plutarch.[2] The work was a long historical exposition of astronomy and navigation that drew mainly on ancient sources but also on the writings of contemporaries, including Jean-Sylvain Bailly and Jérôme de Lalande. Although Ricard's erudition was apparent enough in the verse and notes, he underlined the weight of his scholarship in accompanying 160-page study, in prose, of poetry on astronomy in Greek, Latin, and French.[3] In seriousness as in girth, however, *La Sphère* and all other rivals were soon eclipsed by the *Trois règnes de la nature* by Jacques Delille, a poet and Latinist whose reputation and immersion in the classical tradition were enshrined in descriptions of him as "the Virgil of France."[4] Published in 1808, when Delille was 70, the *Trois règnes de la nature* encapsulated in every respect the fashionable style of didactic poetry inherited from the age of the *philosophes*. It was vast (with nearly six thousand lines of meticulous alexandrines) and dazzling in its intellectual scope. It treated not only the mineral, vegetable, and animal kingdoms but also, in great detail, the elements of earth, air, fire, and water. And, like most works in the genre, it was ostentatiously learned. The notes appended to each of the eight cantos occupied more space than the verse, and they were given deliberate weight not only by the evidence of Delille's vast reading but also by the names of the savants—Antoine Libes, Louis Lefèvre-Gineau, and Georges Cuvier—who had been engaged to advise on them and even, in most cases, to write them. For Delille, the learning of these authorities offered the surest bulwark against the speculative system building that had bred the horrors of Jacobin fervor and led to his own appearance before a tribunal of the people and subsequent enforced residence abroad from 1794 to 1802: it was this same taste for system building that, in another republic, had beguiled Lucretius, a poet despised by Delille as an author who had wasted time and talent on "his absurd universe founded on the fortuitous concourse of atoms."[5]

Even by the time Delille wrote, however, tastes were beginning to change. To readers seduced by the emerging Romantic ideals of spontaneity and sensibility, his verse appeared hidebound, as dated as the prescriptions of Boileau's seventeenth-century *Art poétique* that governed its lapidary form. But literary fashion did not change overnight, and lengthy expository poems in the classical mold retained a significant, if diminishing, following long after their late postrevolutionary heyday during the Directory and Consulate. In 1825, the ceremonial annual meeting of the Institut de France could still be moved to rapturous applause by the "discourse in verse" on "Les facultés de l'homme" by the comte Daru, a military and political figure, man of letters, and (like Delille) eminent Latinist and victim of the Terror.[6] Laplace, who was present, was enchanted. Indeed, it was his praise of "Les facultés de l'homme" and his wish to see poetry advance the task of popularizing science that encouraged Daru to devote the next four years to writing *L'Astronomie*, a work that appeared posthumously in 1830.[7]

The six substantial cantos of *L'Astronomie* were essentially a history of astronomical myth, theories, and observation since the earliest times. It was a dry subject, augmented by the usual panoply of scholarly notes, that Daru himself recognized would probably hold little attraction for the public.[8] Despite the warm response that a reading of most of a preliminary version of the second canto elicited from his peers in the Institute in 1827,[9] Daru's reservations about the wider reception of his work proved well founded. His poem belonged firmly to the old classical tradition in which he had been raised by the Oratorians of Tournon before the Revolution, and by the 1820s younger generations of readers judged it accordingly. Apart from the dated accolade of translation into Latin by Clair-Louis Rohard, a former teacher at the military school of La Flèche in 1839,[10] its publication caused no noticeable stir.

By the time *L'Astronomie* appeared, in 1830, even the man whom Daru recognized as his master,[11] the now long-deceased Delille, was entering the protracted twilight of his popularity. The reputation that Delille had still enjoyed during the Restoration had echoes in a ten-volume *Oeuvres de Delille* in 1832–33, but thereafter editions of his works became conspicuously less frequent. By 1850 neither the *Trois règnes de la nature* nor his once immensely popular poetic exposé on landscape design, *Les Jardins, ou l'art d'embellir les paysages* (first published in 1782 and then frequently reprinted and translated until the mid-1830s), attracted publishers or a significant readership.[12] As the taste for Delille and Daru withered, readers with scientific interests looked elsewhere for information. They were now more likely to turn to journalistic writing with no literary pretensions.

The new genre took the form of reports on public lectures and on communications given in the Académie des sciences and other scholarly institutions. Almost from its foundation in 1819 the monthly *Revue encyclopédique* led the way with a regular account of the doings of the Académie that continued to be published until the *Revue*'s declining years in the early 1830s. The Saint-Simonian periodical *Le Globe* followed, tentatively in its early issues from 1824 and then more systematically from 1827, when it declared a shift in its coverage to the more serious subjects that it felt were increasingly engaging public attention. As an unsigned editorial put it in 1827, "it is no longer enough to preach the cause in favor of serious studies: we must not be afraid to broach them ourselves and to have the public broach them as well."[13] One consequence was a regular science column, with reports on the Académie's proceedings and its elections very much to the fore. Thereafter, through the 1830s and 1840s, similar columns became staple offerings of newspapers and literary and political journals.

While several practicing scientists tried their hand as contributors, the work tended to be left to specialist correspondents. Such men normally had a background in science, but they saw themselves first and foremost as journalists and writers. The impact of the new style of journalism on public interest in science might have been slight had French periodical literature not undergone the same expansion from the 1820s onward that occurred in all industrialized countries. In France as elsewhere, the continuous production of paper by machinery—a technique developed in the first quarter of the century by Louis-Nicolas Robert and the Foudrinier brothers—and the spread of steam-powered cylinder printing led to lower prices, which made publications of all kinds accessible to ever more modest levels of society.

The growth in the market for books and periodicals gathered strength during the July Monarchy, when the spread of primary education began to have a significant effect in expanding the reading public. It was precisely to this new literate market that, starting in 1833, *Le Magasin pittoresque* addressed its eight-page weekly *mélange* of articles on travel, history, art, and snippets of science and natural history, adorned with a modest sprinkling of illustrations. Modeled on the *Penny Magazine* that Charles Knight had launched in Britain in the previous year, the weekly offered a diet that was austere by the standards of the later nineteenth century. But, at 10 centimes a copy (7.50 francs for a handsomely bound edition of a year's issues), it was affordable by all but the poorest working-class families and pitched at a level that appealed to them. In the words of the editorial of the first issue, the *Magasin pittoresque* was to be open to "all inquiring minds and all pockets."[14] In this, its ideals corresponded precisely to those of its founder and leading light, Edouard Charton, whose republican leanings and Saint-Simonian past on the editorial staff of *Le Globe* still fired his commitment to popular education, despite his disenchantment with the turn the Saint-Simonian movement had been taking.[15] Like the very similar *Musée des familles. Lectures du soir*, which was launched in the same year and at the same price, the *Magasin pittoresque* was conceived primarily as family reading; it offered, as it claimed, instruction of quality that would "speak to the heart, imagination, and discernment" of its readers and thereby achieve its overriding aim of enriching the leisure hours of families, rich and poor alike.[16]

Periodicals with more serious intellectual intent than the *Magasin pittoresque* necessarily found fewer readers. But they too fed on the spread of literacy. Two that did so successfully were *L'Institut* and *L'Echo du monde savant*, both launched as weeklies in the early 1830s and both devoted to reports on the world of learned culture, with a clear though not exclusive emphasis on the sciences. The typeface was small and smudgy, and the reporting and editorial material were typical of the period in their dry, respectful style. The content of both *L'Institut* and *L'Echo du monde savant*, in fact, was similar to that of the science columns of newspapers. Occasional supplements or expanded issues that broke with the standard four-page format allowed for some cautious discursiveness and the introduction of obituaries, more detailed information about the work of individual scientists, and extracts from notable lectures. But the core offerings remained news of the Académie and other *sociétés savantes*, and brief reports on their prizes and meetings.

By the standards of later scientific writing, it was tame stuff, conspicuously lacking the spice of controversy. The unsigned prospectus of *L'Institut* in 1833, probably written by the founder, owner, and editor, Eugène Arnoult, underlined the position of detached objectivity that was preserved, as a matter of policy, until *L'Institut* died in 1876. "*L'Institut*," the prospectus stated, "never takes it upon itself to pass judgment on the memoirs or reports read in the meetings of the academies whose work it records. It limits itself to setting down facts, leaving the task of evaluating them to its readers."[17] The *Echo du monde savant* at least began more combatively. According to a punchy editorial in an early issue of 1834, it would do more than merely convey information and do so throughout all the classes of society: "From a position of openness and independence, we shall identify wrongs, support well founded claims, and open our columns to any justified complaints."[18] But by the time of the newspaper's demise in 1846, any pretence of offering even the most benign criticism had disappeared, and the traditional relations of superiority and deference between academic science and the lay public had reasserted themselves and triumphed over the principle of the public accountability of the nation's elite. In this instance at least, the liberalism of the July Monarchy, with its cautious readiness to expose political and intellectual power holders to more rigorous scrutiny, had left little mark.

The persistence of the older style of respectful reporting, until it finally yielded before the more engaged practices of the 1860s, owed much to the opposition of many in the scientific community to any external commentary on their doings. The resistance tended to come from an older generation whose careers had begun in the early years of the century. Cuvier issued his most explicit statement in the prospectus for his great *Dictionnaire des sciences naturelles*: his warning of the dangers that attended the growing popularity of natural history conveyed his fear of a loss of authority when the sober work of dedicated and informed naturalists became the foundation for speculations by outsiders whose partiality threatened the order of society as well as the world of learning.[19] Such anxieties were extravagant. Nevertheless, they demonstrated the cultural chasm that existed between Cuvier the elitist and Bory de Saint-Vincent, a naturalist with a more populist instinct whose aim was precisely to bring natural history within the reach of a nonspecialist public. Bory's response was to dismiss Cuvier's concerns as a "bizarre miscellany"[20] and to publish his own *Dictionnaire classique d'histoire naturelle*, a work whose style and moderate girth contrasted pointedly with Cuvier's sixty-two-volume *Dictionnaire*.[21]

Jean-Baptiste Biot was a slightly younger member of Cuvier's generation for whom even the decision to publish the dry and scientifically uncompromising *Comptes rendus hebdomadaires des séances de l'Académie des sciences,* launched in 1835, brought academicians too much into the public arena.[22] Still more menacing, for Biot, was a series of other decisions, begun in the 1820s, that had resulted not only in the admission of journalists to the Académie's weekly meetings but also in their being given access, after the meetings, to the papers that had been read. A special room was even placed at their disposal, and there for several hours, in what came to be known as a "sixth class" of the Institute, they were free to read and copy the texts in readiness for their reports.[23] Such openness was bad enough. But Biot saw it as just one facet of a broader laxity that had opened sessions of the Académie des sciences to the general public and so made spectacles out of debates once conducted in an enclave of privacy and only accessible to members. In an acerbic reflection in 1842, he described the tone of frivolity that had invaded the proceedings. "Nowadays," he wrote, "people come early to reserve their seats, as if they are going to the theater. Space alone limits the number of people who are admitted. And to keep them quiet, academicians have to interest them. The public does not care for technical details."[24] To cater for the new class of spectator, every subject had now to be prefaced with an elementary introduction, and eloquence became at least as important as content. Sixteen years later, Biot returned to the fray once more, describing the Académie as "a kind of open clearing-house for announcements, accessible indiscriminately to anyone and attended by a captive audience."[25] Things, if anything, had gotten worse.

The Italian-born mathematician Guglielmo Libri was another of the old guard to which Biot and Cuvier belonged. He was just as virulent and even more explicit in his analysis of the vulgarity that he saw as the inevitable curse of exposure. He laid the blame squarely at the door of one man: François Arago, who had succeeded Fourier as the permanent secretary for the mathematical sciences in 1830.[26] In an unsigned article in the *Revue des deux mondes* in 1840, he portrayed Arago as sacrificing the rigor and severity of "high science" in favor of a misguided liberalism that prized exposure to lay comment above the cultivation of expertise.[27] Gradually, as Libri had it, an openness that had initially had some justification as a means of resisting undue governmental influence on the Académie des sciences during the Restoration had become assimilated in a campaign of personal aggrandisement. Arago's overriding concern was now the impression he was making in the public gallery and in the press.[28] Self-indulgently long *éloges* of deceased colleagues, in particular, were pretexts for theatrical displays that skated thinly over technical content. The *éloge* of Ampère that Arago had delivered in the previous year was the most recent case in point, one that had provoked widespread journalistic condemnation.[29] Occupying over a hundred pages in its published form, it had taken up almost three hours of the academy's time, while devoting no more than what Libri saw as a few "unintelligible" words to the subject's work.

Libri judged Arago's behavior to be no less reprehensible on other fronts as well. In the lectures that he had given for the last decade in his capacity as director of the Observatoire de Paris, Arago had gone out of his way to attract a public even more superficial than that which now frequented the Académie; and he had succeeded, shamefully in Libri's view.[30] The lectures had become society events of such allure and easy accessibility that they sustained a large undiminished following until Arago finally ceased lecturing in 1846. Even the monarchist duchesse de Duras overcame her dislike of Arago's liberal politics to hear what the historian and literary scholar Abel Villemain described as a dazzling oratory applied to the presentation of astronomy.[31] Arago plainly was glib and set on using his performances to turn himself into a personality. But the affront to those who shared the conservative disposition of Biot and Libri (and politics as well as temperament colored both their judgments) was not just that Arago was debasing science by drawing on whatever literary device or quip seemed likely to amuse; it was also that his courting of public approval led to a dwindling to almost nothing of his own contribution to science. Arago's public, in fact, was fashioning his life as a savant.

While Arago was a favored target among those who disliked his politics or his showmanship, even his critics recognized that his search for the immediate gratification of applause and adulation reflected a far broader shift that had propelled well-presented popular science to the status of a salable commodity. One academic physicist who followed in Arago's footsteps as a popularizer, Jacques Babinet, was sensitive both to the opportunities and to the snare of triviality. Like Arago, Babinet built his authority on a serious engagement in science at the highest level.[32] Early in the Restoration, after graduating from Polytechnique and a brief military career, he established himself in Paris, where he taught in secondary schools (for most of his career at the Collège Saint-Louis) and pursued research, mainly in optics, for which he was recognized by election to the Académie des sciences in 1840. By the end of the July Monarchy, however, his work as a journalist (with a regular science column in the *Journal des débats* and frequent contributions to the *Revue des deux mondes*) and as a lecturer (notably for working men, under the voluntarist auspices of the Association polytechnique) was beginning to divert him from active science. In 1858, conscious of the effect that his change of focus had had, he came close to apologizing for the lack of seriousness in his writing. "Perhaps," he wrote, "I have been wrong to turn my back on the majesty of science"; the point had apparently been made to him by one of his correspondents.[33] His response, though, was unequivocal: "I had to ensure that I would be read, and in that I succeeded." It was a telling comment by a reflective savant torn between two worlds. While Babinet acknowledged that the limitations of his audiences had obliged him to jettison abstruse mathematics, he maintained that he had never compromised on the rigor of the facts and observations he presented.

One thing that Babinet perceived with overriding clarity was that the days of a polite, uncritical public ready to consume a diet of dry, predigested knowledge delivered by remote sages in the higher echelons of science were coming to an end. A capacity to stimulate and excite were now essential for success, and Babinet was certainly a good performer. Another writer who rode the transition with success was Victor Meunier. From his early days as a contributor to the *Echo du monde savant* in the 1830s, Meunier moved on to the new generation of moderately priced progressive newspapers that began to capture readers during the July Monarchy. His science column in the most successful of all dailies of the period, *La Presse*, for which he wrote regularly until 1855 (when he left to found his popular journal *L'Ami des sciences*), and his occasional contributions to newspapers with a similar liberal stamp, *Le Siècle* and *L'Opinion nationale*, conveyed the gathering tone of engagement that still characterized his work in the 1880s, when he was in his seventies. Driven by an unswerving political commitment and by what he saw as unfair impediments to his intended career as an experimental zoologist, Meunier emerged from the relative anonymity of a jobbing columnist to become a public figure with uncompromisingly leftist credentials during the Second Empire.

Meunier's stock-in-trade was searchingly critical commentary on the professional community of *la science officielle* and hence, by implication, on the imperial regime itself. In a series of five substantial volumes entitled *La Science et les savants*, between 1865 and 1868, he laced well-informed accounts of the previous year's science in France with acid judgments on the way in which influential figures in the Académie des sciences and other seats of scientific power succeeded in suppressing the unwelcome contributions of outsiders. In keeping with his immersion in the socialist wing of the political opposition, he portrayed the Académie's vindication of Pasteur over Pouchet and his allies on the question of spontaneous generation in 1864 as all that was most stifling about the tyrannical rule of the elite of the empire's savants. Pouchet, for his part, was cast as the heroic victim of the nation's "scientific oligarchy,"[34] while no show of authority or supposed peccadillo was now too small for Pasteur to escape Meunier's vituperative treatment. In the same year, the whiff of impropriety that surrounded Pasteur's support for the candidacy of a former pupil of his, Philippe Van Tieghem, for a teaching position in botany at the Ecole normale supérieure was grist to Meunier's relentlessly hostile mill, and he weighed in with characteristic vigor.[35] His accusation that Pasteur had abused his position as head of the school's scientific section, not least by encouraging the students to petition the Minister in Van Tieghem's favor, probably had substance, as did Georges Pouchet's more general but related charge of persistent collusion within the Ministry of Public Instruction that made chairs in effect the "property" of professors.[36] Van Tieghem's doctoral thesis, on a chemical subject related to Pasteur's interest in fermentation, certainly made him a less than obvious candidate for the post, although he did go on to write a second doctoral thesis more closely related to botany and make a distinguished career in the discipline.[37] Whatever wrongdoing Pasteur may have perpetrated, however, the fact remains that politically motivated passion had much to do with the intensity of Meunier's criticism in this as in most of what he wrote.

Another successful exponent of the combative, personal style of the midcentury was the abbé François Moigno. Despite an admiration for the republican Arago's "incomparable" capacity for captivating, lucid exposition, Moigno shared none of his or Meunier's political radicalism. Following an early formation as a Jesuit and his entry into the order in 1822, his vocation as a scientific writer and lecturer grew initially from an association with the ultra-Catholic Augustin Cauchy that took him from the status of a mathematical pupil to that of a friend. It was only after two decades as a priest, however, that he began to devote himself wholly to science, following an obscure conflict with the ecclesiastical authorities, probably occasioned by an unsuccessful industrial speculation and concern within the order about his increasing independence. The result was a thirty-year separation from the Society of Jesus beginning in 1843. In these years Moigno steadily expanded his repertoire. Beginning as an occasional contributor to the Catholic newspapers *L'Univers* and *L'Union* and the author of mathematical textbooks marked by Cauchy's style,[38] he established himself during the 1850s and 1860s as a popular writer committed to combating the ungodly scientism that many at the time saw as the natural corollary of a savant's worldview.

Moigno's career, like that of Meunier, illustrates the importance of the expanding daily press and other cheap publications in the midcentury. His breakthrough into the world of popular science came as the author of regular science columns in *La Presse* and *L'Epoque* (two newspapers closely associated with the press pioneer Emile de Girardin) and *Le Pays*.[39] This experience, allied to the extensive travels to the scientific capitals of Europe on which he reported in his column in *L'Epoque*, provided the secure base from which he could broaden his literary repertoire. While continuing to write for newspapers, he became a translator (notably of all of John Tyndall's main writings and in 1856 of William Grove's *On the Correlation of Physical Forces*), and the author of a rather tough treatise on telegraphy and popular expository works (mainly on optics).[40] Most important, however, he emerged as the guiding spirit of the first in a new generation of ambitious scientific periodicals fashioned for the educated *grand public*.

The weekly *Cosmos*, which Moigno edited from its foundation in 1852 until 1863, before abandoning the journal to found and edit *Les Mondes* (renamed, from 1874,

ΚΟΣΜΟΣ-Les Mondes[41]), was his most remarkable literary achievement. A leading objective of *Cosmos* grew from Moigno's nostalgia for a lost golden age of public debate in science that he situated in the decade immediately after the First Empire. Then, in his words, the *Annales de chimie et de physique* had been the vehicle for criticism that had been "informed, lively, impetuous, even on occasions passionate."[42] It is hard, in retrospect, to see the unfailingly august *Annales*, with its tradition of publishing carefully worked articles at the recondite cutting edge of knowledge, in this light. But Moigno's perception and the point he wished to draw from it were unambiguous. As he believed, even if the old spirit of criticism had become, on occasions, overexuberant, it had generated an excitement conspicuously absent from the drab forms of communication and bland mutual approval that had characterized the journal since the later 1820s. The result, for him, was scientific writing that had degenerated into a "ponderous mass of memoirs printed one after the other, lacking order, lacking intelligence, lacking life, in which error, provided it emanates from a friendly pen, finds its place on the same footing as truth."[43] Moigno, by contrast, sought to inject passion. His *Cosmos* would therefore be a vehicle not just of information but also, and more important, of criticism. He made the point in the first issue, in 1852: "We shall be no chill, stiff echo of information. We shall criticize, since we are profoundly convinced of the necessity, the legitimacy, and the timeliness of criticism in science."[44]

Moigno remained loyal not only to his word but also to the title of his journal, which he took, with due respect and acknowledgement, from Alexander von Humboldt. In the best Humboldtian tradition, *Cosmos* and its successors covered all the sciences. Lively and often difficult reflections on developments in physics comfortably held their own with more immediately accessible material on natural history, anthropology, and archaeology. Behind the diversity, however, there lay a guiding philosophy rooted in Moigno's relentless quest to explore the spiritual dimension in science. It was entirely in keeping with this philosophy that he welcomed as a contributor to *Cosmos* Gustave-Adolphe Hirn, the member of an important Alsatian family in the cotton industry. Hirn commended himself to Moigno by the distance, physical and intellectual, between him and the leading savants of the capital and by the quality of his experimental and theoretical contributions to early thermodynamics. But the greater bond lay in a shared Christian commitment, in Hirn's case deriving from the Protestant traditions in which he was raised and to which he remained loyal throughout his life.[45]

In contrast with Meunier's writing, which bore the stamp of a deep hatred of the empire, Moigno's objective was moral and, though conservative in tendency, only secondarily political. His overriding goal was an alliance in which a science unobjectionable to Catholic teaching would sit easily with an understanding of the material world as rigorous as any demanded by the most hardened rationalist. In the words of *La Foi et la science*, which he published in his own series of Actualités scientifiques in 1875, "Faith has nothing to fear from true science, mature science, science that has reached a state of absolute certainty."[46] Moigno's case for the compatibility of science with Christianity was uncompromising. It faced up not only to the rising tide of republican scientism but also and more specifically to the antireligious cast of John William Draper's *History of the Conflict between Religion and Science*, just translated into French in the year of its original publication in English, and the "false science" of contemporary free-thinkers.[47] His main targets were clear, and he tackled them fiercely in extensive footnotes and other commentary accompanying translations of Tyndall's address to the British Association for the Advancement of Science in Belfast in 1874, addresses by Du Bois-Reymond and Richard Owen, and shorter contributions by T. H. Huxley, Joseph Hooker, and John Lubbock.

An even fiercer onslaught on the misappropriation of science by the enemies of religion was to follow. It began in 1877 with the publication of the first of the five large volumes of Moigno's providentialist work, *Les Splendeurs de la foi*, with its expressive subtitle *Accord parfait de la révélation et de la science, de la foi et de la raison*. Completed in 1882, *Les Splendeurs de la foi* offered a summation of all that Moigno had battled for since the early 1830s. Its appearance could not have been more timely or welcome in Catholic circles. The publication, by 1888, of three editions of a rather costly work (at 40 francs for the five volumes) demonstrated the depth of Catholic anxiety about the implications of science. The same concern was conveyed in the printed endorsement of the book from Leo XIII.[48] Such explicit papal approval signaled a formal recognition that free thought could best be fought with its own weapons of reason and argument; more specifically, it also ended any lingering perception of Moigno as a dangerously aberrant member of the Catholic Church. As at least liberal Catholics recognized, Moigno showed better than any clerical contemporary how judicious accommodation to the tenets of modern science might lead to a strengthening, rather than a weakening, of faith.

In both its spirit and its meticulously scholarly execution, *Les Splendeurs de la foi* responded, to the letter, to Pius IX's encyclical of 1853 urging the clergy of France to protect their flock from the "foul pastures" of materialist writings with their harvest of corrupted morals and attacks on the foundations of religion.[49] It was a compelling exemplar for the men of faith whom the encyclical urged to devote their waking hours to writing books and other materials that would defend and spread Catholic doctrine. From the premise that discord between science (when correctly pursued) and the scriptures (when correctly interpreted) was impossible, Moigno presented miracle stories and Old Testament accounts of the creation and human history as immune from the depredations of German biblical criticism and its insidious progeny.[50] By the same token, as he argued, the favored Catholic doctrine of the unity of the human race emerged triumphantly from the scientific evidence gathered by monogenists, who were at once more convincing and (significantly, in his view) incomparably more numerous than their polygenist adversaries.[51]

Refreshingly, though, Moigno's tone was never overly pious. This left his work with little of the religiosity that characterized most contemporary clerical writing. For this reason alone, his passing left its mark in the pages of *Cosmos,* a new series of which was launched after his death in 1884. Now, in an atmosphere of heightened polarization between science and belief that emerged in the Third Republic, the first editorial of the series (presumably by Moigno's successor as editor, the abbé Henri Valette) promised that *Cosmos* and its contributors would obey the teachings of the Catholic church, "the guardian of truth," and follow the elaborate profession of orthodox faith uniting the prescriptions of the sixteenth-century pope Pius IV and the profoundly conservative Pius IX, who had died six years earlier.[52] Such unquestioning submission would have been inconceivable in the mouth of Moigno.

Masters of the Mass Market: Flammarion and Figuier

The extent to which scientific writing could become involved in the wider political and ideological debates of the midcentury is borne out by the examples of several authors who came to maturity during the Second Empire. Although this generation of writers was drawn into popularization by the burgeoning market for their wares, the motives that guided them frequently transcended commercial interest, and the results were rarely anodyne. Indeed, the public's taste for such writers rested on their readiness to go beyond the confines of academic science, often provocatively. The most successful of them, Camille Flammarion, added spice to his work by presenting himself as a detached and often hostile critic of the professional scientific community. In this he resembled Meunier. But his stance was more personal than political, being virtually determined by the circumstances that diverted him quite suddenly from an early humdrum career in the lower reaches of astronomy.

Formal education had done little to fire Flammarion with high ambitions.[53] A far from privileged childhood strongly marked by a Catholic upbringing in the small town of Montigny-le-Roi (in the eastern department of Haute-Marne) had been followed by two years as an apprentice engraver in Paris, an activity that left him with time for reading and attending the evening lectures of the Association polytechnique. He had entered astronomy by chance, thanks to an encounter with a doctor who recognized the extent of his ability and frustrated ambition and encouraged him to compete for a position as pupil astronomer at the Observatoire de Paris in 1858. The director, Urbain Le Verrier, took to him, and he was appointed. Despite the frugality of life on an initial salary of 600 francs a year, Flammarion enjoyed the freedom for study, leading to *baccalauréats* in both letters and science, and for an immersion in the kind of speculative physical astronomy that was to become his intellectual stock-in-trade. But after four years in the observatory, relations between him and Le Verrier had become intolerably fraught, and his inevitable, though (as it transpired) fortunate, departure from his post followed in 1862, when he was only 20. From that moment, Flammarion viewed Le Verrier as the incarnation of all that was pernicious in official science and the living proof of the need to fashion a new astronomy open to all.

The details of the break between Flammarion and Le Verrier are obscure. Even Flammarion's account in his wonderfully informative autobiography was brief.[54] Certainly, though, there was reciprocal personal animosity. Le Verrier, for Flammarion (as for many contemporaries, especially though not exclusively on the left), was a domineering and intolerant autocrat detested by all who worked with him and the possessor of "the most frightful character that anyone could possibly imagine."[55] Le Verrier, for his part, saw Flammarion as lacking the seriousness of a true astronomer; someone whom he regarded as more of a "pupil poet" than a "pupil astronomer" had no place in the observatory.[56] Le Verrier's perception, in fact, got to the heart of the conflict. His crime, in Flammarion's eyes, was that he ruled French astronomy as the supreme exponent of a crabbed science blinkered by computation and cold abstraction; the result was an astronomy emasculated by mathematics and calculated to exclude rather than embrace the lay public. Flammarion, in contrast, presented himself as a philosopher. For him, mathematics was, at best, a temporary scaffolding: once the figures and formulae had done their job, they should be swept away, leaving the "palace of Urania" (his figurative term for the body of astronomical knowledge as he conceived it) to emerge in all its glory.[57] No one perceived the potential of a genre of open, imaginative astronomy more clearly or pursued it more creatively than Flammarion. His success rested on his vision of the subject as a pursuit that far transcended entertainment. Astronomy in Flammarion's hands was the science that alone could inform man of his true nature and destiny in a universe excitingly conceived as teeming with life and unseen powers.

From the time of his first published writings in the middle years of the Second Empire, Flammarion formulated a scientifically founded spiritualist philosophy to replace his now abandoned Christian faith. Thereafter, he never wavered in his search for the hidden forces of nature and for phenomena that evaded immediate everyday perception. His continuing commitment had its initial inspiration in the work of Jean Reynaud, a leading Saint-Simonian and former secretary to Hippolyte Carnot in the heady revolutionary days of 1848. By the mid-1850s Reynaud was attracting attention through his quest for a secular "religious philosophy" that he felt would break with the vestiges of superstitions inherited from the Middle Ages but do so without falling into the trap of what he saw as the outdated materialism of the Enlightenment. Such a sentiment, combining an aspiration to modernity with a rejection of the boundaries that conventionally separated the realm of science from that of human values, struck an immediate chord with Flammarion. As he put it in his *Dieu dans la nature* in 1867, his aim was "to present a *positive philosophy of* science that will embrace within itself a *non-theological refutation of contemporary materialism*," a statement whose tone and emphases might equally well have found their place in a work by Reynaud.[58]

By this time, the principle underlying Reynaud's doctrines—the existence of life and hidden powers suffused throughout the universe—had become embedded in Flammarion's thinking, and it remained at the heart of the enriched, antimaterialist conception of nature with which he continued to captivate imaginations until his death in 1925.[59] In Reynaud, Flammarion found a worldview that accommodated "living forces" animating the universe and fostered the enlarged conception of science that he began to explore in his earliest work, *La Pluralité des mondes habités*. Appearing as little more than a substantial pamphlet in 1862, *La Pluralité des mondes habités* was enlarged to book-length in 1864 (with Reynaud's encouragement), before going into its nineteenth edition by 1873 and its twenty-fifth only four years later (see figure 7).[60] Flammarion claimed no originality for the book's endorsement of the idea of life on other planets: he acknowledged that he was writing in a tradition of speculation that went back in France to Fontenelle as well as having many contemporary adepts, including David Brewster in Britain. Where Flammarion did break new ground, however, was in the presentation of his argument, which conveyed conviction without overt dogmatism and combined speculation of breathtaking audacity with declarations of profound respect for reason and observation conducted in the Baconian manner.

While admitting that it was impossible to prove that a planet at any particular time was peopled by living beings, Flammarion insisted that the natural state of the universe required that planets should be inhabited. He, for one, needed no observational confirmation of a conclusion that rested on the philosophical "absurdity"[61] of supposing that only our earth harbored life when the conditions on other known planets appeared equally favorable, or even more so. His starting points were the limitless fecundity of nature and the difficulty of believing that the planets had been created without purpose (their only conceivable purpose being, for Flammarion, the sustaining of life, albeit life of a very different kind from that on earth) or that the sun existed to sustain the human race alone. From those principles there followed a vision of Nature as the seat of "universal life"[62] and consequent speculations about the physical attributes of beings elsewhere in our universe.

The assumption that such attributes would be determined by local conditions opened the door to some of Flammarion's most imaginative writing. As he suggested in successive editions of his *Terres du ciel* from 1877 and elsewhere, dwellers on Mars, for example, would be taller than ourselves and might well have developed wings as the most effective means of locomotion on a planet where the force of gravity was only a third of the force on the earth.[63] They might also be expected to have originated earlier than ourselves and to be now a race superior to our own.[64] The inhabitants of the remote and very different world of Jupiter, in contrast, could scarcely be expected to resemble human beings in any respect; theirs was probably an entirely aerial existence spent in the upper reaches of an atmosphere (quite unlike ours) on which they would feed.[65] And if our moon lacked signs of habitation, the most probable conclusion was that life there had come to an end or was dormant; perhaps the moon had simply run its course.[66] In all this, only the particular forms of the inhabitants of other parts of the solar system were presented as matters for conjecture: that there were, had been, or would in the future be such inhabitants was not a matter for any but the most perfunctory doubt.[67]

Flammarion's statements about life on other worlds moved easily between contrived displays of caution, which lent them an air of scientific seriousness, and a freedom made possible by the inaccessibility of the phenomena to normal modes of scientific inquiry. The dual register helped to endow his astronomy with its special character as a science that invited speculation while offering little by way of certainty.[68] With mankind set on a centuries-long path leading from present ignorance to the eventual full exploitation of its reason and spirituality,[69] Flammarion commended astronomy as the surest guide. It was this same sense of perpetual groping for a higher, hidden truth that allowed him to pursue his interest in psychic studies as a perfectly logical element in his worldview. According to his later recollection, the interest began in 1861.[70] Within four years, and under the cloak of a pseudonym (which he soon abandoned), he had made public his tentative belief in a spirit world in a 165-page work entitled *Des Forces naturelles inconnues*,[71] and thereafter he was set on a path that involved him with some of the most notorious scientific charlatans of the day. The main aim of his psychic quest was the defining of the boundaries between truth and deception in reports that had multiplied since the late 1840s, when an American family by the name of Fox—two parents and three daughters—claimed to have experienced unexplained knockings and violent movements of furniture and doors and to have received answers to messages spelled out by raps on tables and elsewhere.[72] The women of the family were especially sensitive to the phenomena, and their experiences soon assumed such a spectacular form that they were able to offer consultations and demonstrations of their powers in their hometown of Rochester in New York state and about the country.

Public interest in the activities of the Fox family was reinforced by the appearance elsewhere in the state, in Buffalo, of two teenage brothers, Ira and William Davenport, who were able to defy gravitation and "fly" through their home and even out into the street.[73] The Davenports' demonstration of a pencil writing independently of any visible control heightened the sensation, and séances at which they flew, to the accompaniment of music from instruments that apparently had no players, drew large audiences. Seemingly reliable witnesses watched in astonishment as the Davenports went through their repertoire, soon extending to questions and answers about future events exchanged with unseen correspondents by raps even when the brothers were bound in sacks. The brothers themselves were reluctant to offer an explanation. But the belief that their powers were the fruit of a contact with the dead in a higher, immaterial form of existence captured imaginations, especially from 1864, when after more than a decade of demonstrations in the United States the Davenports made a first, much-publicized visit to England. It was in the wake of this visit and the brothers' stay in France in the following year (during which they demonstrated their powers before the emperor and his family at the palace of Saint Cloud) that Flammarion wrote *Des Forces naturelles inconnues*. Here, Flammarion stopped short of either endorsing or denying the reality of the demonstrations as psychic phenomena, and he did no more than convey the widely shared view that the case required careful investigation before it could be judged either authentic or the fruit of the Davenports' venality.[74] Where he was unusual, however, was in the firmness of his conviction that, despite the lack of conclusive evidence, communion with the spirit world was, in principle, possible.

A similar tone of scientific rigor, balanced by openness to hitherto unexplained phenomena, colored his attitude to the aims of his older friend and spiritualist mentor who went by the name of Allan Kardec. An admirer of the Davenport brothers, Kardec had been trying for the better part of a decade to establish "*spiritisme*" (his word) as one of a number of new religions that circulated in France during the Second Empire.[75] The comprehensiveness and conviction of his beliefs had made him an object of general fascination since 1857. In that year, newly converted to *spiritisme* by a "guardian spirit" (*esprit protecteur*) by the name of Zéphir, Kardec had published his first book, *Le Livre des esprits*, a compendium of beliefs that he claimed had been communicated to him by departed spirits in the form of answers to his questions.[76] By now he had marked his conversion by abandoning his original name of Hippolyte Léon Denizard Rivail, and it was as Allan Kardec that he wrote *Le Livre des esprits* and in 1858 established the Société parisienne des études spirites and the monthly journal *Revue spirite*. On these foundations he fashioned a campaign of voluminous writing on the philosophical, practical, and moral dimensions of the practices of mediums (of whom he was not one) that ended only with his death in 1869. Kardec's ideas fascinated Flammarion. In fact, it was the chance discovery of *Le Livre des esprits* in 1861, when he was already writing his *Pluralité des mondes habités*, that first drew the young Flammarion's attention to the spirit world. Despite his fascination, however, his position with regard to the particular spiritualist phenomena that Kardec described remained one of detachment, though a detachment that fell far short of rejection. Flammarion's view, like Kardec's, was that the phenomena had to be studied as natural, rather than supernatural and hence inexplicable, events. His position was less that of a skeptic than that of a seeker disappointed by the elusiveness of progress in the empirical exploration of the spirit world, despite a perpetuation of Kardec's mission after his death by Henri Sausse, Gabriel Delanne, and other loyal disciples.

On into the twentieth century, despite the unveiling of countless frauds, Flammarion issued new editions of two earlier works—his *Forces naturelles inconnues* and a collection of studies on *L'Inconnu et les problèmes psychiques*, first published in 1900—in which he reaffirmed his belief in the soul as an entity independent of the body and endowed with powers as yet inaccessible to scientific investigation.[77] Even if the mechanism of communication between the realms of the visible and the invisible remained as obscure as ever, the open-minded scrutiny to which he was committed had to go on, not least in pursuit of his social goal of fashioning a compassionate humanity united in its acceptance of the spiritual as well as the material dimension of its existence.[78] To that end, he was as ready to denounce the extravagant claims of the fraudulent and the gullible as he was to resist the blanket condemnation offered in the name of sober science by Jacques Babinet, in whose eyes all mediums were a party to deception.[79] Where unmasking was necessary, friendship was no protection. Like Kardec, the playwright and medium Victorien Sardou was subjected to Flammarion's firm if understanding criticism. Early in Flammarion's engagement with spiritualist phenomena, Sardou claimed to have communicated with the spirits inhabiting Jupiter and to have been guided by them in his drawings of life on the planet including the houses of Mozart, Zoroaster, and Bernard Palissy. Characteristically, Flammarion took Sardou's claims seriously.[80] But by 1907, twenty years of mounting skepticism had hardened to his dismissal of the claims as the product of unintentional self-delusion, fostered by the widely held contemporary belief (which Flammarion shared) that Jupiter was inhabited.[81]

Far from placing him on an unpeopled eccentric fringe, Flammarion's interest in the spirit world gave him a special bond with some of the leaders of French science, including Pierre Curie and Charles Richet, physiologist, future Nobel Prize winner, and from 1880 (initially with Antoine Breguet) until 1902 editor of the intensely serious *Revue scientifique*. The sense of a quest shared with some of the most eminent of his scientific peers helped to sustain Flammarion's inquiries and to make his salon, close to the observatory in the avenue de l'Observatoire, a meeting place for all the great mediums of the day and the savants who were fascinated by them. Preeminent among the mediums, for Flammarion as for many other spiritualists, was the unlettered Neapolitan Eusapia Palladino, whose demonstrations of the levitation of a substantial table in Milan profoundly impressed Richet and led to her having been invited in 1897 for the first of several visits to Paris (see figure 8).[82] Palladino's repertoire of levitations was impressive: at one séance, a table was raised 15 centimeters off the floor five times in a quarter of an hour. That and other and phenomena such as the movement of objects, knocking, and the playing of a musical box without apparent human agency convinced Flammarion that most of the effects that Palladino produced were genuine and could only be explained by forces hidden from normal sense experience. In a reflection on Palladino published a decade after her first visit, Flammarion expressed his undiminished confidence: "We are surrounded by unknown forces, and there is nothing to prove that we are not also surrounded by unseen beings. Our senses reveal nothing of this reality . . . What matters above all else is for us to affirm that the phenomena of mediums are a reality."[83] Flammarion was unwilling to pronounce on whether the "intelligent entities" behind the phenomena were spirits of the dead. Nevertheless, he believed resolutely in the existence of psychic forces and saw their investigation as the "science of tomorrow" that it was a human duty to pursue. On this, the precept he advanced in *Forces naturelles inconnues*, "Quaere et invenies"—"Seek and you will find"—was unambiguous.[84]

Although Flammarion portrayed his scrutiny of *spiritisme* as part and parcel of his science, it is significant that he tended to discuss the subject in nonscientific periodicals (such as the *Annales politiques et littéraires*) and in writings that were not explicitly didactic. His most extravagant speculations, in fact, appeared in excursions into imaginative literature, beginning with *Lumen* in 1873.[85] In this series of five conversations, subtitled *Les Récits de l'infini*, Flammarion used the figure of Lumen, a spirit traveling on a comet, to give information about the spirit world in the form of answers to questions posed by Quaerens, a mortal seeker after truth. Two later philosophical novels, *Uranie* (1889) and *Stella* (1897), and the collection of light, highly speculative essays that he published as *Rêves étoilés* (1886) were similarly suffused with accounts of séances and of other manifestations of the afterlife, communication with other worlds, and encounters with strange beings on planets within and beyond the solar system.[86]

Through his novels, Flammarion captured imaginations that even his most daring scientific writing might not have touched. It could be left for Lumen to describe the inhabitants of a planet of the double star, Gamma Virginis, where the abnormally dense air allowed creatures (manlike though with just one cone-shaped ear on the top of their skulls) to fly without wings.[87] And it was his "heavenly muse," Uranie, who conveyed the truths of mankind's striving, both in history and in each individual life, for liberation from the weight of materiality and for spiritual union with "the higher Uranic life": only when that state had been achieved, as Uranie told him, would the soul be able to dominate matter and so free itself at once from the fetters of earthly existence and human suffering.[88] It was heady stuff, typical of Flammarion's fiction in the thirty years or so that separated his first drafts of *Lumen* in the mid-1860s from the publication of *Stella* in 1897. While it has been suggested that Flammarion moved, in his later writings, to a slightly less fanciful style,[89] any such change was slight, and it did not affect his readers' undimmed taste for tales of the marvelous. *Uranie* (30,000 copies of which were sold in the four years after its publication) continued to sell, as did translations into English (in several versions), German, Spanish, Italian, Swedish, and even Icelandic.

In popularity, the only contemporary scientific writer who remotely matched Flammarion was Louis Figuier. Like Flammarion, Figuier fashioned his vocation during the empire, establishing himself (in the manner of the time) as someone who drew on "official" science while, as a critic, he stood resolutely apart from it.[90] In two obvious respects, however, he differed from Flammarion. The first was that once Arago had advised him to take up popularization in preference to the academic career in pharmacy on which he had embarked, initially in his native Montpellier and then in Paris, he treated the whole range of science and technology. In this, he was able to draw on his early work in the physical sciences, which earned him a doctorate at Toulouse in 1850 (for a study of the chemical action of light), and his pursuit of both pharmaceutical and medical studies, which entailed his writing three theses, two for the *agrégation* for the faculty of medicine in Paris (1844 and 1853), the other for the *agrégation* in chemistry for the Ecole de pharmacie (1853).[91]

The second difference between Flammarion and Figuier was that Figuier lacked the radical streak that sharpened Flammarion's opposition to the scientific elite. Although outside the elite and fiercely critical of it on occasions, Figuier tended to use an indulgent, detached style in his reports on its doings. His polemical objectives were directed differently, mainly to demonstrating that science and technology provided not only the hope of a better material world but also a bulwark against the anxiety and quackery associated with pseudoscientific fads. Until his death in 1894, Figuier pursued these two goals through newspaper articles (most of them in *La Presse* during the Empire) and books that displayed the abundance of marvels even in the world of everyday experience. His earliest substantial work, a four-volume *Exposition et histoire des principales découvertes scientifiques modernes*, the first three volumes of which appeared in 1851 and 1852 and were already in a fourth edition by 1855, set the tone of almost breathless respect for science that was to mark all his writing.[92] Since the beginning of the century, as he argued, the place of science in society had been transformed: this had left science as an indispensable part of modern life and so as a matter of central concern to all humanity and not just to the once restricted world of the savant.[93]

It is characteristic of Figuier's perception of the beneficence and nobility of science that from the moment American spiritist doctrines began to circulate in France in 1854, he dismissed them, most vehemently in his four-volume *Histoire du merveilleux dans les temps modernes*, a scholarly study of forms of belief in the supernatural since antiquity that he published six years later.[94] There was nothing unusual about an attack on table turning and the attempts to interpret tapping from the spirit world. But Figuier gave his criticisms a distinctive twist by casting the contemporary infatuation as just one illustration of humanity's insatiable and undiscriminating appetite for marvels. As he argued, it was the same appetite that, in ancient times, had bred divination, with its paraphernalia of oracles, sibyls, and miracle workers. In the Middle Ages, the appetite had surfaced again in beliefs about witchcraft and magic. And even in the age of reason and the early nineteenth century, it had encouraged Franz Mesmer to peddle his meretricious doctrine of animal magnetism and allowed a French aristocrat, the marquis de Chastenet de Puységur, to arouse intense public interest in his own version of mesmerism and the cures that he was reputed to have effected in his native Buzancy between the 1780s and his death in 1825.[95] Through the 1830s and 1840s interest had waned, and although hypnotic practices of the kind that Puységur pursued had continued, they had done so inconspicuously. Now, however, charlatanism was rampant once again, a point that Figuier laboured in the fourth volume of his *Histoire* in a litany of the threats from latter-day magicians, table turners, and other peddlers of deceit.

In his interest in the contemporary resurgence of superstition, Figuier shared common ground with Flammarion. But the beliefs to which Flammarion gave at least prima facie credence repelled Figuier. Investigations by Faraday, Chevreul, and Babinet may have sufficed to convince Figuier and other skeptics that table turning had a physical rather than a supernatural explanation.[96] Skepticism, though, was no justification for disengagement with the threat. Figuier's main contemporary target was Louis Alphonse Cahagnet, an unnervingly successful admirer of the Swedish philosopher Emmanuel Swedenborg and an adept of magic of every kind (extending to talismans, cabalistic mirrors, and necromancy, among much else).[97] In books and pamphlets and through his short-lived journal *Le Magnétiseur spiritualiste* (1849–51), Cahagnet had become, from the late 1840s, an object of frightened curiosity, first in England (a country for whose tolerance he felt affectionate gratitude) and then his native France. Denunciations of Cahagnet's notions of animal magnetism and his fanciful experiences by more sober hypnotizers and a ban by the Holy Office served only to heighten his reputation.

In fact, Cahagnet's chilling accounts of séances in which mediums in hypnotic trances had conversed with spirits of the dead, including Swedenborg himself, had no need for the added spice of notoriety that came with condemnation. The three densely anecdotal volumes of *Magnétisme. Arcanes de la vie future dévoilés*[98] in which he gave a definitive statement of his beliefs became required reading for critics and adepts alike, and they both fascinated and polarized opinion. An account of a paradise known only to the mediums with whom he was in communion caused particular skepticism, even among fellow practitioners. This was the reaction of one of the most skilled hypnotists of the day, the baron du Potet, who saw the extravagance of Cahagnet's claims as damaging to the whole noble tradition of animal magnetism. If the charge of charlatanism was not to stick, Cahaganet's "lucubration" had to be exposed as the work of a man who mistook trivial ephemeral reflections for sublime insights.[99]

Less discriminating readers, however, had fewer reservations, and on through the 1850s Cahagnet's accounts of revelations about the past and future, communicated by the spirits of Galileo, Hippocrates, Benjamin Franklin, and others, captivated the reading public.[100] For those with a disposition to believe, credulity survived, although it was constantly strained and eventually (though only slowly) undermined by accounts such as that of the ecstatic experience of Cahagnet's favored medium, his "sister in the eternal" Adèle Maginot. In November 1851 Maginot had been assured by no less a spirit than the astronomer William Herschel's that the missing Arctic explorer Sir John Franklin and several of his crew were still alive, after disappearing without trace four years earlier while in search of the Northwest Passage.[101] The discovery of a bottle containing a message, apparently from Franklin, washed up on the Irish coast in 1853, only added to a sense of confusion and expectation that was not dispelled until the finding of the remnants of Franklin's expedition six years later and the proof that Franklin had died in 1847 irretrievably exploded Maginot's claims.

It was precisely a concern about the level of public interest in spirit phenomena that sustained Figuier's belief in the importance of taking the supposedly supernatural forces seriously. For "official" science to have turned its back on table turning, as Figuier accused it of doing, was a dereliction of the duty of the savant to inform lay minds and to respond critically to new evidence. Danger, not security, lay in the systematic dismissiveness that was all too characteristic of the closed and intensely conservative "intellectual senate" of the Parisian scientific elite when confronted with novelty.[102] If a "discovery" transcended the bounds of orthodox learning, its unorthodox status alone, as Figuier insisted, should not render it immune from rigorous scientific examination, especially in areas that fed, and fed on, the gullibility of the ignorant.

In Figuier's later writing, his interest in realms inaccessible to direct observation assumed a more urgent and, in key respects, less critical character. The most personal, and painful, of his new interests followed the devastating loss of his only child, a son, in 1870. Despite the attacks on the excesses of *spiritisme* that he had delivered over the previous two decades, Figuier now plunged into an account of the universe that had the souls of the dead enduring eternally in "superhuman beings" in the planetary ether beyond the thin layer of air surrounding the earth. In this new departure, Figuier saw himself as entering into contact with the "true principles of nature." The appeal in his judgments was accordingly and unswervingly to science. But *Le Lendemain de la mort*, in which he set out his beliefs in 1871, was in reality an intensely emotional book, rooted both in his personal tragedy and in the social and political tensions that had followed the defeat of 1870 by Prussia.[103] The "moral canker" of German materialistic doctrines, as Figuier saw it, had by now spread menacingly from the educated classes to the people, undermining religion and the social order alike and culminating in the anarchy in which the "madmen" of the Commune had defiled the capital of France. Materialism, for Figuier, was "the scourge of our day, the origin of all the evils of European society," and the main casualty remained morality. Hence society's most urgent need, if the moral vacuum of the materialists was to be avoided, was for a reaffirmation of the boundary between good and evil.

Figuier's speculations on the trajectory of the soul after death combined extravagance with a degree of unsubstantiated detail that even Flammarion would have found startling. For the soul to rise after death to the ethereal spaces, it had to lose the besetting weight that wickedness and impenitence imposed on it. So long as it was trammeled by that weight, it could never be released from the miseries of our earthly existence and could achieve only terrestrial reincarnation in another body, though without memory of its past existence.[104] A soul that had achieved perfection and escaped its earthly bonds, in contrast, would undergo reincarnation in one of Figuier's superhuman beings, of undetermined form but of a lightness that allowed it to float in the ethereal regions. There, such a being would pursue its existence with unimaginably refined mental powers, no reproductive organs, and no need for sleep or sustenance other than the planetary ether itself, which had merely to be breathed.[105] In one essential respect, that of its mortality, the "superhuman being" resembled its earthly counterpart. This death, however, marked a further transition rather than an end. It would provoke yet another reincarnation of the soul, this time in a being of even greater spirituality and with still finer senses and perceptions.[106] The process of death and reincarnation might then be expected to continue, leading the soul from its original location in a being that Figuier likened to an angel, through archangels or archhumans, and on eventually to a final resting place in an "archhuman" that had sloughed off the last vestiges of materiality and become pure spirit. That being would now leave the realm of the planets and find its final home in the sun, where the constant accumulation of souls replenished the radiation that in turn sustained the earth.[107] The sun, Figuier conjectured, might even be composed of nothing more than the souls, the "pure and burning spirits," that had completed their celestial peregrinations.

In their general tone if not in their detailed content, Figuier's fantastic speculations would not have been out of place in the diverse theosophical teachings that were circulating with new intensity about the time he was writing. The founding of the Société théosophique de France in 1875 (the year in which national theosophical societies were founded in both New York and London) was a mark of the contemporary upsurge of interest in human beings' eternal spiritual nature. It was just this interest that ensured brisk sales for *Le Lendemain de la mort*, which appeared immediately in English[108] and, between 1872 and 1878, in no fewer than six other French editions. Yet the notoriety of the book pushed Figuier to the more eccentric, and suspect, fringes of public debate. His only accolade, apart from commercial success, was the perverse one of having his book placed on the Catholic Church's index of prohibited works, a distinction that his condemnation not only of materialism but also of the false *spiritisme* of the séances did nothing to avert. The notion of immortality that suffused *Le Lendemain de la mort* had led Figuier into terrain that the Church considered its own, and in doing so he had cut directly across established doctrine. He could hardly have expected theologians to take kindly, for example, to his assertion that the notion of the reincarnation of souls in other terrestrial bodies sustained a prospect of eventual release from the burden of sin that was denied in the Christian conception of eternal damnation.[109] The successive reincarnations would necessarily take longer in the case of the wicked than they would in that of the virtuous, but they proffered perpetual hope and mercifulness where the Christian could look forward only to a dread judgment day.

In contrast with his views on the afterlife, Figuier's handling of another broad theme that engaged him in his mature years, that of human prehistory and the history of life on earth, was soberly orthodox. As this work reminds us, his independence and fascination for the curious did not make Figuier a free-thinker in the more aggressive sense of the term. His *La Terre avant le déluge* (1863), *L'Homme primitif* (1870), and *Les Races humaines* (1872) and the numerous subsequent editions of these works—all three of which he presented as reading for the young, although they were in fact more widely consumed—elaborated a view of the antediluvian earth and the emergence of the races of mankind that was shot through with conventional piety. He followed Quatrefages, the Belgian geologist d'Omalius d'Halloy, and the English ethnologist J. C. Prichard, for example, in arguing for monogenism—the view that the various human races are varieties of a single species with a common distant ancestry—and then using his discussion as a starting point for an attack on transformist views on (to use Prichard's term) humanity's "physical history."[110] On transformism, Figuier followed Quatrefages, whose *Rapport sur les progrès de l'anthropologie* (1867) had demonstrated, in his eyes, the impossibility of the "strange genealogy" that would portray the human race as descended from the ape.[111] As both men agreed, the human creation and the animal creation were set infinitely apart by man's capacity for abstraction and speech.

The genesis of human beings lay, for Figuier, beyond any possibility of scientific investigation, and (using a device commonly favored by the Christian faithful) he simply dismissed the question as one shrouded in the impenetrable "mystery of creation."[112] Although successive expositions of his opinions on human history changed in emphasis and the evidence he used, the bedrock of his beliefs remained inviolate until his death in 1894: every animal species had emerged from a separate act of divine creation, and it was unnecessary to look for the origin of man any further back than a recent phase of the quaternary, in the period of warming following the last ice age.[113] In this way, humanity was comfortingly confirmed as being of low antiquity, with the gloss that it had spread from a single point on the earth, probably on the "smiling banks" of the Euphrates.[114] Such biblically supported articles of faith, which included the reality of a limited Noachian Asian flood provoked by a volcanic eruption that also gave birth to Mount Ararat, were essentially backward looking. In defending the literal truth of a correctly interpreted Genesis, Figuier was positioning himself close to the self-taught naturalist and devout Catholic Marcel de Serres, who had advanced such views since the 1820s from his chair in the faculty of science at Montpellier (see chapter 4). Taking little account of emerging new perceptions and evidence, Figuier remained equally unmoved by the advancing polygenism of Georges Pouchet and his allies[115] and by Jules Desnoyers's discovery (of 1863) at Saint-Prest, near Chartres, of bones of the *Elephas meridionalis*, from the Pliocene (Upper Tertiary).[116] The bones that Desnoyers unearthed seemed to show the signs of incisions made by flint implements and so (through an extrapolation that Figuier could never accept) took human history back to an unimaginably distant past.[117]

The fact that the great popularizers of the Second Empire and early Third Republic showed such diversity in their scientific, political, and ethical positions left them a disparate group. What Meunier, Figuier, Moigno, and Flammarion did share, however, was a commitment to producing a *littérature engagée* in which the new professionalism of scientific writing for a lay audience was put to purposes that embraced but often transcended the diffusion of conventional academic science. In the words of another successful scientific writer, Henri de Parville, the aims of popular writing on science were of a higher order. They were to "destroy prejudice, counter false opinions, guide inquiry, foster ideas and nurture them . . . in short, to serve where possible as a platform for honest, open discussion for the benefit of society as a whole."[118] Such high-mindedness was characteristic of all the popularizers of the period. Nevertheless, it went hand in hand with a commercial interest capable of yielding rich profit through the Second Empire and on to the end of the century. The result was a constant search not only for exciting subjects but also for new ways of stimulating the market for a product—popular science—that lacked and, in its most successful forms, did not need official subsidy or encouragement.

The Spoken Word

As the market grew, so genres multiplied. Those with the necessary presence and oral skills turned most easily to public lecturing, often as a complement to their writing, in the manner of Comte and Arago, both of whom published their courses of lectures on astronomy.[119] In this, they were participating in a long tradition, though one that acquired new visibility during the Restoration, July Monarchy, and most strikingly the Second Empire. Contemporaries, in fact, had a clear sense of witnessing a shift in public culture. In 1877, the article "Conférence" in the supplement to Pierre Larousse's *Grand dictionnaire universel du XIXe siècle* looked back on the later years of the empire as ones in which public enthusiasm for lectures reached unprecedented heights. In the words of the article: "Lectures became a vogue, a passion."[120] It is impossible to confirm a contemporary estimate that twenty thousand people a day regularly attended lectures in Paris in 1865 and that two hundred thousand did so in France as a whole.[121] Nevertheless, the words "vogue" and "passion" were certainly no exaggeration.

While the Larousse article conceded that the taste for lectures had had its counterpart in Britain in the success of Michael Faraday and John Tyndall in the same genre, it insisted that the French style differed from the predominantly utilitarian or moral tone favored in London. The contrast was drawn, in part, as a side swipe at French *anglomanes* who liked to attribute any worthwhile innovation to the British. But there was indeed something distinctive about lecturing in France. After 1860, lectures there conveyed the same sense of freedom that allowed popular writers on science in the empire's liberal years to become more opinionated and even to criticize the scientific establishment. The trend was manifested in spicier, more exciting offerings and a move toward a greater degree of audience participation. The most prominent catalyst was the writer and literary scholar Emile Deschanel. As a republican whose outspoken views made him unacceptable in the teaching profession of the early empire, Deschanel had been arrested and then exiled from France, and it was while living in Brussels in 1852 that he turned to a new career. In 1860, the first signs of liberalization allowed him to return to Paris, where he lectured and promoted lectures, for an eager paying public. He soon found himself at the heart of an explosion of cultural activity that far transcended his own initiative, encompassing all intellectual levels and audiences across the social spectrum.

What occurred was more by way of a dramatic expansion of existing provision than a wholly new departure. In Metz, the series on industrial subjects that Henri Scoutetten, the head of the city's military hospital, launched in 1861 resurrected the tradition of public lecturing to which Poncelet had contributed in the 1820s.[122] And among the offerings for artisans, the free courses of the Association polytechnique and the Association philotechnique maintained a tradition that had never flagged. When these two associations collaborated in organizing a festive prize giving, under ministerial patronage, in the vast arena of the Cirque Napoléon (the future Cirque d'hiver) in 1860, the attendance for what was in effect a mass celebration of the movement for popular education topped five thousand.[123] Demand on such a scale came with consequences, however. The Association polytechnique saw the demand as an opportunity for expanding its clientele, and soon it was offering an impressive program in Paris and across the country for aspiring, better-educated audiences, for which it had not previously catered. The new audiences included women, and many of the early lectures were published in a new series of books directed primarily at travelers by train, the "Bibliothèque des chemins de fer."[124] An authoritative report in 1867 noted the trend, observing as a matter of concern the Association polytechnique's change of focus from the *ouvriers* for whose benefit it had originally been established to the more skilled members of the industrial workforce.[125] It was a mark of precisely the *embourgeoisement* that afflicted nineteenth-century initiatives in working-class education in Britain and many other countries, and it caused the same misgivings.

Although the public for Deschanel's ventures had much in common with the new audiences that the Association polytechnique was seeking to cultivate, it was at once more affluent and socially more elevated. The locations that Deschanel was initially able to provide certainly had a distinctly makeshift character. Yet it was no coincidence that they were always in the expanding, elegant area close to Garnier's new Opéra, still in a fifteen-year period of construction before its eventual inauguration in 1875. The first lectures, organized on a weekly basis by Deschanel and two associates, took place in temporary premises in the rue de la Paix (see figure 9); thereafter, they moved to a room at 5 rue de la Scribe, then to a newly constructed concert hall in the same street, financed by the banker and generous benefactor of science, Raphaël Bischoffsheim, and finally, in February 1868, to a definitive home in a lecture room at 39 boulevard des Capucines. In all its locations, the venture was seen as, first and foremost, a commercial one. Renting the premises in the boulevard des Capucines, for example, was made possible through the enterprise of one of Deschanel's collaborators, Yves Henry, who assembled a capital of 20,000 francs, in shares of 100 francs each.[126] And the success of the lectures in this setting could be measured financially.[127] A profit of 2,000 francs during the first year indicated that the program—a mixture of history, art, travel, and the wonders of science and technology—was pitched just right.

Among the scientific lecturers whom Deschanel engaged, the "stars" were the old hand Jacques Babinet in physics (despite his age, 74 in 1868), Félix Hément (a writer, teacher, and pioneer of primary education and of lectures for women and working-class audiences) in the life sciences, and, outstandingly, Flammarion, each of them an expert in his own field but an expert above all in the art of communication. Every successful lecturer had a particular way of winning his audience. Babinet was known for a lightness of touch that distinguished him even from Arago. In the words of Pierre Larousse, he was "less methodical, less elevated, less of a professor than Arago"; his style was racier and better adapted to more superficial audiences, which tended to appreciate his anecdotes and humor.[128] Flammarion, too, was an attractive speaker. But he also excelled technically through the use of the magic lantern and projected images. In this, he was a trendsetter, chiefly through his use of high-quality equipment that he developed with the young optical instrument maker Alfred Molteni: the thirty slides of images drawn from his *Merveilles célestes* were soon essential props in his biweekly appearances in the boulevard des Capucines.[129]

Important though eloquence and hardware were, a third element in a lecturer's success was the atmosphere of the room he used. On this score, the 1860s marked a new departure. Deschanel and another great literary speaker of the period, Francisque Sarcey, conceived the lecture as more akin to a public conversation than it was to the more traditional performances that deployed the "big guns of eloquence."[130] When things went well, there would be a constant interaction between speaker and audience, as happened in 1866 when the chemist Edme Fremy addressed a large society audience on oxygen and ozone in the concert hall of the Conservatoire de musique. In one of a series of lectures patronized and attended by the Empress Eugénie to raise funds for the charitable Société des amis des sciences, Fremy's experimental demonstrations were calculatedly grandiose and theatrically presented under the glare of powerful magnesium lamps. The preparation of ozone from oxygen with the aid of an induction coil powerful enough to kill an ox was intended to captivate the eye and the imagination, and journalists, like the audience, were duly impressed. Never, as Victor Meunier put it admiringly, had so many experiments been crammed into one lecture.[131] It was all a far cry from the respectful silence in which scientific information had once been absorbed. It was a far cry too from another in the series, by the mathematician Joseph Bertrand, whose uninspiring style and lack of exciting demonstrations only added to the heaviness of the unpromising subject of the history of the prerevolutionary Académie des sciences.

The accumulating commercial and pedagogical possibilities escaped few of the popularizers of science during the empire. In this as in other respects, Moigno was a pioneer. Indeed, the journal he edited, *Cosmos*, was originally conceived, in 1852, as part of a much larger scheme based on something akin to a modern multimedia science center. From spacious premises in the boulevard des Italiens that the energetic founder and first proprietor of *Cosmos*, Benito de Montfort, had placed at his disposal, Moigno proposed to offer not just the journal but also lectures, demonstrations, a library of recent books and journals, and a "salon photographique" housing a collection of images of great technical originality.[132] Building on the support of de Montfort, who was an early enthusiast for photography and the founder of a short-lived photographic journal (*La Lumière*, to which *Cosmos* was the successor), the "salle du Cosmos" set out to be a pioneer in the art of communication. It prided itself on using the revolutionary wet collodion method for making glass negatives, which had arrived from England and was now sweeping Paris. And further effects were to be obtained by the use of huge projected images and photographs incorporated in the exceptionally fine stereoscopes being produced in the workshop of Jules Duboscq.

Despite the immediate success of *Cosmos* (which achieved a circulation of a thousand within a few months of its launching), the wider scheme got off to a faltering start and soon collapsed. The "salle du Cosmos" seems to have been abandoned by the spring of 1853, the victim of de Montfort's withdrawal, its own complexity, and the difficulty of obtaining police authorization for an activity whose survival depended on its capacity to attract precisely the large numbers that were thought to promote the risk of disorder. Even if the sales of *Cosmos* offered consolation,[133] Moigno was left to look enviously across the Channel to the Royal Panopticon of Science at Leicester Square in London.[134] There, since 1850, an initiative similar to his—incorporating demonstrations of industrial and artistic processes, a publicly accessible chemistry laboratory and cabinet of physical apparatus, facilities for practice and instruction in photography, captivating lectures, and projected images of distant lands—had enjoyed a popularity that his own venture failed, even fleetingly, to match.

Despite the setback, Moigno remained convinced that a substantial lay market for science existed and that it would respond if only the right profile of alluring subjects, imaginative presentations, and congenial premises could be found. Some years later, in 1872, he tried again, this time with lectures, illustrated once more with projections from photographic images on glass slides. The setting, in his "salle du Progrès" in the cité du Rétiro, a courtyard off the elegant faubourg Saint-Honoré, was grandiose,[135] and the seriousness of the venture, which Moigno conceived as the first of many, was reflected in the fifty thousand copies of a prospectus that he had printed.[136] But after three months, public preoccupation with political strife and a consequent neglect of culture led to failure. Even then, Moigno was not discouraged, and within three years the lectures he gave in the working-class suburb of Saint-Denis were regularly attracting audiences of more than two thousand. There, as Moigno recounted with obvious pride, he addressed "all classes of society," including both men and women and even children. The fatal flaw was that Saint-Denis in the early Third Republic was ideologically contested territory in which religious and secular voices were profoundly at odds. After two or three years Moigno's success attracted the attention of the free-thinking Ligue de l'enseignement, which objected to the providentialist tone of his representations of nature (especially in his Sunday evening discourses on his favored theme of the agreement between faith and reason). Moigno finally admitted defeat, and the lectures were abandoned.

The ending of Moigno's activities as a lecturer and the subsequent sale of his slides and projection equipment that followed in 1882—including 4,500 slides that were sold at 1.25 francs each—signaled a personal failure hastened by the aggressive secularism and intolerance of the republican left. It also reflected heightened competition both from other independent initiatives and from *universitaires* who, since the midcentury, had been urged to redouble their efforts to address lay audiences. In promoting lectures by professors, Duruy's personal commitment and his ambitions for his ministry were crucial. In addition to responding to the bourgeois taste for the spoken word, it also had the political objective of curbing the conflicting ambitions of the church, on the one hand, and radical free-thinkers, on the other, to lead in the task of filling minds. In large measure, Duruy succeeded. By a ministerial decision of 1864, he encouraged the professoriate under his jurisdiction to offer proposals for free public lectures that, once approved, could be given in a faculty, taking advantage of equipment and other facilities. Professors throughout France responded to the call. For the academic year 1865–66, Duruy was able to approve just over a thousand lectures and lecture courses, 349 of them given by professors and others teaching in the University.[137] With regard to subjects, while literature and history had pride of place (with 497), pure and applied science (with 223) accounted for a respectable share and did even better in the following year (with 313 out of 893).[138]

For some professors, what was now asked of them implied little change in their established practices. Henri Lecoq, for example, had long been a fixture in the cultural life of Clermont-Ferrand, and he settled easily into the new, public routines, though without ever shaking off the reputation for superficiality that he had in the ministry (see chapter 2). Others had to make a more conscious effort to adapt to the call. The lectures, as Duruy explained to the rector of the academy of Montpellier, should be at once useful and morally uplifting for working-class audiences and "an elegant and beneficial diversion" for the higher classes of society.[139] The balance was a fine one, and the boundaries between what was and was not permissible were precisely drawn. The moral and political criteria for the approval of lectures required that lecturers, whether *universitaires* or not, should be men of "maturity, experience, and ability," while their subjects had to avoid even the most oblique reference to religion or politics.

In this scrutiny, faculty-based proposals tended to fare better than those from independent lecturers. Duruy apparently had no hesitation in authorizing the professor of zoology at the faculty of science in Strasbourg, Dominique-Auguste Lereboullet, to deliver a course on popular zoology destined for women: a report from the prefect of the

Bas-Rhin testified that Lereboullet's character was sufficient guarantee of the soundness of his teaching.[140] Deschanel, in contrast, was less kindly treated when he applied for permission to lecture, also in Strasbourg, in 1866: the prefect and the rector of the academy were as one in presenting lectures by him as a potential focus for disorder at a time of instability following recent elections in the city.[141] Such decisions were guaranteed to provoke indignation among liberals, and protests duly erupted when Duruy intervened to suspend the lectures on the history of the medieval church that the journalist Jules Labbé had already begun in the rue de la Paix.[142] As liberal critics insisted, this was far from an isolated decision: Labbé, in fact, was only the latest in a series of prominent victims.

Although the public lectures that were given in response to Duruy's encouragement were well attended across France, the largest audiences assembled in Paris. The "soirées scientifiques et littéraires" at the Sorbonne, launched with ministerial approval as a rival to the independent lecture courses, marked the pinnacle of the movement. There were to be two lectures a week, Monday evenings on science and Fridays on literary subjects. When Jules Jamin, professor of physics at the Sorbonne and at the Ecole polytechnique, inaugurated the series in March 1864, the Sorbonne's main lecture hall was besieged. It was said that thousands, including many women, spurned the attractions of the theater to hear Jamin discourse on the three states of matter. By the beginning of the lecture, two thousand were inside, with far more outside.[143] With a streak of amused irony, Victor Meunier's account conveyed the excitement:

> By the time I arrived, the efforts of the attendants darting about the edges of the herd had succeeded in forming it into three columns, which were now laying siege to the door of the Sorbonne. Once the door opened, the waves of the triple current merged and surged through the archway. It was a dangerous moment for crinolines . . . I was momentarily raised from the ground, transported four paces like an erratic boulder, and then deposited safe and sound inside the courtyard.[144]

For a quarter of an hour, even the minister himself could not get in.

The struggle was worth it. Jamin's lecture surpassed all expectations, with projection equipment, a Serrin arc lamp that could be directed to illuminate individual instruments, and gas lighting that the lecturer could raise or lower at a touch on the controls and in a way that seems to have overexcited the younger members of his audience. Meunier, again, was the witness: "A younger element in the audience, forgetting where it was, shouted 'Encore !,' a command that an assistant hastened to obey. But a discrete, though expressive gesture by the professor let it be known that such cries were not to be repeated, and the audience heeded the warning."[145] In Meunier's words, Jamin's aim had been to "reach the mind through the eyes," and he had succeeded triumphantly. As with all the most successful popular lectures of the period, the emphasis in his performance had been firmly on the visual rather than on abstraction. The mediatic means, though, did not necessarily eclipse the higher ends. The most famous of all the lectures at the Sorbonne, given by Pasteur on spontaneous generation in April 1864, broached one of the most difficult and contentious areas of science. Pasteur's aim was not just to please his distinguished audience, which included Princess Mathilde, George Sand, and Duruy; it was also, and more seriously, to take the decisive step that it marked on the way to closure in the tiresome public debate between him and Pouchet.

The celebrity of leading savants such as Jamin and Pasteur and the prestige of the Sorbonne gave lectures there a special cachet. Like the lectures in the boulevard des Capucines, they were calculated to appeal first and foremost to cultivated audiences, and their success points to a degree of intellectual engagement (as opposed to more "decorative" cultural attainment) that has tended to be disregarded in studies of bourgeois culture. The mix of gaiety and seriousness so characteristic of this form of public culture in the later Second Empire was a compelling one, and it survived the transition of 1870 without difficulty. A decade into the Republic the lecture room in the boulevard des Capucines was still drawing respectable audiences. Hundreds would gather for a lecture by Flammarion in the 1880s, as they had done twenty years before. And in 1882 the annual late-night "soirée scientifique" at the Observatoire de Paris, organized around a lecture by the astronomer Charles Wolf, drew 1,500.[146] The quietness of Wolf's delivery evidently detracted from his account of the role of spectroscopy in investigating the chemical composition of the stars, and it may well be that the lecture's success owed as much to Dubosq's projections of images as it did to Wolf's exposition. For, as always, audiences for science demanded some element of entertainment and spectacle. Those who attended the soirée of 1882 would certainly have had a more vivid memory of the night of dancing in marquees erected on the terrace of the observatory than they had of what actually was said.[147]

Although the taste for lectures remained strong at the end of the century, things were changing. As in the case of the written word, the accelerating professionalization of science and the associated alliance between the academic profession and the republican regime were working inexorably to marginalize the great independent showmen of the imperial lecture circuit and their successors in the early years of the Third Republic. At the beginning of the twentieth century, most professors in a faculty of science would still see it as their duty and (in response to continued ministerial badgering) in their interest to speak to the wider public as well as to formally enrolled students. But they were now more likely to do so in the context of one of the groups of friends and supporters that most universities founded in the 1890s. In lectures to such audiences and in the prestige-enhancing institutional publications that often went hand in hand with them, they were performing in a formally constituted academic showcase, and their style was fashioned accordingly. The *Annales de l'université de Grenoble,* begun in 1889 as a vehicle for the best in scientific and scholarly work across the university, was typical of house journals in the new style. Authors who wrote for it and for other such publications could never forget, and they never let their readers forget, that they were writing as leading, formally qualified practitioners of their various disciplines.

Despite the sobriety of the new context, skillful presentation remained important. But both style and content changed. Professors now tended to give a higher priority to the interpretation of new developments in the specialized research in which they were engaged than they did to stimulating audiences with the aid of the dramatic demonstrations or the unorthodox speculations that had captivated the lecture-going public of the 1860s. Then, Henri de Parville had set the tone for popular science by a brutal declaration that "it has long been common knowledge that there is nothing so tedious as official science."[148] Within three decades, however, it was precisely the science from which de Parville had distanced himself that academic scientists were routinely purveying, in suitably simplified form, as their offering for at least the more informed end of the popular market.

Broader Audiences, Bigger Stakes

The success of publications and lectures for a broad nonspecialist audience demonstrates the growing commercial potential of public science during the Second Empire. It is no coincidence that some of the most successful popular journals of the century, spanning the whole range of price and sophistication, began in this period. A typical product at the modest end of the scale was *La Science pour tous*. Launched by the anticlerical socialist Henri Lecouturier as a weekly in 1855 with few illustrations (normally two or three in an issue of eight pages) and rather shabby paper and print, *La Science pour tous* adopted a respectful, factual style that would not have been out of place in writing for a popular audience a quarter of a century earlier. Where it broke new ground, however, was in cost: it prided itself on being the cheapest periodical of its kind, at only 10 centimes a copy or, in Paris, 5 francs for an annual subscription (6.50 francs for a bound volume of a whole year's issues). From that base, it successfully carved out a substantial family readership and survived, with only minor concessions to the possibilities of more alluring presentation, until 1901. Another of Lecouturier's weekly publications, *Le Musée des sciences* (founded in 1856 and transformed four years later, into *La Science pittoresque*), was directed at a similar section of the public. So too were Victor Meunier's ventures into this market—*L'Ami des sciences. Journal du dimanche* (1855–62) and *Le Courrier de l'industrie* (1862–63, then transformed into *Le Courrier des sciences, de l'industrie et de l'agriculture*, 1863–65). Like *La Science pour tous*, Meunier conceived these journals as part of his socialist ideal of diffusing scientific knowledge among the working classes—and to that end they remained cheap (generally 10 centimes per copy, 15 centimes for *L'Ami des sciences*), rudimentary in the quality of their physical production and the intellectual demands they made on their readers, and unswervingly confident in science as a fount of material well-being and freedom.

Initiatives that were addressed to more affluent readers could afford to be more adventurous, in both appearance and content. Moigno's *Cosmos*, which cost 50 centimes per (weekly) issue, was preeminent in this superior class of periodical. But it had its rivals. Readers who consumed the rather uncompromising contributions to *Cosmos* might also be tempted by a new category of more or less annual volumes offering miscellaneous articles and reviews of recent science. A pioneer in the genre was Jacques Babinet's *Etudes et lectures sur les sciences d'observation et leurs applications*; in eight volumes between 1855 and 1868, Babinet republished many of his numerous contributions to the *Revue des deux mondes* and the *Journal des débats*.[149] By the time Flammarion followed in Babinet's footsteps (almost certainly with his encouragement) with nine volumes of *Etudes et lectures sur l'astronomie* between 1867 and 1880,[150] the field was becoming crowded. Victor Meunier's *La Science et les savants* for each of the years from 1864 to 1867 was nearing the end of a short, polemical life in which Meunier relentlessly pursued his campaign against *la science officielle*. But other annual publications, such as Pierre-Paul Dehérain's *Les Progrès des sciences* (1861–70), Henri de Parville's *Causeries scientifiques. Découvertes et inventions. Progrès de la science et de l'industrie* (1862–95), and Figuier's *L'Année scientifique et industrielle*, an occasionally opinionated but generally sober review of the year's science that survived from 1856 to 1913, showed no signs of flagging.[151] The articles in all these series were often lengthy and scientifically (though not mathematically) demanding, and the presentation (with only very occasional illustrations) was austere. Austerity, though, was no deterrent. Selling typically at 3.50 francs for a volume of sometimes more than five hundred pages, these journals offered the serious amateur an economical way of keeping up with discoveries and debate.

Another publishing genre that began to proliferate in the later years of the Second Empire was that of the large-format book, published on high-quality paper and incorporating engravings of exceptional quality and occasional color plates reproduced by chromolithography. These were luxury products aimed at a relatively affluent family audience that was at once educated and prosperous, and they were priced accordingly. Typical volumes such as *Le Ciel* (1864) by the writer, journalist, and former mathematics teacher Amédée Guillemin, Félix Pouchet's *L'Univers* (1865), and *L'Espace céleste et la nature tropicale* (1865) by the astronomer Emmanuel Liais were priced at 20 francs. But they sold well, and all went into successive editions: five of *Le Ciel* and three of *L'Univers* by 1877, followed by a second edition of *L'Espace céleste et la nature tropicale* in 1882. It is a mark of French eminence in this type of writing that success was often echoed abroad. No fewer than twelve English editions of *L'Univers* appeared between 1870 and 1895, with three further reprintings following between 1902 and 1909.

The stock-in-trade of these works was a mixture of authoritative exposition and a visual appeal often enhanced by fine bindings and gilt-edged pages that fitted them as accoutrements of the middle-class home. Their broad appeal and handsome presentation, however, generally made them no place for polemic. On this score, Liais's chapter on the discovery of Neptune in *L'Espace céleste* was exceptional. In view of his suffering between 1854 and 1857 as a victim of Le Verrier's rule at the observatory, Liais must have derived satisfaction from being able to use his book to cast doubt on the role that Le Verrier was commonly believed to have played in predicting the location of the planet.[152] In an account of sustained virulence, and with a confidence bred of his recent appointment as head of the imperial observatory in Rio de Janeiro, Liais presented the correctness of Le Verrier's prediction in June 1846 as wholly fortuitous. Moreover, as Liais described it, the prediction was one that Eugène Bouvard had already presented to the Académie des sciences in September 1845, following a much earlier suggestion by his uncle Alexis Bouvard.[153] Hence, while the "telescopic" discovery of Neptune belonged, as was generally agreed, to the German astronomer Johann Gottfried Galle, the "geometrical discovery" was Bouvard's, a judgment that left Le Verrier with the status of a mere "French calculator" whose work was profoundly flawed.[154]

Even if controversial intrusions such as Liais's on Le Verrier were rare, the authors and publishers of books for the luxury market were not indifferent to what a measure of excitement could do for sales and visibility. But generally the riches of nature offered excitement enough: marvels such as those that Liais depicted in his accounts of the tropical forests of South America required no embellishment.[155] Manifestly, the formula worked, and the successes of the pioneering works of the Second Empire continued after 1870. Now, Flammarion's *Astronomie populaire* (1880) and its substantial supplementary volume *Les Etoiles et les curiosités du ciel* (1882) led the way, at least in terms of sales. By the standards of the genre, these volumes were moderately priced, at 12 francs each for works that contained almost four hundred illustrations. And this helped them to become and remain best sellers, *Astronomie populaire* selling seventy-five thousand copies within six years, one hundred thousand by 1892, and one hundred twenty-five thousand by 1925, the year of Flammarion's death.[156] Periodicals too went from strength to strength. The *Revue scientifique* exemplified both continuity and the rewards of adaptability to an evolving public. It had begun life in 1863 as the fortnightly *Revue des cours scientifiques*. Like its companion in the humanities, the *Revue des cours littéraires*, it was devoted mainly to accounts and complete texts of academic and serious popular lectures. In this form, it served from the start as an accessible shop window for the professoriate of higher education and helped to advance the reform movement while serving the growing public of teachers and academic scientists who had professional reasons for wishing to keep abreast of developments in science beyond their own discipline. After being renamed in 1870, it identified even more closely with the world of academic science and edged further away from the popular market. Yet it remained a widely read periodical and survived, somewhat changed but apparently without difficulty, until its demise in 1954.

One measure of the undiminished vibrancy of the popular market is that as the *Revue scientifique* adjusted to cater for its new, more professional readership, a newcomer, the weekly *La Nature*, moved in to fill the gap. Founded in 1873 with the usual fanfare of patriotism and calls for a greater investment in science,[157] *La Nature* soon became the outstanding periodical of its kind. It distinguished itself as an *illustrated* journal, in contrast with the *Revue scientifique*, which had rather few illustrations, especially in its early years. By offering high-quality paper, the same large page format as the *Revue scientifique*, and engravings comparable with those found in luxury books, it set a far higher aesthetic standard than had been the norm in earlier periodicals for the mass market. The result was an unprecedentedly attractive publication for which writers and illustrators of distinction (including the outstanding Louis Poyet among the engravers) were happy to undertake commissions. Essential to success, however, were the publisher Georges Masson and the founder and first editor Gaston Tissandier, a former collaborator of Edouard Charton on the highly successful *Magasin pittoresque*. Masson and Tissandier targeted very precisely a discerning, relatively affluent readership willing to pay 50 centimes an issue (20 francs for a year's subscription) for sixteen pages of articles, reviews, and snippets of news expertly written for the purpose.

In physical quality and pricing, *La Nature* responded to rising bourgeois prosperity and an expanding readership with a level of scientific literacy that rose above simple enthusiasm and a taste for the sensational. It maintained a respect between expert and lay public not only in the authoritativeness of its articles but also in its weekly correspondence column, in which readers' queries were answered in simple but never condescending terms. Some emphasis on accounts of strange aspects of nature and distant lands, a relatively light coverage of theoretical matters, and a virtual absence of mathematics were predictable enough. But few other concessions were made. From a respectable start, when it had a print run of two thousand, *La Nature* advanced to a circulation of fifteen thousand in 1885[158] and only ceased publication in 1964, when it was assimilated into the weekly *La Recherche*, which still survives.[159]

For its publisher, Georges Masson, *La Nature* was no isolated success story. Continuing a tradition inaugurated by his father, Victor, Masson deployed imagination and commercial good sense in exploiting the parallel growth in popular and academic science in the later nineteenth century. His well-stocked stable, which included leading journals at all levels of seriousness and accessibility (except the most trivial), allowed him to compete not only with established specialists in scientific publishing, such as Bachelier (later Mallet-Bachelier), Germer-Baillière (see below), and Gauthier-Villars, but also with the more broadly based houses of Félix Alcan, Hachette, and Flammarion.[160] Although medicine and the life sciences remained Masson's main speciality, his list, extending to over a hundred pages not long after his death in 1900 and including fifty-two journals, came to embrace all the sciences and technology. His adventurousness paid off, especially in his judicious response to the explosion under the Third Republic in the demand for textbooks and for books and learned journals for specialized professional communities. Offerings that included journals of cutting-edge research ranging from the *Annales de chimie et de physique* and the *Bulletin de la Société chimique de Paris* in the physical sciences to specialized medical periodicals such as the *Revue d'hygiène* and the *Annales de dermatologie et de syphiligraphie* marked him as a discerning publisher willing to take risks with productions that only academic institutions and career savants were likely to buy. With Masson, though, risk was always tempered by a solid base of titles of guaranteed profitability. Such works as the standard *Dictionnaire des sciences médicales* (in a hundred volumes), Georges Dieulafoy's prodigiously successful *Manuel de pathologie interne* (with sales of seventy thousand), the many successive editions of Louis Troost's textbook *Précis de chimie*, and the magnificent account of the voyages of the *Travailleur* and the *Talisman* provided a more than comfortable commercial cushion.[161]

While purveyors of the printed word—both publishers and writers—prospered under the Third Republic as they had done during the Second Empire, they were not alone in their recognition of the market for science. A particularly successful Parisian business that exploited the emerging vogue for do-it-yourself natural history was that of Emile Deyrolle fils. The business, in the rue de la Monnaie and, from 1888, in the rue du Bac (where it still exists), had been founded in 1831 by an enthusiast, Jean-Baptiste Deyrolle, and had remained in the family. Deyrolle and his successors maintained a consistently keen entrepreneurial edge. By the Third Republic, the business was well placed to take advantage both of the growing prominence of natural history in education, especially primary education, and of amateur tastes at all levels of competence and expenditure. For primary schools, it produced a remarkable set of twenty wall charts—covering human anatomy, zoology, botany, and geology—all delivered with appropriate specimens for 24 francs (see figure 10).[162] These encapsulated the Deyrolle ideal of instruction, based on observation rather than abstraction, that would respond to the small child's "innate love of natural history."[163] They also helped to instil the lifelong interests for which Deyrolle catered with offerings for adults ranging from the half-hearted amateur to the most discerning collector.

Deyrolle's core trade, in fact, was that of a dealer in specimens and equipment, and he conceived his popular manuals and other publications as props to his main activity. His *Petites nouvelles entomologiques*, a four-page newssheet that appeared from 1869 to 1879, and its more ambitious and wide-ranging successor, *Le Naturaliste* (1879–1910), were directed first and foremost at naturalists of modest means and aspirations. To judge by their longevity and unbroken frequency (two issues a month for more than forty years), they struck a winning formula. Their mixture of entomological gossip, brief accounts of lectures and publications, descriptions of outstanding collections, and advertisements by dealers and by enthusiasts offering specimens or seeking exchanges changed little, and there was no reason to change. Deyrolle's associated trade, in particular at the popular end of the market, in coleoptera and lepidoptera, duly prospered as well. At the bottom of the range, Deyrolle offered a beginner's collection of one hun-

dred French coleoptera for as little as 12 francs, while, higher on the scale of sophistication, a thousand "exotic" (i.e. non-European) coleoptera could be had for 350 francs.[164]
Lepidoptera were somewhat more expensive: one hundred fifty specimens ranging over
a hundred European species were offered for 37 francs, but, for 750 non-European specimens (representing 500 species), the price rose to 900 francs.[165] At prices way above the
few hundred francs for which good collections typically changed hands, there existed
another, more esoteric market, peopled by specialists. For a complete collection of one
hundred thousand specimens, including thirty thousand species of coleoptera in magnificent glazed drawers, the price was 50,000 francs, said to be half of its true value.[166]

The range of these prices underlines one of the strengths of the late-nineteenth-
century boom in participatory science. There was a place for virtually everyone. The
microscopes that Deyrolle advertised as essential for the enjoyment of a collection sold
at between 20 and 450 francs, with the average at around 280 francs for a microscope
offering a magnification of seven hundred diameters that sufficed for most work, at least
in entomology.[167] By comparison, amateur astronomy had a slightly higher monetary
threshold, although here too investments could vary over a wide range. Flammarion's
advice was that an expenditure of 100 francs was the minimum for even the simplest
observations.[168] For this amount, the Parisian instrument maker Bardou offered a simple refracting telescope with an objective lens 57 millimeters in diameter and with a
focal length of 85 centimeters, giving a magnification of thirty-five times. With such an
instrument, it would be possible to see the details of the surface of the moon, the satel-
lites of Jupiter, large sunspots, the more distinct phases of Venus, the nebulae in Orion
and Andromeda, and (just) the rings of Saturn. But models incorporating larger objective lenses and more powerful eyepieces opened the realms of sidereal as well as planetary astronomy. Bardou's top-of-the-line refractor, with an objective of 108 millimeters
in diameter and a focal length of 1.6 meters, would allow observations of stars of the
twelfth magnitude and the separation of double stars down to 1 second of arc. Such an
instrument, with the indispensable sighting telescope and eyepieces allowing magnifica-
tions of up to 250 times, would cost anything between 600 francs and (with a high-
quality equatorial mounting) 1,450 francs. This constituted "the core equipment of any
private observatory,"[169] and its owner would be "the happiest of mortals," a true citizen
of the skies.[170] Used in association with a first-class star map, such as Flammarion's revision of Charles Dien's *Atlas céleste*,[171] it was clearly the telescope to which the ambitious
amateur was expected to aspire. In this, as at all levels, an endorsement by Flammarion
carried immense weight. By October 1881, less than two years after the publication of
Astronomie populaire, at least three hundred readers were said to have equipped themselves with telescopes in accordance with the advice that the book gave.[172]

In a period in which popular science was selling so well, it is easy to overlook the
failures. One such failure was Figuier's venture into writing and producing plays on
themes drawn from the history of science. Figuier's conception of "scientific theater" sat

easily with the goals of all his writing. Theatrical works in the genre were to provide an elevating diet that would simultaneously entertain and instruct in a manner appropriate to an age of science and progress.[173] His own contributions included "Les six parties du monde" (a series of tableaux inspired by Dumont d'Urville's journey to Antarctica in 1839) and a life of Johannes Kepler that used imaginative biographical material and the visual highlight of a projection of astronomical slides to convey an heroic vision of Kepler and his struggle against the superstitions of his age. The tone was relentlessly moralizing and didactic. In "Les six parties du monde," for example, d'Urville's adventures provided a tenuous structure on which snippets of scientific information and *aperçus* of the habits of unfamiliar peoples (including the Chinese and the Mormons) could be hung. Unlike most of the dozen or so plays that made up Figuier's *corpus*, this one had modest commercial success. It was performed on forty-four occasions at the Théâtre de Cluny in 1877, largely, it seems, to captive audiences of schoolchildren, and then in a number of provincial theaters.[174] Through the 1880s, "Le mariage de Franklin" (with an appropriate theatrical storm), "Gutenberg" (a heroic view of the invention of printing), and re-creations of the lives of such inventors as Denis Papin and Samuel Morse were all performed in Paris. But by 1889 even the indomitable Figuier had to admit that he faced obstacles and marks of indifference comparable with those which had clouded his early years as a scientific writer.[175] No amount of visual gimmickry could salvage his dream of integrating science with theater in the way that Jules Verne had already effected the union of science and the novel.

The fact remains, however, that the successes of popular science far outweighed the failures. In this, the five universal exhibitions that took place in Paris between 1855 and 1900 played a role, sometimes providing facilities that became fixtures in Parisian culture. A 3,200 square meter aquarium, constructed under the slopes of the new Trocadéro Palace for the exhibition of 1878, quickly established itself as a popular excursion,[176] as did a seawater aquarium, built for the exhibition of 1900 and equipped (like the aquarium of the Trocadéro) with a laboratory.[177] An equally ambitious venture that grew from the 1878 exhibition was the public observatory installed in the towers and on the terraces of the Trocadéro. Opened in 1880, the observatory offered high-quality instruments for public use.[178] For more than four hours a day, visitors had access to telescopes and projection equipment (including one that gave an image of the sun 3 m in diameter), and there were plans for a school of astronomy, popular lectures, a library, and a range of powerful microscopes. Although the dream of the observatory's entrepreneurial founder, Léon Jaubert, was never fully realized, the Trocadéro survived until the mid-1890s as a Parisian resort for practical astronomy. If it did not have the enduring success that Jaubert hoped for, the explanation does not lie in any lack of support either from users or from a number of eminent savants who backed the venture. The true obstacle was the more banal one of Jaubert's uneasy relations with the management of the Trocadéro and the consequent difficulty of retaining a foothold in the building.

Jaubert's case for staying in the Trocadéro would have been stronger had it not been for the proliferation of alternative ways of learning and practicing astronomy during the 1880s and 1890s. In 1880, Flammarion launched a well-produced monthly journal, *L'Astronomie*, that offered subscribers a sense of closeness to the leaders of the discipline, while remaining accessible and of practical value to the enthusiastic amateur; at a reasonable 12 francs for a year's subscription, within a year it achieved a circulation of six thousand, a remarkable figure for a periodical dealing with a single science.[179] Seven years later, another Flammarion brainchild, the Société astronomique de France, was founded with precisely the same clientele in view. Again it succeeded: by 1894, membership, open to "all soldiers in the cause of progress," stood at six hundred and by 1902 at almost three thousand.[180] Success on that scale helped to make Flammarion an influential advocate. It was a combination of his own authority and the standing of the Societé astronomique de France that made possible one of his most spectacular exploits: the reinstallation of a Foucault pendulum in the Panthéon in 1902, where it had briefly hung more than fifty years before (see figure 11).

Jaubert's enterprise also had to compete with facilities, better than his, that complemented what enthusiasts could do with telescopes for home use. For a favored few, from 1887, there was the possibility of an invitation to the private observatory that Flammarion installed at Juvisy (see figure 12), to the south of Paris, in a house offered to him in 1882 by Eugène Méret, an elderly amateur astronomer who had abandoned the property following its ransacking by Prussian troops in 1870.[181] And by the 1890s members of the Société astronomique de France had access to the society's observatory on the roof of the recently inaugurated Hôtel des sociétés savantes, a building providing facilities for more than thirty societies in the rue Serpente in the Latin Quarter.[182] Here, a 108–millimeter Bardou equatorial refractor (soon to be augmented by an even more powerful Mailhat refractor), a meridian telescope donated by the instrument maker Secrétan, and a library (doubling as a lecture room and improvised sleeping space for observers) were all of a quality to which only an exceptionally well-heeled independent astronomer could aspire.[183] The observatory quickly became popular with the enthusiastic young observers whose nocturnal activities earned them the title of "gnomes," some of whom, notably Henri Chrétien, the future director of the Institut d'optique in Paris, went on to careers in astronomy. But it responded to other needs and levels of competence as well. Talks and courses of practical instruction for beginners always had as high a priority as lectures on spectral analysis and similar more advanced topics: seats for the informal discussions on popular astronomy that Gaétan Blum, one of the ablest of the "gnomes," led on the first Sunday of each month were always in short supply.

For the devotees of museums, an enduring favorite was, as it still is today, the Muséum d'histoire naturelle, which assumed great popularity, especially after the construction of a magnificent new gallery of zoology in 1889 (see figure 13) and a gallery of paleontology in 1898 (replacing a temporary though already much-visited installation of

1885).[184] The historic collection of apparatus, models, and machinery in the Conservatoire national des arts et métiers also attracted visitors once the notoriously dilapidated display dating from the July Monarchy had given way to the more engaging installation in the old abbey church of Saint-Martin-des-Champs, fitted out in the 1850s and further improved in the 1880s and early 1890s.[185] Temporary exhibitions too enjoyed great popularity. The first international exhibition of electricity in 1881 drew 880,545 visitors (88,000 a day) to the Palais de l'Industrie in thirteen weeks.[186] And on a lesser scale, the Muséum's display of the extraordinary creatures that had been dredged from off the west coast of Portugal and of Africa during the expeditions of the *Travailleur* and *Talisman* between 1880 and 1883 was one of the cultural landmarks of 1884.[187] Despite a presentation that sacrificed immediate allure to the exhaustiveness of six thousand specimen jars crammed in a limited space, the exhibition drew a large public. At one level, the fascination in this case lay simply in the weirdness of creatures living at depths of up to 5,000 meters. But scientists and informed lay visitors alike were also excited by the exhibition's visual endorsement of Alphonse Milne-Edwards's long-standing opposition to the prevailing view, originating with Edward Forbes, that life could not exist at depths of more than about 500 meters. Science once again had demonstrated its capacity to set imaginations racing by opening a window on previously hidden mysteries of nature.

The convergence of the vogues for exhibitions, museums, and amateur science pursued at home or on country rambles points to the last two decades of the century as the pinnacle in France's golden age of scientific popularization. One striking indicator was the simultaneous success of a number of book series distinguished by their longevity, relatively modest cost, and wide diffusion.[188] Moigno, once again, was a pioneer: the series of books that he marketed as Actualités scientifiques was both successful and (especially in its early years) important as a conduit for translated texts, most notably works by John Tyndall: between 1866 and 1878 an average of five volumes a year were published at prices ranging from 1 to 6 francs.[189] But despite the guiding hand and prestige of Gauthier-Villars as its publisher, the series always had a miscellaneous quality and it never enjoyed the visibility of the Hachette Bibliothèque des merveilles, which quickly became something of a French cultural institution. In the three decades of its existence, from 1864 to 1895, the series came to include some 130 titles, mainly though not exclusively on scientific subjects. Under the general editorship of Edouard Charton as part of his campaign in favor of popular education, its authors included many of the great popularizers of the age, some (such as Wilfrid de Fonvielle and Amédée Guillemin) offering more than one title.[190] At the price of 2 francs for paperbound and 3 francs for the distinctive blue clothbound edition, the volumes sold spectacularly well: in all, 1,750,000 copies were printed in a uniform pocket-sized format well adapted to, among other locations, train station bookstalls that began to appear in the 1850s.[191]

At the other end of the spectrum of difficulty and cost lay the Bibliothèque scientifique internationale, which continued to offer a regular flow of new titles from its foundation in 1873 until 1919. In doing so, it reflected the trend, characteristic of these decades, toward the reengagement of academic scientists in the task of popularization. Quatrefages, Paul Schützenberger, Marcellin Berthelot, and Adolphe Wurtz were among the leading figures of republican science who contributed substantial volumes to the series, normally selling at 6 francs in the standard clothbound octavo edition (10 francs in leather). Such authors used the series to lay out positions of current scientific import; it was in an early volume, for example, that Wurtz articulated a particularly clear and detailed public statement of support for the atomic theory in 1879.[192] Despite its seriousness, the Bibliothèque scientifique internationale appears to have sold well, helped in this respect by the recommendation of well over half the volumes by the Ministry of Public Instruction.[193] Even if the print runs never matched those of the Bibliothèque des merveilles, the publishers, Germer-Baillière and from 1883 (following a merger) Félix Alcan, put out 122 titles in the forty-six years of the series' existence, almost a hundred of them by 1900.[194]

The resolute internationalism of the Bibliothèque scientifique internationale made it true to its name. Especially in its first decade or so, when more than half of its titles were the translated works of foreign authors, the series promoted science as a pursuit in which geographical boundaries, even those with Germany, counted (or should count) for nothing. It grew from a proposal first broached at the Edinburgh meeting of the British Association for the Advancement of Science in 1871, for an international agreement that would promote the simultaneous publication of scientific books of outstanding importance in several languages. Kegan Paul's International Scientific Series in Britain, the Brockhaus Internationale wissenschaftliche Bibliothek in Germany, and the Fratelli Dumolard Biblioteca scientifica internazionale in Italy were associated collections with the same ideal and, in many cases, the same titles, translated where necessary into the language of the particular series. While the internationalism of all the participating publishing houses was beyond question, that of Germer-Baillière had a special intensity. The fount of the commitment was the head of the company, Gustave-Germer Baillière, who had studied medicine and gone on to promote a particularly strong scientific and medical list, including a significant number of titles translated from German. Baillière's vision was not just commercial. For him, the Bibliothèque scientifique internationale reflected a personal commitment to the struggle against the insularity of French intellectual life that an important dynasty of Baillières had already waged in their publishing ventures since the early nineteenth century.[195] Hence when his publicity literature commended the series as a way of learning about science in America and elsewhere in Europe without leaving France, he was offering far more than a cultural option. Keeping up with developments abroad verged on a patriotic duty, especially for a man of the left who saw the Republic as a new beginning after the humiliation of 1870.

The prominence of acknowledged disciplinary leaders among the authors of the Bibliothèque scientifique internationale is a clear indicator of the wider challenge that academic scientists presented to the old school in the communication of science. When new titles in the series became somewhat less frequent, as they did after 1900, the challenge continued; other series, usually more cheaply produced, provided ample alternative stages for scientists with formal qualifications and appointments. One existing series that welcomed such authors was the Bibliothèque scientifique contemporaine, published by another branch of the Baillière family, J.-B Baillière et fils, since the 1870s at a price that varied little from a standard 3.50 francs: here titles by major university professors (including Claude Bernard, Albert Gaudry, and Emile Duclaux) rubbed shoulders with works by independent savants, such as Antoine de Saporta and the anthropologist and explorer, the marquis de Folin. The mix made for an eclectic list that maintained a significant, if diminishing, place for the scientific amateur as well as the professional, though only for the amateur with demonstrated disciplinary competence.

Among the series that flourished in the new century, none could compete with a vibrant newcomer, the Bibliothèque de philosophie scientifique.[196] Launched in 1902 by the publishing house of Flammarion under the general editorship of the sociologist and psychologist Gustave Le Bon, the Bibliothèque de philosophie scientifique had, as its first two volumes, Le Bon's own *Psychologie de l'éducation* and Henri Poincaré's *La Science et l'hypothèse*. In a striking orange paper cover and selling at 3.50 francs, these and subsequent volumes (112 titles and 650,000 copies by 1914) became a familiar presence in the libraries of educated households on through the 1930s and beyond. No initial print run fell below 1,500, and many volumes had a huge success. *La Science et l'hypothèse* led the way with twenty-four thousand copies printed by 1914 (fifty thousand by 1938), but the sales of many other titles topped ten thousand copies. Although the level was demanding and the style rigorously expository, the subjects, typically at the interface between science and philosophy, placed scientific themes at the heart of general literate culture. Since all but a handful of the authors were French, the books also did much to raise the domestic profile of the nation's universities and *grandes écoles*, where the great majority of the authors taught.

The brisk but changing market for popular science about the turn of the twentieth century made its inevitable mark on the profile of books and periodicals available in a field that publishers seem to have found irresistible. Shortly before the First World War Félix Alcan showed its confidence in the market with a new series of scientific classics with limited commentary and produced at a price that made them accessible to students and secondary school pupils and their teachers. The aim, as the prefatory matter in early volumes had it, was to promote the reading of primary sources as a way of inculcating a "scientific spirit" while also contributing to students' literary culture; a selection of Léon Foucault's papers on the velocity of light and the mirrors of reflecting telescopes occupied a typical early volume.[197] Alcan also engaged the mathematician Emile Borel as the general editor of a new collection of monographs, the Nouvelle collection scientifique, pitched at a level slightly below that of the Bibliothèque scientifique internationale and at a price (the usual 3.50 francs) that made them accessible to students as well as readers with no scientific background.[198] Publishers' enthusiasm makes it hard to evaluate the overall balance between the success and failure of these and other ventures. Certainly there were casualties, some of them significant. The Hachette Bibliothèque des merveilles encountered difficulties after Charton's death in 1890: by 1895 many copies of books in the collection remained unsold and were disposed of at 1 franc each.[199] And even a promising new weekly, *La Science illustrée,* launched by the normally sure-footed Figuier in 1888, died (with Figuier) after only six years. It may be that the failures reflected an incipient trend. As part of a broader argument that journals of popular science began to lose their appeal in this period, Florence Colin has pointed to a reduction by half in the number of such journals between 1895 and 1914.[200] Perhaps fin de siècle questioning of science and scientism played its part, as Colin and Bruno Béguet suggest. But so too did the accelerating pace of academic professionalization, which made a professorial chair an almost essential badge of authority in addressing lay audiences.

The transition in the style and authorship of scientific writing for the general public was dramatic and irreversible. By 1900, Guillemin, Figuier, Moigno, Tissandier, Meunier, Fonvielle, and Parville were either dead or inactive. This left only Flammarion, in his late fifties, as a significant lingering presence. His imaginative style still had its appeal, and his work continued to sell. But he was beginning to seem a slightly incongruous figure when set beside, say, Marcellin Berthelot, whose hard-nosed vision of nature had no place for marvels, whether they were within or beyond the bounds of normal science. When Berthelot wanted to evoke a more exotic dimension as part of his technique of communication, he turned away from the present to a past that he presented as colorful and curious but flawed by gullibility and superstition. His discussion of alchemy was typical. In *Les Origines de l'alchimie* (1885) he gave free reign to his vision of science's history as the story of the long but heroic and eventually successful struggle of true scientific method to combat primitive error. What was left in Berthelot's own day was, as he triumphantly declared, a world "without mystery," a world from which the notion of miracles and the supernatural had "vanished like a vain mirage, an outmoded prejudice."[201] This portrayal of a universe that could, in principle, be fully understood through observation and reason was of a piece with the scientism at the heart of radical republican opinion. To some, Berthelot's relentless militancy in favor of the union of science with the ideals of free thought was exhilarating. But, as the twentieth century dawned, many who wished to retain a place for a spiritual dimension, whether divine or not, in their worldview found Berthelot's an alien vision. In such perceptions, science appeared drab or even (as reflected in the imaginations of Jules Verne or Albert Robida) threatening. At all events, it was hard to square with the overwhelmingly benign conceptions of nature that had inspired the most exciting forms of popular writing since the midcentury.

The passage to an age of popularization dominated by men bearing the formal labels of academic life was encapsulated in the success of one of the most remarkable of all periodicals in popular science, *La Science et la vie*, launched in 1913. *La Science et la vie* presented itself assertively as a new departure. Its small magazine format set it apart from earlier periodicals aimed at the popular market. So too did its heavy use of photography (in contrast with *La Nature's* infrequent flirtations with the medium) and its exploitation of the latest techniques of color printing to create covers in a vivid modernist style in keeping with its subtitle, *magazine des sciences et de leurs applications à la vie moderne*. Most important, its brief, informative articles conveyed a simple image of science as the virtually inexhaustible fount of material improvement and, by an easy though unarticulated extension, as a pursuit whose beneficence could bind together those of all persuasions, political and religious. Though not inexpensive, at 1 franc per monthly issue, *La Science et la vie* set its sights on a large circulation, and soon it had a print run of one hundred thousand.[202] The fact that this figure set *La Science et la vie* so far ahead of the dwindling number of nonspecialist journals that survived on the eve of the war indicates that its mixture of eye-catching presentation and articles by recognized specialists in their various fields fulfilled expectations of what a magazine of popular science should offer.

One thing that *La Science et la vie* did not convey was any sense that the general reading public might have something to contribute to debate about science and its applications. Instead it offered a peek into a world of academic research and industry to which the profane had no right of entry. This distancing of readers was of a piece with the trend to a progressive "black-boxing" of science and technology and the privileging of entertainment over instruction that occurred in successive universal exhibitions between the 1850s and 1900.[203] Most areas of science belonged in 1914 in the realm of barely accessible esoteric knowledge in which the expert ruled. This is not to say that popularization was a dead or dying art. The sales of *La Science et la vie* and of the new generation of cheap books of which Poincaré's *La Science et l'hypothèse* was an outstanding example suggest that on the eve of the war there still existed a substantial public of readers willing to be cast as consumers and not as participants in the life of science. What had changed was that those same readers were now far more likely to put their trust in acknowledged disciplinary leaders than in professional writers and lecturers of the kind who inaugurated the golden age of popularization under the Second Empire.

CHAPTER SIX

The Public Face of Republican Science

• • •

The despair that followed the defeat of France at Sedan and the capture of the emperor in the late summer of 1870 proved to be fertile ground for the new departures in French science that had begun to take root in Victor Duruy's fruitful but difficult years at the Ministry of Public Instruction. One reason for this and for the subsequent rapprochement between science and republican politics was that most of the savants who had campaigned in the 1860s for better conditions for research and an end to the chauvinism of French intellectual life had also been known as opponents of the imperial regime on a broader, political front. Before 1870, Adolphe Wurtz, Claude Bernard, Paul Broca, Georges Pouchet, and Marcellin Berthelot were among the leading champions of an improved provision for science who publicly identified themselves with a band of opinion centered on the moderate left, and they did not hesitate to parade their views, after the war, by standing in local and national elections or even, as in the case of Berthelot, holding ministerial office.

Plainly, declarations of appropriate political sympathies could do much to help both the career and the public standing of a scientist in the years of the Third Republic that separated the fall of Napoleon III from the outbreak of the First World War. By the same token, those who were tarred with the brush of loyalty to the empire had every reason for apprehension. Pasteur, for example, must have viewed the events of 1870 with anxiety. But his intellectual stature, developing image as a public benefactor, and mild gestures of accommodation to the new political order proved sufficient protection. In 1874, the National Assembly, still united in its postwar patriotism (if in little else), voted overwhelmingly to approve the proposal of a committee chaired by the republican Paul Bert to grant an annual pension of 12,000 francs on behalf of a grateful nation.[1] And even in the more aggressively republican and less consensual climate of the 1880s Pasteur continued to soar serenely above party strife, to the point of having his pension raised in 1882 to 25,000 francs, roughly twice the annual salary for the chair at the Sorbonne from which he had retired in 1875. Jean-Baptiste Dumas was less fortunate. His influence in the Ministry of Public Instruction and as a general inspector of higher education and a senator rooted in the center right was abruptly terminated by the war.[2] And he lost a number other offices he had held under the empire, notably as head of the municipal council of Paris and director of the mint. Even if, at the age of 70, Dumas could not have hoped to maintain his leading roles in public life for much longer, he felt his abrupt exclusion from the seats of power keenly and found only moderate consolation in his election to the Académie française in 1879 and in the contributions he was invited to make to occasional international ventures (such as the expeditions to observe the transit of Venus in 1874) and as a senior, rather venerable figure at congresses.

Far more serious was the plight of a number of younger scientists whose political opposition to the Republic was compounded by their resistance to the secularism that successive governments proceeded to espouse once the prospect of a return to a monarchy with predominantly Catholic values (a real possibility in 1873–74, in the days of the religiously inspired right-wing Ordre moral) had receded and paved the way for the definitive securing of the Republic in the senatorial elections of 1879. The threat to this group lay in the all too easy equation between republicanism, good science, and secularization. The fact that the conflation of these disparate elements served the purposes of a prominent majority in the scientific community and continued to be invoked until the First World War should not blind us to the difficulties it created for the community's conservatively inclined minority that struggled to hold a middle ground in which faith and traditional values could coexist with free scientific inquiry. This minority pursued the task as well as it could both in the institutions of the state and in the rather frail independent structures for higher education and research that emerged in the 1870s with Catholic support. But it faced debilitating opposition in the form of ministerial obstruction and the attitudes of secular-minded opinion formers who admired science and technology precisely because they were deemed to symbolize modernity and the antithesis of the old religious and social order.

There was, then, both a lighter and a darker side to the life of the savant in the Third Republic. Most securely in the ascendant were the academic scientists with posts in higher education. These were the many who benefited from the improved facilities, enhanced public approval, and unprecedented growth of career prospects that flowed from the special concern of successive republican administrations to reinvigorate the nation's underused and underperforming network of faculties of science. They were also among those who most enthusiastically applauded the major educational reforms that led, in 1902, to an enhanced place for science in the secondary school curriculum. But there were also those, both scientists and nonscientists, for whom the prevailing scientism—a scientism that they increasingly contested—made the period from 1870 to 1914 a time of trial rather than opportunity. These divergent experiences and the tensions they created in what has generally been regarded as a period of triumph for science provide the main themes of this chapter.

The Savant at War and Peace

The war with Prussia yielded an early, unexpected dividend for science when the siege of Paris allowed savants to appear on a grimly prominent stage as selfless patriotic heroes. From the moment the siege began on 19 September 1870, only two months into the war, until the surrender of the city on 26 January 1871, inventiveness reached new heights, and science was brought to the center of the war effort by the involvement of several of the leading national societies—the Académie des sciences, the Académie de médecine, and the Société chimique de Paris among them—and by the creation of numerous ad hoc committees devoted to encouraging and implementing ideas for either a military riposte or the relief of suffering.[3] Stimulated by the popular science committees that were set up in *arrondissements* across the city, the response of the lay public was impressive, at least in its intensity. Enthusiasts and teenage tinkerers, many of them working through patriotic people's clubs, joined established savants in inundating scientific societies and local committees with proposals that were more often than not bizarrely fanciful. It is hard to imagine that even the most amateurish committee could have taken seriously the suggestion that a 10 million-ton sledgehammer, 6 kilometers across, should be raised by a balloon, maneuvered over Versailles, and then dropped on the Prussian army.[4] And what possible response could be made to the advocates of the resurrection of Greek fire or to the idea that the Prussians might be asphyxiated by chemically decomposing the air?

However far-fetched these ideas may have been, they spread a comforting impression of the usefulness of science, and the kindly silence with which they were received helped to maintain the sense of a broad public participation in the more serious manifestations of the scientific effort. Popular perceptions put the fantasies on a footing, for example, with the undoubtedly beneficial work of the Académie de médecine (chiefly in the disposal of sewage, the concoction of substitutes for bread and milk, and the battle against dysentery, typhoid, and other diseases), the Commission du génie civil (chaired by the professor of mechanical engineering at the Conservatoire des arts et métiers, Henri Tresca),[5] and two committees under the control of the Ministry of Public Instruction, the seven-man Scientific Committee for the Defense of Paris, chaired by Marcellin Berthelot, and an associated Mechanics Committee, chaired by Charles Delaunay.[6] It must be said that even such seriously constituted bodies as those chaired by Berthelot and Delaunay had their failures: these included the attempt of the physicist Charles d'Almeida to restore telegraphic communication, broken by the Prussians, by using the Seine as a conductor of electricity (work that he carried out after escaping from Paris in a balloon) and the vain search for a new stabilizer for dynamite, to replace the now unavailable *Kieselguhr*. Nevertheless, largely at Berthelot's prompting, science did make a justly admired mark, notably on the supply of explosives and other problems of artillery, work that in turn stimulated Berthelot's long-standing interest in explosives.

Parisian science during the siege was by design and necessity a public affair in which the inherent heroism of the response to immediate need was embellished by studied acts of sacrifice and gallantry that smacked more of symbolism than of genuine utility. The decision of the Académie des sciences to continue its weekly meetings throughout the siege and its maintenance of attendances that barely fell below their prewar level succeeded, as was intended, in cloaking academicians in a mantle of patriotic resolve that captured lay imagination.[7] Even more gripping was the readiness of the administrators of the menagerie in the Jardin des plantes (now filled to overflowing by animals brought from the Jardin d'acclimatation in the Bois de Boulogne) to slaughter its animals for food, although it is certain that much-publicized menus offering the meat of wapiti, camels, and elephants were rare gimmicks by comparison with the daily realities of poodle meat (particularly delectable) and rats and the abrupt emergence of horseflesh as a staple component of the Parisian diet.[8] And no citizen could remain unmoved by the technological feat that between 7 October 1870 and 28 January 1871 allowed a total of fifty-four balloons, the famous *ballons montés*, to fly majestically over the Prussian lines, with their cargoes of mail.[9] In this way, almost three million letters were carried out of Paris, to be answered chiefly by messages borne by carrier pigeons that had accompanied the outgoing mail.[10] The ingenuity of this symbolically charged form of pigeon post made the implementer of the system, the chemist, microscopist, and photographer René Dagron, a popular hero.[11] At the central dispatch point in Tours (to which Dagron himself had escaped by balloon, with his equipment) and later at Poitiers messages destined for the besieged capital were printed and consolidated in the form of a news sheet, which was photographed. About twenty negatives, reduced to minute proportions that allowed twenty thousand characters to be reproduced in an image of roughly a quarter the size of a playing card, were then placed in a tube attached to the pigeon, which before release was taken as close to Paris as the military situation would allow. On receipt in Paris after flights of between 70 and 220 kilometers, the negatives were sent to the Central Telegraph Office, where they were enlarged on a screen with the aid of a magic lantern, and finally the messages were transcribed and distributed within the city as normal telegrams (see figure 14).

The conflict also provided the opportunity for scientists to shine, with other intellectuals, in a war of words. Here, they were helped by the extreme declarations of national sentiment that some of their most eminent German colleagues voiced almost as soon as hostilities began. Less than a month into the war, the historian of ancient Rome Theodor Mommsen combined his well-honed sympathy for Prussian militarism with deliberately tendentious observations on the profoundly German character of Alsatian culture. The main observations, which inevitably led on to an argument for the justness of German territorial ambitions in eastern France,[12] were soberly linguistic. But the case was spiced with less disciplined jibes at the Catholic Church, whose snares, as Mommsen put it, a third of Alsatians had managed to resist. A similar, if more measured, attack came from the biblical scholar David Strauss. In two open letters to Ernest Renan, published in August and October 1870, Strauss also insisted on the German character of Alsace, though in the context of a broader, more rhetorical argument that France's time as a preeminent nation had passed, while Germany's had come.[13] Especially in the second letter, the triumphalist notion of a victor's rights to territory that would serve its own interests loomed large,[14] and its implications for any French hope of regaining Alsatian territory were not lost on incensed readers in France.

Even in the German-speaking world, such arguments did not pass uncontested. Karl Vogt—German by birth and culture, though a long-standing resident of Geneva— earned predictable praise in France for his insistence that the plans for the annexation reflected nothing more than German cupidity.[15] Mommsen and Strauss, for their part, immediately became objects of contempt in the French learned community. Still greater opprobrium was heaped on Emil Du Bois-Reymond following the address he gave on 3 August 1870, as Rector of the University of Berlin, at a formal celebration to mark the new academic year. Speaking just a few days before the decisive Prussian victory at Sedan, Du Bois-Reymond could hardly have given his words greater prominence or caused graver offense. The tone of his address, on the "German war," was crudely abusive: he portrayed France as having succumbed to insularity, moral degradation, and a Catholic tradition that had impeded political development and stultified the Gallic mind.[16] The point struck home all the more tellingly since, as a German of Huguenot descent who had studied in France in the 1850s, Du Bois-Reymond knew his adversary well. Moreover, his attack rested on points that had much in common with those made by many a domestic French critic during the 1860s. Only months before the war, in fact, the reforming *Revue des cours scientifiques* had published a stinging address in which Du Bois-Reymond had contrasted the free, flexible, and intellectually vital German universities with the stiflingly bureaucratic university system of France.[17] The absence of editorial comment can only be taken as a measure of the journal's approval of Du Bois-Reymond's position. But once France and Prussia were at each other's throats, what might have passed in peacetime for just comment on the deficiencies of the centralized Imperial University was read as an affront to the French academic world.

In the trading of insults that went on through the autumn and winter of 1870–71, French scientists contributed prominently to fashioning a new image of a coarse, brutal Germany drunk with military success. The scantiest evidence of vandalism was proffered as a mark of depravity. Only the barbaric descendants of Attila could have shelled the Muséum d'histoire naturelle (as happened during the siege of Paris) or wilfully damaged the cathedral and rich municipal library of the "studious learned city" of Strasbourg.[18] And, as the French saw it, only ignorance or prejudice could account for the gloating enthusiasm with which the embodiments of German science and scholarship proceeded to analyze the weakness of France. A particularly cruel example was the charge, made in February 1871 by the Bavarian doctor Carl Stark, that the French people were prey to collective psychological degeneration.[19] The nation's symptoms, in Stark's analysis, were those of an individual racked with delirium and illusions of grandeur. They reflected a form of advanced national senility, fading into the lunacy of a pathetically aging civilization. Faced with such abuse and with the bitter implications of defeat (culminating in the signing of the preliminaries of peace in late February 1871), the French could only fulminate with an impotence that was soon exacerbated by the insurrection and Commune in Paris in the spring of 1871. In August 1871 the *Revue politique et littéraire* returned to the fray, publishing a long extract from Stark's book, which it then peremptorily dismissed as the rambling of a madman.[20] The assertion, by the anonymous author, that Stark had leveled charges at the French that rang far truer when applied to Stark himself and his fellow countrymen lacked even the merit of subtlety. But the manifest poverty of these and other exchanges did nothing to stop the mutual ascription of demeaning national characteristics, which remained common currency, albeit with diminishing vehemence, for several years after the war.

A particularly rich opportunity for French invective lay in the racial and cultural arguments that had been used to justify German territorial aggrandisement. On this point, Renan was both vociferous and well primed. In September 1871 he used his delayed response to the second of Strauss's two public letters of a year earlier to criticize the anthropological and linguistic considerations on which Strauss had rested much of his case, especially for the annexation of Alsace.[21] As Renan claimed, centuries of migration had undermined any idea that racial similarity or dissimilarity could be invoked as criteria in determining boundaries and territorial claims. The very fact that the names of the Prussian provinces of Pomerania and Silesia, like that of Berlin itself, were unmistakably of Slav origin suggested that if language and place-names were decisive evidence, as Strauss maintained, the logical destiny of much of Prussia must lie, not in a unified Germany with Bismarckian borders, but rather in a pan-Slav movement and hence in a detachment from the rest of the German Empire.

By the time Renan formulated his position, he was entering territory on which anthropologists on both sides of the Rhine had been venturing, in the name of patriotism, for some months. The main French source for this turn was the clear-eyed analysis of the concept of a "Prussian race" that Quatrefages published in February 1871.[22] With the authority of a senior professor at the Muséum d'histoire naturelle, Quatrefages bolstered his argument with elaborate scholarly detail. His core contention, though, remained clear: cranial and other anthropological evidence demonstrated that the case for the racial unity of the people of Germany was flawed. The German case stumbled, above all, on the sharp demarcation that he drew (and had already drawn in the 1860s, though not then with polemical intent) between the mixture of Finnish and Slav characteristics prevailing in all but the highest classes in the heart of Prussia, in particular in Pomerania and Brandenberg, and the nobler German features that were common in the south and west of the German-speaking region. As Quatrefages argued, the Finnish and Slav characteristics (essentially the brachycephalic, short-headed skull), though of ancient origin, still marked the Prussian race. Even if they were now concealed beneath a veneer of cultivation and refinement (assimilated mainly from the French since the seventeenth century), they left their heritage in an almost medieval proclivity to vengeful, combative dealings with other nations. The tragedy, for Germany as for France, was that the racially defined *true* German people had allowed Prussian brutishness to pass as a shared characteristic of the German Empire as a whole.

At a time of heightened sensitivity on both sides, Quatrefages's verbal aggression profoundly offended German readers, and Rudolf Virchow's reply, delivered before the Berlin Anthropological Society, was vehement.[23] Virchow rejected out of hand Quatrefages's vision of a Prussia peopled originally, and still in significant measure, by an indigenous Finnish race of small stature, whose coarseness and violence had been reinforced by subsequent Slav migration, while France prided itself on being a nation of superior Celts. Quatrefages, he argued, had fabricated his "psychological reveries"[24] from generalizations about Prussia based on craniometric and other evidence that in reality could only be applied to the early inhabitants of the Baltic provinces, in particular Estonia. In reality, as Virchow (correctly) observed, the prehistoric specimens found in Prussia and the northeast of the German Empire seldom displayed the brachycephalic and other characteristics of the Finns or the craniologically similar (though not identical) Estonians. If the evidence was accepted, Quatrefages's depiction of the two distinct racial groups of the Empire—Finno-Slav and German—was left without foundation. So, too, and possibly more important, was his gratuitous corollary that Prussia should be regarded as no more than a "colony" in which the original brachycephalic Prussian race had fallen under the sway of the superior dolichocephalic (long-headed) Teutonic peoples of German origin whose knights had asserted their political and military authority and imposed their language between the twelfth and fifteenth centuries.

The confrontation between Virchow and Quatrefages turned on complex, contestable evidence. There was a strong case for believing that—if the distinction between two precisely bounded regions of Germany, dominated respectively by brachycephalic and dolichocephalic peoples, had ever been as sharp as Quatrefages maintained—it had been so only in the mists of the distant quaternary. With regard to more recent times, the distinction rested on the most fragmentary remains and was confused by subsequent migrations and conquests. The Slav arrivals in the fifth and sixth centuries AD followed by the later, Teutonic infiltration after the twelfth century had resulted in a complex mixing of the races of Prussia and so fed Virchow's counterargument that Germany, far from being the home of two distinct racial types, had in reality been a melting pot of races as far back as evidence existed. In their inconclusive character, the exchanges were typical of the patriotic effusions of scientists and scholars on both sides who used a veneer of learned authority to overlay arguments primarily rooted in prejudice. The ostensibly dispassionate address by Virchow to the annual Versammlung deutscher Naturforscher und Aerzte at Rostock in September 1871 encapsulated the tone. The air of even-handedness in Virchow's acknowledgment of the debt that the German scientific tradition owed to France[25] served only to give force to his implied contrast between the importance of France's contribution in science up to the 1830s and a decline since that date—a decline that his withering attack on the intellectual shackles imposed by the Catholic Church associated by implication, though unmistakably, with religion.[26]

Such talk of decline presented a particular challenge for those in France who had been so excoriating in their analysis if the nation's failings during the Second Empire. The 120–page essay on "La réforme intellectuelle et morale de la France" that Renan wrote in the dark hours of defeat exemplified the difficulty.[27] Renan had long argued that one of France's greatest failings lay in a lack of faith in the power of science and in a tendency to allow undisciplined verbosity to pass for serious intellectual endeavor.[28] He had also seen French loyalty to the Catholic tradition, with its hierarchical structures and preoccupation with the mysterious and the supernatural, as a source of weakness, much as Virchow argued.[29] Now, without withdrawing his earlier charges, he insisted on an irreducible core of subtlety and inspiration in French culture that contrasted with Prussian arrogance and gracelessness. The conclusion was clear. A culture, like Prussia's, that was founded on a peculiarly German conception of science and scientific method was at least as flawed as that of France. Renan's ideal, therefore, lay not in adherence to one or other tradition but rather in a complementarity that would balance science with spirituality and reason with imagination; it was only in this meeting of opposites that German culture could be prevented from becoming ever cruder and that of France from becoming more "superficial and backward-looking."[30]

The vituperation of the early 1870s made the immediate resumption of good academic relations between France and Germany after the war virtually impossible, and as the learned communities of both nations became obsessed with the desire to display their patriotic credentials, any pretence of the universalism of knowledge was abandoned. Pasteur's contemptuous return of the diploma he had received to mark his honorary doctorate of medicine from the University of Bonn attracted particular attention in January 1871, the more so as the mutually insulting correspondence between him and the Bonn faculty of medicine was published in the press.[31] But Pasteur's was by no means the only such gesture. In the same month, the Société zoologique d'acclimatation in Paris announced its decision to remove the names of the sovereigns and princes of the various German states that had appeared hitherto in the list of "protectors" of the society.[32] Two months later, in March 1871, two sessions of heady debate in the Académie de médecine stopped short of a decision to expel all Prussians and other Germans sympathetic to Prussia (as the Académie des sciences, belles-lettres et arts de Clermont-Ferrand had already done, with respect to its sole German foreign associate).[33] But patriotic indignation boiled furiously in the announcement by several members of their decision to withdraw from German societies to which they belonged and of their hope their fellow members would do likewise.[34]

As the 1870s wore on and the futility of open abuse became apparent, animosity began to assume a less overt form, softened by a veneer of strained respect. French savants in particular had to aim any jibes with special care. Faced with the growing opulence of German laboratories and libraries and the distinction of the research schools that worked in them, they tended to fall back on Quatrefages's strategy of contrasting the quality of their own achievements with the sheer volume of publications emerging from Germany.[35] Where straightforward abuse seemed inappropriate, faint praise remained the order of the day. The judgment of Georges Pouchet, who genuinely admired and envied many aspects of the German university system, was typical. Writing in 1881, Pouchet insisted that the essential advantage of Germany in the life sciences lay not in the highest reaches of original research, where he believed the French had the edge, but in the number and broad geographical distribution of German universities of the second rank.[36] Paris remained, for him, the leading university city in the world, endowed with incomparable facilities.[37] In Germany, in Pouchet's own discipline of comparative anatomy, only Berlin could boast a collection remotely comparable with that of the Muséum d'histoire naturelle; the Berlin collection, moreover, was in far worse condition. Hence where French professors had failed was in not realizing their potential for international leadership. It was only when the shackles binding them to their ministerial paymasters were broken that France's scientific community would be able to display the eminence for which, as Pouchet insisted, it was destined.

By the time Pouchet wrote, the belief that France had been defeated less by military tactics, manpower, or weapons than by the science or, as some put it, the "scientific spirit" of Prussia had circulated strongly for a decade. It was an opinion shared across a spectrum of opinion extending even to such prominent literary figures as Flaubert and Zola, both of whom looked to science or a greater public confidence in science as a remedy for French ills. Flaubert had already stated his position before the war, when he wrote to George Sand on 5 July 1869: "It is no longer a question of imagining the best form of government, since no one form is better than any other; what matters is to ensure the victory of science . . . politics will remain in a state of perpetual fatuousness so long as it does not draw on science."[38] Zola made his judgment retrospectively but no less categorically in a celebrated paean to science in his "Lettre à la jeunesse" of 1880. As he wrote there, "One thing we have to acknowledge loud and clear is that in 1870 we were defeated by the scientific spirit."[39]

The most vociferous champions of such opinions, however, were scientists, who understandably anticipated an opportunity of wringing some benefit from the darkness of defeat. As they knew, any case had to be couched in terms that avoided even the slightest suggestion of intrinsic French inferiority. It was this that bred the common defensive caveat that the quality of the nation's scientific community left nothing to be desired and that what had condemned France to its scientific decline and to a state of crippling disadvantage in the war were the conditions in which savants had had to work. In a frequently cited article in the Lyon newspaper *Salut public* in March 1871, Pasteur harped repeatedly on this distinction. If the French had failed to find the "superior men" who might have exploited their country's vast resources at its moment of need, the explanation lay in half a century of neglect of the "great labors of the mind," especially in the exact sciences.[40]

A similar tone pervaded a collective outburst in the Académie des sciences. At a meeting on 6 March 1871, apparently without warning, member after member rose to deplore the lack of funds, excessive centralization, and stifling bureaucracy that had shackled education and impeded generation after generation of savants since the early years of the century.[41] The case was introduced in tones of ringing patriotism by Henri Sainte-Claire Deville and supported by academicians ranging from those of a predominantly academic cast, such as the mathematicians Joseph Bertrand and Charles Hermite, to those, like Arthur Morin (the director of the Conservatoire des arts et métiers) and the pathologist and veterinary scientist Henri Bouley, who were more actively engaged in technical and professional education. With such backing, the cries of alarm could not pass unnoticed, and a committee was immediately established to advise on further action. But all too soon far weightier events overtook the cause of scientific reform. On 18 March, a left-wing insurrection culminated in the declaration of the Paris Commune, and academicians had their attention diverted from the quest for unity and national renewal to the alarming immediate realities of civil conflict that made Paris a setting for violence and disorder for more than two months. Protecting buildings and instruments against the ravages of vandalism became the overriding concern. In this, the observatory was particularly unfortunate. In late May insurgents of the Commune took refuge in the building and started a damaging fire before being expelled: despite the efforts of the staff, fine geodesic instruments were destroyed and the Gambey equatorial damaged.[42] Although the Académie, by contrast, survived unscathed, the committee on reform does not appear to have met, and its functions were not renewed.

In failing to build on its declaration of outrage, the Académie missed an opportunity of reaffirming its right to speak to the nation in the name of science. At a time when respect for traditional institutions could no longer be taken for granted, the failure had the effect of encouraging the emergence of other voices. Soon the complaints about the conditions for research and teaching in science were taken up in less venerable circles. Most of the voices came from a younger generation that worked largely outside the Académie and had few of the bonds with the past that inevitably tempered the vehemence of the more established academicians. These "new" men could denounce the empire and endorse the republic without embarrassment. Above all, they could pursue the reform of science as part of a wider program of modernization that would achieve the longed-for resurrection of the soul of France by freeing the country from the shackles of a political regime and a culture that had both been found wanting.

The proliferation of calls for new structures for research and its dissemination after 1870 bears witness to the unprecedented will for change and the conviction that something could be done. The resulting innovations had two recurring characteristics. First, they served, almost without exception, to assert the primacy of specialized disciplinary expertise among the criteria for professional advancement. And secondly, they broke with the notion of a single elite of science defined by membership of the Académie and the near-oracular authority that academicians enjoyed. In both respects, the Société mathématique de France, founded in 1872, was a typical new departure. Its priority was service to an open but informed public defined by its mathematical competence rather than by academic seniority or a position in a particular institution. To that end, membership was open equally to the acknowledged leaders of French mathematics and to the wider community of teachers in the faculties, technical *grandes écoles, classes préparatoires*, and *lycées*.[43] The nature of mathematics, however, meant that the society offered little for the dilettante or the idly curious and nothing for those who might have seen the value of belonging to a *société savante* in purely social terms. Hence when the statutes defined the purposes of the society as "the advancement of science and the spread of pure and applied mathematics,"[44] they meant what they said, and the society's *Bulletin* fulfilled the purposes to the letter. It offered no easy reading. Instead it devoted its monthly pages to the original papers of members pursuing academic careers in which success was coming increasingly to depend on a strong record in research. In the process, the task of diffusing mathematical knowledge on a broader plane devolved to other publications. For digests of recent work and reviews of books and journals, readers had to turn to the *Bulletin des sciences mathématiques et astronomiques*, an independent monthly journal that Gaston Darboux had launched shortly before the war, in January 1870.[45]

The profile of constraints and opportunities within which a scientific society had to work varied greatly with the discipline concerned. In the case of the Société mathématique de France, the strongly academic tone was reinforced by the society's growing attractiveness to former *normaliens*, virtually all of them engaged in higher or secondary education, who made up 14 percent of the membership in 1874 and 35 percent by 1914.[46] In other societies such narrowing of the occupational profile and educational background was less marked. The Société française de physique, for example, never compromised its disciplinary focus, but its still higher priority was a measure of controlled openness. Like many other innovations of the early Republic (it was founded in 1873), it had roots in the Second Empire, in this case in an informal group of academic physicists who had met since the late 1860s in Pierre-Auguste Bertin's laboratory at the Ecole normale supérieure.[47] But the collapse of the empire marked a hiatus, of which Bertin and the cosiness of his group were casualties. His election as vice president of the new society and then, for 1874, its president did not assuage his nostalgia for the very different circle over which he had presided just a few years earlier. A petulant protest against the inscription on a bust of Charles d'Almeida that, not unreasonably, identified d'Almeida as the founder of the society was one of his last acts, and thereafter he simply withdrew.[48]

In return for an entry fee of 10 francs and an annual subscription of 20 francs, the

Société française de physique offered a diet of twice-monthly meetings and a bulletin of experimental physics of uncompromising technicality. It proved a successful package. By 1880, a membership that topped 550 ensured financial stability, if not opulence, and numbers continued to grow, to 880 in 1900 and almost 1,600 ten years later. Although the figure fell short of the almost two thousand members of the Société de géographie, this made the society the largest of the disciplinary societies in the mainstream sciences, rivalled only and at some distance by the Société chimique de France.[49] The success owed much to the core of Parisian academic scientists who provided the society with its early leading lights. With consummate skill, d'Almeida (the first secretary), Alfred Cornu, Jules-Antoine Lissajous, Eleuthère Mascart, and Désiré Gernez (an unusual instance of someone who held posts simultaneously in a *lycée* and at the Ecole centrale des arts et manufactures) tended the common ground between themselves and those for whom physics was a realm of practice and application. To this end, members of the various state corps of engineers and industrial engineers, as well as academic physicists, were welcomed and catered for in meetings, publications, a modest lending service for periodicals, and, above all, the society's important annual exhibitions devoted to applied physics. Utility, as much as competence in the discipline, was seen as yet another way of displaying the contrast with the effete and outmoded past.

Journals too contributed to the postwar *revanchiste* zeal by declaring goals that were at once scholarly and patriotic. The rhetoric varied little from journal to journal. The *Revue scientifique* took an early lead when it signaled its relaunching, on 1 July 1871, with a new title (it had been founded as the *Revue des cours scientifiques* in 1863), an increased size, and a resolve to become even more international in scope. In a belligerent editorial, one of the two coeditors, Emile Alglave, set the tone with the familiar refrain that Germany's strength lay in her universities and that the "scientific spirit" which fired them also inspired the army and the nation as a whole.[50] It followed that if regeneration and revenge were to be achieved, the only course was for France to take up arms once again but to do so now on the battlefield of science. As was made very plain, however, victory on that field would only be achieved if the past shortcomings of the country's provision for scientific education were addressed and if science and scientists tempered their remoteness by fostering closer bonds with society and the economy. More moderately but no less cogently, the first issue of the sober *Journal de physique théorique et appliquée* in the following year carried a declaration by d'Almeida, who was both founder and editor, that the journal's purely scientific aim of advancing teaching and research in physics was part of a wider program rooted in "love of country" and directed at fostering the "intellectual and moral" forces of the nation.[51]

A common theme of the scientific rhetoric of the early years of the republic was a declaration of selflessness. Men of science portrayed themselves as ready for any sacrifice in order to serve the nation, although it was not always clear just what service was on offer or how the nation would benefit. In the *Journal de physique théorique et appliquée*, d'Almeida promised to work for greater openness and cooperation not only among teachers of physics (of whom he was one, at the prestigious Lycée Henri IV) and practitioners across the spectrum of the scientific professions but also, by implication, in French society generally. The talk was always of an effort that would unite the nation and transcend divisions bred of religion and politics. But no amount of conciliatory language could obscure where the driving force of the quest for national renewal through science really lay. Whether it was explicit or not (and it usually was not), the emphasis on resurrection and a fresh start implied a distancing from the empire and the bourgeois, predominantly Catholic values that had sustained it. The corollary that the new republican liberal order offered a far better future hardly needed to be spelled out.

When the political dimension of the revivified reform movement was so blatant, only a contemporary of extreme insensibility could have seen it as a coincidence that the first two meetings of the most explicitly patriotic of the new scientific organizations of the 1870s, the Association française pour l'avancement des sciences, or AFAS, were held in Bordeaux and Lyon. In both cities opposition to Napoleon III had been keen, and now strongly republican local administrations were more than ready to be seen as the patrons of an institution whose motto, "Par la science pour la patrie"—by science for the fatherland—associated the welfare of France indissolubly with that of science. The tone of the association was set when it was welcomed to its inaugural meeting in September 1872 by the mayor of Bordeaux, Emile Fourcand. A pharmacist by training and a resolute champion of the moderate anticlerical left (which he supported as a deputy for the department of the Gironde in the National Assembly, as well as in Bordeaux), Fourcand exemplified the republican resolve to appropriate science as the embodiment of progressive culture and a bulwark against war and social strife. His characterization of science as "a great sovereign whose domain is humanity" and a pursuit whose nobility rested on its having the whole of nature as its object exuded a thoroughly republican conception of secular virtue.[52] As such, it elicited an enthusiastic response from the audience that packed the newly opened headquarters of the Société philomathique de Bordeaux, a building whose sumptuousness and availability for an impressive range of popular vocational courses stood itself as a symbol of national rebirth.[53]

The vision that Adolphe Wurtz presented at the Bordeaux meeting, of science as "one of the levers of modern civilization,"[54] was similar in its inspiring optimism and would certainly have been shared by Fourcand. But whereas Wurtz saw the work of the savant as, first and foremost, a fount of material well-being, Fourcand credited it with equally important, moral virtues: science had been not only the motor of all human progress but also the guarantor of peace and the force that had created civilization out of barbarism. Science, as he put it, "is in every age the divine ray, the seat of all experience and all truth."[55] The affront to traditional religious values that was implied in Fourcand's choice of such a spiritual metaphor was plain. It is significant that the local figure who might have reacted most forcefully, the Cardinal Archbishop of Bordeaux,

Ferdinand Donnet, was absent from the galaxy of dignitaries who heard Fourcand speak, in contrast with the Congrès scientifique in Bordeaux eleven years earlier, when he had been elected president of the congress and taken a leading part in the proceedings, notably with an address on the alliance between science and religion.[56] Quite apart from his position in the Church, Donnet's acquiescence in the empire, compounded after the war by his open espousal of the royalist cause, would have made his an uncomfortable presence at the association's meeting.

The discomfort that Donnet would have experienced at a meeting of the AFAS points to the limitations of the association's declared aim of uniting the nation. Catholic participation was certainly not excluded: the prominence of the abbé Antoine Ducrost in discussions of the prehistoric site of Solutré at the association's second congress, in Lyon in 1873, and the election of the chemist Edme Fremy, a declared Catholic, to the presidency of the Paris congress of 1878, are evidence enough.[57] But a Catholic member would soon recognize that the AFAS embodied a belief in the unaided power of reason that the more conservative elements in his Church had traditionally seen as a threat. Hence while there was no question of overt anticlericalism, the AFAS rejected any notion that Catholic doctrine might be allowed either to impinge on or even constrain the course of science. The point, though, was always subtly made. In so far as the association had a religious policy, in fact, the thrust of it was conveyed by the mere presence, among its early leaders, of three active Protestants (Wurtz, Quatrefages, and Charles Friedel), a declared free-thinker of Protestant descent (Broca), a militant champion of anticlericalism (Gariel), and a Jew (Adolphe d'Eichthal). The watchword was openness, though an openness that spoke volumes and that some would have found disconcerting.

The tolerant liberalism of the AFAS made its congresses a natural forum for scientific opinions that would have been at least muted in the more conformist atmosphere of a Congrès scientifique. It was entirely in keeping with the association's spirit that it invited Broca, whose advocacy of polygenism had long made him a marked man in conservative eyes, to deliver a plenary lecture from the highly visible platform of its inaugural congress. Given the opportunity, Broca did not hold back. His lecture combined patriotism and a sense of hurt national pride with a discrete but unmistakably provocative discussion of Cro-Magnon man and the other troglodytes of the region of the Vézère river in southwest France between Le Moustier and Les Eyzies.[58] According to Broca, the recently discovered remains of these long-headed (dolichocephalic) cave dwellers of the quaternary (dating, as we now know, from roughly 30,000 BC), pointed to the existence of an indigenous people with a range of physical characteristics, including large cranial cavities and tallness, quite different from those of the shorter, round-headed (brachycephalic) Celtic peoples of the kind still dominant in nineteenth-century Brittany and much of central France. The evidence undermined any argument in favor of the racial unity of the people of France, which would have required either the brachycephalic (Celtic) or the dolichocephalic (Cro-Magnon) races to have been the only ones to inhabit the country in pre-Neolithic times. Broca's point, however, was that the coexistence of Celtic and Cro-Magnon cultures did not make anthropology the enemy of national sentiment. For him, the integrity of a nation, whether the French nation or any other, did not stand or fall by its racial homogeneity.[59]

A discussion of anthropological evidence as novel as that of the Cro-Magnon skeletons and artifacts would have provoked an interested response at any time or place. But in 1872 the gloss that informed Broca's lecture to the Bordeaux congress added a spice perfectly adapted to the occasion. As Broca admitted, the people living around the Vézère may have been savages. Yet they had progressed steadily, refining remarkable artistic and technological skills (in their cave drawings, carvings, and weapons) alongside their prowess as hunters. By the end of the quaternary (in the millennia, between 10,000 and 5,000 BC, in which Europe emerged from the grip of the ice age), they were on the threshold of civilization, only to have the cultural torch they bore snuffed out by the invading Neolithic race of barbarians. The conquering race was in due course to have its achievements in agriculture, the erection of megaliths, and the deployment of polished stone axes far superior to the simple tools of "our peaceable reindeer-hunters." But in the course of those technological advances and the transition to a culture rooted in crop growing and herding, the vigor of the Cro-Magnon peoples had been sacrificed. Broca's concluding flourish—"It was evident then, as it has been since, that might prevails over right"[60]—only labored a parallel with recent painful events that no one could have missed. It is not hard to imagine the enthusiasm of the applause that followed.

Countercurrents: Science in the Catholic Tradition

When Broca presented his patriotically embellished interpretation of Cro-Magnon culture before the AFAS in 1872, he was careful to invoke the nuances and cautious evaluation of evidence that were required by the conventions of scholarly debate. Some others who, like him, saw science as the bedrock of republican culture and a key to national salvation felt no such constraint. They preferred the more directly confrontational advocacy of the physiologist and partisan of the far left, Paul Bert. In his addresses to the National Assembly in the early and mid-1870s, Bert displayed his credentials as a political *protégé* of Léon Gambetta by elaborating in an extreme form the view that a republican education had to give pride of place to science. Rejecting the utilitarian argument and hence a call to which citizens across a wide band of political opinion (including many on the right) could have rallied, Bert took a more divisive course. His insistence on the role of science in fostering the ideal of intellectual freedom went hand in hand with an acknowledgment that the Christian tradition had wrought moral benefits in the past. The acknowledgment, though, was perfunctory, and it did nothing to veil the anticlericalism that suffused Bert's case. As he put it in a parliamentary debate in January 1873, in words calculated to appal the opponents of the secular republic, science deserved greater prominence in education "since it destroys superstitions . . . eliminates capriciousness from nature and sets immutable law in its place . . . since it is the queen of modern societies and the liberator of human thought."[61]

The Bonapartists and monarchists who made up the broad conservative block in the Assembly could not fail to be incensed by Bert's vision of a systematic "scientific doubt" doing battle with error and prejudice. They saw it as a microcosm of republican scientistic arrogance and responded with predictable hostility. The protests from conservative quarters of the Chamber were soon reinforced by a wider response from within the Catholic Church. The clerical case was not easily formulated. For the church had always declared itself the friend of science, so that now as always any attack had to be mounted as an attack not on science as a whole but on a flawed science that taught falsehoods at odds with faith. A key voice was that of the elderly philosopher Augustin Bonnetty. As editor of the monthly *Annales de philosophie chrétienne*, a bastion of traditional, dogmatic Catholicism that he had founded in 1830 to answer the liberalism of *Le Globe*, Bonnetty had already articulated his position with aggressive clarity. Writing immediately after the defeat of 1870–71, he argued that the advance of rationalist thought, far from being a springboard for national renewal, should be seen as the cause of the moral and material decay of France that had reached its dismal culmination in the Commune and the murder of several priests, including the Archbishop of Paris, Monsignor Georges Darboy.[62] For Bonnetty, the fault lay with modern philosophers, besotted with the power of reason, who had encouraged citizens to fashion their own conclusions and thereby (when those conclusions did not accord with the authority and ancient teachings of the church) fostered ignorance of the nature of God. The remedy lay in a return to revelation, which alone provided a guarantee of certainty.

In Bonnetty's attack, as in some others from the Catholic side, the pagan classics came under the same withering fire as rationalism; they were seen as equally menacing founts of the secularism that the state had long disseminated through its schools and faculties, and as provocations that demanded a reassertion of Christian principles. Twenty years earlier, the Papal Protonotary Apostolic, Jean-Joseph Gaume, had already expressed his views on the dangers of a "gnawing worm" (*ver rongeur*) founded on the works of Greek and Latin authors.[63] And he now returned to the charge. In an analysis that breathed precisely the same spirit as Bonnetty's, he presented the ancient classics as no less culpable than the unfettered use of reason in fashioning a modern educated class of "Voltaireans and sensualists" whose attitude to religion was, at best, one of indifference.[64] Both threats called for action. It was time to resurrect the uncompromising Catholic arguments against an exclusively secular education and to rally to a position of the kind that the bishop of Langres, Pierre-Louis Parisis, and other senior bishops had articulated in opposing the state monopoly of secondary education in the parliamentary and extra-parliamentary debates of 1849–50.[65]

The strength that such broadsides gathered in the early 1870s did more to reinforce than to allay prejudices in republican circles: unsympathetic eyes inevitably saw the rhetorical flourishes of a Bonnetty and a Gaume as caricatures of the backward-looking Catholic response to modernity. In fact, such a judgment was harsh. Gaume, in particular, acknowledged the need for the clergy to reflect on their own failings as teachers of the young: they should not forget that the age that had preceded and in important ways spawned the French Revolution had been one in which Catholic pedagogy had held sway. In a similar spirit of compromise, other prominent Catholic thinkers too argued for cautious accommodation with the modern world and with the traditions of non-Christian thought. In educated Catholic circles, their position won much support. The attempts to reconcile science with religion that the Bishop of Orléans, Félix Dupanloup, had pursued since the early 1850s, for example, represented a rallying point for many middle-of-the-road believers who refused to turn their faces totally against industrial society and its scientific underpinnings and who, like him, saw the invective that Bonnetty and Gaume poured on the pagan classics as extravagant.[66]

In the aftermath of war, moderation of the kind that Dupanloup had articulated struck many responsive chords in all but the most conservative sectors of the Church. And it gained additional favor from July 1875, when the long-awaited victory in the struggle for the liberty of higher education left the Catholic interest with a new level of responsibility in the fashioning of young minds.[67] Catholic intellectuals could no longer allow themselves to be seen as indifferent to science, and within two years, a full range of scientific studies had been established in the four Catholic universities that were established in Lyon, Paris, Lille, and Angers, only Toulouse of the new universities remaining without a faculty of science (until 1882).[68] The beginnings were encouraging. Donations from leading figures within the church and the Catholic nobility provided for immediate material needs and augured well for the future. Professors, too, were found, often but by no means exclusively through transfer from Catholic secondary schools (a route that could entail attendance at a state faculty in order to acquire the necessary *licence* or doctorate). If only because of the haste with which the new faculties had to be put in place, the standard of the professorial appointments was mixed. But at least a few stars were recruited. Lyon could boast as its dean the mathematician Claude-Alphonse Valson (a man of militant piety and the biographer of Augustin-Louis Cauchy, one of the heroes of Catholic science) and, in physics, the greatest of all authorities on the measurement of high pressures and the compressibility of liquids, Emile-Hilaire Amagat.

Paris, predictably, had more than its share of distinguished teachers. A faculty launched with the physicist Edouard Branly, the chemist Georges Lemoine, and the geologist Albert de Lapparent among its professors was, at least potentially, a serious rival to the faculties of the state. But if the primacy of Paris among the Catholic universities was inevitable, the success with which this position was sustained and exploited could not have been achieved without the sustained commitment of the abbé Maurice d'Hulst. A scholarly theologian and authority on canon law, d'Hulst quickly emerged as the most persuasive and combative of all spokesmen for the cause of Catholic higher education.[69] In association with the bishops who established the Université catholique de Paris in 1876 and then, starting in 1880, as rector of what had now become the Institut catholique (in accordance with the law of 18 March 1880 forbidding the Catholic institutions to use the title university), d'Hulst was responsible for recruiting the majority of the first professors. At a time when careers in a structure unrecognized by the Ministry of Public Instruction were fraught with personal risk and when growing republican confidence from the late 1880s found expression in heightened ministerial opposition to the new Catholic institutions, d'Hulst's role was crucial. As he saw, the battle could only be won with strong lay support, and he duly took the lead in attempts to persuade the public, and Catholic families above all, of the merits of an independent alternative to the state system.

In pamphlets and public addresses (in particular those he delivered beginning in 1881 at the ceremonies of the Institut catholique at the start of each academic year), d'Hulst sharpened the by now familiar distinction between the "false" science that had spread its materialist tentacles in the state faculties, and "true" science that could never conflict with faith.[70] As he insisted, it was precisely this distinction that provided Catholic initiatives in education with one of their main missions as a bulwark against the malign tendencies of many modern interpretations of science. In the conflict to come, parents could be expected to play their part by entrusting their children, wherever possible, to the founts of Christian science that the church alone could provide. Only with that support would it be possible to resist the tide of secularism emanating from ministerial circles and, even more provocatively, from the Ligue de l'enseignement, for whose members d'Hulst reserved two of his most withering denunciations, in addresses to the faithful at Rouen and Evreux in 1883.[71] Since its foundation by the militant republican Jean Macé in 1866, the Ligue had associated its campaign for primary and, more generally, popular education with an uncompromising scientism founded on the principle of free inquiry. The reality, for d'Hulst, was that the Ligue, far from promoting freedom of thought as it claimed to do, was setting up a new authority, that of science, in opposition to the authority of the church. In doing so, it was going far beyond its brief of popularization, which would have been unobjectionable. Its real objective was the diffusion of a science "poisoned by atheism"[72] that encapsulated the worst evils of the time. The sole remedy was for the church to take possession of science, to cultivate it in its faculties, and to convey it, divested of its atheistic connotations, to the people. Once that was done, but only then, would it be possible to combat the common scientistic contention that science and faith were irreconcilably at odds.[73]

In giving concrete expression to the view that science should not be allowed to become the preserve of the secular republic, the new Catholic universities achieved an important objective. But the contribution they could make to research, or even teaching, was always muted by their precarious status and material deficiencies. Amid the initial euphoria, diocesan, industrial, and private benefactions flowed encouragingly, and some of them could boast real achievements. In 1879, the claim of the rector of the Catholic university in Lille—that the collection of apparatus for physics placed the university among the European leaders in the discipline—was not entirely the fruit of wishful thinking.[74] And at Lyon the professor of zoology, Adolphe-Louis Donnadieu, could exult in 600 square meters of space, extensive collections, and excellent facilities for microscopy and other specialist activities that, in his view, many professors in the state's provincial faculties would have envied.[75] Donnadieu, though, was speaking as one whose previous teaching experience had been in a *lycée* and to whom even his faculty's modest facilities must have appeared luxurious by comparison with what he had enjoyed earlier in his career. In truth, the facilities in the Catholic sector only allowed serious competition with the state faculties in a few very limited instances.

Onerous regulations governing academic careers contributed to an additional drain, one that affected morale as well as recruitment. When, in 1880, heightened republican hostility toward the Catholic universities forced Lapparent to choose between his chair of geology and mineralogy and service in the state Corps des mines (in which he had hitherto been able to retain his coveted rank of *ingénieur d'Etat*), a secure personal financial position made the decision easy: abandoning the Corps, he stayed on at the Institut catholique until his death, after thirty-two years of service, in 1908.[76] Faced with a similar choice, Georges Lemoine took a different route: he had no such private resources to fall back on and in 1881, after six years of apparent contentment, he resigned his chair to devote himself more fully to the promising career as a Ponts et chaussés engineer (with the rank by now of *ingénieur en chef*) that he had interrupted on his appointment to the Institute.[77] Resignation also allowed him to pursue his long-standing ambitions at the Ecole polytechnique, where he had been a pupil in the late 1850s. As a teaching assistant (*répétiteur*) at Polytechnique, he had become especially close to Fremy and had emerged by 1881 as a natural candidate for one the school's two chairs of chemistry when Auguste Cahours retired in that year. Despite Lemoine's being the school's first choice, however, the Ministry of War did not endorse his appointment. Instead, it appointed Edouard Grimaux, so delivering a blow that must have weighed heavily in Lemoine's decision to put an end to his career in the Catholic system. Consolation came in the form of appointment to one of the prestigious positions of examiner in the final examination in 1884. But it was not until 1897 that a chair at Polytechnique came his way, as successor to Henri Gal.

The difficulties of the Catholic institutions of higher education multiplied, along with those of individual Catholic scientists, as republican attitudes toward them hardened after 1880. Apart from withdrawing the right to use the title university, the most vexing measure was a ban on the participation of professors from the Catholic faculties in the juries for the state examinations, which henceforth became the exclusive preserve of the professoriate of the state faculties. The signs of harassment and accumulating deprivation, accentuated by the state's enhanced investment in its own faculties, now became even more marked. In 1891, the once euphoric Donnadieu resigned from his post at Lyon in the face of the Catholic faculty's inability adequately to support his research and teaching.[78] In the following year, Lyon suffered a further blow when the meager level of funding for research loomed large in Amagat's decision to leave for the relatively modest position of *répétiteur* at the Ecole polytechnique.[79] At Lille too the strain began to show once the exhilaration and relative opulence of the early years had passed. There, the brightest scientific star was Antoine Béchamp, who became the first dean of the Catholic faculty of medicine and pharmacy when it opened in 1877. Béchamp had been recruited from the state faculty of medicine at Montpellier, where he had been professor of medical chemistry and pharmacy and become a respected authority on *microzymas* (which he discovered) and on the diseases of silk worms (which brought him into conflict with Pasteur). On his appointment in Lille, he served the faculty for ten years. But then he and his son Joseph, who also held a post in the faculty, as professor of analytical chemistry and toxicology, left in circumstances of some acrimony. The conflict, precipitated by wrangling and culminating in a lawsuit over the permanence of their positions (which the Béchamps lost),[80] was a symptom of the internal stresses and diminishing financial resources that afflicted virtually all of the Catholic institutions. Through the 1880s and 1890s, a mixture of republican repression and (to a degree that threatened the very survival of the Institut catholique in Toulouse) flagging support by regional Catholic interests took a heavy toll.

Of all the scientists whose careers were marked by their Catholicism, however, the best-known and most controversial was the first professor of physics at the Université catholique in Paris, Edouard Branly. Things had begun well enough for Branly. As Christine Blondel has observed, a description of his laboratory dating from 1879 suggests that the initial material provision for his research had been rather generous. He enjoyed the exclusive use of several rooms for different types of experiment and the services of a *préparateur* and a "lab boy," and the inventory of his apparatus, which included a 1.2-ton electromagnet, one of the first absolute Thomson electrometers in any French institution, and a gas-powered Gramme generator, showed the benefit of an allocation for equipment of 80,000 francs over two years.[81] But the deteriorating financial circumstances of the Catholic faculties after 1880 soon left him a poor relation of his peers in most state institutions and unprotected against the policies of the secular republic. By 1900, the meanness of the conditions in which he had to work had become legendary, at least in Catholic eyes, and it took another quarter of a century before he was given a laboratory commensurate with his reputation, thanks mainly to the generosity of the newspaper owner François Coty.[82]

From the time Branly's position began to worsen, his loyalty both to his institution and to his discipline was sorely tried. A mixture of insecurity and gathering disappointment led him to work for the doctorate in medicine that he took in 1882 and then to compete for the *agrégation* in medical physics that might have been expected to lead to his appointment to a chair in the subject at the faculty of medicine in Paris. In fact, the anticlericalism of the time put such a position beyond Branly's reach, as a chilling conversation with the engineer and dean of the faculty (and long-serving secretary general of the Association française pour l'avancement des sciences) Charles-Marie Gariel made all too plain. When Branly paid Gariel the customary visit, to introduce himself as a candidate for an academic career in medicine, the reply was of brutal frankness. Gariel's words, as recounted by Branly's daughter, put an abrupt end to Branly's hopes: "You are a professor at the Catholic university? Then, stay there!"[83] The advice, though accepted with resignation, was profoundly distressing for Branly, the more so as his motives for taking a chair in the Catholic system in the first place had been mixed. A reluctance to leave Paris for an appointment in a provincial faculty had certainly contributed to that decision. So, too, according to his own testimony, had a desire to escape from a delicate personal situation involving his relations with Paul Desains, the professor of physics at the Sorbonne. Desains, it seems, saw Branly (who held a junior post under him) as an all too eligible suitor for his daughter.[84]

The fact remains that, whatever his reasons for entering the system, Branly's religious conviction drew him into an association with the Catholic cause that was anything but ephemeral. It was a conviction that allowed him to endure, albeit with less than equanimity, a long history of indignities. Salary was one problem, as it was for most professors in the Catholic system: by setting professional salaries at 8,000 francs, d'Hulst had hoped to place the Institut catholique de Paris on a footing with the state faculties, but after 1881 the disparity between pay in the Catholic and state sectors began to widen. It was therefore to meet the needs of a family man of his standing that Branly augmented his income by practicing medicine for more than twenty years, from the mid-1890s to 1917.[85] To his other, greater burden of exclusion from the impenetrable republican establishment of savants there was no such straightforward solution. His public Catholic commitment always loomed large in the eyes of the free-thinkers and Protestants who dominated the academic elite. It did much to stoke the opposition when he was a candidate for election to the Académie des sciences in 1911. This anti-Catholic fervor helps to explain the anger that greeted his narrow victory over Marie Curie, by thirty votes to twenty-eight on a second ballot and in the face of the physics section's decision to recommend Curie as its first choice and to list Branly as just one of five candidates in the "deuxième ligne."[86] The mathematician Gaston Darboux (a Protestant) and the Swiss physicist Charles-Edouard Guillaume (at the time deputy director of the Bureau international des poids et mesures) were especially incensed by the result, but their indignation had echoes across a broad spectrum of liberal and secular opinion.

Branly's defeat of a rival who had won a Nobel Prize for physics eight years earlier would have been inconceivable but for a determined campaign in his favor by his ideologically sympathetic scientific friends (especially Amagat, Arsène d'Arsonval, and Jules Violle), the antirepublican writer and journalist Paul de Cassagnac, and the Catholic press (which almost unanimously and very vociferously spoke up for Branly). Supporters made much of the fact that Curie's Nobel Prize had been shared with her husband, Pierre Curie. They also invoked the Académie's unwritten norms with regard to elections. Whereas Curie had never been up for election, Branly had been a serious candidate twice in 1908, on the second occasion gaining a respectable eighteen votes against the thirty-four for the successful candidate, Paul Villard.[87] Hence it could be argued that by 1911 Branly's time had come. But this was no ordinary election. It was an exemplary episode in the confrontation between religiously and scientifically informed worldviews at a time when political and ideological tensions were at their height.

It is less clear whether Branly's Catholicism also contributed to the reluctance of the French scientific community to support his claims to a more prominent place in accounts of the origins of wireless telegraphy. But in Catholic eyes at least, he should have shared in the Nobel Prize that went to Guglielmo Marconi and Ferdinand Braun in 1909.[88] In the charged atmosphere of the time, the belief that the Académie des sciences simply failed to respond to the Nobel committee's request for an endorsement of Branly's case slipped easily into Catholic lore. Categorical denial of the rumor never entirely removed the suspicions of Branly's admirers, and the belief about the Académie's negligence in the matter resurfaced as recently as the 1960s, following a repetition of the story in the medical journal *Presse médicale*. The insistence of its permanent secretaries that the Académie would not have been consulted about this or any other Nobel Prize conveyed what must surely be the case as far as formal exchanges are concerned. However, we simply cannot know to what extent (if at all) the well-known coolness of certain members of the French scientific community toward Branly may have contributed to impeding his chances.[89]

The impediments that even a savant of Branly's eminence had to endure underlines the limitations of what the movement in favor of Catholic science had achieved by 1914. The republic had fashioned a public culture of science that had no place for religion (even if the two could, in principle, exist side by side but separately), and it had then appropriated that culture for its own ideological ends. In such circumstances, Catholic higher education was condemned to a perpetual struggle in which the personal sacrifices of professors and the expenditure of what, for private organizations without state backing, were large sums bore disappointingly little fruit. Upbeat addresses marking each new academic year sat uneasily with modest enrolments and chronic financial deficits that conveyed the Catholic predicament all too eloquently.[90] Parents and pupils were understandably reluctant to opt for institutions that, especially in the sciences, were seen to be materially disadvantaged and marginal to the well-oiled state machinery for the conferring of diplomas and the fashioning of careers. In its first ten years, to 1886, the Institut catholique in Paris produced only fifty-three *licenciés* in *sciences physiques*, fifteen in *sciences mathématiques*, and three in *sciences naturelles*.[91] A report seven years later showed that the pace had increased: by 1893, a total of 141 *licenciés* in science had emerged since 1876.[92] But the increase was achieved at a time when the state faculties were undergoing expansion on a scale that the Catholic system could not hope to match. In 1903, for example, the total number of students enrolled in the four Catholic faculties of science was 173, a figure eclipsed by well over four thousand students in the sixteen corresponding state faculties.[93] The disparity in the size of the scientific professoriate was somewhat less marked but still great, with only thirty chairs in the Catholic faculties, compared with 154 in the faculties of the state.

As Harry Paul has argued, it is important to set the Catholic faculties' rather muted achievements against successes in the more vocational aspects of education.[94] In this respect Lille was exemplary in ways that Paul and André Grelon have elaborated.[95] The jewel in Lille's Catholic crown was the country's only Catholic faculty of medicine, opened in 1877 and embracing pharmacy as well. The faculty surmounted some predictably hostile interventions by the municipal council in establishing an excellent working relationship with the city's Sainte-Eugénie hospital, where it had access to three hundred beds, in addition to overseeing two dispensaries, a maternity hospital, and a number of other services.[96] The city also became, by any standards, a major Catholic center for industrial and agricultural training, following the inauguration in 1885 of an Ecole des hautes études industrielles and in 1886 of an Ecole des hautes études agricoles, both of them associated formally with the Catholic faculty of science. The preparation in the industrial school was innovatively broad, including law, geography, foreign languages, history, and French composition as well as the core technical subjects. And its aim was clear. In a program normally lasting two years for candidates holding the *baccalauréat*, it was to fashion an employee who would be "a man of distinction" in addition to being a competent practitioner and who would in due course occupy a position of seniority from which he would be able to promote "the reign of truth, along with that of justice and peace."[97]

More than any other city, Lille was the setting for decades of rivalry between the competing forces of religious and secular education. Leading lay Catholics in the city and the surrounding department of the Nord, such as the architect Louis Cordonnier, the manufacturer Léon Harmel, and the brothers-in-law Philibert Vrau and Camille Feron-Vrau dominated the Catholic professional schools. And, as employers, they gave practical expression to their support by favoring graduates of the schools, whom they saw as being at once better trained than their counterparts from institutions under state or municipal control and also less vulnerable to the siren calls of socialism, strikes, and workers' unrest. Ministers, for their part, watched Catholic initiatives closely and on occasions were stimulated by them. Jostling, in this case involving the Ministry of Public Instruction, began in the mid-1870s, when the founding of a state faculty of medicine responded to plans for the Catholic faculty that opened shortly afterward.[98] In a similar spirit of rivalry, the opening of a Jesuit-controlled trade school, the Institut catholique des arts et métiers, in 1898, spurred the Ministry of Commerce to implement a long-delayed plan to make Lille the location for an addition to its network of *écoles d'arts et métiers*.

Elsewhere in France, the Catholic cause in education could seldom draw on the degree of lay support that made Lille and the Nord such a special case. Nationally, money and services in kind that the structures of Catholic education and research received from sympathetic industrialists, bankers, and diocesan authorities were never sufficient to allow competition across the board with the institutions of the ministries that provided for education and research within their various domains. The recruitment of pupils too was a source of unremitting anxiety and a spur to vigilance. On this score, the contrast was routinely drawn between the weaknesses of Catholic provision in higher education and its strength in primary education and (albeit more patchily) in the secondary sector, in which a Catholic school of the stature of the Collège Stanislas in Paris was quite the equal of the best state *lycées*. As even the most loyal champions of Catholic higher education had to recognize, their own faculties remained perpetually vulnerable to the capacity of the state faculties to draw pupils toward them once young men and women acquired the independence that came with the passage from secondary to university studies.[99]

Catholic ventures in all sectors and at all levels, therefore, were beset with impediments that made such successes as they had all the more remarkable. This was true not only of educational achievements such as those in Lille but also of contributions to the five international Catholic scientific congresses that took place, at three-year intervals, between 1888 and 1900, twice in Paris and thereafter in Brussels, Fribourg, and Munich (see chapter 4). Catholic learned journals too played their part, none more so than the Belgian *Revue des questions scientifiques*, whose Jesuit editor, Ignace Carbonnelle, proclaimed its tone from the start (in 1877) with an attack on the materialism and atheistic tendencies of contemporary science, as represented by John Tyndall's address to the British Association for the Advancement of Science in Belfast in 1874.[100] On these international stages, the French Catholic community was not found wanting. Nowhere, in fact, did Catholic resolve flag: if anything, the rising tide of laicization served to strengthen the resolve to resist secular browbeating. Nevertheless, the constraints and indignities extracted a heavy price. Even if they failed to stifle a distinctively Catholic approach to science and learning, they certainly diminished its influence and, by constantly stressing its separateness from the main stream of republican academic life, undermined public perceptions of its worth in the eyes of any but the most faithful.

The Republic of the Savants

While most citizens of a conservative disposition found the intellectual tone of the early Third Republic hard to swallow, they had to acknowledge that the provision for higher education and research improved significantly in the years of soul searching that followed the defeat of 1870. Predictably, the main benefits accrued to science. But the spokesmen for nonscientific disciplines too strove to present themselves as essential to the renewal of France. No less than scientists, such spokesmen had support in high places. Among their most influential champions was the philosopher and republican politician Jules Simon. As Minister of Public Instruction in the new Government of National Defense of September 1870 and then, until May 1873, in the moderate republican administration of Adolphe Thiers, Simon fought against the underfunding that had left its mark in laboratories of grotesque inadequacy and in a state of abandon that had crippled the whole world of learning in France. The series of ministerial awards that enabled the young Gabriel Lippmann to work for a doctorate at Heidelberg during a two-year stay there between 1872 and 1874 and then to move on for a semester at Berlin were pushed through, with characteristic resolve, by Simon.[101] In all he did, however, Simon did not seek to accord either science or the faculties of science a privileged position. When he spoke at the annual prize-giving for the members of the provincial *sociétés savantes* in the Sorbonne in April 1873, his criticism extended to the meagerness of governmental provision for all branches of higher education.[102] His policies for the reform of primary education (for which he is, rightly, best known) and secondary education conveyed the same breadth of vision.[103] At the secondary level, science was to be promoted as part of a wider program aimed at promoting a modern "literary culture" that also embraced geography, history, and modern languages.[104]

Whether science was assimilated within a more general program, such as Simon's, or given special status, it found its way into the rhetoric of virtually every corner of the reform movement. Of the disciplines that were not traditionally associated with the sciences, geography was readier than most to combine the usual assertions of national fervor and utility with an aspiration to scientific status. The specter for modern-minded geographers was a retreat to the old world of gentlemanly tale-telling that they associated with the venerable Société de géographie, founded in Paris in 1821. But things were changing. Geographers of every persuasion applauded an influx that saw the membership of the society double (to more than 1,350) between 1871 and 1875.[105] They also welcomed the launching of new periodicals, such as *L'Exploration*, *L'Explorateur*, the *Revue géographique internationale*, and the *Revue de géographie* (in an ascending order of seriousness) in the mid-1870s, expeditions,[106] and the founding of geographical societies, bringing the number of such societies across the country to twenty-six by 1884. But while such initiatives reflected and fostered a swell of public support both for academic geography and for the Republic's colonial adventures and oceanographic expeditions, the leaders of the discipline—Ludovic Drapeyron (from a rather unsatisfactory Parisian base in the Lycée Charlemagne) and Charles Hertz (the founder of *L'Explorateur*) in particular—had their sights set on more specific objectives.[107] Chief among these was the advancement of geography as a quintessentially patriotic discipline that offered benefits ranging from promoting international trade to remedying the notorious inadequacy of the maps the French had used in the war of 1870.[108] If geography was to serve the nation in this way, its association with the literary cast of the "old" geography had to be tempered by the new scientific spirit that Drapeyron thought to be a cornerstone of France's revival. Only if that were done would the pattern of cause and effect be perceived beneath otherwise inexplicable superstitions and the caprices of human history.[109]

In important ways, the advance of geography in the 1870s and 1880s was favored by its freedom from the constraints of a long academic tradition. By comparison, historians were trammeled by their discipline's past. Gabriel Monod was typical of those who came to the Republic marked by an outspoken admiration for German scholarship that he had acquired as a young scholar under the Second Empire. After studying at the Ecole normale supérieure and then attending the seminar of the medievalist Georg Waitz at Göttingen in 1867–68, he returned to France intent on recreating, at the newly formed Ecole pratique des hautes études, the atmosphere of rigorous critical analysis of evidence that he saw as a main cause of Germany's lead in historical research. In his response to the war, Monod maintained a measured, unrancorous view of Germany and poured his energies into the campaign in favor of regional universities and the modernization of historical and (with his slightly older contemporary and fellow veteran of Göttingen, Gaston Paris) philological scholarship along German lines.[110] Like so many of his and the next generation of scholars bent on reform, Monod couched the case for innovation—notably for the journal *Revue historique* that he was instrumental in founding (with his pupil Gabriel Fagniez)—in patriotic terms. An introduction to the first issue of the journal, published in 1876, advanced the study of France's past as a means of restoring the unity and moral strength that the nation needed in its time of crisis.[111] The *Revue historique*, as Monod conceived it, was to be disinterested and "scientific" in its approach. It would eschew political and philosophical theorizing, speak to a public broader than would be attracted by a journal of pure erudition, and, above all, present an understanding of the past as a force for advancing the progress of humanity. It was a tough program and yet another one rooted in the perception of science as the touchstone of intellectual rigor.

No discipline remained unmarked by the mixture of patriotic sentiment and sympathy for the Third Republic that bound most of those working in the institutions of the Ministry of Public Instruction. After 1870, even in a pursuit as seemingly detached from the present as ancient history, the favor accorded to Gaston Boissier's *Cicéron et ses amis*, which went through eleven editions between 1870 and 1902, owed much to its perceived contemporary relevance. The book's depiction of Cicero as a defender of democratic republican virtues (in contrast with the despotic Caesar, widely seen as a surrogate for Napoleon III) had unequivocal meaning for readers in the early days of the republic. Boissier voiced a gentle protest against the fashion for quarrying the past in search of "arms for fighting today's battles."[112] But reminders of relevance in the unlikeliest corners of scholarship served an essential rhetorical purpose if scholars in the humanities and the social sciences were to stand shoulder to shoulder with scientists in the cause of reform. The meetings in Renan's Parisian apartment in the early 1870s encapsulated the reform movement's broad base. They engaged the mathematician Joseph Bertrand, the chemist Marcellin Berthelot, and the physician Henri Liouville in discussions with literary figures, historians, and jurists (including Hippolyte Taine, Emile Boutmy, Michel Bréal, and Ferdinand Hérold, as well as Monod, Paris, and Boissier).[113] And the door to political and economic reality was kept open by the presence of Armand du Mesnil, the director of higher education in the Ministry of Public Instruction.

The union of modern-minded aspiration, political adroitness, and administrative expertise helped to create an unstoppable momentum for change that assumed concrete form, in 1878, in the Société de l'enseignement supérieur and, from 1881, in the society's *Revue internationale de l'enseignement*. Here, in one of the cradles of what Albert Thibaudet was to characterize, with the benefit of a half century of hindsight, as the "République des professeurs,"[114] the powerful corporate spirit of the professoriate of higher education and the professoriate's overwhelming conformity to republican values found easy reinforcement. Through the 1880s and 1890s, the reforms of higher education came thick and fast, pushed ahead by committed Ministers of Public Instruction—especially Jules Ferry (1879–83, with two relatively short gaps), René Goblet (1885–86), and Léon Bourgeois (1890–92)—and equally committed administrators, among whom Louis Liard stood out as the ministry's director of higher education from 1884 to 1902.[115]

Underlying the reforms were two enduring strands of policy. One was a belief in the virtues of a controlled devolution of power that would give the provincial faculties a measure of independence. The other, related strand was a belief that faculties should have the right, indeed the obligation, to raise a significant part of their funding in their various regions. A powerful motive for these policies, from 1882, was a deepening economic crisis that caused public spending on education to slow and then to stagnate. Also driving the policy, however, was a conviction that a greater dependence on local support—from municipal and departmental councils and commercial and industrial interests—would strengthen faculties, especially science faculties, by obliging them to respond to concrete needs. The process of devolution got under way in the 1880s, aided by Goblet's granting of "civil status" to the faculties in 1885. The measure—in French terms, a recognition as bodies of "public utility" —authorized the faculties to keep whatever funds they could raise, independent of the state, and marked an important step on the road to their grouping first as *corps de facultés* in 1893 and then as largely self-governing universities in 1896. The creation of universities fulfilled a number of Liard's long-standing dreams, by fostering an emphasis on fiscal responsibility and responding to academic demands for greater freedom from the overbearing Parisian bureaucracy.

The first fruits were encouraging. In places where town-gown alliances were strong, professorial morale was boosted: the science faculties in Bordeaux, Lille, Lyon, Nancy, Paris, Toulouse and (late in the day) Grenoble all benefited in this way, enjoying a new flexibility and enhanced income and enrolments. And even in such smaller faculties as Caen, Clermont, Dijon, Poitiers, and Rennes the number of enrolled students rose steadily.

As events were to show, a preoccupation with satisfying local demands for manpower and technical expertise ate menacingly into professorial time and energy, something that became particularly irksome as the demands grew after 1900. An even more threatening intrusion arose from the decision, taken in 1893, to make science faculties responsible for teaching basic physics, chemistry, and natural history to future medical students. Such teaching, for the *certificat d'études physiques, chimiques, et naturelles* (always known as the PCN), met its objectives by filling hitherto underused lecture theaters and laboratories and serving as a gateway to more advanced study not only in medicine but also in vocational disciplines relevant to industry and agriculture.[116] Especially in the provinces, the presence of the PCN students, as of those in the specialized technical institutes that most faculties had created by 1900, reinforced perceptions of universities as sources of publicly accountable services rather than (as many had seen them) the refuges of a distinguished but remote professoriate. For some professors, burdened by increased teaching loads and an uninspiring, elementary curriculum, the transition was painful. But even they enjoyed the fruits of the faculties' greater responsiveness to the needs of society and the economy. Across France the improvements in the conditions in which professors and their younger colleagues and pupils worked were a reality.

In some faculties, existing buildings were simply rehabilitated: this occurred in Montpellier, where the faculty was rehoused in a former hospital, the Hôtel-Dieu Saint-Eloi, along with the faculties of letters and law.[117] Elsewhere, a mixture of civic and regional pride and ministerial encouragement resulted in purpose-built structures of outward classical magnificence if not always of genuine practicality. In Rennes, the premises that the faculty of science had occupied in the purpose-built but cramped Palais universitaire since the 1850s (along with the faculties of law and letters, the secondary school of medicine, and the civic museums of art and natural history) were abandoned in favor of a dedicated new building begun in 1889 on an initial budget of a million francs.[118] Here, a familiar story unfolded as the original estimate receded into irrelevance; by the late 1890s, the cost stood at almost 1.5 million francs (including 500,000 francs from the state), though still without satisfying the escalating demands for space and facilities for research and teaching.[119] A similar voraciousness and diversity of sources of funding were evident in Bordeaux, where local support was showered even more liberally than in Rennes. Between 1880 and 1894 2,615,436 francs were spent on an imposing new building to be shared by the faculties of science and letters. Of this, only 300,000 francs were contributed by the state, the rest (almost eight times as much) coming from the city.[120]

The changes of the last two decades of the century left professors in the sciences busier and more prominent in public life. They also made for more complex, tension-filled lives. For while academic scientists were seen locally as the providers of useful knowledge and props of the economy, they remained the administrators of the national examination system and, hence, in ministerial eyes, the public face of central authority and the nation's intellectual tradition. As Marcellin Berthelot made clear in a sketch of the worldly existence of the modern savant in 1905, the age of the ivory tower was definitively over. Speaking with the wealth of personal experience he had gathered during the more than three decades in which he had come to embody all that was best or worst (according to taste) about the public man of science, Berthelot observed:

> The life of today's savant is a many-sided one, calling upon him to direct his activity in many different directions. It is not that he is driven to this by some vain desire for excitement or popularity. Perhaps he would prefer to remain shut up in his laboratory and to devote all his time to his favorite studies. But he is not permitted to confine himself in this way. . . . He is sought out, and his services are requested. Often they are even urgently solicited in the name of the public interest, in the most diverse spheres: specific applications in industry or national defense, public education, or even general politics.[121]

Although Berthelot was one of those who revelled in the multifaceted life that he described, he knew better than anyone that a multiplicity of extraneous distractions posed a threat to research and high-level publication. Yet he coped, remaining a major figure in science until his death in 1907. In fact, it was one of the signal achievements of the educational administrations of the early Third Republic that the heightened external demands on faculties of science coincided with a surge in intellectual productivity in the last quarter of the nineteenth century. In making this point, Terry Shinn has pointed to a contrast between the doldrums of the singularly unproductive period from 1846 and 1875 and the far more vibrant years from 1876 to 1900: an almost fourfold per capita increase in the number of articles published annually by professors in what Shinn defines as "fundamental research" set the two periods poles apart.[122] The circumstances that favored the surge are not easy to unravel. But they included the improvement in the provision for laboratories (tentatively reinforced from 1901 by the creation of a Caisse des recherches scientifiques for funding research[123]), the growing emphasis on publication (rather than an honourable record as a teacher in a *lycée* or faculty) in making appointments, and the shift in the ambitions of the ablest pupils at the Ecole normale supérieure, whose trajectories increasingly led on to higher rather than secondary education.[124] At all events, the indicators of the new prominence that came to be given to research are unmistakable. Quite apart from Shinn's evidence about productivity, there are significant pointers in the increasing size of doctoral theses and the tendency for the doctorate to be awarded at an ever later age. As Victor Karady has shown, the proportion of nineteenth-century theses in science of fewer than forty pages fell from 63 percent for the years up to 1850 to 16 percent between 1850 and 1869 and finally to 7 percent after 1869, while that of doctorates obtained before the age of 30 fell from 65 percent (until 1850) to 49 percent thereafter.[125]

In republican circles, blending intellectual endeavor with practical services and administrative functions was generally deemed to be desirable and to have been implemented successfully. And that perception never wavered. Once the Republic was definitively confirmed by the victory of the left in the legislative elections of October 1877 and then in the senatorial elections of 1879, science and its values became more central than ever to republican ideology.[126] One mark of the respect that the nation's great savants came to enjoy, both in society at large and in government, lay in granting state funerals.[127] Between 1878 and 1907, five scientists were granted such funerals, compared with only two writers (Hugo and Renan) and one musician (Charles Gounod).[128] Of the five, Paul Bert (1890) and Marcellin Berthelot (1907) were pillars of the Republic who had enjoyed political as well scientific eminence, and their funerals were seen as occasions for flaunting an easily concocted brew of scientism and republican virtue that set Berthelot in particular so far above the normal run of humanity that he also received the exceptional honor of burial (even more exceptionally, with his wife) in the Panthéon. Granting a state funeral to Michel-Eugène Chevreul was seen in a similar light, thanks to a mixture of his distinction and great age: dying as the Nestor of French science in 1889 in his 103rd year, he had transcended the allegiance to any one of the many regimes through which he had lived, and no patriot could forget his bitter outburst against Prussia's act of vandalism in bombarding the Muséum d'histoire naturelle during the night of 8–9 January 1871.[129] Claude Bernard's known closeness to the emperor and his service as a senator at the end of the empire might have created rather greater problems when, in 1878, he became the first scientist to receive a state funeral. However, a portrayal of him as a man of genius and simplicity who had never sought honors but on whom they had been thrust—"a man at once so great and so simple, one who never sought fame but whom fame sought out," in the words of Paul Bert, his pupil and successor in the chair of general physiology at the Sorbonne—did enough to efface uncomfortable memories of his imperial past.[130]

The case of the fifth of the savants whom the republic honored, Pasteur, also required delicate handling. His adherence to the Catholic Church, however formal, and his indifference in politics made him a less than ideal republican hero.[131] But the reverence with which he was regarded across the whole social spectrum by the time of his death in 1895 made his a special case, one moreover that had potential political benefits. Pasteur could be lauded by the scientifically inclined left as a symbol of the beneficence of science, though without risking the provocation to bourgeois sensibilities that savants of an openly secular cast would have caused. When he received the impressive honor, rare for a scientist, of being admitted to the Académie française in 1882, he declared his position unambiguously in an address to his fellow "immortals" that marked out the ideological space separating him from the free-thinking Emile Littré, whom by a piquant irony he succeeded, and Auguste Comte, whose positivism had captivated Littré but left Pasteur defiantly "mistrustful."[132] For Renan, whose task it was formally to welcome Pasteur, fashioning a response to words that defined such a telling divide between Pasteurian science and the scientism at the heart of liberal ideology was far from straightforward. But he performed unflinchingly, accompanying the customary pleasantries with the spice of a statement of his own intellectual affinities to Littré (though not to Comte).[133]

In the polarized political atmosphere of the early Third Republic, such exchanges were milked for every drop of significance. Pasteur's standing, though, withstood the scrutiny and continued to grow, albeit under a Republic whose ideals he scarcely shared. In December 1892, his seventieth birthday was marked with a glittering celebration in the great amphitheater of the new Sorbonne at which he entered on the arm of the president of the Republic, Sadi Carnot.[134] And when he died almost three years later, the ceremonies were worthy of one of the nation's greatest heroes. In huge numbers, the public filed respectfully past the casket containing his body in the Institut Pasteur, and the Mass in the Cathedral of Notre Dame was attended by foreign royalty as well as Félix Faure, who had succeeded Carnot and Jean Casimir-Périer as president. It was only at the family's insistence that a place in the Panthéon was declined in favor of burial in a sumptuously decorated crypt at the Institut Pasteur. A great French man of science who was also a benefactor of all humanity, Pasteur had assumed the aura of a secular saint, which he retained (though not uncontested) a century on, when the centenary of his death was commemorated in 1995.[135]

For the majority of lay people, both science and the republic as its patron emerged well from the three universal exhibitions that took place in Paris in 1878, 1889, and 1900. The manner in which this was achieved rested on vague associations rather than on explicit demonstrations of the contributions that science and savants had made to the material blessings of technology. But the strategy, if such it was, worked. As the first, rather austere declaration of France's rebirth, in the hastily organized and exhibition of 1878, gave way to the more sumptuous festivities to mark the centenary of the Revolution in 1889 and finally in 1900 to the celebration of the century that had just ended, the popular appeal of the exhibitions grew. A broad swath of the public was beguiled by an ever more ostentatious style of presentation that succeeded increasingly by spectacular novelties (not least those made possible by the *fée électricité*, a benign "electricity fairy" with unmistakable republican credentials; see figure 15). This was in contrast with the displays of the bare bones of machinery that had tended to be uppermost in the universal exhibitions of 1855 and 1867. The fact that the great popular successes of the later exhibitions—Bell's telephone and Edison's phonograph in 1878, the colored fountains and Eiffel Tower in 1889, and the moving walkways and "grande lunette" in 1900—rested on techniques more sophisticated than those manifested in even the most striking exhibits of 1855 and 1867 escaped no one. And the public images of science and science-based technology (as the embodiments of progress) were the beneficiaries, despite the ever thicker layer of exuberance and gimmickry that, by 1889 and 1900, tended to obscure the underlying principles.[136]

Official attendances reflected the success of the formula that set science and technology in the context of a larger package of spectacle and the celebration of modernity. The attendance of the 1878 exhibition, 16,156,626, rose to 32,350,297 in 1889,[137] and then to a gigantic 50,860,801 in 1900, far in excess of the number achieved at any other exhibition until well after the Second World War.[138] With rising numbers of visitors and increasingly prudent management, there came financial dividends too: a loss of more than 28 million francs in 1878 (caused mainly by building costs, especially for the Trocadéro Palace) was transformed into operating profits of 8 million and 7 million francs, respectively, in 1889 and 1900. The most striking indicator of success, however, remained the growth of public interest, which fed and was fed by an explosion of accompanying literature, official and unofficial, to which men of science made a conspicuous contribution. The published reports of the specialist juries, finely printed in many bulky volumes, conveyed the ministerial sense of a job well done and of public money judiciously expended. Only occasionally were they the setting for serious reflection on France's place in the modern world. On this count, Albin Haller's analysis of the eclipsing of the French chemical industry by that of Germany in the official report on the 1900 exhibition stood out, as did the comments of the physicist Alfred Cornu on the state of the French instrument-making industry.[139] Weighed against the celebratory literature of guides, reviews, and ephemeral publications for popular consumption, Haller's writings must have appeared to some as rather curmudgeonly departures from the converging tides of republican self-congratulation and belle époque gaiety. And Cornu could be seen to be reacting in a spirit of barely concealed pique to the triumphalist tone of the catalogue of the German display of instruments for optics and precise measurement.

The sad fact was that the judgments of Haller and Cornu had substance. But they were easily lost from view in the euphoria that characterized all the exhibitions, especially those after 1870. When Sadi Carnot, in his capacity as president of the republic, opened the 1889 exhibition, he expressed the overriding sentiments, both official and popular, in the ringing terms appropriate to an official inauguration: "In the invigorating atmosphere of liberty, the human spirit is rediscovering its initiative, science is taking flight, steam and electricity are transforming the world. A century that has witnessed such miracles had to be celebrated."[140] Twelve years later, Alfred Picard, a former *polytechnicien* and Ponts et chaussées engineer who had served for seven years as the senior administrator for the exhibition of 1900, mixed the same laudatory gravitas with an even more lyrical response. His work had been, as he described it, an inspiring experience for him. The exhibition had bathed the most pessimistic spirits in a new optimism and done so by laying bare the progressive nature of human history. In his *Bilan du siècle*, a sumptuous six-volume review of the century, inspired by the exhibition, he made his boldest claim: "The sense of a relentless onward march and of humanity's incessant advance puts despondency to flight and brings profound comfort. It affords consolation in old age, inspires resolve in the prime of life, and instills trust and a competitive spirit in the young."[141]

The source of Picard's ennobling vision lay unambiguously in science in its most disinterested theoretical form, the form that made it a life-giving spring, a perpetual fount of renewal for the human mind.[142] In science, the nineteenth century had been not only "a truly heroic age" but also one in which France had played an outstanding role: "Noblesse oblige! Our country will maintain its standing by preserving as a sacred legacy the traditions of the nation's genius, that is to say the spirit of synthesis, simplicity, and lucid precision."[143] Thirty years on from the searing experiences of the Franco-Prussian war and the Commune, the exhibition brimmed with patriotic significance. There, to a degree unmatched in earlier exhibitions, the universal benefits of scientific and technological progress had combined with France's unique cultural and intellectual traditions to set the course for a century of unprecedented achievement to come. Such, at least, was the dream.

Fin de Siècle: From Inspiration to Anxiety

By the time Picard wrote his fulsome reflections on the 1900 exhibition, the conflation of human well-being with the onward march of science and the identification of science as one of the glories of France's past had been commonplace for a quarter of a century. But the admiration for science and the benefits that material progress were assumed to bring had never gone unchallenged, and since the 1880s the disquiet had gradually assumed a coherent voice. As plans for the 1889 exhibition took shape, opposition to republican conceptions of modernity boiled over in a debate about the rising Eiffel Tower. For the spokesmen and former pupils of the Ecole centrale des arts et manufactures, where Eiffel had studied in the 1850s, the "tour de 300 mètres" (as contemporaries called it) was a potent symbol: as a feat of engineering that allowed France to boast a structure almost twice the height of its closest rivals, the Washington Monument and the Mole Antonelliana in Turin (both of them just under 170 meters high), it celebrated the nation's renewal and the eminence that Centrale enjoyed in the elaboration of the most advanced technologies for the erection of large iron structures.[144] Aesthetic considerations, however, opened the door to a more critical stance. In February 1887, the month after Eiffel's plan for a tower had been approved by the Minister of Commerce

(Edouard Lockroy) and the prefect of the department of the Seine (Eugène Poubelle),[145] a group of intellectuals describing themselves as "writers, painters, sculptors, architects, passionate lovers of the hitherto unsullied beauty of Paris"—including such major cultural figures as Charles Garnier, Ernest Meissonier, William Bouguereau, and Antonin Mercié—delivered a public protest to Adolphe Alphand, the director of construction for the forthcoming exhibition.[146] The group's core question encapsulated their hostility. Was Paris, of all cities, to allow the "commercial fantasies of a machine-builder" to set it on the path of "ugliness and dishonor"?[147]

In literature, reservations about the tower were reinforced by the Parnassian poet François Coppée in his derision of the "monstrous belvedere," topped with a restaurant.[148] Coppée's outpouring of more than thirty stanzas of invective against a structure whose hideousness was matched, for him, only by its futility was the *cri de coeur* of a conservative who perceived all too clearly the gathering threat to the traditions of old Paris. Other reactions to the tower had deeper political or ideological motives. Among those who regarded with suspicion everything that smacked of republican breast beating, opposition to the structure was never far from the surface. The first issue of the new series of the now assertively clerical *Cosmos* in 1885 had already poured vitriol on Eiffel's preliminary plan for a construction that the anonymous author portrayed as an overblown Tower of Babel, a "folly."[149] Decorum decreed that *Cosmos* should reserve its final judgment on the project. But no amount of rethinking was going to change the journal's opinion of what it saw as a detestable exercise in pointless, grandiose frivolity.

The onslaughts did nothing to impede the tower's completion. Banks, companies, and individuals eagerly subscribed to create a share capital of 5.1 million francs, and popular adulation once the tower was in place demonstrated how little impact the criticism had on the mass of the population. As Minister of Commerce, Lockroy urged Alphand to stand firm in the face of the jibes, while visitors voted supportively with eager feet, more than 23,000 of them on one spring holiday alone.[150] The elevators and staircases took large crowds to the four restaurants, with combined accommodation for almost four thousand people, on the first level, then on to the second level, with its kiosks, a post office, and a printing office fitted out to produce a daily edition of *Le Figaro*, and finally to the third level, where an enclosed gallery allowed up to five hundred visitors at a time to obtain an incomparable view of Paris. The novelty never faded, and in all, in the six months of the exhibition, just under two million people ascended to at least the first level, at a cost of 2 francs each.[151] The overwhelming public favor for the tower and other material manifestations of the modern world made head-on confrontations with the modernizing philosophy underlying them difficult to sustain. Criticisms, such as Coppée's, that were presented as a declaration in favor of enduring human and religious values in the face of the forces of materialism could all too easily be read as retrograde obscurantism and a blanket rejection of modernity in all its forms.

Even so, by 1890, the questioning of the many variants of scientism was coalescing into a dispersed sense of anxiety. It took as its targets the twin (and frequently confused) dangers of blind optimism in the beneficence of technology and the bankruptcy of systems of morality and thought based on science. The most engaging of the writers who took technology as their target, to the extent of presenting it as a two-edged weapon, was Albert Robida, a prolific book illustrator and author who came to prominence in the 1880s. His stock-in-trade, apart from the illustration of children's books, was the nostalgic evocation of the "old France" of regional diversity and human types unsullied by industrial modernity.[152] But in two volumes published in 1883 and 1892—*Le Vingtième siècle* and another vision of the twentieth century, also bearing the main title *Le Vingtième siècle* but always known by its subtitle *La Vie électrique*[153]—he turned his attention to the future. The stories, both of them imaginative portrayals of life in the 1950s, were light-hearted tales of bourgeois society into which technology had intruded irreversibly. In *Le Vingtième siècle*, Robida cast his futurist speculations around the family of the wealthy banker Raphaël Ponto and his orphaned niece and pupil, Hélène Colobry, through whose eyes mid-twentieth-century Paris was observed; in *La Vie électrique*, the family was that of Phyloxène Lorris, a businessman and former *polytechnicien*. But the perceptions that Robida conveyed through his characters were essentially the same.

Despite the frailty of Robida's stories as a literary vehicle, their evocation of the impact of science and its applications was at once powerful and prescient. The civilization of his mid-twentieth century would be gripped by gadgetry. A vehicle in what Robida seems to have conceived as a sort of pneumatic tube would put Paris less than an hour from Madrid; pictures conveyed through the *téléphonoscope* to screens in public places and the home (by means that Robida never specified) would allow Parisians to follow events in distant lands and places of entertainment closer to home; and sound transmitted to controllable speakers would bring news directly to the breakfast table. Thus far, so good. But few of the anticipated inventions would be without cost to the quality of life, and here Robida's anxieties surfaced. Often the cost, measured in the waning importance of personal interactions, would be grave. News arriving by a loudspeaker at the breakfast table would kill conversation, and the speed and convenience of travel in the flying machines, the "aéronefs," that Robida expected to become a normal means of domestic transport would be offset by the inevitable congestion of the skies. And everywhere pollution would be a threat. Air and water would be spoiled by the effluents of industrial society, and a mixture of noise and dirt would make it necessary to exclude the steam engine and other symbols of the modern world from specially protected parks. For Robida, the locomotive—"the beast of iron and fire"—had already left its melancholy mark on a nineteenth century that had lost its quaint and lively coaching inns,[154] and ahead he saw only more of the same dehumanization and threat to individuality. By 1892, the electricity fairy of *La Vie électrique* was depicted as a slave who bore the threat of rebellion rather than the promise of serene beneficence that dwellers in the electric age had come to expect of the new technology (see figure 16).

Those who exulted in the technological achievements of the age would probably have seen Robida's reservations as an essentially harmless conservative response to change and one whose influence was restricted by the cost and small print runs of *Le Vingtième siècle* and *La Vie électrique*.[155] It would have been harder for them to treat Villiers de L'Isle Adam's *L'Eve future* (1886) so lightly. Few readers could have avoided an uncomfortable *frisson* on reading the novel's portrayal of Thomas Edison as a sinister wizard who responded to a friend's disappointment in love by constructing an ideal mechanical woman moved by electricity and with phonographs for lungs.[156] In response to such disquiet, the only weapon was reassurance of the kind that Berthelot's admirer and former pupil, Charles Richet, delivered in his prophetic book, *Dans cent ans*, in 1892. The world that Richet foresaw a hundred years on would be marked profoundly and, on the whole, for the better by technology and science. The metric system would have triumphed, even in Britain; life expectancy would have risen to over fifty; a tunnel would connect France and England; work in the home would be helped by electric sewing machines; steam-driven vehicles would be commonplace; and steam tramways would link previously isolated communities.[157] There would be huge economic, social, and political changes too: household goods would have fallen dramatically in price, and the balance of power would have shifted to the two great nations of the United States and Russia, whose combined population, promoted by medical advances and plentiful space and food, would have risen to 600 million.[158] Richet's vision of progress extended still further to changes on a higher plane. Some years later he looked forward to the gradual victory of peace and solidarity between the peoples of the world over war and misplaced national sentiment.[159] In that better world, science and technology would show their colors as quintessentially the pursuits of peace and the property of all nations, irrespective of where a discovery or invention originated. War and military organizations, in contrast, would be seen as the "remnants of barbarity of which we all bear traces within ourselves."

It is hard to evaluate the impact that such speculations had on readers. But the ideological undertow that they carried with them—broadly reactionary, backward-looking, and pessimistic in Robida's case and progressive, modern-minded, and optimistic in Richet's—conveyed the tension between conservatism and modernity that racked late nineteenth-century French society. Even the most widely read and superficially least engaged of the authors who used science and technology as a resource for imaginative writing—Jules Verne—articulated the same tension. Although Verne's emphasis evolved in the course of a long literary career extending from the early 1860s until his death in 1905, the theme of human responses to the power of science and, even more so, of technology remained at the heart of his work. Among his early novels, *Cinq semaines en ballon* (1863), *De la terre à la lune* (1865) and its sequel *Autour de la lune* (1869–70),

Voyage au centre de la terre (1864), and *Vingt mille lieues sous les mers* (1870) all evoked the extraordinary possibilities for travel that the modern engineer had placed at the disposal of nineteenth-century society. But they were also a paean to the old-fashioned qualities of human bravery and resourcefulness. For in realizing the fruits of technology, Verne's travelers had to display courage of barely imaginable proportions.

With regard to modernity, the technology on which Verne's tales rested displayed few flights of fancy: it was, almost without exception, the technology of Verne's day. And while scientific allusions were plentiful (especially in astronomy and geography, in which Verne was most at home), the science too was conventional: the sequence of animal and vegetable species observed by Pierre Aronnax (Verne's imaginary naturalist from the Muséum d'histoire naturelle) and his two companions on their journey to the center of the earth, for example, was drawn straight from Cuvier. In fact, there was nothing in Verne's stories that called for extravagant departures from the body of contemporary knowledge. What distinguished the most adventurous and scientifically and technologically informed of his characters (all of them male—there were no women among Verne's heroes) was a capacity for leaps of perception that fully revealed the potential of nature. The leaps were made to particularly dramatic effect by the five Americans in what is generally regarded as Verne's most engaging novel, *L'Ile mystérieuse* (1875). The heroes, significantly possessing the adventurousness that Verne admired in Americans, were the engineer Professor Cyrus Smith, the young naturalist Harbert Brown, and their three companions. The five were distinguished by an ability to exploit their knowledge that allowed them to survive following the crash of their balloon on the seemingly inhospitable island of the novel's title. As Jean Chesneaux has put it, the island was a microcosm, the setting for "a symbolic recapitulation of all the animal, vegetable and mineral resources of the earth,"[160] and Verne's reworking of the Robinson Crusoe story can be read as casting humanity's accumulating technological mastery over nature in the form of a heroic exercise in survival against the odds. That the nature in question was essentially nature as it was understood in the late nineteenth century and not in some imaginary future made the message all the more compelling.

Such a vision lent itself to restrained optimism, and generally Verne's endings were happy. Readers of *De la terre à la lune* who at the end of the book had left the three travelers drifting in their projectile helplessly and out of sight behind the moon were reassured, four years later, by their reappearance and eventual safe landing on earth in *Autour de la lune*: the characteristic Vernean combination of human resolve and technological ingenuity had borne its predictable fruit. In *Hector Servadac* (1877) even the terrifying events that accompanied the collision between the comet Gallia and the earth and the resulting enforced tour of the solar system by Servadac and his fellow passengers on a part of the earth carried off by the comet were made to appear controllable by the calmness of Servadac's investigations of the scientific dimensions of the experience. Men who could indulge in observations and calculations of the weight of the comet[161] as they hurtled through space and eventually returned to earth (in an implausible final descent from Gallia by balloon) had evidently conquered fear and drawn on science to do so.

Menace, however, was a no less essential ingredient of Verne's narratives, and it is this that makes him a telling witness to contemporary concerns about the unsure dehumanized world that science and technology seemed to promise. At a purely literary level, the menace served to heighten suspense, especially at the end of each of the episodes of the stories that appeared (as most originally did) in the periodical press. More profoundly, though, it gave a brooding tone to some of his most memorable works.[162] In *Vingt mille lieues sous les mers* Captain Nemo and his crew on board the sinister vessel *Nautilus* were rebels against society. Who else but an anarchist would have raised a black flag at the South Pole to assert his claims to land untouched by the taint of colonization? The point would not have been lost on readers who remembered 1848 or, within a few months of the book's publication, witnessed the ravages of the Commune. The fact remains, however, that if menace lurked in Nemo, it did not triumph. This leaves us with a Janus-faced Verne in whom what Michel Serres has called "a mildly progressivist, scientistic confidence" vies with "a relatively catastrophist caution."[163]

More than a century after Verne's death, it remains as difficult as ever to determine how far his insistence on Nemo as a disturbing but in the end impotent threat conveyed his own view of the dangers that beset modern society. One reason for this is that his writings were strongly colored from the beginning by his relations with his strong-willed publisher Jules Hetzel, always sensitive (as Verne was himself) to the effect that an excess of foreboding might have on sales. As a convinced republican who had suffered exile in Brussels for his political views during the early years of the Empire (before returning to Paris after the amnesty of 1859), Hetzel was attracted by the modernity of Verne's conception of human progress founded in the material world of technology. Far less acceptable to him was the darker side of Verne's worldview, and, for commercial reasons if no others, that darker side only intermittently found its way into print. In one case, that of the posthumously published futuristic novel, *Paris au XXᵉ siècle*, it seems probable that Hetzel, with an eye on Verne's predominantly bourgeois family readership, rejected the text not just for its literary deficiencies but also for its portrayal of the harm that might come in the train of such exciting developments as underground rail travel, gas-powered automobiles, and fax-like methods of communication (many of them broached in embryo by the time Verne first wrote about them in the early 1860s).[164] If this is so, the Verne we have in the recognized canon of his published writings emerges as a figure whose attitudes to modernity have to be teased out with due attention both to their ambiguities and to the filtering mechanisms affecting their entry into the public domain.

The technology that so fascinated and, in different degrees, alarmed Robida and Verne was only one source of fin de siècle anxiety about science and scientific ways of thinking. Even more disturbing, for many contemporaries, were the moral and philosophical implications of scientism. It was at this ideological level that Marcellin Berthelot aroused the radically conflicting reactions that made him simultaneously one of the most admired and one of the most hated public figures of the republic. Paul Painlevé's sympathetic appreciation, written after Berthelot's death in 1907, portrayed him as the very embodiment of scientific reason, "that calm, powerful faculty of the human mind that views nature with lucidity, builds with facts not words, and brooks no check on its labors, no impediment to its inquiries."[165] Berthelot was indeed that representative, but it was precisely his confidence in reason and (as the unique embodiment of reason) science that also made him an archvillain in the eyes of any who sought to defend traditional modes of thought. Such critics would have found little to admire in the four widely read volumes of essays and addresses in which, in the two decades before his death, he brought together a lifetime of preaching of the universal applicability of science to the moral as well as to the natural world.[166] And they would have abhorred the extravagant celebration of his half century as an active scientist in 1901, when the Sorbonne's main amphitheater (which "had never housed a larger or more select audience," including the president of the republic and a galaxy of ministers and academic dignitaries) was the setting for admiring speeches and addresses from academies throughout the world.[167] In his speech, Berthelot excelled himself. His message—that science should be seen as society's material, intellectual, and moral guide and the motor of modern civilization's accelerating onward march—was familiar enough.[168] But the style, vehemence, and sense of occasion were exceptional.

The public adulation that accompanied Berthelot's state funeral six years later compounded the impression of greatness. And it drove home, to supporter and critic alike, just how much the ideals of secular modernity owed to his tireless rhetoric and the influence he wielded close to and far beyond the seats of power. When in his popular *Science et libre pensée* (1901), he eulogized "the inexorable ascendancy of science, secular solidarity, and free thought in the intellectual, moral, and political guidance of human societies,"[169] he was voicing what had become not only the approved sentiment of the republican elite but also an opinion that (for those with the ears to hear) transcended boundaries of class and age. The 750 admirers who gathered in the working-class Parisian district of Saint-Mandé in April 1895 for a dinner organized in his honor by the Union de la jeunesse républicaine listened to him with minds already attuned to his message. They did not need Berthelot to persuade them that science was "the source of all progress achieved by the human race from its most distant past";[170] that it had liberated thought; or that, as an essential element in the union of the beautiful and the true, it had even inspired art and poetry, neither of which could achieve perfection in the absence of an understanding of nature and the scientific method that made understanding possible.[171] In their speeches at the dinner, Raymond Poincaré, Emile Zola, Edmond

Perrier, Auguste Comte's executor Ernest-Pierre-Julien Delbet, the radical president of the Chamber of Deputies Henri Brisson, and Charles Richet duly added their predictable mite to the eulogy of Berthelot and the scientistic orthodoxy he articulated.

Such unanimity bred anxiety as well as reassurance. For it was precisely the unquestioning dogmatism of the claims circulating at Saint-Mandé that alarmed those who saw the world differently. No serious conservative response denied the importance of the advances in understanding and the material benefits that science had bestowed. But what conservative could accept Berthelot's characteristically sweeping statement that the modern world had no place for mysteries, still less for miracles,[172] or the declaration of his loyal pupil, Richet, that science was the fount of a "moral ideal superior to any in the past, one founded on respect for human life and for the suffering of others, solidarity among men, and fraternity between nations"?[173] Such categorical sentiments opened a window on a future stripped of human values, as these had conventionally been understood.

Another author whose rhapsodic vision of modernity sowed alarm, especially among Catholics (including even those of a liberal inclination), was the classical and literary scholar, free-thinker, and professor in the Ecole d'anthropologie in Paris, André Lefèvre. By the time he delivered his most explicit broadside, in *La Religion* (1892), Lefèvre had been on the anticlerical barricades for over a decade. Two earlier books, *La Philosophie* (1879) and *La Renaissance du matérialisme* (1881), had left no doubt as to where his sympathies lay: invoking the by now well-worn antithesis between two main poles of thought, he argued that materialism had advanced inexorably against spiritualism and that hope for the future of both philosophy and humanity as a whole rested on the continuation of the process.[174] In *La Religion*, he repeated his Manichean view of the contemporary polarization of philosophy.[175] But here any residual objectivity was cast aside. In a work of unyielding scientism that was intended to shock and offend, Lefèvre dismissed religions as "the purged remnants of superstitions": faced with these pared-down vestiges of a primitive past, civilizations could only progress to the extent that they restrained religious fervor and promoted the retreat of the supernatural.[176] The means to that end was science, a judgment that Lefèvre encapsulated in his uncompromising concluding words, "The future lies with science."[177]

Among the many who were provoked by the aggressiveness of such assertions, in particular in the form in which Berthelot articulated them, was the literary critic Ferdinand Brunetière, *maître de conférences* at the Ecole normale supérieure and, as chief editor of the *Revue des deux mondes*, the master of one of the most influential tribunes of the day. The failing of science, in Brunetière's eyes, lay not (as some argued) in its total bankruptcy but rather in its limitations and failure to fulfil its extravagant promises: it had simply not delivered.[178] At the heart of the argument lay science's inability to penetrate such areas of darkness as the origin and fate of the human race (both matters on which Darwinian theory was mute) or the nature of man (in the fullest sense, not just of man the animal).[179] When the limitations of science were so palpable, what grounds were there for believing that other studies, such as philosophy, history, language, and (most importantly, for someone of Brunetière's increasing closeness to the Catholic Church) religion would benefit from being recast in accordance with the principles of scientific method?[180] As any Christian knew, reason and experience would never throw light on the divinity of Christ.

There are obvious dangers in taking any individuals, but especially Berthelot and Brunetière, as the representatives of the complex tendencies in thought and morality that did battle for French hearts and minds at the end of the nineteenth century. Berthelot was extravagant, overbearing, and categorical in his claims. And Brunetière, for his part, became an increasingly controversial, not to say eccentric figure from the mid-1890s until his death in 1906, even among those who broadly sympathized with him. His wordy, oracular style and the rigidity of his literary preferences, which led him to an unswerving admiration for Bossuet and the French classical tradition of the seventeenth century and a contempt for much contemporary writing, in particular for the naturalism of Zola, allied him with the most traditional canons of taste. In religion too, his inclinations were to conformity and obedience. While he was sympathetic to the injunction of the cautiously liberal pope Leo XIII that Catholics should face the modern world, that world was one that he could never wholly embrace: his natural intellectual allies were the "green cardinals,"[181] an informal network composed mainly of fellow members of the Institut de France (he had been elected to the Académie française in 1894) who, like him, maintained a powerful lay voice in support of Catholic orthodoxy. It was a mix that led the writer and critic Jules Lemaître, in a characteristically trenchant cameo, to acknowledge Brunetière (with whom he had collaborated on the *Revue des deux mondes*) as a "master" but also to present him as a public figure who inspired little affection and whose works were disdained equally by the young, by university professors (on the grounds that he was too pedantic), and by women: "Such sympathies as he inspires are few and without warmth" was an accurate comment.[182] Brunetière and Berthelot, in short, occupied the extremes, leaving between them a huge middle ground that harbored all possible shades of confidence and unease about the modern world.

Among the more sophisticated reading public, conservative sentiment that fell short of Brunetière's intransigence was manifested in the popularity of the twenty-six volumes of *Les Oeuvres et les hommes* by the reactionary Catholic writer and dandy Jules Barbey d'Aurevilly. In this huge work, published between 1860 and 1909, Barbey d'Aurevilly drew on articles, most of them originally contributed to Catholic newspapers, that formed an "intellectual inventory" of the nineteenth century. In an inflated imitation of Jean François La Harpe's vignettes of the philosophers of eighteenth-century France, historians, poets, novelists, women writers (his derided "blue-stockings of the nineteenth century"), travelers, critics, philosophers, and theologians all found their place in Barbey d'Aurevilly's scheme. It was the four volumes on "Les philosophes et les écrivans religieux," though, that attracted most attention. Published at intervals throughout the composition of the larger work, these volumes made much of the Catholic tradition as a force for good, scorned Comtean positivism as the aberration of a troubled mind, treated Renan roughly, and castigated Hippolyte Taine's *De l'intelligence* (1870), a study of human psychology in the sensationalist tradition of Condillac that Barbey d'Aurevilly dismissed as a tedious book offering nothing but "lifeless words and scientific abstractions."[183] As a battering ram against scientism, such a profile of opinions would have evoked supportive murmurs in any educated *bien pensant* bourgeois home.

In *bien pensant* minds, a similarly favorable chord would have been struck by Paul Bourget's novel *Le Disciple*, published in 1889 as an attack on the damaging effects of the scientism that Taine's work, in particular his psychological and literary theories, enshrined. The story was of an austere Parisian philosopher, Adrien Sixte, whose unswerving teaching of determinism had led his besotted pupil, Robert Greslou, to seduce a young woman, Charlotte de Jussat, who in due course took her own life. It was a horrifying story with an appropriately horrifying ending in which Greslou also died, shot by Jussat's father. In Bourget's view, justice had been done, in contrast with the justice of the court, which had found Greslou not responsible for his actions but rather the hapless and hence innocent victim of the teaching he had received from Sixte.

Informed readers would have seen the novel as a natural sequel to earlier attacks on the evils of scientism that Bourget had published in the mid-1880s. In these, Bourget had castigated not only Taine, with whose ideas he had been briefly infatuated in his student days in Paris in the 1870s, but also Flaubert. Despite an admiration for Flaubert that he continued to feel in many respects, Bourget found both him and Taine simplistic in their conception of human nature and in the associated hollow claim to objectivity that underlay their writing. For the imaginary "young man of my country" to whom he addressed *Le Disciple*, the moral implications of Bourget's literary judgments were spelled out in an admonitory preface. The pitfalls of indifference and self-indulgent aestheticism lurked at every turn in the modern world.[184] More urgent and immediate, however, was the menace of nihilism and amorality that came of supposedly scientific conceptions of human nature. It was these conceptions above all that threatened the souls of the young, leaving them perilously exposed, "less capable of love, less capable of resolve."[185]

As a sustained cautionary tale, *Le Disciple* made dry reading. Sixte and Greslou were characters so overdrawn in their obsessions that they appear unpersuasive to modern eyes. But the enthusiastic response to the book (twenty-two thousand copies of which were sold in the first six weeks) showed that readers recognized its significance for contemporary ideological debate, and they took sides. In literary circles, the book's admirers proclaimed it to be another nail in the coffin of the whole school of writers—Zola chief among them—whose sterile naturalistic aesthetic (with Taine as its main inspiration) Bourget despised. Critics, though, saw the book as an unfair and unsubstantiated attack on science: how could the teachings of science be held responsible for the acts of a man who was quite simply a born criminal? Admirers and critics agreed, however, that *Le Disciple* touched on issues at the heart of the divide among the intellectuals of the Republic. Their responses were predictable. Brunetière, in the *Revue des deux mondes*, took Bourget's side in an admiring review.[186] Against him, Anatole France, in the liberal newspaper *Le Temps*, displayed his loyalty to the Enlightenment tradition by rejecting any notion that Greslou's behavior and fate could be seen as the consequence of a deterministic chain of cause and effect with its origin in the scientism of current psychological theory. For France, to see "pure thoughts" as a fount of impiety was both illogical and, for the freedom of intellectual life, dangerous.[187] While the exchanges between the camps that Brunetière and France represented helped to clarify the points at issue, they did nothing to resolve them and probably did little to shift opinion. Most French people who cared about such things knew where they stood and they were not to be moved. The role of Bourget's supporters and detractors, like that of *Le Disciple* itself, was rather to stoke the fires of debate than to light them.

The polemical dimension that characterized Bourget's writing made its mark in poetry too. Here, the position that Charles Baudelaire had articulated in the 1850s, in favor of the poetic ideal of "art for art's sake" and a distancing of poetry from discussions of morality and scientifically inspired notions of truth, probably represented the prevailing view. Yet in the later nineteenth century there were poets, including some much admired at the time, who used the medium of verse to engage with the implications of science and technology. The positions taken could be appreciative as well as critical. Against a Coppée excoriating the Eiffel Tower we can set Louise Ackermann exploring the implications of science for a committed free-thinker, like herself, whose outlook had been profoundly affected by the doctrines of evolution and the transformation of forces;[188] her short poem "A la comète de 1861," with its reflection that the human race might have disappeared by the time of the comet's return, gives a flavor of the philosophical thrust of her work.[189] Less combatively, in Sully Prudhomme France had a leading writer who used both poetry and prose in search of the common ground where science would not necessarily threaten human values. For Prudhomme, whose work earned him the first Nobel Prize for literature in 1901, science and poetic imagination were easy bedfellows, and he made many sorties into scientific subjects. These drew both on his background in science (he had begun his working life as an engineer at Le Creusot) and on a sympathy for the objectivity of the scientific worldview that allied him, through his detached evocation of events and places, to a central aesthetic principle of some of the leaders of the circle of Parnassian poets to which he belonged. His poem of 1876, "Le Zénith," expressed the nobility of science in its evocation of the heroism of a high-altitude balloon ascent undertaken in the previous year by three men of whom only one (the most famous of all balloonists, Gaston Tissandier) survived.[190] In "Le Zénith," as in his later poems, Prudhomme's portrayal of science was uplifting. It offered the reassurance of an eternal, open-ended quest for understanding rather than the dogmatism and bleakness that many found distasteful in Ackermann's work.

Prudhomme's ideal of uniting humanistic and scientific mentalities and of bringing scientific modes of thought into the mainstream of literate culture found even more elaborate expression in his prose. In 1869, quite early in his literary career, he had entered the lists in the debates on materialism with a preface composed for his verse translation of the first book of Lucretius's *De rerum natura*. Here, Prudhomme argued that life could not be explained solely by the action of inert matter as physicists or chemists conventionally understood it; equally, though, he rejected the notion of a spiritual entity lying outside matter and giving it a capacity for action. Once matter was truly conceived, according to Prudhomme, the dichotomy between passive matter and a separate active spirit disappeared. Matter was force, and force matter, as the preface made clear:

> We must understand that matter is not separate from force and that all that exists in nature is substance in action [*substance active*]. Far from being defined by qualities of mass and inertia, matter has activity as its essential characteristic. The different forms of activity go under the names of properties, powers, or forces. A force is matter manifested through one of its properties. Matter is the very substance of these forces.[191]

Although Prudhomme's philosophical proclivities hovered about his work from its beginnings in the 1860s, it was not until the mid-1890s that they emerged as central to his writing. Now, in a series of articles, he used a rather rambling statement of his perplexity on the conditions that had preceded the emergence of life on earth to affirm his belief in scientific method as the source of true understanding and to ally himself firmly with Berthelot in rejecting any idea of the moral bankruptcy of science.[192] In 1899, he returned to philosophical debate, this time in seven published letters on the concept of final causes in science stimulated by a contribution to the *Revue scientifique* by Charles Richet.[193] The letters were essentially a friendly commentary in response to Richet's analysis of a "striving for life" that expressed the characteristic will of all individuals and species to survive. Richet, in fact, was arguing for a scientifically informed use of the concept of final cause, one far removed from providentialist teleologies in the manner of the abbé Pluche or William Paley. He insisted that to state, as he did, that the eye served the purpose of vision was not to imply any transcendental, still less a divine, plan of which the eye was the culmination. With that proviso, final causes had their place in science. Love, the fear of heights, and the sensations of pain and hunger could all be interpreted as having purpose and so contributing to the good of survival, but there the veil descended. For Richet, the fact that everything happened *as if* Nature had wished life to be preserved did not justify the ascription of such a desire to Nature or to any other entity.

Prudhomme opposed Richet's assertion on the grounds that it blocked discussion at the very threshold of understanding. It was too easy for Richet to hedge his conclusion with a statement of the limitations of scientific inquiry and to proclaim his impotence. But had he really rid himself of anthropomorphic conceptions of the causes at work in the universe when the notions of "will" and "effort" remained, to all appearances, so central to his vision of Nature? While Richet took the objections seriously, the modifications he made in the light of Prudhomme's comments were slight. As he admitted, the imagery he had used in speaking of Nature had been too colorful. But even if the anthropomorphic connotations of the word "effort" did leave him vulnerable to Prudhomme's questioning, was there any better way of expressing a manifest tendency in Nature that had hitherto remained unexplained? Moreover, as he insisted, an acknowledgment of human ignorance concerning the causes of the will to survive (as of animal intelligence or the complex structures of living creatures) certainly did not impugn the power of scientific method. Still less did they undermine those science-based teachings, such as Darwin's theory of evolution by natural section, which, as he saw it, had advanced to the status of certainties. In this way, with a twist that sat well with his current excursions into the arcane world of spiritualism, Richet associated his brand of unrepentant scientism with the conviction, essentially Henri Poincaré's (but by now widely shared among the reading public) that an element of the mysterious and the unknown lay behind even the most solid scientific theories.

Although Prudhomme's letters of 1899 had elicited this elaboration of Richet's position and brought points at issue between the two men to public attention, they were a somewhat damp squib that had no real impact. At best, they contributed to softening the image of arrogance and dogmatism that many saw as characteristic of the scientific mind, and they demonstrated the possibility of at least some level of mutual understanding between scientific and humanistic conceptions of the universe. The possibility was one to which Prudhomme attached particular importance, and it loomed large in his quest for a cultural middle ground. He presented science as a force for uniting human beings in the pursuit of truth, while distancing himself from any notion of science's infallibility[194] and insisting that both science and imagination were essential for an understanding of human nature and the relations between human beings.[195] But by the turn of the twentieth century, events beyond the realm of science were pulling toward heightened tensions rather than reconciliation, and making *rapprochement* difficult.

No episode did more to divide opinion than the Dreyfus affair. Between 1894, when the Jewish army captain, Alfred Dreyfus, was found guilty of the unfounded charge that he had betrayed his country to Germany, and 1899, when he received a pardon from the president of the republic, the hostility between the broadly liberal camp that supported Dreyfus and the predominantly Catholic, anti-Semitic elements in society who believed him guilty reached unprecedented levels. In the process, families and friendships were split asunder, and political sensibilities were sharpened in ways that drove the extremes of secular and religious modes of thought further apart than ever before. Pope Leo XIII's hopes of cautious accommodation with the republic were one victim of the strife (see chapter 4). So too was the Napoleonic Concordat between France and the Holy See, broken as part of the formal separation between church and state for which the Chamber of Deputies voted, after several months of acrimonious debate, in December 1905.

While the tensions that culminated in the separation of 1905 did not turn centrally on science, the nature of republican cultural orthodoxy meant that science could not remain untouched. Events in Nantes in the summer of 1898 made the point. The chemist Edouard Grimaux had made no secret of his support for "J'accuse," the public letter in which in January 1898 Emile Zola attacked the condemnation of Dreyfus. He would therefore have known that he was entering a lions' den when he visited Nantes, seven months later, to preside at the annual meeting of the Association française pour l'avancement des sciences. The city had been the scene of troubles since January, mainly provoked by anti-Dreyfus sentiments of the kind that took root easily in a region of strongly Catholic sympathies. As soon as Grimaux arrived, demonstrators renewed their activity, forcing him to leave a café in which he was eating and then shouting him down as he attempted to give his opening address in the théâtre Graslin.[196] The hostility forced the removal of the session to a nearby *lycée*, and throughout the week of the congress, the presence of a man whose sympathies for Dreyfus and outspoken support for the anticlerical cause had already cost him his chair at the Ecole polytechnique continued to enrage conservative opinion.[197] By the end of the congress, Grimaux had bowed to the inevitable and entrusted the task of presiding over the closing ceremonies to his vice president, the medical professor Paul Brouardel.

The local and national press followed the tumultuous events in Nantes with a keen and invariably partial interest. As both sides in the debate agreed, the target of the demonstrations was not the AFAS, a small number of whose senior members even urged Grimaux to resign his presidency as a way of demonstrating the body's detachment from politics. Nevertheless, science was exposed. To observers already uneasy about the prevailing scientism of the fin de siècle republic, it was both unsurprising and significant that what was essentially an episode in a political war should have been played out in and around a scientific congress. Even some of those who felt disquiet about the cultural tone of the republic judged that the hounding of Grimaux had gone too far. But such reservations did nothing to alter conservative perceptions of the dangers inherent in a culture founded on science. To that extent the battle lines between right and left, church and state, and religious and secular ideologies remained unchanged.

The frothy spectacle of the Exposition universelle of 1900 did no more than temporarily gloss over the divisions that were exposed by the Dreyfus affair. The fact that a deep-rooted struggle for hearts and minds was being pursued with escalating ferocity escaped no one. French society at the beginning of the twentieth century was at a turning point, divided more profoundly than ever in its attitudes to the republican ideal of a secular society. Anxieties about materialism and the uncertain future of a modern, technological world were not the leading issues, but they had become inextricably bound up in the debate. In the process, science had come to matter to citizens far beyond the higher reaches of government, education, and the economy, in which it had always been important. And it mattered at the most visceral level.

In exchanges that invited the flaunting of extreme positions, spectacular clashes have tended to attract most attention. Yet it is important not to overlook the middle ground. There was always a strong streak of liberally inclined Protestant opinion that had no difficulty in reconciling faith with modernity and republican values. And Catholic universities and scientific congresses brought together many within the Catholic tradition who had no fear of science, rightly conducted. But those who sought reconciliation and compromise rather than confrontation found their calls for moderation all too often drowned out by the louder voices on the extremes. For them, the accession of the anti-modernist Pope Pius X to succeed Leo XIII in 1903 was a dismal turning point. In an atmosphere that was to be irretrievably poisoned by the final break between church and state two years later, moderate French Catholics who aspired to the coexistence of secular and religious thought knew that their cause was all but lost. For them, the decade preceding the First World War was a difficult one, irredeemably scarred by the intransigence on both sides that had done so much to undermine their ideals.

Conclusion

● ● ●

The core of my argument is that science mattered, across the board of French society and throughout my slightly displaced nineteenth century. It would require detailed comparative studies to confirm that, as I suggest in the introduction, it mattered more in France than elsewhere. But the argument does not stand or fall by the truth or falsity of that belief. The unifying thread of *The Savant and the State* is the evidence that science had a consistently prominent place in French public debate and that it penetrated exchanges far beyond its own particular realm. This was most obviously, and unsurprisingly, true of the late nineteenth and early twentieth century, by which time international exhibitions, newspapers, popular magazines and books, and the proliferation of educational institutions, public lectures, and science-based industries had brought science closer than ever before to the experience and concerns of the nonscientific public. But the earlier chapters of the book have shown that this heightened public exposure toward the end of my period was the culmination of a long history going back (for my immediate purposes) to the Bourbon Restoration or even (as a rich secondary literature has taught us) to more distant roots in the ancien régime. What occurred in the second half of the nineteenth century certainly represented an acceleration, but it was part of a process rather than a fundamentally new departure.

The ever greater visibility of science and its practitioners in the nineteenth century had parallels in all scientifically developed nations. The same is true of other broad trends during the same period. The growing emphasis on qualifications and formal training, for example, and the associated marginalization of the self-taught devotee went hand in hand with the increasingly sophisticated character of scientific knowledge and practice everywhere. The enhanced investment in laboratories, particularly marked in France from the 1880s, also had its counterparts internationally. The obvious fact remains, however, that despite a measure of similarity, science developed in distinct ways in different national contexts. It is that distinctiveness which justifies a study, such as this, focused on a single country.

The point about the particularity of the course of science in France was not lost on scientists there, many of whom in the later nineteenth century were convinced that French science was set not just on a distinctive track but on one leading inexorably to a loss of status among the world's scientific nations. Assertions that all was not well in either research or education converged to a common view that France needed to break out of what critics saw as unduly narrow national preoccupations and to compete more vigorously on the world stage, especially in response to the rise of Germany as a scientific and economic power. Pasteur's evocation, in March 1871, of a golden age in French science half a century earlier, when "all foreign nations acknowledged our superiority,"[1] may be regarded as an emotional response to the immediate circumstances of defeat at the hands of Prussia. But the contrast that Pasteur drew between a glorious past and a present characterized by hardship and the sacrifice of international leadership was to have its echoes in statements by many French savants up to the First World War.

There is irony in the fact that such pessimistic judgments proliferated at a time when, after the long struggle for better facilities and public recognition, science and scientists enjoyed unprecedented favor in the eyes of the governments of the Third Republic. Even those working in the faculties and other institutions in the domain of the now conspicuously indulgent Ministry of Public Instruction routinely expressed discontent, as they looked with envy across the Rhine to a world of scientific plenty, openness, and functionality. One unhappy professor, Pierre Duhem, was seen by many in and around the Parisian administration as a tiresome malcontent. But he may well have had a point when he criticized the new buildings of the faculties for their tendency to showy opulence: his comment "What we needed was a workshop, a factory; what has been provided, or is said to have been provided, is a monument" had, to some extent, the ring of truth.[2] In addition, especially for professors in the faculties who spoke for the experimental and applied sciences, there was always the blight of a system of examinations and rewards, social and material, that channeled generations of the ablest pupils in secondary education into the ferocious mathematical grind of preparation for entry to the Ecole polytechnique. Balzac's portrayal of the emotionally and intellectually stunted graduate of Polytechnique, Grégoire Gérard, in the *Curé de village* bears classic witness to perceptions of the damaging personal effects of the grind (see chapter 1). By the 1830s, when Balzac wrote, such perceptions were common, and they remained so to the end of my period. Observers from Théodore Olivier in the mid-nineteenth century to Henry Le Chatelier during the First World War continued to blame the selection procedures and the educational experience that followed at Polytechnique both for promoting an unduly abstract cast in French science and for leaving all but the most resilient *polytechniciens* unfit either for creative scientific work or for economically relevant employment.

In such criticisms, self-interest and special pleading played their part. Duhem's comments drew on a well of personal indignation at the impediments he believed he had suffered as an ambitious but frustrated professor in a provincial faculty (Bordeaux) and as a declared Catholic in an age of rampant republican secularism. The resentment of an outsider also had much to do with the ferocious campaign that the physicist Henri Bouasse waged against the Parisian scientific establishment, embodied in the "crowd of mediocrities and ignoramuses" in the Académie des sciences (to which he was never elected).[3] Personal rancor likewise fired Bouasse's other complaint, against a pattern of dispersed funding that condemned individual faculties, such as his own in Toulouse (where he taught from 1892 to 1937) and a fortiori the smaller faculties in Clermont-Ferrand, Poitiers, Dijon, Caen, and Rennes, to impotence.[4] His remedy, that faculties should be allowed to specialize rather than trying to cover the whole range of the sciences, never won favor. Yet even Bouasse, who was no friend of republican secularism, would have had to recognize that in the relatively benign atmosphere of the Third Republic, the life of a professor in a provincial faculty at the beginning of the twentieth century was far less frugal than it had been in the first fifty years of the University's existence. Still, the life was not easy, and it became even less so when the introduction of low-level teaching for the Certificat d'études physiques, chimiques, et naturelles, or PCN, and in the proliferating institutes of applied science through the 1880s and 1890s converged to raise teaching loads and eat into the time and energy available for research. Better facilities and increased student numbers were a welcome reality in these years, after decades in which faculties of science had been notoriously underused and poorly equipped. But they had been bought at a price that many academic scientists now found too high.

Such conflicting perceptions point to tempting questions. Put most succinctly, after the long struggle to improve the position of science in higher education and to commend science to educational administrators, politicians, and public alike, had things really gone so badly wrong? Are we to believe, on the one hand, the frustrated Duhem and the notoriously cantankerous Bouasse or, on the other, the political figures and senior officials in the Ministry of Public Instruction who prided themselves on their achievements in advancing science as the foundational culture of the progressive modern society they wanted France to become? Here, we immediately face the difficulty of determining what constitutes success or failure. The leading role that France took in mounting universal exhibitions and in the rising tide of scientific congresses between the 1860s and 1914 has surely to be accounted a success, a mark of the respect that the nation enjoyed in science and technology as in other areas of culture and learning. On this increasingly visible front, Germany and the United States presented serious challenges by the turn of the twentieth century. But at least until then, France was a leader, arguably *the* leader, in the securing of international agreements on electrical units and standards, nomenclature in the natural history sciences, and the promotion of ventures in scientific cooperation of the kind that resulted in the founding (in Paris) of the Association internationale des académies in 1900.[5] Of course, the accumulation of high offices and influence in international organizations is just one among many touchstones of a nation's eminence. Yet the fact remains that, in setting its sights on this particular form of leadership, the French did reap significant rewards. It was an aspect of competition in science in which the French could plausibly be accounted victors.

It might be argued—and I have some sympathy for this view—that the preoccupation with exhibitions and congresses encouraged complacency, even some diversion of energy from other areas of science. But I stress what occurred less for the real or potential weakness that it may have been than as a facet of a particular French way of science, manifested in the strategic choices about investment of time and resources by the nation's scientific leaders. As such, it invites explanation. It unquestionably had something to do with wounded national pride in the aftermath of the defeat of 1870, perhaps also with a legacy of internationalism bred of the traditional standing of France and French culture in the world of diplomacy. Whatever the explanation, it was a choice that the French made, and in the circumstances of the time it was an understandable one. It was a product of a chronologically and geographically rooted interplay between political, economic, and social context and the individual and collective aspirations of scientists in pursuit of intellectual and career goals. It was precisely the type of interplay that I have tried to elucidate, for different periods and different settings, in *The Savant and the State*.

Throughout the century that I discuss, the interactions between the broader context and the style and practices of French science were kaleidoscopic, and they can only be reconstructed with reference to specific times and places. Take, for example, the very different profiles of public opinion and politics within which science had to make its way during the Bourbon monarchy under Louis XVIII and Charles X (chapter 1) and in the Second Empire of Napoleon III (chapters 3 and 4). Under the Bourbon kings, public perceptions of the savant and his work were strongly colored by memories of the Enlightenment and the years of revolutionary and Napoleonic rule, which engendered a suspicion of reason and reason's most visible embodiment, science. It followed that the champions of science had to cope with currents of deeply conservative opinion that circulated even in the more liberal phases of the Restoration. My argument in chapter 1 was not that science was fatally crippled by this; albeit with some difficulty, leading men of science, with Cuvier as a supreme case, managed to preserve the freedom and material support that allowed them to work at the highest level. Nevertheless, the clerically suffused atmosphere of the Bourbon period encouraged a particular orientation, notably toward ideologically neutral goals, including industrial and agricultural utility. By the same token, it also trammeled the initiatives of those—Dupin and Raspail in their different ways being conspicuous examples—who chose the more challenging course of opening science to the people. It was not that the context fashioned the style of Restoration science. Its effect was simply, though no less important, to define the boundaries of possible actions.

Under the Second Empire, the relations between context and scientific practices were even clearer. Even in the improving climate of the later, comparatively liberal phase of the empire, a mixture of peculiarly French administrative and political circumstances inhibited discussion of a clutch of radical ideas, including not only Darwinism but also spontaneous generation, polygenism in anthropology, and a new wave of materialist interpretations of the working of the human body. The inhibition, as I describe it in chapter 4, arose essentially from Duruy's need to retain bourgeois support for the moderately liberal educational reforms he wanted to implement. With the Catholic Church on edge about the moral implications of German materialism and intent on storming the last bastion of state monopoly in education (limited after 1850 to higher education), Duruy had no choice but to assert his authority and moral rectitude by subjecting his professors to rigid discipline. When discipline was seen to have slipped, as happened after Charles Robin publicly espoused materialism in his lectures in the Paris faculty of medicine in 1868, the result was religiously motivated outrage and a debate in the Senate that came close to removing Duruy from office. The furor served as a warning, to academic scientists, of the perils of deviance and, to Duruy, of the difficulty of promoting intellectual freedom in a still profoundly conforming society.

Duruy's travails during the Second Empire and his handling of them relate to what I have presented as a central theme of this book, that of bureaucracy. As I describe it, bureaucracy was not the same thing as centralization; it was rather the tool of centralization. At all events, the two had many common manifestations, especially in the recurring tensions (of the kind I describe in chapter 2) between the center, defined by the national network of professors and educational administrators, and the periphery, where independent savants and scholars pursued research and publication without regard to career advancement. On the periphery, devotees typically achieved independence by combining their scientific vocation with a successful professional life (usually as a doctor or a lawyer), while some, like Arcisse de Caumont, did so by simply drawing on a private fortune. But most French savants, certainly the ones whose work appears most prominently in accounts of French science, were employees of the state. Whether they held posts under the Ministry of Public Instruction or one of the other ministries with responsibility for education, they were public servants, in French terminology *fonctionnaires*. As such, they were answerable ultimately to an ever vigilant Parisian administration and through that administration to the senior scientists and others whose advice helped to form ministerial policy. With respect to the Second Empire and the early Third Republic, the effect was generally conservative, manifested in a pattern of authority that helps to account for French slowness in taking up the ideas of Charles Gerhardt, Auguste Laurent, and Adolphe Wurtz on atomism, Charles Darwin on natural selection, Rudolf Clausius and William Thomson in thermodynamics, or James Clerk Maxwell in electromagnetism. The pattern of reluctance to adopt new theories did not pass unnoticed by commentators outside France, while within the French community internationally minded critics were already identifying indifference or hostility to developments abroad as a French weakness in the 1860s and contrasting it with the freer way of things in German universities. As one observer put it, the intellectual consequences for chemistry were a discipline being taught in 1864 much as it had been taught in the heyday of Thenard decades earlier.[6]

The criticisms, usually reinforced in France with a strong dash of liberal political sentiment, rang true. Of course, it would be wrong to suggest that ideas from abroad made no mark in France before the midcentury; there is much evidence of the impact of foreign concepts and practices in this earlier period, as Pietro Corsi has shown with reference to natural history between 1800 and 1830.[7] Yet there was still much ground to make up when the *Annales de chimie et de physique* launched a long-running series of translations of mainly British and German articles in chemistry (by Wurtz) and physics (by Emile Verdet, later by Pierre-Augustin Bertin) in the 1850s. The consequences of the translations were a mixed bag. Certainly, this and similar attempts to open up the world of French science were welcomed and exploited by individual savants; the young Gabriel Lippmann, for example, had his horizons enlarged by the translations from German and English that he undertook at Bertin's request.[8] But the new openness to cutting-edge work abroad had limited impact on ministerial strategy. In 1870, despite Duruy's best efforts, the priorities of educational administrators remained less with fostering creativity and a critical engagement with developments in the wider world of science than with rewarding mental and oral dexterity and the mastery of a vast body of received knowledge.[9] It is telling that the supreme test at the threshold of an academic career remained the *agrégation*, an examination whose brutal competitiveness bred admiration (for its rigor) and criticism (for its failure to reward originality) in equal measure. Despite greater openness during the Third Republic (though an openness tempered by ambivalence with regard to Germany's rising star in science), the effect of such preferences was long-lasting. Dominique Pestre's demonstration of the slowness with which textbooks assimilated new departures in physics in the early years of the twentieth century, and even after the First World War, is of a piece with this observation.[10]

Since this is a book about the public face of science, I have left much unsaid about the relations between the public and the private. The fact remains, however, that the long-running debate about the supposed decline of French science from the high point it reached in the first quarter of the nineteenth century has frequently invoked aspects of public science that I discuss. Hence a brief comment seems necessary. For some years now, commentaries on the reality or otherwise of a decline beginning in about 1830 have been moving uncertainly toward a revisionist consensus, and I find myself venturing on the same course. Essentially, I feel that we may have been unduly influenced by the declinist rhetoric of scientists themselves, especially after 1870, and that we have yet to fashion the nuanced view of French science that reliable judgment—discipline by discipline and period by period—requires. A major comparative study by Paul Forman, John Heilbron, and Spencer Weart has shown that a typical French physicist in a university or institution of higher technical education at the turn of the twentieth century was not greatly less productive than his German counterpart; and that conclusion squares with Terry Shinn's demonstration of an increase in the number of publications (in all the sciences) emerging from science faculties in France between 1876 and 1900.[11] Other research has drawn attention to the vigor of Catholic activity, especially in training scientists and engineers for industrial employment: here, as I comment in chapter 6, the work of Harry Paul and André Grelon, in particular, has opened important new perspectives.[12] Mary Jo Nye's revisionism has focussed mainly on the provincial faculties of science, specifically Nancy, Grenoble, Lyon, Toulouse, and Bordeaux. As she demonstrates, during the Third Republic compromise between a professor's status as an obedient functionary and as a free citizen of the world of learning was possible. Her accounts of the path-breaking work of the joint Nobel Prize winners for chemistry in 1912, Paul Sabatier in Toulouse and Victor Grignard in Lyon, show how committed professors could fashion international reputations in seemingly unpromising provincial settings and so go beyond their limited, bureaucratically defined roles.[13]

The conclusions that Nye draws are properly cautious. She insists on the specific and very different circumstances of Sabatier's and Grignard's implantation in provincial academic life and avoids any simple generalization about a flowering of intellectual endeavor in the provinces.[14] But what her work and an accumulating body of other studies have done, and have done irreversibly, is to point to a far greater diversity of pedagogical and research practices than was once thought to exist in the French system. No account of the supposed backwardness of French physics, for example, can now ignore the evidence of numerous settings in which by the 1860s and early 1870s energy conservation and the second law of thermodynamics were already taking significant strides toward general acceptance through the lectures and textbooks of adventurous professors who have only attracted the serious attention of historians in recent years. These more progressive spirits included Verdet, whose premature death in 1866 deprived France of a conspicuously outward-looking professor who, as John Herivel has argued, might well have gone on to establish the Ecole normale supérieure (where Verdet taught from 1863) as a leading center for theoretical physics.[15] They also included such understudied figures as Athanase Dupré in the faculty of science at Rennes and Jules Moutier, a former *polytechnicien* who held a variety of posts—in the state telegraph service, as a teaching assistant and examiner at Polytechnique, and for more than twenty years as a teacher in the *classes préparatoires* at the Collège Stanislas—that served to keep him in Paris, with time for his pioneering work in thermodynamics. Even if the leading innovations of the midcentury were made in Germany and Britain, there were always those in the French physics community who followed and, in some cases, built on developments abroad. A bibliography in Moutier's *Eléments de thermodynamique* (1872) bears striking testimony to the number of early books on the mechanical theory of heat by French physicists, including not only Verdet, Dupré, and Moutier himself but also Gustave-Adolphe Hirn, Charles Combes, and Charles Briot.[16]

With regard to electromagnetism, Michel Atten has identified the previously disregarded world of telegraph engineers as one in which Maxwell's theories won acceptance long before they were generally incorporated in national curricula. As Atten shows, it was in the Ecole supérieure de télégraphie in the 1880s that the theories were first taught systematically, and it was only the appointment of three teachers fashioned in telegraphy (Ernest Mercadier, Jules Raynaud, and Aimé Vaschy) that finally brought Maxwell to Polytechnique.[17] As it transpired, the exposure of *polytechniciens* to Maxwellian ideas was even then a slow process: one of the school's two professors of physics, Alfred Potier, did not introduce the new electromagnetism in his teaching until 1893, while the other professor, Alfred Cornu, persisted in his view that Maxwell's *Treatise* lacked the "order and clarity" of his own favored speciality of optics. The important point to make, however, is that the reticence evident at Polytechnique should not be taken as symptomatic of attitudes in the French scientific community as a whole. The reticence may have been widespread but it was not universal.

Similar patterns can be discerned in chemistry, where atomism had its dispersed, somewhat marginal champions before it entered slowly into the approved canon of the discipline. Advocates from Gerhardt in the 1850s to his disciple Wurtz and other members of a predominantly Alsatian network of chemists much later in the century had to face resolute resistance from supporters of the rival idea of equivalents, led by the powerful Berthelot, and the tide only began to turn when some crucial professorial appointments were made in the early 1880s: the arrival of the militant atomist Edouard Grimaux at Polytechnique in 1881, the more retiring Edmond Willm at the faculty of science in Lille in 1882, and Charles Friedel at the Sorbonne (as Wurtz's successor in the chair of organic chemistry) in 1884 laid the foundations for the eventual routing of equivalentist orthodoxy.[18] But by then damage had been done, resulting in what Alan J. Rocke describes as "a dramatic decline, vis-à-vis Germany, in both quality and quantity of published research."[19] As in the case of electromagnetism, even when things improved, the victory of the new ideas was by no means assured. As late as the early 1890s Auguste Béhal, a pupil of Friedel and a committed champion of the atomic theory, still encountered fierce opposition when his five-year appointment as a visiting lecturer at the Ecole de pharmacie, where he had taught the atomic theory since 1888, came up for renewal.[20]

It is significant that the life sciences appear to have gone through a weak period at about the same time as happened in physics and chemistry. The first half century of the existence of the Muséum d'histoire naturelle, from its foundation in 1793 has generally been regarded as a golden age in which the museum was widely regarded as the leading institution of its kind in the world. In these years, emerging disciplines, such as geology and crystallography, found a welcoming home and a degree of freedom that allowed them, as well as new departures in the more established sciences, to flourish. But from the death of Etienne Geoffroy Saint-Hilaire in 1844 the institution entered three decades of financial hardship and confused internal management that provoked the external inquiries to which I refer in chapter 3. For Jules Marcou, writing in the late 1860s, the professors' reluctance to follow the recommendations of the inquiries was a symptom of the conservatism that afflicted the museum, as well as French science more generally.[21]

In these years of "stagnation" (the word is Camille Limoges's), change certainly did occur: several new chairs were added, most of them in experimental disciplines, and the profile of research edged away from its traditional emphasis on descriptive natural history. As Limoges has argued, however, divergences between the newer experimental sciences and the older collection-based work in taxonomy made for destructive internal conflict rather than the productive modernization that might have been expected.[22] An eventual swing back toward natural history, allied from the 1890s to a new colonial mission, did little to improve the museum's fortunes. By that time, also, the best of the revitalized faculties of science had stepped in to occupy territory (notably in physiology) that had once been the preserve of the museum. Such a shift in institutional focus was not in itself a bad thing. But, along the way, opportunities for leading professors to engage with new theories, most conspicuously Darwin's theory of evolution by natural selection, had slipped by. The dismissive response to Darwinism by Flourens, the professor of human anatomy from 1832 to 1868, had become symptomatic of a sclerosis that affected the whole institution, even after his death (see chapter 4).

All this said, enduring French excellence in mathematics and in meticulously pursued experimental physics of the kind in which Victor Regnault excelled reminds us how important it is to distinguish between performance in different disciplines. So too does the tradition of mechanical engineering that prepared the country for a position of leadership in early motor car production. At the very least, such examples make it hard to sustain a declinist interpretation as uncompromising and all-embracing as the ones that Robert Gilpin and Joseph Ben-David advanced in classic studies some forty years ago.[23] What has to be said, however, is that a pattern of resistance to change in certain key areas of science seems to have characterized the middle decades of the nineteenth century, with consequences, in those areas, that hindered attempts to reengage with the cutting edge of world science from Victor Duruy onward.

To return to some of the leading themes of *The Savant and the State*, I believe that what historians have commonly identified as midcentury lethargy or insularity can be related to handicaps under which the French scientific community labored over a much longer period. One such handicap was the generally poor quality of laboratories and equipment. Even if the disparity between French and German provision was not so great as we once believed, French laboratories in 1914 remained, by and large, less well equipped than the most opulent of their German equivalents; the rare islands of generous provision, such as Marie Curie enjoyed in the early years of the twentieth century, do not invalidate that general point. Similarly, the fact that resolute professors throughout my period managed to fashion a measure of intellectual autonomy should not blind us to their reliance on paymasters committed to the priority of imposing the overriding bureaucratic ideals of efficiency and conformity. Even after 1880, when the Ministry of Public Instruction instituted a policy of limited devolution of authority, freedom was limited, and opportunities for exploiting it had to be sought and actively exploited.

And there remains the question of the quantity, as opposed to the quality, of research. Effective though Pasteur's attempts to foster a research culture at the Ecole normale supérieure from the late 1850s undoubtedly were, we have still to ask whether the number of students who benefited and went on to academic careers was sufficient for a nation competing at the highest level in science. Was it sufficient, for example, to compensate for the drain on scientific talent resulting from the lure of Polytechnique and the well-paid and largely administrative careers to which most *polytechniciens* aspired?[24] Might relative size, in fact, go a long way to explaining the sense that French scientists shared by 1914 of being swamped by the publications of their German peers? The French could and often did insist that their work had a finesse that German work lacked. But they could not deny the sheer volume of publications emerging from universities and Technische Hochschulen across the Rhine. For this reason, it is important to read the evidence of Forman, Heilbron, and Weart about the respectable level of productivity of individual physicists in France in conjunction with their equally compelling evidence about the disparity between the size of the physics community there and in Germany. There were simply far more teachers and students of physics in German institutions of higher and advanced technical education than there were in the comparable institutions in France, as would be expected in a country whose population by 1900 was at least half as large again as that of France.[25] When that disparity is taken into account, France's combined total of eleven Nobel Prize winners in physics, chemistry, and medicine or physiology between 1901 and 1914, compared with Germany's fourteen, should be regarded as perfectly honorable. It certainly exceeded Britain's five and would suggest that by the early twentieth century, if not before, French science was more competitive than the observations of its severest contemporary critics would lead us to believe.

The identification of such open-ended questions directs attention to the concrete, the particular, and the local, precisely the focus I have adopted in this book. The history that results from such a focus will always be rich in contingency. And it will necessarily be sensitive to the interplay between individual aspirations and the initiative that scientists in all ages and all places have had to display in making the most of opportunities for research and the advancement of their careers. The history of science in France that has yet to be written will also be a more broadly cast history than was once thought possible or even necessary. It is not just that we have now to incorporate actors and settings beyond the better-studied world of the great institutions of the capital. No less important, we have to direct systematic attention to individual disciplines and specializations within disciplines, and to walk, as historians, the precarious interfaces between them. It is tempting to ask, for example, whether France might have occupied a far more prominent place in theoretical physics in the age of Maxwell if the interaction between the experimental and mathematical sides of physics had been as strong at the midcentury as it had been, with such distinguished results, until the 1820s.[26]

The historiographical challenges involved in the detailed study of disciplines, locations, and personal achievements and, above all, in trying to bring these into a coherent whole are formidable. They are ones that we will have to face if our understanding of science in nineteenth-century and early-twentieth-century France is to progress beyond a macro-history focused exclusively on national administrations, metropolitan elites, and large-scale clashes between ideologies and programs. I see, and seek, no alternative. I believe that this is the only way of digging, as I have tried to do, into the fine structure of science in the life of a nation.

The French System of Education and Research

The complexity and changing character of the French system of education and research in the sciences between 1814 and 1914 make it hard to offer a succinct résumé. What follows is therefore far from comprehensive. It merely summarizes the main institutions and institutional structures that bear on the themes treated in this book.

The University

The institution that I variously designate as the "Imperial University," the "Royal University," or simply the "University" was created by Napoleon I, as the Université impériale, in 1808. It was not a single institution, as is generally associated with the word university, but rather a national organization, embracing all places of learning under state control and the teachers who were employed in them. In higher education the system was made up of university-level faculties dispersed in major towns throughout the country. At the secondary level, it included *lycées* (secondary schools that were the Napoleonic successors to the *écoles centrales* of the Directory) and *collèges communaux* (lesser secondary schools, administered and financed through local authorities). Primary education was also overseen by the central administration of the University, although with a far greater measure of delegation of control and financial responsibility to municipalities and departments. A consequence of this unified structure was that the term *universitaires,* as it was commonly used in the nineteenth century and as I use it in this book, included those who taught in *lycées* as well as those who taught in institutions of higher learning.

At the head of the Napoleonic University was a grand master (*grand-maître*), with responsibility for the quality of education at all levels. It was he, along with his senior administrators and advisers on the Conseil de l'Université, who established curricula and standards for instruction. In the major reorganization of August 1815, his powers devolved to a five-man Commission of Public Instruction (Commission de l'instruction publique). In the reorganization of 1820, the Commission was renamed the Royal Council of Public Instruction (Conseil royal de l'instruction publique), and it survived as such in the further reorganization that followed in June 1822, with its chairman now holding the resurrected title of grand master. During the July Monarchy, power increasingly lay with the Ministry of Public Instruction, which was established as an independent ministry in 1832. From 1850, the Higher Council of Public Instruction (Conseil supérieur de l'instruction publique) served, under the Ministry of Public Instruction, as an influential committee with responsibility for all aspects of the University's work.

While overall control remained with the Ministry of Public Instruction, local responsibility was deputed to the academies into which France was divided for the purposes of administering educational policy, each of them headed by a rector. There was a short period (1850–54) when the geographical areas of the academies became those of the country's eighty-six departments— administrative units created after the Revolution. Otherwise, academies typically embraced a number of departments. There were between twenty-six and twenty-nine academies before 1850 and eighteen (including Algiers and, until 1870, Strasbourg) after 1854. These academies, which I conventionally spell with a lowercase *a*, are not to be confused with the many learned academies throughout France that existed independently of ministerial control.

After a century of many changes, the structure of the educational system was broadly as follows around 1900.

PRIMARY EDUCATION

The cursus of primary education was divided between the elementary *enseignement primaire*, for pupils to the age of 14, followed by the optional *enseignement primaire supérieur*, commonly dispensed in an *école primaire supérieure*, which offered more advanced instruction, including options in vocational subjects.

SECONDARY EDUCATION

A recurring debate throughout the nineteenth century turned on the relative importance of the humanities and the sciences in secondary education. By the 1890s, these cultural traditions were enshrined in secondary education in two distinct streams, the *enseignement classique* and the *enseignement moderne*.

In the last three years of *enseignement classique*, known collectively as the *division supérieure*, pupils passed through the classes of *troisième, seconde,* and *rhétorique*, taking the first of the two parts of the *baccalauréat* after the *classe de rhétorique*. Pupils then moved on to a year of either *philosophie* or *mathématiques élémentaires* during which they specialized for the first time. The *baccalauréat* that they took at the end of this year, in either *lettres-philosophie* or *lettres-mathématiques*, reflected their choice of options.

The *enseignement moderne* succeeded the *enseignement secondaire spécial* that Victor Duruy had introduced in 1865 and was organized in 1891 as a vocationally oriented alternative to the *enseignement classique*. Crucially, it did not require Latin. Candidates took the first part of the "modern" *baccalauréat* at the end of the *classe de seconde*; the second part being taken a year later, at the end of what in this stream was designated the *classe de première*. In this year, candidates oriented their studies toward science, letters, or mathematics and completed a *baccalauréat* designated accordingly in *lettres-sciences, lettres-philosophie,* or *lettres-mathématiques*.

The absence of Latin from the *enseignement moderne* condemned the program to an inferior status. The major reforms of 1902 sought to tackle the problem by integrating both the classical and the modern curricula within a unified *enseignement secondaire* leading to a *baccalauréat* whose standing was in principle the same, irrespective of a candidate's choice of options. The aim was to diminish the distinction between classical and modern, but the prestige of curricula involving Latin remained virtually intact until long after the First World War.

The *baccalauréat* allowed candidates (whatever options they had chosen) to proceed from their secondary school to higher education in a faculty of their choice. But if they wished to prepare for entry to a *grande école* (see below), they had to continue their studies in one of the advanced *classes préparatoires* that were attached to leading *lycées*. These classes specialized in either letters or science and mathematics. Candidates in a scientific *classe préparatoire* began with a year of *mathématiques élémentaires supérieures* and then moved on to one or two years of *mathématiques spéciales* before attempting the competitive entrance examination for their chosen school.

HIGHER EDUCATION

Until the last years of the nineteenth century, the national network of faculties of science, letters, medicine, law, and theology continued to constitute the University's sector of higher education. Each faculty was headed by a dean (chosen from among the professors) and was answerable to the rector of the academy in which it was located and, through the rector, to the Minister of Public Instruction (who at various periods during the nineteenth century also held the title of grand master of the University). Under the Third Republic, the diffused structure of faculties and their subjection to centralized ministerial control came increasingly to be criticized. Moves to group faculties into institutions with some measure of autonomy culminated in legislation, promoted by Louis Liard, as the ministry's director of higher education, which led to the creation of sixteen quasi-independent universities in 1896. All except one of these universities (Algiers) had a faculty of science.

The University retained its monopoly in higher education until 1875, when Catholic universities were established in Paris, Lyon, Lille, Angers, and Toulouse. By the original legislation of 12 July 1875, these institutions had the right to the title "University" wherever at least three faculties (in addition to a faculty of theology) were created. The hardening of republican attitudes led, in 18 March 1880, to the withdrawal of that right and to the replacement of the title "University" by either "Institut catholique" or "Facultés libres."

The seven faculties of medicine that existed in 1900 (some of them offering training in pharmacy as well) were important locations for scientific expertise. In addition to these faculties, a number of lower-level "secondary" schools of medicine (renamed in 1840 écoles préparatoires de médecine or écoles préparatoires de médecine et de pharmacie) offered the more elementary part of the medical syllabus. A small number of two-year écoles préparatoires à l'enseignement supérieur des sciences et des lettres, founded in the 1850s, fulfilled a similar role in the sciences and humanities.

Degrees and Qualifications

THE *BACCALAURÉAT*

By the legislation founding the Imperial University in 1808, two separate *baccalauréats* were created: a *baccalauréat-ès-lettres* for the humanities and a *baccalauréat-ès-sciences* for the sciences and mathematics. The *baccalauréat-ès-sciences* could only be taken by candidates who already held the *baccalauréat-ès-lettres*. This system existed until 1852, when Hippolyte Fortoul established the *baccalauréats* as examinations to be taken independently of each other rather than consecutively. The measure, part of Fortoul's system of *bifurcation*, was conceived as a way of giving the sciences a status equal to that of the humanities. But from the start, the *bifurcation* had its critics, especially among teachers in the humanities. During the 1860s it began to be dismantled, as part of a reaction against Fortoul's legacy. In ways I outline in chapter 3, the dismantling reaffirmed the cultural primacy of literary over scientific studies.

THE *LICENCE*

The examinations for the *licence* were normally taken, in either science or letters, two or three years after the *baccalauréat*. Teaching for the *licence* was provided in the faculties and at the Ecole normale supérieure. The qualification was seen primarily as a means of entry to careers in secondary education.

THE *DOCTORAT*

The *doctorat-ès-sciences* and *doctorat-ès-lettres* were awarded for theses in, respectively, science and letters. Although the doctorate was formally required for appointment to a chair in a faculty, in reality this regulation was only enforced rigorously from the 1850s (see table 1 in chapter 1). A doctorate was not required for the essentially interim positions of *chargé de cours* or *suppléant* or at the rank of lecturer (*maître de conférences*). The universities that were created under the legislation of 1896 had the right to award a new *doctorat d'université*, mainly intended for foreign students, and many did so. This doctorate carried with it neither the privileges nor the prestige of the existing doctorates.

THE *AGRÉGATION*

Except where stated, the *agrégations* to which I refer in this book were the fiercely competitive annual examinations that selected candidates for the most prestigious posts in secondary education, while also helping the ablest *agrégés* to move on to posts in higher education or research, either directly or via posts in a *lycée*. The Ecole normale supérieure was particularly effective in preparing candidates for these *concours*. A quite different system of *agrégations* had an important role in appointments to the faculties of medicine and law and schools of pharmacy. But there was no equivalent in the faculties of science or letters.

The Grandes écoles

The term *grandes écoles* came into common use in the later nineteenth century to designate a group of advanced vocational schools (also known as *écoles spéciales*) administered by a variety of ministries. The most celebrated school in the sciences was the Ecole polytechnique, founded in 1794 as the Ecole centrale des travaux publics and renamed the Ecole polytechnique in the following year. A two-year program at Polytechnique qualified the ablest *polytechniciens* (as the pupils were known) for further study in one of a number of specialized *écoles d'application*. Of these schools, the Ecole des ponts et chaussées (founded in 1747) and the Ecole des mines (1783) were the most sought after, because of the prestigious careers in state service to which they led. New schools, such as the Ecole supérieure de télégraphie (founded in 1878) and the Ecole supérieure d'électricté (1894), were created as new demands emerged during the nineteenth century.

The nineteenth century saw increasing intellectual rivalry between the Ecole polytechnique and the Ecole normale supérieure. The Ecole normale (renamed the Ecole normale supérieure in 1845) was established in 1808, following an earlier short-lived institution of 1794, commonly known as the "Ecole de l'an III." Divided into sections for science and letters, it prepared men for the various *agrégations* (see above) and hence for teaching posts in the secondary system, as did the corresponding school for women, the Ecole normale supérieure de jeunes filles, established at Sèvres in 1881. The Ecole normale supérieure is traditionally associated with the rue d'Ulm in Paris, where it occupied new premises from 1847.

The Institut de France and Other Research Institutions

The Institut de France was created in 1795. It brought together the successors of the national academies of the ancien régime, redesignated as classes, the First Class corresponding to the old

Académie royale des sciences. As part of the reorganization of 1816, the classes of the Institute regained the title of academies. The Académie des sciences was typical in undergoing only modest expansion during the nineteenth century, from a complement of seventy-five resident members (in addition to one hundred nonresident correspondents and eight foreign associates) in 1815 to one of seventy-eight resident members, 122 correspondents and other nonresident members, and twelve foreign associates by 1914.

Three great research institutions of the ancien régime survived the revolutionary period and continued to play leading roles in the sciences. These were the Collège de France, the Paris Observatoire, and the Muséum d'histoire naturelle.

The Collège de France was founded in 1530 to advance and (through its public lectures) diffuse science and learning. Among the forty-five chairs that it had in 1914, roughly a third were in science, mathematics, or medicine. Although the Collège conferred no diplomas, its laboratories (which included that of Claude Bernard in experimental medicine) were the source of much important research.

The premises of the Observatoire de Paris, inaugurated in 1671, underwent significant expansion during the nineteenth century. During the 1850s and 1860s, under the directorship of Urbain Le Verrier, the observatory developed a meteorological service, as an adjunct to the core work of positional astronomy. An observatory for physical astronomy was established at Meudon just outside the capital in 1876.

The Muséum d'histoire naturelle—known over time as the royal, imperial, and (at present) national museum of natural history—was founded in 1793 as the revolutionary successor to the Jardin du roi of the ancien régime. On the eve of the First World War, it had eighteen chairs, with specialities across the whole range of natural history, including anthropology, and relevant areas of physics and chemistry. Professors were responsible for collections in their fields as well as for their various laboratories and the museum's popular menagerie.

An important innovation among research institutions in both the sciences and the humanities was the founding in 1868 of the Ecole pratique des hautes études, also commonly known by its initials, EPHE. In the five sections of the EPHE (including sections for mathematics, the physical sciences, and natural history and physiology), students were free to fashion their own programs, with a strong element of individual research. The EPHE was a federation of existing laboratories and research facilities (mainly in Parisian institutions) and had no facilities of its own. Its creation was one of Victor Duruy's most notable achievements during his time as Minister of Public Instruction.

Technical and Other Vocational Education

Responsibility for technical and other vocational instruction was divided among several ministries. Although the Ministry of Commerce was particularly active, the Ministry of Public Instruction launched a number of important initiatives, notably in the creation of specialized institutes attached to most faculties of science from the 1880s. In this way, the faculties in Nancy (with an institute of applied chemistry) and Grenoble (with its institute of electrical engineering) assumed special prominence. The most important technical schools that I discuss were the Conservatoire des arts et métiers, the Ecole centrale des arts et manufactures, and the *écoles d'arts et métiers*.

The Conservatoire was created in 1794 as a publicly accessible depository for machines and models, mainly intended for the instruction of artisans. The establishment of chairs of mechanics, applied chemistry, and industrial economy in 1819 marked a turn toward the provision of public lectures in vocational subjects, and this aspect of the Conservatoire's work expanded thereafter, as did research and teaching in the laboratories attached to the various chairs. The Conservatoire offered no degrees or other qualifications.

Founded in Paris as a private initiative in 1829, the Ecole centrale came under the control of the Ministry of Commerce in 1857. Its scientifically rigorous three-year course provided a high-level preparation for civil careers in industry. The school is not to be confused with the *écoles centrales*, the secondary schools (with a strong scientific orientation) that existed from 1795 until their replacement by the more classically oriented *lycées* in 1803.

The *écoles d'arts et metiers* were a network of technical schools, with a strong emphasis on workshop practice, that was developed from the trade school established by the duc de La Rochefoucauld-Liancourt at Liancourt in Picardy in 1788. Two such schools existed in 1815 (at Châlons-sur-Marne and Angers), and new schools were inaugurated at Aix (1843), Lille (1881), Cluny (1901), and Paris (1906). Pupils normally entered at the age of between 14 and 17 and stayed (as boarders) for four years.

Exchange Rates and Incomes in Nineteenth-Century France

Exchange Rates

Exchange rates between the French franc and other major currencies changed little during the nineteenth and early twentieth centuries. Approximate equivalents are as follows:

Great Britain	Pound worth 25 francs
German Empire (from 1870)	Mark worth 1.2 francs
United States	Dollar worth 5.4 francs

Incomes

During the First Empire, the practice of *cumul* allowed a few leading savants to achieve very high incomes. The cases of Claude-Louis Berthollet and Pierre-Simon Laplace, each of whom had an annual income in excess of 100,000 francs in 1814, were exceptional (see chapter 1). But, even as a younger member of the elite community of Parisian science, the chemist Louis-Jacques Thenard was able to accumulate a total of over 17,000 francs in 1811, mainly through the chairs he held at the Collège de France (6,000 francs p.a.) and the Sorbonne (4,500 francs). An even more striking case is that of Joseph-Louis Gay-Lussac, whose income exceeded 40,000 francs in the 1830s and 50,000 francs in the 1840s, thanks not only to his academic appointments but also to a judicious engagement in industrial activities and consultancy that I discuss in chapter 1.

Generally, however, professors had to live on far less, in most cases on the income from a single chair. A typical salary of between 4,000 and 5,000 francs in a provincial faculty in the 1840s (rather more in Paris) allowed for a comfortable but far from elegant lifestyle. And even the rector of an academy, with responsibility for several departments, earned no more than 6,000 to 7,200 francs, significantly less than senior tax collectors or the prefect of all but the smallest departments. Salaries in secondary education were less, rarely more than 3,000 francs for a senior professor in a Parisian *lycée* and between half and two-thirds of that in the provinces. The pay of teachers in primary schools was lower still. Many of them earned no more than 500 francs, corresponding broadly to the income of an unskilled workingman, although the amount increased significantly about the midcentury and was often augmented by the benefit of accommodation in the school and income from private lessons and other paid work.

As Minister of Public Instruction between 1863 and 1869, Victor Duruy made a determined effort to increase salaries at all levels. By the end of the Second Empire, professors in the provincial faculties of science had basic salaries of 4,000 francs, in addition to which they received a *traitement éventuel* of up to 2,400 francs drawn from fees from examining. The eighteen chairholders at the Sorbonne were better paid, with basic salaries of 7,500 francs (equal to that of a professor at the Collège de France or the Muséum d'histoire naturelle) and a supplementary *traitement éventuel* for examining that could rise to almost 4,000 francs. The salaries of senior educational administrators were generally higher than those of serving professors and other teachers, though often not significantly so. The salary of an inspector-general of higher education, for example, was 12,000 francs, while rectors earned between 10,000 and 15,000 francs, depending on the size of their academy. While such incomes compared "at least respectably" (in George Weisz's words) with those of other professional groups, they lagged well behind the salaries for top positions in the state administration.

Under the Third Republic, professorial salaries in the faculties of science, as in other faculties, continued to grow. On the eve of the First World War, they stood at between 12,000 and 15,000 francs at the Sorbonne and between 6,000 and 12,000 francs in the provinces. The incomes of *lycée* professors increased also, although they were always lower than those in higher education: in 1913, they were between 6,000 and 9,500 francs for an *agrégé* teaching in Paris and between 4,200 and 6,700 francs in the provinces, in both cases roughly the amounts earned by *maîtres de conférences* and other junior teaching staff in the faculties.

Sources: Statistique 1865–68 and *Statistique 1876.* Antoine Prost, *Histoire de l'enseignement en France 1800–1967* (Paris, 1968), 356–60, and the table on p. 372. George Weisz, *The Emergence of Modern Universities in France, 1863–1914* (Princeton, NJ, 1983), 56–60 and 328–33. On Thenard's income: Anne-Claire Déré and Gérard Emptoz, *Autour du chimiste Louis-Jacques Thenard (1777–1857). Grandeur et fragilité d'une famille de notables au XIXᵉ siècle* (Chalon-sur-Saône, 2008), 121–22.

Libraries, Archives, and Institutions

AAS	Archives of the Académie des sciences, Paris
AD	Departmental archives, indicated by AD followed by the name of the department concerned
AFAS	Association française pour l'avancement des sciences
AN	Archives nationales, Paris, followed by the number of the file and document, e.g., F^{17} 3719
BAAS	British Association for the Advancement of Science
BCMHN	Bibliothèque centrale of the Muséum d'histoire naturelle, Paris
BIF	Bibliothèque de l'Institut de France, Paris
BL	British Library, London
BnF	Bibliothèque nationale de France, Paris
CNAM	Conservatoire national [royal, impérial] des arts et métiers, Paris
PCN	Certificat d'études physiques, chimiques, et naturelles

Periodicals and Other Publications

ACP	*Annales de chimie et de physique*
Actes Acad. Bordeaux	*Actes de l'Académie des sciences, belles-lettres et arts de Bordeaux*
AFAS	*Association française pour l'avancement des sciences. Compte-rendu de la . . . session.* In the abbreviated form, I use *AFAS, 14e session. Grenoble 1885*, for example, to refer to the proceedings of the fourteenth session, published in two volumes as *Association française pour l'avancement des sciences. Compte-rendu de la 14me session. Grenoble 1885.* From the mid-1880s, publication in two volumes became the norm. As was usually the case, the *Compte-rendu* did not appear until the year after the meeting.
AIP	*Annuaire de l'Institut des provinces [et des congrès scientifiques]*
AMHNM	*Annales du Musée d'histoire naturelle de Marseille*

APC	*Annales de philosophie chrétienne*
AS	*Annals of Science*
ASI	*Année scientifique et industrielle ou exposé annuel des travaux scientifiques . . . par Louis Figuier*
BAIP	*Bulletin administratif [du Ministère] de l'instruction publique*
BAM	*Bulletin de l'Académie de médecine*
BASF	*Bulletin [hebdomadaire] de l'Association scientifique de France*
Beauchamp, *Lois et règlements*	Arthur Marais de Beauchamp et al., *Recueil des lois et règlements sur l'enseignement supérieur comprenant les décisions de la jurisprudence et les avis des conseils de l'instruction publique et du Conseil d'Etat*, 7 vols. (Paris, 1880–1915)
BJHS	*The British Journal for the History of Science*
BSGF	*Bulletin de la Société géologique de France*
BSLN	*Bulletin de la Société linnéenne de Normandie*
BSZF	*Bulletin de la Société zoologique de France*
CR	*Comptes rendus hebdomadaires des séances de l'Académie des sciences*
CRSB	*Comptes rendus hebdomadaires des séances et mémoires de la Société de biologie*
CSF	Proceedings of the annual *congrès scientifiques de France*, published under slightly varying titles. In references I use the abbreviated form *CSF, 6e session. Clermont-Ferrand 1838*, for example, to refer to the proceedings of the Clermont-Ferrand congress of 1838, published as *Congrès scientifique de France. Sixième session, tenue à Clermont-Ferrand, en septembre 1838*. In this as in virtually every other case, the proceedings appeared in the year after the congress, usually in one volume though occasionally in two and once (following the Bordeaux congress of 1861) in five.
DBF	*Dictionnaire de biographie française*, ed. J. Balteau et al., 19 vols. to date (Paris, 1933–)
ED	*Enquêtes et documents relatifs à l'enseignement supérieur*. The 124 volumes of the *Enquêtes et documents* were published by the Ministry of Public Instruction between 1883 and 1929.
HSPS	*Historical Studies in the Physical Sciences*
JAPNP	*Journal de l'anatomie et de la physiologie normales et pathologiques de l'homme et des animaux*
JGIP	*Journal général de l'instruction publique [et des cultes]*
JPTA	*Journal de physique théorique et appliquée*
JS	*Journal des savants*
MAS	*Mémoires de l'Académie des sciences de l'Institut de France*
Mém. Acad. Caen	*Mémoires de l'Académie [royale] des sciences, arts et belles-lettres de Caen*

Mém. Soc. Lille	*Mémoires de la Société [royale, impériale] des sciences, de l'agricuture et des arts de Lille*
MS	*Le Moniteur scientifique. Journal des sciences pures et appliquées à l'usage des chimistes, des pharmaciens et des manufacturiers . . . fondé et dirigé par le Dr Quesneville*
MSLC	*Mémoires de la Société linnéenne du Calvados* [from 1827 *Société linnéenne de Normandie*]
MU	*Le Moniteur universel [Journal officiel de l'Empire]*
NRRS	*Notes and Records of the Royal Society [of London]*
Phil. Trans.	*Philosophical Transactions of the Royal Society*
PNE	*Petites nouvelles entomologiques*
PP	*La Philosophie positive*
PV	*Académie des sciences. Procès-verbaux des séances de l'Académie tenues depuis la fondation de l'Institut jusqu'au mois d'août 1835*, 10 vols. (Hendaye, 1910–22)
RCL	*Revue des cours littéraires de la France et de l'étranger*
RCS	*Revue des cours scientifiques de la France et de l'étranger*
RDM	*Revue des deux mondes*
RG	*Revue de géographie*
RH	*Revue historique*
RHS	*Revue d'histoire des sciences [et de leurs applications]*
RIE	*Revue internationale de l'enseignement*
RMC	*Revue maritime et coloniale*
RO	*Revue occidentale philosophique sociale et politique*
RPI	*Revue positiviste internationale*
RPL	*Revue politique et littéraire*
RQS	*Revue des questions scientifiques, publiée par la Société scientifique de Bruxelles*
RR	*Revue de Rouen*
RS	*Revue scientifique de la France et de l'étranger [Revue rose]*
RSI	*Revue scientifique industrielle. Sous la direction du docteur Quesneville*
RSS	*Revue des sociétés savantes* [then variously *de la France et de l'étranger*, or *des départements*] *publiée sous les auspices du Ministre de l'instruction publique [et des cultes]*
SSS	*Social Studies of Science*
Statistique 1865–68	*Statistique de l'enseignement supérieur. 1865–1868* (Paris, 1868)
Statistique 1876	*Statistique de l'enseignement supérieur. Enseignement, examens, grades, recettes et dépenses, en 1876. Actes administratifs jusqu'en août 1878* (Paris, 1878)

Introduction

1. For more on the unique relationship between the state and education at all levels, see appendix A.

2. Picard, "La science et la recherche scientifique, *RS* 50, 2(1912): 577–81, esp. 581.

Chapter 1 · Science and the New Order

1. Ambroise Fourcy, *Histoire de l'Ecole polytechnique* (Paris, 1828), 323, and Gaston L. E. Pinet, *Histoire de l'Ecole polytechnique* (Paris, 1887), 75.

2. Cited in Edouard Estaunié et al., *Société des amis de l'Ecole. L'Ecole polytechnique* (Paris, 1932), 11.

3. My figures are for France alone. They do not take into account faculties in parts of the empire outside France that were lost after 1814.

4. Pierre-Simon Laplace, *Traité de mécanique céleste* (Paris, 1798–1827), 3:v (an XI, 1802).

5. Georges Cuvier, *Rapport historique sur les progrès des sciences naturelles depuis 1789, et sur leur état actuel, présenté à Sa Majesté l'Empereur et Roi, en son Conseil d'état, le 6 février 1808, par la Classe des sciences physiques et mathématiques* . . . (Paris, 1810), 2–3.

6. I say "learned world" here, since scientists were not alone in their declarations of loyalty to the passing regimes. Indeed, no major savant quite matched the ill-timed effusiveness of Jean-Baptiste Say's dedication of the second edition of his *Traité d'économie politique* to the emperor of Russia, Alexander I, in 1814; see Say, *Traité d'économie politique, ou simple exposition de la manière dont se forment, se distribuent et se consomment les richesses*, 2nd ed. (Paris, 1814), 1:v.

7. Nicole and Jean Dhombres, *Naissance d'un nouveau pouvoir. Sciences et savants en France (1793–1824)* (Paris, 1987), passim, but see 7–10 for a succinct statement.

8. Charles Coulston Gillispie, *Science and Polity in France: The Revolutionary and Napoleonic Years* (Princeton, NJ, 2004), 695. The word encapsulates the thrust of Gillispie's analysis of a process that saw the modernization of both science and politics during the revolutionary and Napoleonic periods.

9. *Dictionnaire des girouettes, ou nos contemporains peints d'après eux-mêmes par une société de girouettes* (Paris, 1815), 102–3 and 225. My references are to the first edition. Other editions followed quickly (a third by the end of 1815), each with significantly different pagination. On the *Dictionnaire*, a multi-authored work in which le comte César Proisy d'Eppe and the publisher Alexis Eymery seem to have played a leading role, and more generally on the wave of criticism of the shifting political allegiances of leading public figures in 1814–15, see Pierre Serna, *La République des girouettes* (Seyssel, 2005), 194–250.

10. Laplace's strategies for survival are discussed in Roger Hahn, *Pierre Simon Laplace, 1749–1827: A Determined Scientist* (Cambridge, MA, 2005), 190–97.

11. On Cuvier's conduct between the collapse of the Napoleonic Empire and the definitive return of the Bourbon line, see Dorinda Outram, *Georges Cuvier: Vocation, Science and Authority in Post-Revolutionary France* (Manchester, 1984), esp. chaps. 4 and 5.

12. Cuvier, "Analyse des travaux de l'Académie royale des sciences, pour les années 1813, 1814, 1815. Partie physique," *Mémoires de la Classe des sciences mathématiques et physiques de l'Institut de France, Années 1813, 1814, 1815* (published 1818), cxcv. Note Cuvier's use of the First Class's restored title of Académie royale des sciences in the title of his report.

13. Ibid., cxcvi.

14. Cuvier made similar points in his address to the combined academies of the Institute in April 1816. See Georges Cuvier, "Réflexions sur la marche actuelle des sciences, et leurs rapports avec la société," in *Recueil des discours prononcés dans la séance publique annuelle de l'Institut royal de France, le mercredi 24 avril 1816.*

15. Adolphe Robert and Gaston Cougny, *Dictionnaire des parlementaires français... depuis le 1er mai 1789 jusqu'au 1er mai 1889* (Paris, 1889–91), 4:53–54.

16. *Institut de France... Le mercredi 11 mai 1814, l'Institut a eu l'honneur d'être présenté à Sa Majesté. M. Lefèvre-Gineau, président de la Première Classe, portant la parole, a dit . . .* (Paris, 1814). I have used the copy of this report in the British Library (BL 733 g.11 (35)).

17. A valuable guide through the institute's transformations is Maurice P. Crosland, *Science under Control: The French Academy of Science 1795–1914* (Cambridge, 1992), esp. (on this period) 50–90. The legislation, including lists of members, is in Léon Aucoc, *L'Institut de France. Lois, statuts et règlements concernant les anciennes académies et l'Institut, de 1635 à 1898. Tableau des fondations* (Paris, 1889), 108–17.

18. On this last phase of Carnot's career, see Jean and Nicole Dhombres, *Lazare Carnot* (Paris, 1997), chaps. 21 and 22.

19. Francoeur fils, *Notice sur la vie et les ouvrages de M. L.-B. Francoeur* (Paris, 1853), 8–9.

20. Crosland, *Science under Control*, 304–5. Hachette was finally elected to the Académie in 1831, when (in the more liberal atmosphere of the July Monarchy) a vacancy arose in the mechanics section.

21. J. P. F. Deleuze, *Histoire et description du Muséum royal d'histoire naturelle* (Paris, 1823), 126–27. After two years, the normal budget was restored, and the museum embarked on a program of building and refurbishment that improved conditions both for the collections and, notably in the menagerie, for the visiting public.

22. Abel Lefranc, *Histoire du Collège de France depuis ses origines jusqu'à la fin du Premier Empire* (Paris, 1893), 338–39 and 372.

23. Antoine Picon, *L'Invention de l'ingénieur moderne. L'Ecole des ponts et chaussées 1747–1851* (Paris, 1992), 317–18 and 403–4, and Louis Aguillon, "L'Ecole des mines de Paris. Notice historique," *Annales des mines*, 8th ser., 15 (1889): 519–41.

24. Among the plentiful accounts of the heroism of *polytechniciens* during the defence of Paris, see Fourcy, *Ecole polytechnique*, 322–27; Pinet, *Histoire de l'Ecole polytechnique*, 72–78; and, above all, Maurice Sautai, *L'Ecole polytechnique pendant la campagne de France (1814)* (Paris, 1910), 8–29.

25. Pinet, *Histoire de l'Ecole polytechnique*, 96–102.

26. For a particularly virulent retrospective view of Laplace's role, see Théodore Olivier's unfinished text "De l'Ecole polytechnique," published as a preface to his *Mémoires de géométrie descriptive, théorique et appliquée* (Paris, 1851), esp. xi–xii. Olivier saw Laplace, along with Poisson

and Cauchy, as responsible for an exclusive emphasis on algebra that had turned Polytechnique into a school for mathematicians, an "Ecole monotechnique."

27. On the origins of these schools, see below.

28. It was the requirement for pupils at the Ecole centrale des travaux publics (as Polytechnique was called at the time) to swear their hatred of royalty that led Liautard to withdraw from the school. See *Mémoires de l'abbé Liautard, fondateur du Collège Stanislas, mort archiprêtre, curé de Fontainebleau . . . précédés d'un essai biographique . . . par M. l'abbé A. Denys* (Paris, 1844), 1:29–33.

29. [Liautard], *Mémoire sur l'Université* (Paris, 1814), 45–56. The *Mémoire* appeared with another text, *De l'Université*, the pagination of which it continued. Although both texts were published anonymously, the preface to a reissue of the *Mémoire* (Lyon, 1845), v–vi, puts the identity of the author beyond reasonable doubt.

30. François-René Chateaubriand, *De Buonaparte et des Bourbons et de la nécessité de se rallier à nos princes légitimes, pour le bonheur de la France et celui de l'Europe* (Paris, 1814), 17.

31. On the attempts of Royer-Collard and Guizot to refashion the University, see Louis Liard, *L'Enseignement supérieur en France, 1789–1889 [1893]* (Paris, 1888–94), 2:129–32.

32. Frayssinous's time at the head of the University between 1824 and 1828 is well treated in Antoine Roquette, *Monseigneur Frayssinous, Grand-Maître de l'Université sous la Restauration (1765–1841). Evêque d'Hermepolis ou le chant du cygnet du trône et de l'autel* (Paris, 2007), 135–68.

33. The Royal Commission of Public Instruction decided on the closure of the faculties on 31 October 1815, and the measure received royal assent on 18 January 1816. See Beauchamp, *Lois et règlements*, 1:387–88 and 1:402.

34. Cuvier had been elected a member of the First Class of the Institute on its creation in 1795 and as permanent secretary for the physical sciences in 1803.

35. The legislation of 1 November 1820 changed the title of the Commission but, apart from a small increase in size, barely altered its functions. See Beauchamp, *Lois et règlements*, 1:452–54.

36. Outram, *Georges Cuvier*, 106–7.

37. Ibid., 93–117.

38. On the reaction against Laplace's conception of physics and Berthollet's associated views on chemical affinities, see Robert Fox, "The Rise and Fall of Laplacian Physics," *HSPS* 4 (1974): 109–32.

39. On the program, see Fox, "Laplacian Physics," 91–109, and Ivor Grattan-Guinness, *Convolutions in French Mathematics, 1800–1840: From the Calculus and Mechanics to Mathematical Analysis and Mathematical Physics* (Basel, 1990), 1:436–517.

40. Fox, "Laplacian Physics," 112–27; Robert H. Silliman, "Fresnel and the Emergence of Physics as a Discipline," *HSPS* 4 (1974): 137–62; and Eugene Frankel, "Corpuscular Optics and the Wave Theory of Light: The Science and Politics of a Revolution in Physics," *SSS* 6 (1976): 141–84; and Jed Z. Buchwald, *The Rise of the Wave Theory of Light: Optical Theory and Experiment in the Early Nineteenth Century* (Chicago, 1989).

41. In a letter to Macvey Napier, the Scottish physicist John Leslie expressed his surprise at finding Laplace and Berthollet with annual incomes of between £5,000 and £6,000 each when he visited Paris in 1814. This suggests my figure of c. 100,000 francs. The letter is cited in J. B. Morrell, "Science and Scottish University Reform. Edinburgh in 1826," *BJHS* 6 (1972–73): 51.

42. Berthollet's income quickly fell to about 24,000 francs. See his letter to Charles Blagden, 3 February 1817, Blagden Letters, Royal Society, B143.

43. Laplace's failure to deliver to the king a petition of the Académie française against the

government's plan to restrict press freedom in January 1827 aroused liberal resentment. The explanation, however, may lie in his illness (he died only a few weeks later) or in the king's prior refusal to receive the petition, as Roger Hahn has observed; see Hahn, *Pierre Simon Laplace*, 196.

44. Outram, *Georges Cuvier*, 109.

45. A full and perceptive biography of Raspail is Dora B. Weiner, *Raspail: Scientist and Reformer* (New York, 1968). Among older studies, Raphaël Blanchard, "François-Vincent Raspail," *Archives de parasitologie* 8 (1903–4): 5–87, is especially helpful.

46. Just four volumes of the *Annales* appeared before it finally succumbed to its financial difficulties in June 1830.

47. Raspail, "Nécrologie," *Annales des sciences d'observation* 3 (1830): 159–60.

48. Raspail, *Nouveau système de physiologie végétale et de botanique, fondé sur les méthodes d'observation, qui ont été développées dans le nouveau système de chimie organique* (Paris, 1837), 1:v–xxiv and 2:609–31. Raspail took up the theme with undiminished venom in 1838 in the "Avertissement historique" to the second edition of his *Nouveau système de chimie organique fondé sur des méthodes nouvelles d'observation* (3 vols. and 1 vol. of plates), a manual of organic chemistry that emphasized experimental procedures, notably with the microscope, in the study of the phenomena of life; see Raspail, *Nouveau système de chimie organique*, 1:xvii–lxiv.

49. Raspail, *Nouveau système de physiologie végétale*, 2:610, 2:612, 2:624, 2:630, 2:611, 2:629.

50. On Biot's perception of the political and religious prejudices (especially Arago's) that weighed against him, see Francisque Lefort, "Un savant chrétien. J. B. Biot," *Le Correspondant* n.s., 36 (1867): 982–83.

51. Arago's decision to give his support to Fourier after being himself a candidate seems to have been the main cause of Biot's resentment. The continuing multifaceted hostility between Biot and Arago is a core theme of Theresa Levitt, *The Shadow of Enlightenment: Optical and Political Transparency in France, 1789–1848* (Oxford, 2009).

52 Toby A. Appel, *The Cuvier-Geoffroy Debate: French Biology in the Decades before Darwin* (New York, 1987), passim but esp. 1–10 for a helpful summary. See also Henri Le Guyader, *Geoffroy Saint-Hilaire. A Visionary Naturalist*, trans. Marjorie Grene [from the original, French edition of 1998] (Chicago, 2004), chaps. 5 and 6.

53. Franck Bourdier, "Le prophète Geoffroy Saint-Hilaire, George Sand et les Saint-Simoniens," *Histoire et nature* 3, no. 1 (1973): 47–66.

54. A still standard source on nineteenth-century debates concerning the state's monopoly in education is Louis Grimaud, *Histoire de la liberté d'enseignement en France depuis la chute de l'ancien régime jusqu'à nos jours* (Paris, 1898); see pp. 195–375 on the July Monarchy. Although Liautard's most vehement outbursts date from the 1820s, he remained a vigilant critic of the University and its secularism and illiberal centralizing tendencies until his death in 1842; see *Mémoires de Liautard*, 2:7–144.

55. *Rapport sur les besoins du Muséum d'histoire naturelle, pour l'année 1835, et sur la Bibliothèque royale, présenté au Ministre de l'Instruction publique* (Paris, 1834), 2. The Muséum was seeking an increase of a sixth in its budget for 1835 to help to strengthen its support for the foreign missions of its mid-level staff of naturalists.

56. For this recollection of the condition in the Grenoble faculty, by the long-serving professor and dean, François Raoult, see "Centenaire de la faculté des sciences de Grenoble. Discours de M. le doyen J. Collet," *RIE* 63 (1912): 405–6.

57. Ibid., 404.

58. A. de Saint-Germain, "Recherches sur l'histoire de la faculté des sciences de Caen de 1809 à 1850," *Mém Acad. Caen* (1891): 65–68.

59. Beauchamp, *Lois et règlements*, 1:249–55, esp. 250 and 253 (clauses 10–11 and 55–56).

60. Ibid., 1:903–4.

61. *AFAS, 16ᵉ session. Toulouse 1887*, 760.

62. In the more limited period discussed in this chapter, 1811–48, the disparity (seventy-seven theses in the provinces compared with 155 in Paris) was less marked. But the thinness of doctoral activity in the provincial faculties was masked by the exceptional vigor of Strasbourg, where thirty-four of the seventy-seven provincial theses were written.

63. Beauchamp, *Lois et règlements*, 1:171–88; decree of 17 March 1808, clause 143 (188).

64. Laurent to Gerhardt, 12 June 1845, in Edouard Grimaux and Charles Gerhardt, *Charles Gerhardt. Sa vie, son oeuvre, sa correspondance, 1816–1856. Document d'histoire de la chimie* (Paris, 1900), 96. The volume evokes both the difficulties and the resolution of Laurent and Gerhardt in their struggles against "the formidable coalition of the savants of the official school" (91).

65. Correspondence in his personal file in AN F¹⁷ 21091/A reveals the growing resentment of the rector of the academy of Bordeaux and of Laurent's colleagues at his repeated requests for leave. The resentment contrasts with leniency within the ministry that allowed him to retain half of his salary for at least the first two years of his absence.

66. Biot, "Appréciation des titres de M. Laurent et M. Balard," *RSI* 3rd ser., 6 (1850): 441–45; reproduced in Grimaux and Gerhardt, *Charles Gerhardt*, 588–91.

67. *CR* 32 (1851): 46 (13 Jan. 1851).

68. Evidence of Gerhardt's frustration and the precariousness of his attempts to establish his school is in Grimaux and Gerhardt, *Charles Gerhardt*, esp. 208–54. His personal file in AN F¹⁷ 20829 endorses the point, showing that he again sought to absent himself when he was appointed as Pasteur's successor in the chair of chemistry at Strasbourg in 1855.

69. Georges Hervé and L. de Quatrefages, "Armand de Quatrefages de Bréau, médecin, zoologiste, anthropologiste (1810–1892)," *Bulletin de la Société française d'histoire de la médecine* 20 (1926): 318–30.

70. Rémusat, *Mémoires de ma vie*, ed. Charles H. Pouthas (Paris, 1958–68), 1:242.

71. J. N. P. Hachette to Michael Faraday, 9 July 1832, in Frank A. J. L. James, ed., *Correspondence of Michael Faraday* (London, 1991–2008), 2:64–68.

72. Charles-Augustin Sainte-Beuve, *Volupté* (Paris: Eugène Renduel, 1834), 2:285 (in chap. 25). Amaury's comment is discussed in Robert Fox, "Scientific Enterprise and Patronage of Research in France, 1800–70," *Minerva* 11 (1973) : 452–58.

73. See chap. 5.

74. Maurice P. Crosland, *Gay-Lussac: Scientist and Bourgeois* (Cambridge, 1978), 228–34.

75. Anne-Claire Déré and Gérard Emptoz, *Autour du chimiste Louis-Jacques Thenard (1777–1857). Grandeur et fragilité d'une famille de notables au XIXe siècle* (Chalon-sur-Saône, 2008), 242–63.

76. Henri Tribout, *Un grand savant. Le général Jean-Victor Poncelet 1788–1867* (Paris, 1936), 101–18. Cf. Félix Dujardin's similar preoccupation with securing election as a resident member of the Académie; see chap. 2.

77. Babbage, *Reflections on the Decline of Science in England and on Some of Its Causes* (London, 1830), 22–26.

78. The point, illustrated by Humphry Davy's visit to France in 1813–14, is made in Gavin de Beer, *The Sciences Were Never at War* (London, 1960).

79. On the visit, which was financed by the Restoration government, and Say's disappointingly fragmentary diary of his travels, see André Tiran, "Préface historique," in Jean-Baptiste Say, *Manuscrits sur la monnaie, la banque et la finance (1767–1832)* (Lyon, 1995).

80. Ernest Teilhac, *Œuvre économique de Jean-Baptiste Say . . . Thèse pour le doctorat (Sciences politiques et économiques) . . . Faculté de droit de Bordeaux* (Paris, 1927), 7–9.

81. Say, *De l'Angleterre et des Anglais* (Paris, 1815), 16–27.

82. Ibid., 30.

83. Ibid., 30–31.

84. The numerous older accounts of Dupin's life have now been augmented by the important collection of studies in Carole Christen and François Vatin, eds., *Charles Dupin (1784–1873). Ingénieur, savant, économiste, pédagogue et parlementaire du Premier au Second Empire* (Rennes, 2009).

85. Robert Fox, "From Corfu to Caledonia. The Early Travels of Charles Dupin, 1808–1820," in John D. North and John J. Roche, eds., *The Light of Nature: Essays in the History and Philosophy of Science, Presented to A. C. Crombie* (Dordrecht, 1985), 303–20; Margaret Bradley and Fernand Perrin, "Charles Dupin's Study Visits to the British Isles, 1816–1824," *Technology and Culture* 32 (1991): 47–68; and Benoît Agnès, "Le passeur des deux rives? La Grande-Bretagne dans l'action politique de Charles Dupin, 1814–1835," in Christen and Vatin, *Charles Dupin*, 53–65.

86. Dupin, *Mémoires sur la marine et les ponts et chaussées de France et d'Angleterre* (Paris, 1818), 72.

87. Dupin, "De la structure des vaisseaux anglais, considérée dans ses derniers perfectionnements," *Phil. Trans.* 107 (1817): 86–135, esp. 89–92.

88. Dupin, *Réponse au discours de milord Stanhope sur l'occupation de la France, par l'armée d'Angleterre,* 2nd ed. (Paris, 1818). The episode is described in the article on Dupin in Victor Lacaine and H. Charles Laurent, *Biographies et nécrologies des hommes marquants du XIXe siècle* (Paris, 1844–66), 4:284–85.

89. Dupin, *Forces productives et commerciales de la France,* 2 vols. (Paris, 1827). Dupin's admiration for the improvements that France had made, extending even to morals, is most explicit in the introduction to the first volume, "Introduction, montrant la situation progressive des forces de la France, depuis 1814," 1:ix–xxxx, also printed separately as Dupin, *Situation progressive des forces de la France* (Paris, 1827).

90. Dupin, *Forces productives et commerciales,* 1:xvii.

91. Ibid., 1:xvii–xviiii and 1:xxxvi–xxxviii.

92. Say, *De l'Angleterre et des Anglais,* 28.

93. Chaptal, *De l'Industrie française,* 2 vols. (Paris, 1819).

94. Ibid., 2:38, 2:102.

95. Among numerous studies of La Rochefoucauld-Liancourt, a succinct modern account is the article on him, by Charles Rodney Day, in Claudine Fontanon and André Grelon, eds., *Les Professeurs du Conservatoire national des arts et metiers. Dictionnaire biographique, 1794–1955* (Paris, 1994), 2:41–49. See also the classic biography by Frédéric Gaétan, *Vie de La Rochefoucauld-Liancourt (François-Alexandre-Frédéric)* (Paris, 1827).

96. Georges-Albert Bouty et al., *Cent-cinquante ans de haut enseignement technique au Conservatoire national des arts et metiers* (Paris, 1970), 3–18.

97. Correspondence dating from 1816–17 in the archives of the CNAM, especially in file 10.594, conveys the pressures on the institution from the Ministry of the Interior, notably for the completion of a proper catalogue of the collections. The appointment of Gérard-Joseph Christian as the Conservatoire's director, following the retirement in 1816 of its long-serving and now ailing administrator, Claude-Pierre Molard, seems to have allowed this project to move ahead quickly. The result was the 167-page *Catalogue général des collections du Conservatoire royal des arts et métiers* (Paris, 1818), with Christian's interesting "Notice sur le Conservatoire royal des arts et métiers," i–xx.

98. Chaptal's difficult passage from the Napoleonic Empire to the Restoration is described in the contribution to "La vie et l'oeuvre de Chaptal" by his great-grandson, the vicomte Antoine Chaptal, in *Mes Souvenirs sur Napoléon par le Cte Chaptal, publiés par son arrière petit fils* (Paris, 1893), 136–46, and Jean Pigeire, *La Vie et l'oeuvre de Chaptal* (Paris, 1932), 329–52.

99. On the exhibition, see Louis Costaz, *Rapport du jury central sur les produits de l'industrie française, présenté à S. E. M. le comte Decazes* (Paris, 1819), esp. the administrative documents predating the exhibition (379–94).

100. Chaptal, *Industrie française*, 2:307–9.

101. On the reorientation of the Conservatoire toward a primary engagement in advanced science-based teaching, see Bouty et al., *Cent-cinquante ans de haut enseignement technique*, 19–33, and the brief but helpful comment in René Tresse, "J. A. Chaptal et l'enseignement technique de 1800 à 1819," *RHS* 10 (1957): 172–73.

102. Chaptal, *Souvenirs sur Napoléon*, 91.

103. Dupin, *Mémoires sur la marine*, 67–69.

104. Dupin appears to have withdrawn his protest from publication at Carnot's insistence.

105. Dupin, *Essai sur les services et les travaux scientifiques de Gaspard Monge* (Paris, 1819). Dupin had dedicated his *Développements de géométrie* (Paris, 1813) to Monge, "Mon illustre maître."

106. Dupin expressed fulsome gratitude to Decazes in the inaugural lecture that he delivered in January 1822: Dupin, *Discours d'inauguration de l'amphithéâtre du Conservatoire des arts et métiers; prononcé le 8 janvier 1821* [corrected to 1822 in a copy, signed by Dupin, in BnF 8–Z Le Senne-10498] (Paris, 1822): 1–4. The sum of 85,000 francs is mentioned as an estimate of the likely cost in documents in AN F^{13} 881 and F^{13} 670.

107. Paul Dupuy, "Notice historique," in *L'Ecole normale (1810–1883). Notice historique. Liste des élèves par promotions. Travaux littéraires et scientifiques* (Paris, 1884), 23–28; also the updated version of Dupuy's text in *Le Centenaire de l'Ecole normale. 1795–1895. Edition du bicentenaire* (Paris, 1895), 219–22.

108. Auguste Corlieu, *Centenaire la faculté de médecine de Paris (1794–1894)* (Paris, 1896), 222–25.

109. Adrien Garnier, *Frayssinous. Son rôle dans l'Université sous la Restauration (1822–1828)* (Paris and Rodez, 1925), 84–236, is informative on the measures taken against the Ecole normale and the faculty of medicine and more generally on student disorder during the Restoration. See also Roquette, *Monseigneur Frayssinous*, 135–68.

110. Gaétan, *Vie de La Rochefoucauld-Liancourt*, 66–72. The account of the loss of these and other positions by Gaétan, who was La Rochefoucauld-Liancourt's son, is especially authoritative, although not necessarily impartial.

111. Dupin, *Bien-être et concorde des classes du peuple français* (Paris, 1840) and *Enseignement et sort des ouvriers et de l'industrie, avant, pendant et après 1848. Leçon donnée au Conservatoire des arts et métiers* (Paris, 1848), esp. 71–72.

112. *Histoire de l'Association polytechnique et du développement de l'instruction populaire en France... 1830–1880* (Paris, 1880) and François Vatin, "L'Association polytechnique (1830–1900). 'Education' ou 'instruction'? ou la place des sciences sociales dans la formation du peuple," *Management et sciences sociales. Revue scientifique et industrielle* 3 (2007): 245–96.

113. On the Association philotechnique, see Antoine Pressard, *Histoire de l'Association philotechnique* (Paris, 1899), esp. 7–23, on its break with the Association polytechnique, which it saw as pitching its courses at an unduly high scientific level. See also ibid., 30–37, on the initiatives of the Association philotechnique in providing courses for women and a growing number of mixed gender courses.

114. On Péclet's lectures, offered under the auspices of the municipality, see Léon Guillet, *Cent ans dans de vie de l'Ecole centrale des arts et manufactures 1829–1929* (Paris, 1920), 48.

115. The lectures, organized by the Société académique de Metz, were published as Poncelet, *Cours de mécanique industrielle fait aux artistes et ouvriers messins pendant les hivers de 1827–1828 et de 1828–1829*, 3 parts (Metz, 1827–30). The strong tradition of public lecturing in Metz began in 1816 and was sustained over the next twenty years not only by Poncelet but also by other former *polytechniciens* garrisoned in the city. The tradition flagged after the creation, in 1833, of one of the country's first *écoles primaires supérieures*, which offered a vocationally oriented extension of primary-school education under municipal patronage; see *BAIP* n.s., 6 (1866): 885–89.

116. My comments on the *écoles d'arts et métiers* draw on Charles Rodney Day, *Education for the Industrial World: The Ecoles d'arts et métiers and the Rise of French Industrial Engineering* (Cambridge, MA, 1987), chaps. 3 and 5; also on Day, "The Making of Mechanical Engineers in France: The Ecoles d'arts et métiers, 1803–1914," *French Historical Studies* 10 (1978): 439–60.

117. As Anne-Françoise Garçon shows, however, even a location in an area as heavily industrialized as Saint Etienne could not protect the Ecole des mineurs from phases of difficulty in placing its former pupils. See Garçon, *Entre l'état et l'usine. L'Ecole des mines de Saint-Etienne au XIXe siècle* (Rennes, 2004), passim but esp. 307–11.

118. Day, *Education for the Industrial World*, 80–82. For a comment on the continued success of the schools under the Second Empire, see also chap. 3.

119. Balzac, *Curé de village* [1841], Pléiade edition, Honoré de Balzac. *La Comédie humaine*, ed. Pierre-Georges Castex et al. (Paris, 1976–81), 9:794–95.

120. I stress "minority" here, following Antoine Picon's argument that the secondary literature has tended to exaggerate the Saint-Simonian leanings of *polytechniciens*. See below, note 129.

121. Olivier, *Mémoires*, vi and xi–xviii.

122. Biot described the school as "descending every day to a state of uniform mediocrity, which engenders neither resistance nor noise" in an undated letter of c.1822 to Thomas Chalmers; see William Hanna, *Memoirs of the Life and Writings of Thomas Chalmers, D.D. LL.D.* (Edinburgh, 1849–52), 2:15. Cf. the similar point he made in his review of Charles Babbage's *Reflections on the Decline of Science in England*, in the *Journal des savans*, January 1831, 47. According to Biot, the same governmental suspicion of intellectual activity that had been at work in the replacement of the scientifically oriented *écoles centrales* by the Napoleonic *lycées* in the early years of the century had also progressively weakened Polytechnique's position as a school of advanced mathematics.

123. Stendahl, *Lucien Leuwen* [written 1834, published 1894], Pléiade edition, Stendhal, *Romans et nouvelles*, ed. Henri Martineau (Paris, 1952), 1:768–72.

124. Terry Shinn, *Savoir scientifique & pouvoir social. L'Ecole polytechnique 1794–1914* (Paris, 1980), 60–64.

125. Robert Fox, "Regards étrangers sur l'Ecole polytechnique, 1794–1850," in Bruno Belhoste, Amy Dahan-Dalmedico, Dominique Pestre, and Antoine Picon, eds., *La France des X. Deux siècles d'histoire* (Paris, 1995), 63–74.

126. On the Ecole centrale, Charles de Comberousse, *Histoire de l'Ecole centrale des arts et manufactures depuis sa fondation jusqu'à ce jour* (Paris, 1879) is a thoroughly documented history and my main source. Also valuable are Francis Pothier, *Histoire de l'Ecole centrale des arts et manufactures d'après des documents authentiques et en partie inédits. XIXe siècle à nos jours* (Paris, 1887); Guillet, *Cent ans de la vie de l'Ecole centrale*, 7–186; and John Hubbel Weiss, *The Making of Technological Man: The Social Origins of French Engineering Education* (Cambridge, MA, 1982).

127. Guillet, *Cent ans de la vie de l'Ecole centrale*, 87.

128. Weiss, *Making of Technological Man*, 183–87. According to Weiss (70–87 and 256), two-thirds of *centraliens* between 1830 and 1847 came from *haut bourgeois* backgrounds, much as happened at Polytechnique. Despite Weiss's warning of the difficulty of setting the boundaries between *haute bourgeoisie, moyenne bourgeoisie,* and *petite bourgeoisie,* the conclusion that I cite seems unassailable. So too does that, drawn from Day, concerning the distinctly humbler social origins of the pupils of the *écoles d'arts et métiers,* almost one-fifth of whom between 1830 and 1860 were the sons of peasants and factory or other miscellaneous workers, compared with a proportion of barely 2 percent at Centrale.

129. Antoine Picon has shown convincingly that, despite their prominence and influence in the secondary literature, the number of former *polytechniciens* who subscribed to Saint-Simonian ideas was less than has commonly been supposed. This was especially so from the mid-1830s. See Picon, *Les Saint-Simoniens. Raison, imagination et utopie* (Paris, 2002), 102–12.

130. Comberousse, *Histoire de l'Ecole centrale*, 126.

131. Quoted by Weiss, *Making of Technological Man*, 240.

132. Olivier, *Mémoires*, vi.

133. Ibid., iii.

134. The closeness of the relations between Centrale and the state under the July Monarchy is evident from documents in two rich files on the school in AN F$^{17\mathrm{bis}}$ 7233 and 7234.

135. Ministère de l'Instruction publique, *Rapport sur l'enseignement scientifique dans les collèges, les écoles intermédiaires et les écoles primaires* (Paris, 1847).

136. Correspondence, reports, and newspaper cuttings in the Dumas papers in the Archives of the Académie des sciences, esp. boxes 16–18, bear witness to Dumas's resolve and the eventual failure of his schemes.

137. Saint Simon, *Lettres d'un habitant de Genève* [1803], in *Oeuvres de Claude-Henri de Saint-Simon* (Paris: Editions Anthropos, 1966), 1:49.

138. Saint-Simon, *Lettres au Bureau des Longitudes* [1808], in ibid., 6:231–32 and 6:268–69 (passim for the sustained criticism of Laplace and 268 for the quoted judgement).

139. Saint-Simon, *Nouveau Christianisme. Dialogues entre un conservateur et un novateur. Premier dialogue* (Paris, 1825).

140. On Comte's early life and his links with Saint-Simon, I draw on Henri Gouhier, *La Jeunesse d'Auguste Comte et la formation du positivisme,* 3 vols. (Paris, 1933–41); also on Gouhier's briefer *La Vie d'Auguste Comte* (Paris, 1931) and Mary Pickering, *Auguste Comte. An Intellectual Biography* (Cambridge, 1993), vol. 1.

141. Saint-Simon, *Introduction aux travaux scientifiques du dix-neuvième siècle* [1808], in *Œuvres de Saint-Simon*, 6:172.

142. Comte described his quest for an independent philosophy retrospectively in a letter of 1 May 1824 to his young follower Gustave d'Eichthal. Reporting his final break with Saint-Simon, he stated that it was four or five years since he had had anything to learn from his former master. See *Auguste Comte. Correspondance générale et confessions,* ed. Paulo E. de Berrédo Carneiro et al. (Paris, 1973–90), 1:80.

143. The "Prospectus" was originally published as a separately authored part of Saint-Simon, *Du Contrat social. Suite des travaux ayant pour objet de fonder le système industriel* (Paris, 1822). It was reissued, with some additional pages, as Comte, *Système de politique positive,* vol. 1, part 1 (Paris, 1824). In this version, which formed the independently paginated third *cahier* of Saint-Simon, *Catéchisme des industriels* (1823–24), the authorship ("par Auguste Comte, ancien élève de l'Ecole polytechnique, élève de Henri Saint-Simon") was again clearly indicated.

144. For a succinct definition of the three states, as Comte understood them in the early

1820s, see Comte, *Système de politique positive* (1824), 63–73. Similar, though not identical, brief statements also appeared in his "Considérations philosophiques sur les sciences et les savants" (November 1825), in *Appendice general du Système de politique positive contenant tous les opuscules primitifs de l'auteur sur la philosophie sociale*, an independently paginated appendix to Comte, *Système de politique positive* (Paris, 1854), 4:137, and in Comte, *Cours de philosophie positive* (Paris, 1830–42), vol. 1, 1st lesson (1–56, esp. 2–20)

145. Comte recounted the successes he had had with his lectures in Comte, *Cours de philosophie positive*, 1:v (1830).

146. Fourier, *Théorie analytique de la chaleur* (Paris, 1822), i.

147. See especially Comte, *Cours de philosophie positive*, 1:16–17 and 2:453–54. Lessons 29–31 (2:465–593) illuminate the criteria that led Comte so to admire Fourier.

148. Ibid., 2:549.

149. Ibid., 2:505 and 1:12.

150. Lamé, *Cours de physique de l'Ecole polytechnique*, 2 vols. [the second in two separately paginated parts] (Paris, 1836), 1:ii–iii.

151. On Dumas's work and its bearing on his views of atoms, see Robert Fox, *The Caloric Theory of Gases from Lavoisier to Regnault* (Oxford, 1971), 282–95.

152. Dumas, *Traité de chimie appliquée* (Paris, 1828–46), 1:xxxiii–li (1828).

153. Dumas, *Leçons sur la philosophie chimique. Professées au Collège de France. Recueillies par M. Binau [sic]* (Paris, 1837), 270. Another edition of the same year give the name of the compiler, Armand Bineau, correctly.

154. Ibid.

155. This trend is discussed in Fox, *Caloric Theory of Gases*, 248–70. Lamé was still pursuing the ideal of an analytical theory of heat in the manner of Fourier (allied now to a related theory of elasticity) in the 1860s; see Lamé, *Leçons sur la théorie analytique de la chaleur* (Paris, 1861), esp. the "Discours préliminaire," v–xx. A paper of 1828 on the conduction of heat in solids exemplified Duhamel's principle of working, like Fourier, without reference to any assumptions about the nature of heat; see Duhamel, "Mémoire sur les équations générales de la propagation de la chaleur dans les corps solides," *Journal de l'Ecole polytechnique* 13, no. 21 (1832): 356–99, esp. 356–57.

156. Gouhier, *Vie d'Auguste Comte*, 179–80.

157. The lectures—Comte's contribution to a movement for the diffusion of scientific knowledge through lectures organized in the town halls of Parisian *arrondissements*—were eventually published as Auguste Comte, *Traité philosophique d'astronomie populaire ou exposition systématique de toutes les notions de philosophie astronomique, soit scientifiques, soit logiques, qui doivent devenir universellement familières* (Paris, 1844). The title conveyed the priority that Comte gave to the philosophical—in practice, the positivist—thrust of the lectures.

158. Comte to John Stuart Mill, 27 February 1843 and 6 February 1844, in *Auguste Comte. Correspondance générale*, 2:139 and 2:238.

159. Comte's warm recollections of the two months of his interim appointment reflect his pride at the appreciative response of the students and the favorable impression he had made on the school's director studies, Dulong. See Comte, *Cours de philosophie positive*, 6:xxiii–xxiv and 6:470n (1842).

160. Comte to the president of the Académie des sciences, 13 July 1840, in *Auguste Comte. Correspondance générale*, 1:345–50.

161. Gouhier describes the episode in his *Vie d'Auguste Comte*, 182–84. The evidence of what occurred in the Académie is in *CR* 11(1840): 210. Comte's fullest expressions of indignation ap-

peared in a letter of 26 August 1840 to his long-standing friend Pierre Valat and two years later in the sixth volume of the *Cours*; see *Auguste Comte. Correspondance générale*, 1:353–59, and Comte, *Cours de philosophie positive*, vol. 6, "Préface personnelle," xix (n) and (in the 57th lesson) 267n.

162. Comte to Blainville, 17 March 1833, in *Auguste Comte. Correspondance générale*, 1:243. In the election, on the following day, Libri was elected with thirty-seven votes, against Duhamel's sixteen, and one for Liouville; see *PV*, 10:227, entry for 18 March 1833.

163. Comte to Blainville, 21 June 1832, in *Auguste Comte. Correspondance générale*, 1:238.

164. When voting took place, on 9 July 1832, the candidates were all members of the Académie. Dulong was elected on the second round with thirty votes, against ten for Pierre Flourens, three for François Beudant, and two for Geoffroy Saint-Hilaire; see *PV*, 10:84, entry for 9 July 1832. Following Dulong's resignation, Flourens was elected and served until his death in 1867.

165. Comte to Blainville, 27 January 1843, in *Auguste Comte. Correspondance générale*, 2:134–35.

166. Comte, *Cours de philosophie positive*, 4:539–70, and Comte, *Système de politique positive*, 1:669–90 and 1:701–30.

167. Comte would have found much to draw upon in Franz Joseph Gall [and Johann Gaspar Spurzheim], *Anatomie et physiologie du système nerveux en général et du cerveau en particulier* (Paris, 1810–19), 4:315–85.

168. In the autumn of 1821, elections of both a full and a corresponding member of the medicine and surgery section took place. Without French nationality, however, Gall could only compete for corresponding membership, despite his residence in Paris. His name did not appear even on the long list of eight candidates whom the section proposed as candidates on 15 October 1821; see *PV*, 7:235–36, entry for 15 Oct.1821, and, for the election, *PV*, 7:236, entry for 22 Oct. 1821.

169. Johann Josef Gall, *A Messieurs les membres de l'Académie royale des sciences*, a three-page pamphlet, signed "Votre très humble serviteur" and dated 15 October 1821. Here Gall used an alternative German form of his forenames.

170. Outram, *Georges Cuvier*, 129–34.

171. Comte, *Cours de philosophie positive*, 6:v–xxxviii, esp. 6:xxiii–xxxviii; reproduced in *Auguste Comte. Correspondance générale*, 2:439–55. Cf. the no less blistering words that concluded the volume, where Comte referred to his suffering from "the blind or malicious impulse of prejudices and passions that go with the deplorable regime that rules over our science" (6:895).

172. Comte, *Cours de philosophie positive*, 6:xvi.

173. In copies of the sixth volume in which it appears (BnF, R-10118, for example), Blanchard's disclaimer occupies a page just before Comte's "Préface personnelle." Facing it in this BnF copy is a printed extract from the judgment of the Tribunal de commerce de Paris, dated 29 December 1842. The text of the "Avis de l'éditeur" is in Gouhier, *Vie d'Auguste Comte*, 203–6.

174. Comte, *Discours sur l'ensemble du positivisme, ou exposition sommaire de la doctrine philosophique et sociale propre à la grande république occidentale* (Paris, 1848), v. The *Discours* was reproduced three years later, with only minor changes, as the "Discours préliminaire sur l'ensemble du positivisme," in Comte, *Système de politique positive* (1851), 1:1–399.

175. For his bitter reflections on the loss of his post, see his "Appel au public occidental," published as a "post-scriptum" in Comte, *Discours sur l'ensemble du positivisme*, 395–99.

176. Comte voiced his regret in one of his public lectures on positivism on 27 February 1848 and had his retraction published in a number of newspapers in France and abroad; see Comte, *Discours sur l'ensemble du positivisme*, xi–xii, where he reproduced a letter to Emile Littré, also dated 27 February, in which he expressed his change of heart. It is unlikely that Comte's regret diminished Arago's contempt for him. At the time of the publication of the sixth volume of the

Cours, Arago had described Comte as an ill-tempered man "in whose mathematical work I saw no merit of any kind," a description that Blanchard quoted in his "Avis de l'éditeur" (cited above in note 173).

177. Comte, *Discours sur l'ensemble du positivisme*, xii–xiii. The attack on Poinsot that Comte now regretted had appeared in the *Cours de philosophie positive*, 6:471–72n. Comte formally broke off relations with Poinsot in a letter of typical frankness accompanying his gift of a copy of the sixth volume of the *Cours*; Comte to Poinsot, 21 August 1842, in *Auguste Comte. Correspondance générale*, 2:69–70.

178. Comte, *Discours sur l'ensemble du positivisme*, vi. Annie Petit's discussion of Comte's heightened emphasis on the importance of "diffusion" (a favored term in his writings) to the less educated orders of society is especially helpful; see Petit, "Heurs et malheurs du positivisme. Philosophie des sciences et politique scientifique chez Auguste Comte et ses premiers disciples (1820–1900)," thesis for the *doctorat d'Etat*, Université de Paris-Sorbonne (1993), 2:369–95, esp. 2:382–92.

179. Petit, "Heurs et malheurs du positivisme," 2:383.

180. The list first appeared in October 1851 as an eight-page pamphlet *Bibliothèque du prolétaire au dix-neuvième siècle*. It was reissued in the following year as an unpaginated appendix to the preface of Auguste Comte, *Catéchisme positiviste* (Paris, 1852) and in 1854 as an appendix, also unpaginated, to the fourth volume of Comte, *Système de politique positive*.

181. In the version of the list that appeared in the *Système de politique positive*, Comte renamed this last category "Synthèse."

182. Petit, "Heurs et malheurs du positivisme," 2:386–92.

183. Comte, *Cours de philosophie positive*, 6:xxxvi.

Chapter 2 • Voices on the Periphery

1. Edmond Goblot, *La Barrière et le niveau. Etude sociologique sur la bourgeoisie française moderne* (Paris, 1925).

2. Gustave Rouland, "Des revues de province, et de la Revue de Rouen," *RR* 3 [incorrectly printed 5 on the title page] (1835): 7.

3. Charles Louandre, "De l'association littéraire et scientifique en France. II. Les sociétés savantes et littéraires de la province," *RDM* n.s., 16 (1846): 792–818.

4. So too the unsigned dismissal of the provincial academies as "stricken with lethargy" that appeared in the *Revue de Rouen* in the following year. See "Chronique," *RR* 4 (1836): 139.

5. Nicolas Bignon, "Rapport," *Précis analytique des travaux de l'Académie royale des sciences, belles-lettres et arts de Rouen, pendant l'année 1824* (Rouen, 1825), 123–35. Bignon, a priest who had taught in the *école centrale* and then the *lycée* in Rouen, was speaking as secretary of the Rouen Academy.

6. Ibid., 124.

7. François-Gabriel Bertrand, "Discours d'ouverture" [26 Nov. 1840], *Mém. Acad. Caen* (1840): xxvi–xxvii.

8. Ibid., xxxiv.

9. The figures in table 2 for other years show the continued pattern of expansion through to 1914, with growth in the provincial faculties slightly exceeding that of Paris, especially after 1870.

10. The practice had virtually ceased by the end of the July Monarchy. By then, only two of the sixty chairs in the faculties of science were occupied by professors who also taught in a *collège royal*.

11. On Eudes-Deslongchamps, see A. de Saint-Germain, "Recherches sur l'histoire de la faculté des sciences de Caen de 1809 à 1850," *Mém. Acad. Caen* (1891): 42–104 (97–99).

12. On Lamouroux, see J.-P. Lamouroux, "Notice biographique sur J. V. F. Lamouroux," in

Jean-Vincent-Félix Lamouroux, *Résumé d'un cours élémentaire de géographie physique*, 2nd ed. (Paris, 1829), vii–xxxii, and Jacques-Armand Eudes-Deslongchamps, "Notice sur la vie et les ouvrages de M. J. F. V. Lamouroux," *Mém. Acad. Caen* (1829): 357–83.

13. The origins and early history of the Caen museums are described in Eugène Eudes-Deslongchamps, "But et plan de l'Annuaire," *Annuaire du Musée d'histoire naturelle de Caen* 1 (1880): v–xvi. On the travels and Norman roots of Dumont d'Urville, see Jacques Guillon, *Dumont d'Urville 1790–1842. La Vénus de Milo. Les épaves de La Pérouse. L'Antarctique et la Terre Adéline* (Paris, 1986). On Roussel, see Grégoire Jacques Lange, *Notice historique sur H.-F.-A. Deroussel, ancien professeur royal de médecine à l'université de Caen, professeur d'histoire naturelle de l'Académie impériale . . .* (Caen, 1812).

14. Flaubert, *Madame Bovary. Moeurs de province* [1857], ed. Pierre-Marie de Biasi (Paris: Imprimerie nationale, 1994), part 3, chap. 11, 526.

15. Eudes-Deslongchamps to Geoffroy Saint-Hilaire, 13 July 1833, in Eudes-Deslongchamps's file in the Archives of the Académie des Sciences (I2763c).

16. Eudes-Deslongchamps too held a doctorate but, in a manner common at the time, he took the degree (as well as the necessary preliminary of the *licence*) after his appointment, in 1826. See Albert Maire, *Catalogue des thèses de sciences soutenues en France de 1810 à 1890 inclusivement* (Paris, 1892), 132, for the titles of Eudes-Deslongchamps's two theses that were required for the doctorate—one in geology and palaeontology, the other in zoology.

17. For more information on the Institut des provinces, see later in the chapter.

18. On Baudrimont's evolving attitude to the institutions of Bordeaux, see the obituary notices of him by Léopold Micé: "Discours d'ouverture de la séance publique du 19 mai 1881," *Actes Acad. Bordeaux* 3rd ser., 42 (1880): 730–66, and "Notes complémentaires," ibid. 3rd ser., 44 and 45 (1883): 557–624.

19. Pasteur, "Mémoire sur la fermentation appelée lactée" [presented 3 Aug. 1857], *Mém. Soc. Lille* 2nd ser., 5 (1858) : 13–26.

20. *CR* 45 (1857): 913–16, and *ACP*, 3rd ser., 52 (1858): 404–18.

21. See the bibliographical note for information on the secondary literature on nineteenth-century *sociétés savantes*.

22. Eugène Lefèvre-Pontalis, *Bibliographie des sociétés savantes de la France* (Paris, 1887), 727. Of these twenty-six societies, eleven were concerned, at least partially, with science, medicine, or technology.

23. Ibid., 10–12, 34–38, and 119–27. On the growth in Rouen, to a total of some thirty societies by the beginning of the twentieth century, see Jean-Pierre Chaline, *Les Bourgeois de Rouen. Une élite urbaine au XIXe siècle* (Paris, 1982), 231–47.

24. *Annuaire des sociétés savantes de la France et de l'étranger, publié sous les auspices du Ministère de l'instruction publique. Première année. 1846* [cited hereafter as *Annuaire 1846*] (Paris, 1846), 1013–14 and 1016. Though launched as an annual publication, only the edition of the *Annuaire* for 1846 ever appeared.

25. Ibid., 465–71, 729–32, and 920–25.

26. Ibid., 430.

27. *Mém. Acad. Caen* (1855) : 527.

28. *Annuaire 1846*, 481.

29. Ibid., 567 and 575. In the mid-1840s, the Académie des sciences, belles-lettres et arts de Bordeaux received 2,000 francs from the General Council of the Gironde and 1,500 francs from the municipality. The Académie royale des sciences, inscriptions et belles-lettres de Toulouse received its 3,000 francs from the municipality.

30. *Annuaire 1846*, 566.

31. Ibid., 545–72. In fact, the growth of societies in Toulouse, as traced in Caroline Barrera, *Les Sociétés savantes de Toulouse au XIXe siècle (1797–1865)* (Paris, 2003), seems to have been less marked than in most other large towns.

32. *Annuaire 1846*, 1007–10, for a convenient listing by region.

33. Lefèvre-Pontalis, *Bibliographie des sociétés savantes*, v–vi.

34. Ibid.

35. Robert de Lasteyrie et al., *Bibliographie générale des travaux historiques et archéologiques publiés par les sociétés savantes de la France*, 6 vols. (Paris, 1888–1918).

36. Gustave Regelsperger, "Le XXXIᵉ Congrès des sociétés savantes de géographie," *Bulletin de la Société de géographie de Rochefort (agriculture, lettres, sciences et arts)* (1913): 158–9. For a helpful brief account with a list of societies, see also Philippe Pinchemel, "Les sociétés savantes et la géographie," in *Les Sociétés savantes. Leur histoire. Actes du 100ᵉ congrès national des sociétés savantes. Paris 1975. Section d'histoire moderne et contemporaine et Commission d'histoire des sciences et des techniques* (Paris, 1976), 69–78.

37. Cuvier, "Prospectus," in Cuvier et al., *Dictionnaire des sciences naturelles, dans lequel on traite méthodiquement des différens êtres de la nature*, 3 vols. and 2 vols. of plates (Paris, 1804–6), 1:v– xx (1804). The *Nouveau dictionnaire d'histoire naturelle appliquée aux arts, principalement à l'agriculture et à l'économie rurale et domestique, par une société de naturalistes et d'agriculteurs*, 24 vols. (Paris, 1803–4) was marked by Virey's more utilitarian priorities. Yet it too reflected the growth of interest in natural history, not least among women ("those radiant flowers of the human species"). Passages in Virey's "Discours préliminaire," vol. 1, esp. v–viii and xliii convey the tone. On the initiatives of Cuvier and Virey and Bory de Saint-Vincent's rival *Dictionnaire classique d'histoire naturelle* (1822–31); see Pietro Corsi, *The Age of Lamarck: Evolutionary Theories in France 1790–1830,* trans. Jonathan Mandelbaum (Berkeley, CA, 1988), esp. 218–22.

38. Virey similarly advocated natural history as a pursuit that allowed the soul, "wearied by the tribulations of life," to find rest in "the calm bosom of nature." See Virey, "Discours préliminaire," 1:viii.

39. Cuvier, "Prospectus," 1:v. The same statement (with only minor typographical changes) appeared in 1816, when the uncompleted project for a *Dictionnaire des sciences naturelles* of 1804–6 (see above, note 37) was relaunched after collapsing ten years earlier.

40. As in *La Peau de chagrin*, where Cuvier is lauded not only as "our immortal naturalist" but also as "the greatest poet of our century," a man whose sublime understanding has given life to a geological past laid down over millions of years. See Balzac, *La Peau de chagrin* [1831], Pléiade edition, Honoré de Balzac, *La Comédie humaine*, ed. Pierre-Georges Castex et al. (Paris, 1976–81), 10:74–76.

41. On this episode, see Gabriel Dardaud (annotations by Olivier Lebleu), *Une Girafe pour le roi. Ou l'histoire de la première girafe en France* (Paris, 1985), and Olivier Lagueux, "Geoffroy's Giraffe. The Hagiography of a Charismatic Animal," *Journal of the History of Biology* 36 (2003): 225–47.

42. Correspondence between Cuvier and the prefect of the Bouches du Rhône (AD Bouches du Rhône 4.T.53) reflects the uncertainty about the best method of getting the giraffe to Paris. The prefect favored starting the journey by boat on the Rhône but was evidently overruled.

43. [Balzac], "Etudes de philosophie morale des habitans du Jardin des plantes," *La Silhouette. Journal des caricatures, beaux-arts, dessins, mœurs, théâtres, etc.*, 2, 13ᵉ livraison (17 June 1830): 101–2. Although the article is simply signed "B," there seems no reason to doubt its authorship, not least because Balzac was a frequent contributor to the magazine.

44. *La France littéraire*, 2nd ser., 5, no. 1; Pierre Flourens, "Eloge historique de Benjamin Delessert," *MAS* 2nd ser., 22 (1850): cxix–clxiv; and Charles Dupin, *Travaux et bienfaits de M. le baron Benjamin Delessert* (Paris, 1847).

45. *Icones selectae plantarum quas in systemate universali . . . descripsit Aug. Pyr. De Candolle*, ed. Benjamin Delessert, 5 vols. (Paris, 1820–46).

46. Descriptions of Delessert's collections include Jean-Charles Chenu, *Sur le musée conchyliologique de M. le baron Benjamin Delessert* (Paris, 1844) and Antoine Lasègue, *Musée botanique de M. Benjamin Delessert. Notices sur les collections de plantes et la bibliothèque qui le composent* (Paris, 1845). Chenu and Lasègue were the last in a succession of curators whom Delessert had employed since 1817. Some of the specimens in his collection came from the herbarium of Jean-Jacques Rousseau, who had introduced him to botany.

47. Flourens, "Eloge historique de Delessert," cxxxi, and, for a modern study, Thierry Hoquet, "La bibliothèque botanique de Benjamin Delessert," *Bulletin du bibliophile* 1 (2002), 100–141.

48. Albert Gaudry, "Notice sur les travaux scientifiques de d'Archiac," *BSGF* 3rd ser., 2 (1874): 230–44.

49. Etienne Jourdan et al., "Notice sur la vie et les travaux de A.-F. Marion," *Annales de la faculté des sciences de Marseille* 11 (1901): 1–26; Albert Vayssière, "Notice biographique sur A.-F. Marion," *AMHNM* 6, Section de zoologie (1901): 7–9; G. Petit, "A. F. Marion (1846–1900)," *Bulletin de la Société d'histoire naturelle de Marseille* 1 (1941): 5–12.

50. *Notice sur les travaux scientifiques du Cte G. de Saporta* (Paris, 1875).

51. A. Maud'heux (fils) and Etienne Lahache, "Notice biographique sur M. le docteur Mougeot père," *Annales de la Société d'émulation du département des Vosges* 10, no. 1 (1858): 398–417.

52. On Dufour's life as a savant and physician, see his *Souvenirs d'un savant français. A travers un siècle 1780–1865. Science et histoire* (Paris, 1888) and the correspondence he maintained with other naturalists, notably Bory de Saint-Vincent and Henri Milne Edwards, rich traces of which survive in the Archives of the Académie des sciences and the Bibliothèque centrale of the Muséum d'histoire naturelle. A valuable modern study is Pascal Duris and Elvire Diaz, *Petite histoire naturelle de la première moitié du XIXe siècle. Léon Dufour, correspondant de l'Institut (1780–1865)* (Talence, 1987).

53. Claudine Cohen and Jean-Jacques Hublin, *Boucher de Perthes 1788–1868. Les origines romantiques de la préhistoire* (Paris, 1989).

54. Louis Joubin, "Félix Dujardin," *Archives de parasitologie* 4 (1901): 5–57, esp. 8–31.

55. On the financial difficulties and procrastination that delayed the project for the rehousing of the faculties of science, letters, and law and the secondary school of medicine, along with the city's collections of art and natural history, see Jean-Yves Veillard, *Rennes au XIXe siècle. Architectes, urbanisme et architecture* (Rennes, 1978), 312–18.

56. Dujardin, *Histoire naturelle des helminthes ou vers intestinaux* (Paris, 1845).

57. Jean Dhombres et al., *Un Musée dans sa ville. Le Muséum d'histoire naturelle. Sciences, industries et société à Nantes et dans sa région. XVIIIe–XXe siècles* (Nantes, 1990), 201–21 and 442–43. The museum was not formally established until 1810, and it was only then that Dubuisson began regular lecturing. The story of municipal patronage in Nantes had a happy ending in the installation, in 1875, of the now much enlarged collection in a handsome former school of surgery, where it remains; ibid., 244–48 and 356–60.

58. Robert Bourgat, "Perpignan Museum: From Natural History Cabinet to Municipal Institution," *Journal of the History of Collections* 7 (1995): 73–79.

59. The printed announcement of the lectures, which were given on Mondays, Wednesdays, and Fridays at 6.30 a.m., is in AD Bouches du Rhône 1.T.166.

60. The attendances are recorded in the annual reports on the botanical garden, in AD Bouches du Rhône 4.T.52.

61. Albert Vayssière, "Historique du Musée d'histoire naturelle de Marseille," *AMHNM* 6, Section de zoologie (1901): 14–43, and L. Laurent, "Les grands musées d'histoire naturelle de province. Le Muséum de Marseille," *La Terre et la vie* (March 1933): 1–15.

62. Correspondence and other materials in AD Bouches du Rhône 4.T.52 demonstrate the enthusiasm of both municipal and departmental authorities for public lectures on science in Marseille, including a course on physics and chemistry inaugurated in 1829. Although the support for science continued through the century, priority was increasingly given to vocational teaching. An illustration of this is the municipal council's contribution of 3,000 francs in 1894 toward the creation of a chair of industrial physics, essentially applied electricity, in the faculty of science (AD Bouches du Rhône 1.T.166).

63. AD Nord 1.T.19.1–4 and 1.T.277.12–13.

64. Louis-Benoît Guersent, "Notice biographique sur M. Varin," *Précis analytique des travaux de l'Académie des sciences, belles-lettres et arts de Rouen pendant l'année 1808* (1809): 67–80. On these lectures and their revival after a brief suspension in 1789, see B********* [Jules-Etienne Bouteiller], *Le Jardin des plantes de Rouen* (Rouen, 1856), esp. 32–39.

65. Jean-Baptiste Vitalis, *Manuel du teinturier, sur fil et sur coton filé* (Rouen, 1810) and *Cours élémentaire de teinture sur laine, soie, lin, chanvre et coton, et sur l'art d'imprimer les toiles* (Paris, 1823). Translations of the *Cours* into both German and Spanish bear witness to the international recognition of its importance.

66. Pierre Bizeau, "De l'Académie de Rouen à la cure de Saint-Eustache de Paris. Le chimiste Jean-Baptiste Vitalis (1759–1832)," in *La Normandie et Paris. Actes du XXIe congrès des sociétés historiques et archéologiques de Normandie. Paris, 3–6 septembre 1986. Cahiers Léopold Delisle* (Paris, 1986–87), vols. 35–36, 183–204.

67. Jules de la Quérière, *Notice sur la Société libre d'émulation, du commerce et de l'industrie de la Seine-Inférieure* (Rouen, 1884), 26–32, and the unsigned *Notice sur les cours publics & gratuits professés sous le patronage de la Société libre d'émulation, du commerce et de l'industrie de la Seine-Inférieure* (Rouen, 1889). The account in Théodore Licquet, *Rouen. Son histoire, ses monuments et ses environs. Guide nécessaire,* 8th ed. (Rouen, 1871), 160–64, conveys the range and seriousness of the lectures available in Rouen by the early 1870s.

68. *Notice sur les travaux de zoologie et de physiologie de M. F.-Archimède Pouchet* (Rouen, 1861), 6.

69. Ibid., 4.

70. Ibid., 6.

71. On Lecoq's life and work, Pierre Pénicaud, *Henri Lecoq. Les fortunes d'un naturaliste à Clermont-Ferrand* (Clermont-Ferrand, 2003), stands as an up-to-date comprehensive source. On the commonly told story about the letter intended for another Henri Lecoq, see Pénicaud, op. cit., 37–39, and Jean-Marc Drouin and Robert Fox, "Corolles et crinolines. Le mélange des genres dans l'œuvre de Henri Lecoq (1802–1871)," *Revue de synthèse* 4th ser., 120 (1999): 585. Details of Lecoq's career in the faculty at Clermont-Ferrand can be found in his unusually full personal file in AN F[17] 21116.

72. Promotion to the rank of a professor of the first class in 1869 brought his core salary to 5,500 francs, with an additional payment (*traitement éventuel*) of about 500 francs for examining. Until then, he had earned the basic professorial salary of 4,000 francs, plus a *traitement éventuel* of between 300 and 500 francs. It is easy to see why Lecoq, who was said to have a private income

of 30,000 francs per annum between 1858 and 1860, frequently stated that he was tempted to abandon his chair.

73. Dhombres, *Musée dans sa ville*, 218–21.

74. *Lettre sur la création d'une faculté des sciences, adressée à Monsieur le Maire de la ville de Rouen par MM. Dainez, Girardin, Pouchet et Preissen* [dated Rouen, 15 July 1847]. The authors of this eight-page printed letter would all have benefited financially if they had been appointed to chairs in the faculty whose creation they advocated. As lecturers attached to what had now been formalized as municipal school, they each earned 2,400 francs a year (ibid., 3n); as professors in a faculty, their incomes would have been roughly twice this amount.

75. Maire, *Catalogue des thèses*, 139.

76. Again, Lecoq's personal file in AN F^{17} 21116 is a rich source.

77. The phrase appears in the rector's confidential annual report, dated 9 June 1863, to the Ministry of Public Instruction, AN F^{17} 21116.

78. Files of the academy of Rennes (AD Ille-et-Vilaine 11.T.253 and 10.T.578A) and his personal file in AN F^{17} 20654 convey a vivid impression of Dujardin's desperation and the irritation that his requests for leave of absence caused. Between 1846 and 1848, in return for freedom from teaching duties, Dujardin accepted a reduction of his professorial salary by half (from 4,000 francs to 2,000 francs), the other half going to his assistant and keeper of the collections of both the faculty and the city of Rennes, Hyacinthe Pontallié, who replaced him. Likewise, in letters of 2 December 1844 and 12 April 1852 to the President of the Académie des sciences (AAS, in Dujardin's personal file) Dujardin expressed his determination to establish himself in Paris. In both letters, he stated that, in the event of his being elected to the Académie, he would give up all posts and other obligations that would prevent him from residing in the capital. In fact, he was not elected until 1859, less than a year before his death, and then only as a corresponding member.

79. Dufour to Cuvier, 10 May 1830, in the personal file "Léon Dufour," AAS.

80. Eudes-Deslongchamps to Flourens, 12 December 1849, in the personal file "J. A. Eudes-Deslongchamps," AAS. The letter to Boussingault is in the same file.

81. The figures are in the annual reports from the rector of the academy of Clermont-Ferrand to the Ministry of Public Instruction, AN F^{17} 21116.

82. Lecoq, *La Vie des fleurs* (Paris, 1861), 326; also the revised, illustrated edition of the work, published as Lecoq, *Le Monde des fleurs. Botanique pittoresque* (Paris, 1870), 455. The chapter, entitled "De la toilette et de la coquetterie des végétaux," developed a labored analogy between flowers and women, who shared the "desire to please" and the "certainty of doing so."

83. A typical illustration of his style is the preamble to his will, written in 1864 in his remote country chalet near Menat, thirty miles or so to the north of Clermont-Ferrand:

> While the nightingale and warbler sing their songs of love amid the budding foliage, and the morning breeze bears the fragrance of spring, I here set down, in tranquility and in full possession of my faculties, the desires and hopes that I ask to be regarded as my last wishes.

The passage is cited in Antoine Vernière, *Henri Lecoq. Notice biographique* (Clermont-Ferrand, 1907), 17, and commented on in Pénicauld, *Fortunes d'un naturaliste*, 160–63.

84. The comments are from the rector's confidential reports on Lecoq, 9 June 1863 and 14 June 1864, AN F^{17} 21116.

85. Letters of 17 July 1854 and 18 November 1856 to Brongniart (BCMHN MS 2358) show Lecoq asking Brongniart to propose him as a corresponding member of the botany section. Brongniart's notes in the same file indicate that he considered Lecoq's contributions to hybridity and

other aspects of agricultural botany as lacking the precision of the best recent work. The coolness of attitudes toward Lecoq within the Académie des sciences (especially Brongniart's) is discussed in Pénicaud, *Fortunes d'un naturaliste,* 116–17.

86. Antoine Vernière, "Les voyageurs et les naturalistes dans l'Auvergne et dans le Velay," *Revue d'Auvergne* 17 (1900): 270–90.

87. For a fulsome reference to the first volume of Lecoq's *Etudes sur la géographie botanique de l'Europe,* 9 vols. (Paris, 1854–58), see the "Historical Sketch of the Recent Progress of Opinion on the Origin of Species," in Darwin, *On the Origin of Species by Means of Natural Selection,* 4th ed. (London, 1866), xx.

88. Maud'heux and Lahache, "Mougeot," 404.

89. Dufour, *Souvenirs,* 313–42.

90. See, for example, "Des églises gothiques," chap. 8 of book 1 ("Beaux-arts") of the third part of Chateaubriand, *Génie du Christianisme, ou beautés de la religion chrétienne* (Paris, 1802), 3:23–28.

91. Anon., "Normandy Architecture of the Middle Ages," *Quarterly Review* 25 (1821): 147.

92. Licquet's indignation boiled over in the preface to his French translation of Thomas Frognall Dibdin's *Bibliographical, Antiquarian and Picturesque Tour in France and Germany* (London, 1821). See Dibdin, *Voyage bibliographique, archéologique et pittoresque en France,* trans. Théodore Licquet (Paris, 1825), 1:vii–x, where Licquet expressed bitter resentment at the disdain of British visitors for the work of French antiquarians, which they nevertheless plagiarized.

93. Michel Guibert, "Arcisse de Caumont et Charles de Gerville," in Vincent Juhel, ed., *Arcisse de Caumont (1801–1873). Erudit normand et fondateur de l'archéologie française* (Caen, 2004), 66–79.

94. The point is explored in François Guillet, "Arcisse de Caumont, un archéologue provincial," in Juhel, *Arcisse de Caumont,* 81–93.

95. Emile Viellard and Gustave Dollfus, "Etude géologique des terrains crétacés, et tertiaire du Cotentin," *BSLN* 2nd ser., 9 (1874–75): 5–181, esp. 9–12.

96. Older biographical studies of Caumont tend to be lengthy and generous. Among more modern sources, Marcel Baudot, "Trente ans de coordination des sociétiés savantes (1831–1861)," in *Colloque interdisciplinaire sur les sociétiés savantes. Actes du 100e congrès national des sociétiés savantes (Paris, 1975)* (Paris, 1976), 7–40, is particularly helpful on Caumont's relations with successive Ministers of Public Instruction, while the collective volume Juhel, *Arcisse de Caumont,* is now the standard source on most aspects of Caumont's life.

97. Hufton, *Bayeux in the Late Eighteenth Century* (Oxford, 1967), 208.

98. For the granting of *lettres de noblesse* to François Caumont, on 16 December 1815, see vicomte Albert Révérend, *Les Familles titrées et anoblies au XIXe siècle. Titres, anoblissements et pairies de la Restauration 1814–1830* (Paris, 1901–6), 2:52–53. Curiously, Révérend gives the date of Arcisse de Caumont's birth as 26 August 1802, rather than the usual 26 August 1801.

99. On Caumont's debt to Hervieu, see Guy Héraud, "La jeunesse d'Arcisse de Caumont: l'écolier, l'étudiant, le provincial à Paris," in Juhel, *Arcisse de Caumont,* 44–46.

100. Laspougeas, "La Normandie au temps d'Arcisse de Caumont," ibid., 14–15.

101. On Lamouroux, see note 12 above.

102. Claude Pareyn, "Arcisse de Caumont. Acteur du renouveau des sciences naturelles," in Juhel, *Arcisse de Caumont,* 49–58, esp. 51–52.

103. Desnoyers's work as a geologist is evoked briefly in Albert Gaudry's presidential address to the annual general meeting of the Société géologique de France, in *BSGF* 3rd ser., 16 (1887–88): 455–56. Longer appreciations of his antiquarian work appear in *Bibliothèque de l'Ecole des chartes* 48 (1887): 612–16.

104. Jules Morière wrote substantial obituaries of both men: on Lenormand, "Notice biogra-

phique sur Sébastien-René Lenormand," in *Mém. Acad. Caen* (1874): 49–78, and on Brébisson, "Notice biographique sur Alphonse de Brébisson," *BSLN* 2nd ser., 8 (1873–4): 3–27.

105. Caumont, "Mémoire géologique sur quelques terrains de la Normandie occidentale," *MSLC* (1825): 447–597, read to the society on 8 November 1824 and 3 January and 7 February 1825. A separately published copy of the "Mémoire" (Paris, 1825) in the Bibliothèque centrale of the Muséum d'histoire naturelle is inscribed on the flyleaf with a warm dedication to Cuvier.

106. *Carte géologique du département du Calvados. Dressée en 1825 par M. de Caumont, dédiée à Mr de Gerville, membre de plusieurs sociétés savantes. Lithographiée par M. C. L. Maufras, membre de plusieurs sociétés savantes,* a single printed, hand-colored sheet on the scale 1:200,000. The map was reissued to accompany Arcisse de Caumont, *Essai sur le topographie géognostique du département du Calvados,* 2nd ed. (Paris, 1867).

107. His *Carte géologique du département de la Manche,* prepared with the aid of a grant of 800 francs from the General Council of the Calvados, appeared in 1833.

108. Caumont, *Cours d'antiquités monumentales professé à Caen [en 1830],* 6 vols. and one vol. of plates (Paris, Caen, and Rouen, 1830–41). The "Avertissement" to the first volume (v–ix) explains how the lectures of 1830 came to be given.

109. A brief but authoritative history of the society, by Jeanne Grall, is reproduced as an introduction to the typewritten handlist "Répertoire numérique détaillé de la sous-série 83F. Fonds de la Société des antiquaires de Normandie," 1–10, in the reading room of the AD Calvados, Caen.

110. The *Revue normande rédigée par une société de savants et de littérateurs de Rouen, de Caen et des principales villes de la Normandie, sous la direction de M. de Caumont* was launched in September 1831, with a three-page list of collaborators, the most important of whom were Gerville in Valognes, Le Prévost in Rouen, and Pierre-Jacques Féret, archaeologist, polymath, and head of the public library in Dieppe since its foundation in 1827.

111. I say "supposed" here since Caumont's rather bleak view invites the kind of qualification that is offered in Laspougeas, "La Normandie au temps d'Arcisse de Caumont," 14. As Laspougeas observes, for example, Normandy had a higher level of literacy than any other region of France.

112. The *Annuaire,* also commonly known as the *Annuaire normand,* offered a compendium of articles on the economy and culture of Normandy without interruption from its foundation in 1835 until the Second World War.

113. Bernard Huchet, "Les origines de la Société française d'archéologie," in Juhel, *Arcisse de Caumont,* 165–90.

114. [Anon.] "Un congrès fini," *Le Charivari,* 11, 16 Oct. 1842. The satirized meal took place in Mulhouse on an excursion from Strasbourg.

115. Dominique Branche, "Compte-rendu de l'excursion que le congrès a faite à Vienne (Isère), le 7 septembre 1841," in *CSF, 9e session. Lyon 1841,* 2:417–27.

116. *CSF, 12e session. Nîmes 1844,* 84–86.

117. On Caumont's admiration for the German congresses, see Richelet, *Notice sur M. de Caumont,* 13 and *CSF, 1ère session. Caen 1833,* vii–viii.

118. Jack Morrell and Arnold W. Thackray, *Gentlemen of Science: Early Years of the British Association for the Advancement of Science* (Oxford, 1981), 132–33.

119. Attendances at the meetings of Versammlung der deutscher Naturforscher und Aertze in the 1830s, as recorded in the annual reports, varied considerably. But a list of about 350 subscribers was normal, the number of those attending being certainly rather less than this.

120. See *CSF, 1ère session. Caen 1833,* 7. An authoritative study of Jullien, with translated extracts from his writings, is Robert R. Palmer, *From Jacobin to Liberal: Marc-Antoine Jullien, 1775–1848* (Princeton, NJ, 1993).

121. See de Vaudoré's lengthy address at the opening session of the Congress, in *CSF, 2ᵉ session. Poitiers 1834*, 16–17.

122. Chatelain, *Arrêt de mort du Congrès scientifique de France. Aux membres du congrès de Metz* (Paris, 1837).

123. *CSF, 5ᵉ session. Metz 1837*, 257–59.

124. *CSF, 6ᵉ session. Clermont-Ferrand 1838*, xl–xliv.

125. "Mes souvenirs des deux congrès scientifiques de France et d'Italie," in Jullien, *Petit code philosophique et moral. Exposé sommaire de douze lois générales . . . et souvenirs des deux congrès scientifiques de France et d'Italie* (Paris, 1844), 14–15.

126. Caumont, "Rapport sur les travaux de la Société linnéenne du Calvados, depuis son origine jusqu'au 24 mai 1824," *MSLC* (1824): lx.

127. Caumont, *Cours d'antiquités monumentales* (1830–41), 1:vii–viii (1830).

128. *CSF, 9ᵉ session. Lyon 1841*, 1:244.

129. *CSF, 6ᵉ session. Clermont-Ferrand 1838*, 19–29.

130. Descended from an old aristocratic family of the Auvergne, de Parieu was the father of Félix Esquirou de Parieu, who as Minister of Public Instruction carried through the Falloux law on the freedom of secondary education in 1850.

131. Chatelain, *Arrêt de mort du Congrès scientifique*, 5 and 8–9.

132. Paul Léon, "Les principes de la conservation des monuments historiques. Evolution des doctrines," in *Congrès archéologique de France. XCVIIe session tenue à Paris en 1934* (Paris, 1935–36), 1:24–28, and Arlette Auduc, "Arcisse de Caumont et le Service des monuments historiques," in Juhel, *Arcisse de Caumont*, 181–90.

133. *CSF, 1ʳᵉ session. Caen 1833*, 6–7 and *CSF, 3ᵉ session. Douai 1835*, 13.

134. *CSF, 9ᵉ session. Lyon 1841*, 1:liii.

135. AD Bouches du Rhône 4.T.57 and *CSF, 14ᵉ session. Marseille 1846*, 1:5–40 (Roux's opening address to the congress) and 95–96.

136. For an account of these associations, see "congrès régionaux," in the *Annuaire de l'Institut des Provinces et des congrès scientifiques* for 1846, 112–47. Plans for the formation of an Association des rives de la Loire are referred to in *Annuaire normand*, 14 (1848), 571. Those for a similar association for Provence seem to have come to nothing.

137. Salvandy to Caumont, 28 August 1846, in AN F[17] 3026.

138. Caumont to Salvandy, 6 July 1847, in AN F[17] 3026.

139. Charlotte Robert, "L'Institut des provinces de France, le détonateur d'une lutte pour la décentralisation intellectuelle," in Juhel, *Arcisse de Caumont*, 191–95.

140. A first volume of the *AIP* appeared in 1846, and annual volumes were published thereafter from 1850 to 1880. The *Mémoires* included one volume devoted to science (1859), and two to historical studies (1845–69).

141. See the "Notice sur l'Institut des provinces de France" (unsigned but almost certainly by Caumont), *AIP* (1846): 5.

142. Copies of Salvandy's instruction of 4 January 1847 and associated correspondence survive in many departmental archives.

143. For Dupanloup's address, see *CSF, 18ᵉ session. Orléans 1851*, 1:32–34. Rivet's comments are in *CSF, 21ᵉ session. Dijon 1854*, 24.

144. "Loi sur l'instruction publique," art. 41, in Léon Aucoc, *L'Institut de France. Lois, statuts et règlements concernant les anciennes académies et l'Institut, de 1635 à 1898. Tableau des fondations* (Paris, 1889), 65.

145. The Minister of Justice made the point in a letter to Fortoul, 3 March 1852, in AN F[17]

3021. Correspondence in the same file and in AN F^{17} 3026 reveals the Ministry of Public Instruction's undimmed determination to impose a change of title on the Institut des provinces as late as the 1870s. Legal obstacles, however, repeatedly prevented the move, leaving the Institut to die a natural death.

146. *Bulletin bibliographique des sociétés savantes des départements*, no. 12 (Dec. 1853): 321–22. Although the title page bears the date December 1853, Du Chatellier's note is dated "Versailles avril 1854."

147. The proceedings of the congresses were reported in the *AIP* for the year concerned.

148. On the ministerial intervention, of which no mention appears in the volume for the Toulouse congress, see Jean-Michel Renault, "Notice biographique sur M. de Caumont," *Annuaire des cinq départements de l'ancienne Normandie*, 40 (1874): 491–92.

149. Caumont made the contrast in these terms in his introduction to *AIP* 13 (1861): vi.

150. *CSF, 36ᵉ session. Chartres 1869*: 4–5.

151. A letter from the Division de la Sûreté publique of the Ministry of the Interior, 4 June 1859 (AD Ille-et-Vilaine 4.T.57) expressed the minister's determination to act against the association. For a brief account of the furiously contested closure, which put an end to the fifteen annual congresses that the association had organized since its foundation at Vannes in 1843, see "Avant-propos," in *Association bretonne. Seizième session, tenue à Quimper en 1873. Comptes-rendus et procès-verbaux publiés par les soins de la rédaction* (Saint-Brieuc, 1874), i–iii. The association was reconstituted in 1873.

152. The firmness of Druilhet-Lafargue's resolve is evident in the drafts of letters and correspondence between him and the Ministry of Public Instruction in AN F^{17} 3026.

153. *CSF, 44ᵉ session. Nice 1878*, 2 vols. (Nice, 1878–79).

154. Docteur Guétrot, *Le Quarantenaire de la Société mycologique de France (1884–1924)* (Paris, 1934), 1–132.

155. Jean-Loup d'Hondt, "Histoire de la Société zoologique de France. Son évolution et son rôle dans le développement de la zoologie," *Revue française d'aquarologie. Herpétologie* 16, no. 3 (1989): 65–100, and Robert Fox, "La Société zoologique de France. Ses origines et ses premières années," *BSZF* 101 (1976): 149–67; translated as "The Early History of the Société zoologique de France," item IV in Fox, *The Culture of Science in France, 1700–1900* (Aldershot, 1992).

156. Correspondence in AN F^{17} 2912 throws light on Bouvier's earlier career, first as a medical student and, from 1864 to 1866, as a volunteer naturalist with the Mexican expedition. The quality of the specimens, especially of birds, that he sent back to the Muséum d'histoire naturelle was praised by Firmin Bocourt, then a senior naturalist with the expedition.

157. The figures are calculated from the records of meetings in the MS volumes of "Procès-verbaux" in the archives of the Société entomologique.

158. *BSGF*, 4th ser., 10 (1910): 336.

159. Mary Jo Nye, *Science in the Provinces: Scientific Communities and Provincial Leadership in France, 1860–1930* (Los Angeles, 1986).

Chapter 3 · *Science, Bureaucracy, and the Empire*

1. Renan was suspended following the excited applause of students at his inaugural lecture at the Collège de France on 26 February 1862. He was removed from the chair on 11 June 1864, by which time his controversial *Vie de Jésus* had been published.

2. The text of the law of 15 March 1850 and associated documents are in Beauchamp, *Lois et règlements*, 2:85–122. A classic study of the law's impact on the University is Paul Gerbod, *La Condition universitaire en France au XIXe siècle* (Paris, 1965), 237–82. Although the law was the

creation of the comte de Falloux during his time as Minister of Public Instruction between December 1848 and October 1849, it was passed into legislation under his successor at the ministry, Félix Esquirou de Parieu (see chap. 2).

3. For the reorganization into eighty-six academies and the composition of the departmental academic councils, see chap. 2 of the Falloux law, in Beauchamp, *Lois et règlements*, 2:87–89.

4. Ibid., 2:108–20: "The law demonstrates that its aim is to make the wisest and most enlightened elements of society itself once again responsible for the ultimate control of teaching" (112).

5. Ibid., 2:209–11. The University was now to be represented on the council by eight senior academics with the rank of "general inspectors." Although other places were reserved for five members of the Institut, no provision was made for the appointment of representatives of the rank and file of the professoriate.

6. Jourdain, *Rapport sur l'organisation et les progrès de l'instruction publique* (Paris, 1867), 7–13, and, in a later, less critical vein, Jourdain, *Les Conseils de l'instruction publique* (Paris, 1879), esp. 12–27.

7. The legislation, finally approved on 14 June 1854, is reproduced in Beauchamp, *Lois et règlements*, 2:316–23 (with important explanatory material on 2:324–36). In his "Rapport à l'Empereur," dated 4 April 1854, Fortoul noted the need for a constant improvement of the ministry's institutions of secondary education in the face of the growing popularity of independent schools under ecclesiastical control; see *JGIP* 23 (1854): 261–65.

8. For the legislation, dated 10 April 1852, see Beauchamp, *Lois et règlements*, 2:219–22. Also, with a brief commentary and some passages deleted, in Belhoste, *Les Sciences dans l'enseignement secondaire. Textes officiels réunis et présentés par Bruno Belhoste avec la collaboration de Claudette Balpe et Thierry Laport*, vol. 1, *1789–1914* (Paris, 1995), 251–57.

9. Fortoul showed his suspicion of the *agrégation* in philosophy in his decision, taken immediately after the coup d'état, to cancel the competition for the current year, on the grounds that the University already had a sufficient supply of teachers in the discipline.

10. André Chervel, *Histoire de l'agrégation. Contribution à l'histoire de la culture scolaire* (Paris, 1993), 149–53.

11. The quotations in this and the following paragraph are from Fortoul's preamble to the decree of 10 April 1852. See Beauchamp, *Lois et règlements*, 2:216–19.

12. On this period in the Ecole normale's history, see Paul Dupuy, "Notice historique," in *L'Ecole normale (1810–1883). Notice historique. Liste des élèves par promotions. Travaux littéraires et scientifiques* (Paris, 1884), 59–68, and the identical passage in Dupuy's enlarged text (extending now to 1895) in *Le Centenaire de l'Ecole normale. 1795–1895. Edition du bicentenaire* (Paris, 1994), 241–47.

13. Gerbod, *Condition universitaire*, 309–53 (a chapter "L'université sous le joug").

14. Liard, *L'Enseignement supérieur [en France, 1789–1899 [1893]* (Paris, 1888–94), 2:241.

15. Belhoste, *Sciences dans l'enseignement secondaire*, 258–301, gives the essentials of the the syllabuses for the *baccalauréats* in science and letters, laid down in ministerial documents of 30 August and 5 and 7 September 1852. Dumas's report of 23 July 1852 on the Thenard committee had evidently played a crucial role.

16. The combining into a single *baccalauréat-ès-sciences* of the previously separate *baccalauréats* in mathematics and in the physical and life sciences created additional pressure for more time to be made available for science in the later years of secondary education.

17. Charles Jourdain, *Le Budget de l'instruction publique et des établissements scientifiques et littéraires depuis la fondation de l'Université impériale jusqu'à nos jours* (Paris, 1857), 128–30, and the statistical tables in Jean-Benoît Piobetta, *Le Baccalauréat* (Paris, 1937), 304–8. The income

generated by the faculties of science tripled in a year, leaving the corresponding figure for the faculties of letters substantially diminished.

18. Nicole Hulin-Jung argues persuasively for the role of Saint-Simonian ideas, especially those of Chevalier, in the fashioning of Fortoul's policy. See Hulin-Jung, *L'Organisation de l'enseignement des sciences. La voie ouverte par le Second Empire* (Paris, 1989), 121–28.

19. Laprade, a Catholic legitimist, was dismissed from his chair of French literature in the faculty of letters at Lyon in 1861 for the poem, which satirized the servile, regimented character of "official" literary life under the Second Empire. The chief butt was Charles-Augustin Sainte-Beuve, whose scorn for bohemianism fired his aspiration for the empire to support a network of approved men of letters. See Edmond Biré, *Victor de Laprade. Sa vie et ses oeuvres* (Paris, 1886), 219–51.

20. Dupuy, "Notice historique," in *Ecole normale (1810–1883)*, 66, and *Centenaire de l'Ecole normale*, 245–46. For the legislation, dated 22 February 1855, see Beauchamp, *Lois et règlements*, 2:430.

21. Zwerling, "The Emergence of the Ecole normale supérieure as a Centre of Scientific Education in the Nineteenth Century," in Robert Fox and George Weisz, eds., *The Organization of Science and Technology in France, 1808–1914* (Cambridge and Paris, 1980), 41–50; also Dupuy, "Notice historique," in *Ecole normale (1810–1883)*, 66–68, and *Centenaire de l'Ecole normale*, 246.

22. *Normaliens* who completed their two years in the "division supérieure" but did not take the doctorate had to serve for only two, instead of the normal three, years in a *lycée* before they could be confirmed as *agrégés*. Those who took the doctorate had their status as *agrégés* confirmed immediately.

23. Zwerling, "Emergence of the Ecole normale supérieure," 44–45; also Dupuy, "Notice historique," in *Ecole normale (1810–1883)*, 70, and *Centenaire de l'Ecole normale*, 248. The positions, initially five in number, were created by reallocating the existing appointments at the level of laboratory assistant (*préparateur*), hitherto not restricted to either *agregés* or *normaliens*.

24. The *Annales* was restricted almost exclusively to contributions by present or former *normaliens*, including a high proportion by *agrégés-préparateurs*; see Zwerling, "Emergence of the Ecole normale supérieure," 49–50.

25. Nisard to the Minister of Public Instruction (Gustave Rouland), 5 November 1861, reproduced in *JGIP* 30 (1861): 723.

26. Zwerling, "Emergence of the Ecole normale supérieure," 47–48.

27. These attendances contrast sharply with enrolments of ten (normal for Bordeaux, but above the average for a provincial faculty) for a course in chemistry for candidates for the *licence*.

28. Integration did not proceed easily in the newly founded Marseille faculty. In a letter to the rector of the academy, dated 11 October 1856 (AD Bouches du Rhône 1.T.1159), Dean Charles Person reported the opposition of his fellow professors to instruction in applied subjects. Their main objection seems to have been the appointment of *lycée* teachers (including Claude Valson, who became *professeur adjoint* in 1858 and taught pure and applied mathematics in the faculty for over a decade) to help with the teaching.

29. This said, any evaluation of the success of the lectures, which drew audiences of up to 200, has to be qualified by the evidence of their modest intellectual level and the fickleness of the audiences, referred to in *L'Université de Nancy (1572–1934)*, preface by Charles Adam (Nancy, 1934), 36.

30. See the extract from a speech by Godron in Ernest Bichat, "L'enseignement des sciences appliquées à la faculté des sciences de Nancy," *RIE* 35 (1898): 299–300. Metz's claims to be considered the natural location for a faculty of science in Lorraine continued for some years after the

faculty's establishment in Nancy, as correspondence and related documents in AD Meurthe-et-Moselle T. 265 and T. 2125 show.

31. Denise Wrotnowska, *Louis Pasteur, professeur et doyen de la faculté des sciences de Lille (1854–1857)* (Paris, 1975), passim but esp. 122–26.

32. On Mahistre's career and engagement in the work of the Lille society, see the address at his funeral by the dean and professor of chemistry in the faculty, Jean Girardin: *Mém. Soc. Lille* 2nd ser., 7 (1860): xxxvi–xxxix. I comment on the publications by the professors in the Lille faculty in Fox, "The Savant Confronts His Peers: Scientific Societies in France, 1815–1914," in Fox and Weisz, *Organization of Science and Technology*, 249–52.

33. Fortoul's comment to the rector of the academy of Douai that "M. Pasteur increasingly justifies the hopes that I invested in him when I appointed him to lead the faculty of science in Lille" (8 Feb. 1856) is typical of several expressions of ministerial approval in correspondence in AD Nord T. Rectorat 182.

34. My information about Abria and the publications of other professors in the Bordeaux faculty is taken largely from Georges Rayet, "Histoire de la faculté des sciences de Bordeaux (1838–1894)," *Actes Acad. Bordeaux* 3rd ser., 59 (1897): 5–369, esp. 155–363.

35. Gerbod, *Condition universitaire*, 286–92. The relevant passage in the decree of 9 March 1852 is in *BAIP* 3 (1852): 35 and *JGIP* 21 (1852): 159.

36. The point is frequently made in John M. Merriman, ed., *French Cities in the Nineteenth Century* (London, 1982), especially in Merriman's introduction and the contributions by Charles Tilly, Ted. W. Margadant, and Michael P. Hanagan.

37. On these and all aspects of industry in the Haut-Rhin, I draw heavily on the magnificent *Histoire documentaire de l'industrie de Mulhouse et de ses environs au XIXe siècle (Enquête centennale)* (Mulhouse, 1902). See also Fox, "Science, Industry, and the Social Order in Mulhouse, 1798–1871," *BJHS* 17 (1984): 127–68, and, on the crucial role of the Société industrielle de Mulhouse, Florence Ott, *La Société industrielle de Mulhouse 1826–1876. Ses membres, son action, ses réseaux* (Strasbourg, 1999).

38. Fox, "Science, Industry, and the Social Order," 167.

39. "Séance d'installation de l'Ecole gratuite des chauffeurs," *Mém. Soc. Lille* 2nd ser., 5 (1858): v–viii.

40. The classic study of education in Mulhouse remains Raymond Oberlé, *L'Enseignement à Mulhouse de 1798 à 1870* (Paris, 1961), esp. (on the new schools of the 1860s) 227–31. On the schools, see also *Histoire documentaire de l'industrie de Mulhouse*, 91–92, and Ott, *Société industrielle de Mulhouse*, 468–74. The creation of a commercial school in 1866 was part of the same Mulhousien response to heightened British competition provoked by the Cobden-Chevalier treaty.

41. Charles Rodney Day, *Education for the Industrial World: The Ecoles d'arts et métiers and the Rise of French Industrial Engineering* (Cambridge, MA, 1987), 31–38.

42. Paul Raphael and Maurice Gontard, *Un Ministre de l'instruction publique sous l'Empire autoritaire. Hippolyte Fourtoul 1851–1856* (Paris, 1975), 10–59.

43. The report and decree establishing the new teaching in applied science formed part of the major reform of 22 August 1854. See Beauchamp, *Lois et règlements*, 2:349–64.

44. Oberlé, *Enseignement à Mulhouse*, 215–17.

45. Ibid., 197–205; *Histoire documentaire de l'industrie de Mulhouse*, 87–88; Ott, *Société industrielle de Mulhouse*, 465–66. A pamphlet published to mark the opening of the school on 17 November 1855, *Inauguration de l'Ecole préparatoire à l'enseignement supérieur des sciences et des lettres de Mulhouse* (Bibliothèque de l'Arsenal, BR-22568), is especially informative.

46. Beauchamp, *Lois et règlements*, 2:459.

47. Ibid., 2:453.

48. Ibid., 2:455 and 2:457.

49. Ibid., 2:459.

50. *Statistique 1865–68*, 382. On Pasteur's enthusiasm for industrially related instruction and its legacy after his departure from Lille in 1857, see Wrotnowska, *Louis Pasteur*, esp. 57–79, and Harry W. Paul, "Apollo Courts the Vulcans. The Applied Science Institutes in Nineteenth-Century Science Faculties," in Fox and Weisz, *Organization of Science and Technology*, 156–58.

51. *Statistique 1876*, 580.

52. See Charles Rodney Day, "Technical and Professional Education in France. The Rise and Fall of l'Enseignement secondaire special, 1865–1902," *Journal of Social History* 6 (1972–73): 177–201, as well as Duruy's own account of the initiative in his *Notes et souvenirs (1811–1894)* (Paris, 1901), 1:252–75.

53. Duruy's address at the inauguration is reproduced in Victor Duruy, *L'Administration de l'instruction publique de 1863 à 1869. Ministère de S. Exc. M. Duruy* (Paris, 1870), 329–40. The school was created from an existing *collège communal*.

54. Duruy, "Exposé de la situation de l'instruction publique en 1868," ibid., 861.

55. The reform of 1902 has been treated authoritatively by Nicole Hulin in works cited in the bibliographical note.

56. Day, "Technical and Professional Education," 186–87, and Day, *Education for the Industrial World*, 35–36.

57. Pigeonneau, *La Réforme de l'enseignement secondaire spécial* (Paris, 1891).

58. Francis Pothier, *Histoire de l'Ecole centrale des arts et manufactures d'après des documents authentiques et en partie inédits. XIXe siècle à nos jours* (Paris, 1887), 231–55, and, for the relevant correspondence and official documents, 481–528.

59. Inaugurated in 1843 as a mainly municipal initiative, the Ecole des maîtres mineurs in Alais quickly came under the control of the administrators of the Ecole des mines in Paris and hence of the Ministry of Public Works. See André Thépot, *Les Ingénieurs des mines. Histoire d'un corps technique d'état,* vol. 1, *1810–1914* (Paris, 1998), 233–35.

60. The Ecole spéciale d'horlogerie at Cluses (in Haute-Savoie) and the Ecole muncipale d'horlogerie at Besançon had been opened in 1849 and 1861, respectively. In 1890–91 both schools were given the status of "national schools," under the Ministry of Commerce as part of a policy of promoting high-level apprenticeship schools.

61. Day, "Education for the Industrial World," in Fox and Weisz, *Organization of Science and Technology*, 143.

62. Day, "Education for the Industrial World," 90–107.

63. Rouher's report to the Emperor commending the idea of an inquiry is in *BAIP* 14 (1863): 111–15, and Thérèse Charmasson, Anne-Marie Lelorrain, and Yannick Ripa, eds., *L'Enseignement technique de la Révolution à nos jours. Textes officiels avec introduction, notes et annexes,* vol.1, *De la Révolution à 1926* (1987), 173–77.

64. *Ministère de l'agriculture, du commerce et des travaux publics. Commission de l'enseignement technique. Enquête sur l'enseignement professionnel,* 2 vols. (Paris, 1864–65). For an account of the inquiry and its consequences, see Charmasson et al., *Enseignement technique,* 26–31.

65. Inaugurated as the Ecole municipale d'apprentis, the school was renamed the Ecole municipale professionnelle Diderot in 1884.

66. Morin and Tresca, *De l'organisation de l'enseignement industriel et de l'enseignement professionnel* (Paris, 1862) and Michel Chevalier, ed., *Exposition universelle de Londres 1862. Rapports*

des membres de la section française du jury international (Paris, 1862), 6:186–242. Morin and Tresca (*De l'organisation*, 6) calculated that roughly six hundred pupils a year emerged from schools offering a preparation for industrial careers at all levels (from the Ecole centrale des arts et manufactures down to specialized trade schools), compared with a total of 1.2 million people employed in industry. The figures pointed to precisely the shortage of qualified recruits that Duruy's *enseignement secondaire spécial* was intended to rectify.

67. Jourdain, *Budget de l'instruction publique*, 227–29. Expenditure on the Muséum d'histoire naturelle in 1855 stood at 479,779 francs, compared with figures of 178,920 francs and 105,886 francs for the Collège de France and the Paris Observatoire respectively. By 1869, the budgets had risen to 678,180 francs for the museum and 280,500 francs and 152,460 francs respectively for the Collège de France and the observatory (excluding the Bureau des Longitudes). Over the same period, the budget for the Institut de France rose by a lesser proportion, from 573,514 francs to 661,200 francs; see *JGIP* 38 (1868): 397, 410–11, and 425.

68. Limoges, "The Development of the Muséum d'histoire naturelle of Paris, c.1800–1914," in Fox and Weisz, *Organization of Science and Technology*, 221–29.

69. Corne, *Rapport au nom de la commission spéciale instituée par M. Le Ministre de l'instruction publique pour étudier les questions qui se rattachent soit à l'administration, soit à l'enseignement du Muséum d'histoire naturelle* (Paris, 1850), and Allard, *Rapport présenté à Son Exc. Le Ministre de l'instruction publique et des cultes, par la commission chargée d'étudier l'organisation du Muséum d'histoire naturelle. Arrêté du 21 mai 1858* (Paris, 1863). On the reports and their context and effect, see Limoges, "Une 'République de savants' sous l'épreuve du regard administratif. Le Muséum d'histoire naturelle 1849–1863," in Claude Blanckaert, Claudine Cohen, Pietro Corsi, and Jean-Louis Fischer, eds., *Le Muséum au premier siècle de son histoire* (Paris, 1997), 65–84.

70. *Observations en réponse au rapport de la commission spéciale . . .* (Paris, 1851), and *Mémoire des professeurs-administrateurs du Muséum d'histoire naturelle, en réponse au rapport fait en 1858 . . .* (Paris, 1863).

71. "Rapport de S. Exc. M. le Ministre à S. M. l'Empereur, précédant le décret du 29 décembre 1863, portant réorganisation de l'administration du Muséum d'histoire naturelle," in Duruy, *Administration de l'instruction publique*, 48–51; also in Beauchamp, *Lois et règlements*, 2:633–34.

72. Limoges, "République de savants," 67–81.

73. Three pamphlets by Fremy summarized his position: *L'Organisation des carrières scientifiques* (Paris, 1866), *Les Volontaires de la science* (Paris, 1868), and *L'Abandon des carrières scientifiques* (Paris, 1870).

74. On Fremy's laboratory and his work as professor of inorganic chemistry at the Muséum, see Danielle Fauque, "Organisation des laboratories de chimie à Paris sous le ministère Duruy (1863–1869). Le cas des laboratoires de Fremy et de Wurtz," *AS* 62 (2005): 501–31, esp. 513–20 and 526–31.

75. Edme Fremy, *Les Savants délaissés* (Paris, 1884).

76. The Société de secours des amis des sciences was founded in 1857 by the chemist Louis-Jacques Thenard, mainly to provide financial support for the families of deceased savants.

77. Limoges, "Development of the Muséum," 212.

78. Joseph Bertrand, "Eloge historique de Urbain-Jean-Joseph Le Verrier," *Annales de l'Observatoire de Paris. Mémoires* 15 (1880): 14; also in Bertrand, *Eloges académiques* (Paris, 1890), 177–78.

79. Beauchamp, *Lois et règlements*, 2:306–11.

80. Le Verrier, "Rapport sur l'Observatoire impérial de Paris et projet d'organisation," *Annales de l'Observatoire de Paris* (1855): 1–68.

81. Le Verrier's contributions to meteorology are discussed in John L. Davis, "Weather Fore-

casting and the Development of Meteorological Theory at the Paris Observatory, 1853–1878," *AS* 41 (1984): 359–82.

82. On France's loss of the dominant position in terrestrial magnetism to Germany and Britain from the 1830s (a process for which Arago's ordering of his political and scientific priorities seems to have been largely responsible), see John Cawood, "Terrestrial Magnetism and the Development of International Collaboration in the Early Nineteenth Century," *AS* 34 (1977): 583–87.

83. "M. Le Verrier présente un travail fait à l'observatoire impérial, par M. Liais, sur la tempête de la Mer noire, en novembre 1854," *CR* 41 (1855): 1197–1204.

84. Le Verrier gave generally fair summaries of his scientific achievements in the Observatory in the justificatory texts that he published during the attacks on his regime in the later 1860s. See, for example, *Travaux des treize dernières années. Abstention fatale de l'administration supérieure. Exécution des projets de la ville de Paris* and *Observatoire impérial. Notes administratives*, undated but almost certainly published in 1869.

85. The observations eventually appeared as Liais, *Influence de la mer sur les climats, ou résultats des observations météorologiques faites à Cherbourg en 1848, 1849, 1850, 1851* (Paris, 1860), after publication in *Mémoires de la Société impériale des sciences naturelles de Cherbourg* 7 (1859): 171–238.

86. Jacques Ancellin, *Un Homme de science du XIXe siècle. L'astronome Emmanuel Liais (1826–1900)* (Coutances, 1985), 65–81, 109–10, and 133–57.

87. Liais, "De l'état de l'astronomie en France," *L'Ami des sciences. Journal du dimanche* 4, no. 18 (2 May 1858): 273–76, and no. 19 (9 May 1858): 291–93.

88. Ancellin, *Emmanuel Liais*, 19–63.

89. Davis, "Weather Forecasting," 366. For the views of Regnault and Biot, see "Opinion de M. Biot sur les observatoires météorologiques permanents que l'on propose d'établir en divers points de l'Algérie," *CR* 41: 1177–90, esp. 1186–90 (31 Dec. 1855) on a proposal for the establishment of permanent meteorological stations in Algeria.

90. "Report of the Council of the British Association, Presented to the General Committee at Aberdeen, September 14, 1859," in *Report of the Twenty-ninth Meeting of the British Association for the Advancement of Science; Held at Aberdeen in September 1859* (London, 1860), xxviii–xl.

91. *BASF* 1 (1865–7): 51–53, and Duruy's address at the annual prize-giving for the nation's *sociétés savantes*, 22 April 1865, in *RSSD* 4th ser., 1 (1865): 276; the latter also in Duruy, *Administration de l'instruction publique*, 213–25.

92. Le Verrier, *Rapport fait au Ministre de l'instruction publique sur la situation des travaux météorologiques au 1er août 1865* (Paris, 1865), 4–5.

93. The structure that the Ministry of Public Instruction set up in 1864–65, with the Paris Observatoire as the national coordinating body, is described in Le Verrier's introduction to the *Atlas des orages de l'année 1865* (Paris, 1866), vii–viii.

94. Guy Caplat, ed., *Les Inspecteurs généraux de l'instruction publique. Dictionnaire biographique 1802–1914* (Paris, 1986), 195–97, 318–20, 471–72, and (on the role of the "inspection générale"), 3–116.

95. On the history of the Comité, see Xavier Charmes, *Le Comité des travaux historiques et scientifiques (Travaux et documents)* [cited hereafter as Charmes, *CTHS*] (Paris, 1886), 1:i–ccxxv. See also the administrative and other documents covering the period 1834–85 in vol. 2.

96. Ibid., 2:376–81. The figure for 1874 appears in *Rapports au Ministre sur la collection des documents inédits de l'histoire de France et sur les actes du Comité des travaux historiques* (Paris, 1874), 29.

97. Guizot, "Circulaire et instructions relatives à la recherche et à la publication de documents inédits," in Charmes, *CTHS*, 2:28–37.

98. Charmes, *CTHS*, 1:cxlv–cxlvi, and Salvandy, "Arrêté portant organisation de cinq comités historiques," ibid., 2:60–66.

99. Cousin, "Arrêté réunissant en un comité unique les comités de littérature et de sciences sous le titre de Comité pour la publication des documents inédits de l'histoire de France. 30 Août 1840," in Charmes, *CTHS*, 2:97–98. The new consolidated committee was to exist in parallel with the Comité des arts et des monuments, which had quite separate responsibility for historic monuments and archaeology.

100. Rouland, "Arrêté réorganisant le Comité de la langue, de l'histoire et des arts de la France sous le titre de Comité des travaux historique et des sociétés savantes," in Charmes *CTHS*, 2:184–87. The committee retained its new title until the reorganization of 1881, when it became the Comité des travaux historiques et scientifiques.

101. Charmes, *CTHS*, 2:380–81.

102. Rouland, "Circulaire relative à la préparation d'une Description scientifique de la France. 1er juin 1860," in Charmes, *CTHS*, 2:207–10.

103. The legislation concerning the *Répertoire archéologique de la France* and the *Description topographique* [originally *géographique*] *de la France* is in Charmes, *CTHS*, 2:199–200 and 2:202–6.

104. On these and other displays of opposition within the Institut de France, see Raphael and Gontard, *Hippolyte Fortoul*, 273–96.

105. The comte de Viel-Castel's contempt for the excessive zeal of Fortoul's campaign for election in 1854 is conveyed in *Mémoires du comte Horace de Viel-Castel sur le règne de Napoléon III (1851–1864)*, ed. Pierre Josserand (Paris, 1942), 1:204 and 1:207–9. The victory of his friend Adrien de Longpérier gave Viel-Castel much satisfaction.

106. Napoleon III, *Histoire de Jules César* (Paris, 1865–66), 2:521–53. A collection of correspondence, notes, and proofs dating from 1863–66, the years in which the *Histoire* was being written, is in the library of the Institut de France (BIF MS 3717).

107. On Duruy's eventually warm friendship with Dumas and his conflict with Nisard, see Duruy, *Notes et souvenirs*, 2:238–42, and, for the reminiscences of his adversary, Nisard, *Souvenirs et notes biographiques* (Paris, 1888), 2:61–75. The literature in question at Saint-Etienne included Voltaire's *Dictionnaire philosophique*, Rousseau's *Confessions*, works by the socialist Pierre-Joseph Proudhon, and the novels of George Sand. Nisard's sense of scandal contrasted sharply with Sainte-Beuve's "laxity."

108. On the conflicts between Le Verrier and his colleagues within the Observatoire de Paris and with Duruy, see James Lequeux, *Le Verrier. Savant magnifique et détesté* (Paris, 2009), chap. 6 ("Le dictateur"), 123–69.

109. For an illustration of the intense personal animosity that fired the conflict, see (among the many exchanges recorded in the *Comptes rendus*) *CR* 50 (1860): 510–31.

110. "Note de M. Delaunay sur le degré d'importance des erreurs qu'il a signalées dans le tome II des *Annales de l'Observatoire*," *CR* 51 (1860): 735–40; also the heated debate between Le Verrier and Delaunay that followed these papers, in *CR* 51 (1860): 740–46.

111. Newspaper clippings in AN F^{17} 3719 give a flavor of press criticism of the observatory. Volumes 3 and 4 of "Documents divers sur l'Observatoire de Paris 1854–1872" in the archives of the observatory and correspondence and other materials in the Institut de France (BIF MS 3716) also throw light on the Le Verrier affair; see Davis, "Weather Forecasting," 376n.

112. William Tobin, *The Life and Science of Léon Foucault: The Man Who Proved the Earth Rotates* (Cambridge, 2003), 204–6 and 282.

113. Ibid., 282.

114. Georges Pouchet, "Situation de l'Observatoire de Paris," *Cosmos* 3rd ser., 1 (1867–68):

6–7. According to Camille Flammarion, no fewer than twenty-eight members of the observatory's staff were affected by Le Verrier's aggressive behavior in 1866–67 alone, when the conflicts reached their height. Flammarion made the charge in 1867 in an article in *Le Siècle* reproduced in his *Mémoires biographiques et philosophiques d'un astronome* (Paris, 1911), 508–23, esp. 514–15. A list of those who resigned is in AN F^{17} 3719 and is reproduced in Tobin, *Life and Science of Léon Foucault,* 282.

115. Flammarion, *Mémoires d'un astronome,* 515.

116. This was just one of the complaints that Marié-Davy articulated in the letter that he sent to the emperor later in the year; see below, note 118.

117. Marié-Davy, *Météorologie. Les mouvements de l'atmosphère et des mers considérés au point de vue de la prévision du temps* (Paris, 1866), ii–iii. It is significant that a decade later Marié-Davy (now in charge of the meteorological observatory in the parc Montsouris) used the preface to the second edition of his book to comment on the scant progress in French meteorology since 1865; see Marié-Davy, *Météorologie. Les mouvements de l'atmosphère et les variations du temps* (Paris, 1877), preface (dated 16 April 1876), i–ii. The assumption, by Niels Hoffmeyer in Copenhagen, of responsibility for the atlas of movements in the atmosphere that had begun life under Léon Sonrel in the Observatiore de Paris was, for Marié-Davy, a sign of France's declining international role in the field.

118. Nine copies of the twelve-page letter were prepared (BnF 4 Ln^{27} 23602), as the *Rapport à l'Empereur, par M. Marié-Davy astronome à l'Observatoire de Paris,* for the members of the inquiry that Duruy instituted in the autumn of 1867 (see below, note 123). The letter was then printed more formally and (apart from an explanatory introduction) unchanged as *Mémoire adressé à Sa Majesté l'Empereur par M. Marié-Davy astronome à l'Observatoire* (Paris, 1868).

119. See, for example, the particularly virulent "Décadence de l'Observatoire de Paris et de l'administration astronomique en France," reproduced in Flammarion, *Mémoires d'un astronome,* 514–17, from *Le Siècle,* 30 November 1867.

120. I have not seen the letter, dated 21 August 1867, but it is extensively quoted in the manuscript "Personnel de l'Observatoire Impérial au 31 Août 1867" in AN F^{17} 3719. In the letter, Le Verrier cited Marié-Davy as "the cause of all the disorder" in the observatory and sought his dismissal after what he described as a year of inactivity on the part of Marié Davy.

121. Duruy to the Emperor, 4 October 1867, draft in AN F^{17} 3719. The text, with minor changes and omissions, is in Emile Ollivier, *L'Empire libéral. Etudes, récits, souvenirs* (Paris, 1894–1917), 12:527; also in Flammarion, *Mémoires d'un astronome,* 519. Duruy described his deteriorating relations with Le Verrier in his *Notes et souvenirs,* 2:242–46.

122. The bitterness of his eight-page *Observatoire impérial de Paris. Travaux des treize dernières années. Abstention fatale de l'administration supérieure. Exécution des projets de la ville de Paris* (Paris, 1867) was capped by that of the thirty-eight pages of his *Observatoire impérial de Paris. Notes administratives,* addressed to the Council of State and undated, though probably written soon after the institution of the committee.

123. The manuscript proceedings of the committee and the drafts and final printed report are in AN F^{17} 3719.

124. The point emerges strongly from the correspondence, much of it heated, in AN F^{17} 3719 and from the minutes of meetings of the committee (at which Le Verrier usually presided) between 28 April 1868 and 2 June 1870 in the library of the Observatoire de Paris (MS X.31–182).

125. Duruy to the Empress Eugénie, 16 March 1868, in Duruy, *Notes et souvenirs,* 2:243–45.

126. Delaunay to the Minister of Public Instruction, 9 December 1869, BIF MS 15446.

127. The vehemence of Delaunay's letter of December 1869 contrasts with the reserve of his

public statements. In his *Rapport sur les progrès de l'astronomie* (Paris, 1867), he gave respectful prominence to Le Verrier's work. Only his insistence (p. 14) that Adams in Cambridge had performed the same calculations that had led Le Verrier to predict the existence of Neptune in 1846 hinted at the latent animosity.

128. Lequeux, *Le Verrier*, 172–77.

129. The memorandum was published as *Mémoire sur l'état actuel de l'Observatoire impérial présenté par les astronomes à son Exc. le Ministre de l'Instruction publique* (Paris, 1870) and reproduced as "Affaire de l'Observatoire," *JGIP* 40 (1870): 88–93. The preamble in the latter version (87–88), dated 9 February 1870, summarizes the history of the conflict and the final sequence of events that had now led to Le Verrier's dismissal.

130. Ollivier, *Empire libéral*, 12: 527.

131. The decision was reversed in the very different circumstances of November 1870, when the wartime government of national defense restored Renan to the chair, which he held for the rest of his life.

132. Perrine Simon-Nahum, "La scandale de la Vie de Jésus de Renan. Du succès littéraire comme mode d'échec de la science," *Mil neuf cent. Revue d'histoire intellectuelle* 25 (2007): 63–64.

133. Duruy, *Histoire des Romains et des peuples soumis à leur domination*, 2 vols. (Paris, 1843–44).

134. Duruy did not complete his *Histoire des Romains depuis les temps les plus reculés jusqu'à l'invasion des barbares*, 7 vols. (Paris, 1870–79) until long after his departure from ministerial office in 1869.

135. Nisard, *Souvenirs et notes biographiques*, 1:88–97; Duruy, *Notes et souvenirs*, 1:80–86; and Ernest Lavisse, *Un Ministre. Victor Duruy* (Paris, 1895), 26–27.

136. For the relevant decree, dated 11 July 1863, see Beauchamp, *Lois et règlements*, 2:628–69.

137. For more on salaries in the educational field in France at this time, with a guide to exchange rates between francs and other currencies, see appendix B.

138. George Weisz, *The Emergence of Modern Universities in France, 1863–1914* (Princeton, NJ, 1983), 57–59, and *Statistique 1865–68*, 216–33.

139. On this, as on all aspects of secondary education in the sciences in this period, Hulin-Jung, *Enseignement des sciences*, esp. 285–95, is a sure guide. See also the texts and commentaries in Belhoste, *Sciences dans l'enseignement secondaire*, 383–88.

140. Duruy, "Rapport de S. Exc. M. le Ministre à S. M. l'Empereur, précédant le décret du 4 décembre 1864," in Duruy, *Administration de l'instruction publique*, 126–28. Also (with a helpful commentary) in Belhoste, *Sciences dans l'enseignement secondaire*, 388–90.

141. Duruy, *Administration de l'instruction publique*, 126–27, and Belhoste, *Sciences dans l'enseignement secondaire*, 389.

142. Duruy, *Administration de l'instruction publique*, 128, and Belhoste, *Sciences dans l'enseignement secondaire*, 390.

143. Hulin-Jung, *Enseignement des sciences*, 207–12, and Chervel, *Histoire de l'agrégation*, 157. It was Bonnier who laid the foundations for the reforms of 1880 in the curricula for the *enseignement classique* and the establishment of the Section des sciences naturelles at the Ecole normale supérieure (in which he taught).

144. The collection was intended to include more volumes than the twenty-nine mentioned here. For a list of the twenty-seven titles that had appeared by March 1870, when he wrote, see Duruy, *Administration de l'instruction publique*, 882–83. Only two of the eleven other volumes that Duruy listed as being in preparation (including six said to be already in press) were ever published.

145. *Statistique 1865–68*. Equally full reports on primary and secondary education were published in the same year.

146. *BAIP* n.s., 1 (1864): 284–85, and *JGIP* 33 (1864): 242. Copies of the circular are in many departmental archives.

147. Jaccoud, *De l'organisation des facultés de médecine en Allemagne. Rapport présenté à son Excellence le Ministre de l'instruction publique le 6 octobre 1863* (Paris, 1864).

148. See, for example, Jean-Magloire Baudoüin, *Rapport sur l'état actuel de l'enseignement spécial et de l'enseignement primaire en Belgique, en Allemagne et en Suisse* (1865), and Jacques Demogeot and Henry Montucci, *De l'enseignement secondaire en Angleterre et en Ecosse. Rapport adressé à son Exc. M. le Ministre de l'Instruction publique* (Paris, 1868).

149. Jules Marcou, *De la science en France. Première partie* (Paris, 1869), 101–324.

150. As occurred in Duruy's speech at the annual meeting of the representatives of the provincial *sociétés savantes* in April 1868; *RSSD* 4th ser., 7 (1868): 327–28. While dissociating himself from any idea that France had been overtaken by Germany or Britain in certain areas, Duruy saw it as the government's duty to act as if this were so.

151. Lorain, *De la réforme des études médicales par les laboratoires* (Paris, 1868), 5.

152. Alan J. Rocke, *Nationalizing Science: Adolphe Wurtz and the Battle for French Chemistry* (Cambridge, MA), 282–83.

153. Ibid., 284–87. Rocke speculates, plausibly in my view, that Hofmann (newly appointed to the chair of chemistry at Berlin) may have acted as a consultant to the French government on the reinvigoration of academic science.

154. Chevalier, "Introduction. Quatrième partie. Observations sur les premiers ressorts de la production," in *Exposition universelle de 1867 à Paris. Rapports du jury international publiés sous la direction de M. Chevalier* (Paris, 1868), 1:cccxi–cccxvi.

155. Wurtz, *Les hautes études pratiques dans les universités allemandes* (Paris, 1870).

156. Lorain, *Réforme des études médicales*, 23–24.

157. Wurtz, *Hautes études pratiques*, 12.

158. Quoted in Rocke, *Nationalizing Science*, 283–84, from a letter of 18 October 1864 from Milne Edwards to the vice rector of the academy of Paris.

159. Bernard, *Rapport sur les progrès et la marche de la physiologie générale en France* (Paris, 1867), esp.143–49 and 230–37.

160. Ibid., 147.

161. On Pasteur's intervention, see Rocke, *Nationalizing Science*, 288–92; also René Vallery-Radot, *La Vie de Pasteur* (Paris, 1900), 160–63, and Pasteur's letters of 5 September 1867 to the emperor and Duruy, in *Correspondance de Pasteur 1840–1895* (Paris, 1940–51), 2:345–51.

162. Pasteur had first submitted his text to the semiofficial administrative newspaper of the empire, *Le Moniteur universel*. Although the editor thought the *MU* was inappropriate for such trenchant criticism of governmental policy, he sympathized with the case and was apparently instrumental in passing Pasteur's article to the emperor's secretary, who in turn (at the request of the emperor) encouraged its publication elsewhere. As a result, before its appearance as the pamphlet to which I refer, it was published in February 1868 as Pasteur, "Le budget de la science," *RCS* 5 (1867–68): 137–39.

163. Pasteur, "Visite de l'Empereur au laboratoire de chimie de l'Ecole normale supérieure et à la Sorbonne," *MS* 10 (1868): 108–10. It is a mark of the emperor's sympathy that the article was published originally in the *Moniteur universel* (28 Jan. 1868).

164. Eventually published as "Suppression du cumul dans l'enseignement des sciences physiques et naturelles," in Pasteur, *Quelques réflexions sur la science en France* (Paris, 1871), 15–24.

165. Vallery-Radot, *Vie de Pasteur*, 218–21; Sir Ashley Miles, "Reports by Louis Pasteur and Claude Bernard on the organization of scientific teaching and research," *NRRS* 37 (1982–83): 101–18; Rocke, *Nationalizing Science*, 291–92. Duruy's confidence in the emperor's support for his initiative is evident in his letter to the emperor, 12 April 1868, in Duruy, *Notes et souvenirs*, 1:312.

166. *RSSD* 4th ser., 7 (1868): 322–30; also in Duruy, *Administration de l'instruction publique*, 592–603.

167. Hillebrand, *De la réforme de l'enseignement supérieur* (Paris, 1878), 83.

168. William R. Keylor, *Academy and Community: The Foundation of the French Historical Profession* (Cambridge, MA, 1975), 19–54.

169. The title of this last section, "sciences historiques et philologiques," indicates the rigorous, quasi-scientific methodology that Duruy wished to foster.

170. The foundation of the Ecole and the first ten years of its existence are well documented in *Statistique 1876*, 707–27. The text of the decree of 31 July 1868 and accompanying documents are also in Beauchamp, *Lois et règlements*, 2:746–61, and (slightly abbreviated) *RCS* 5 (1867–68): 587–92.

171. Duruy, *Administration de l'instruction publique*, xix n1. The provincial laboratories were in Caen (agricultural chemistry), Marseille (physiology, geology, and chemistry) and Montpellier (physiology). For a convenient list, see *Statistique 1876*, 707–9.

172. Louandre's editorial, in *JGIP*, 38 (1868): 529–30, reflected a broader tendency of liberal journalists on the left and center left to turn against Duruy from the early months of 1868; see Charles Dejob, "Le réveil de l'opinion dans l'Université sous le Second Empire. La *Revue de l'instruction publique* et Victor Duruy," *L'Enseignement secondaire. Organe de la Société pour l'étude des questions d'enseignement secondaire* 35 (1914): 65–68, 82–85, 107–10, and 120–24, esp. 123–24.

173. *RCS* 5 (1867–68): 585–87.

174. Duruy, *Notes et souvenirs*, 1:301–24.

175. By 1878, the section for history and philology had published thirty-four volumes. See *Statistique 1876*, 727.

176. See Duruy's address, cited in note 150 above, 326.

177. Most of the information that follows on the laboratories of Desains and Jamin is drawn from the published annual *Rapport sur l'Ecole pratique des hautes études*.

178. Jamin's laboratory was certainly better equipped than the other facilities for physics at the Sorbonne until the major rebuilding of the 1880s, a point made by Emile Picard in his account of the time that Gabriel Lippmann spent working with Jamin between 1875 and 1878. See "La vie et l'oeuvre de Gabriel Lippmann," in Picard, *Discours et notices* (Paris, 1936), 11.

179. Octave Gréard, "Introduction," in Henri Paul Nénot, *Monographie de la nouvelle Sorbonne* (Paris, 1903), 20.

180. Schuster, *Biographical Fragments* (London, 1932), 198.

181. For Duruy's own account of his inhibiting and often turbulent relations with a number of senior Catholic clergy, in particular the bishops of Orléans (Félix Dupanloup) and Montauban (Jean-Marie Doney), and the cardinal archbishop of Rouen (Henri Marie Gaston de Bonnechose), see Duruy, *Notes et souvenirs*, 1:325–90. The many texts and commentaries in the *Journal général de l'instruction publique*, esp. vol. 38 (1868) vividly record the confrontations.

182. Duruy, *Administration de l'instruction publique*, xviii.

183. On the negotiations leading to the establishment of the observatory, see Duruy, *Notes et souvenirs*, 1:309–12.

184. Although the astronomical observatory was intended primarily for training naval offi-

cers (who also received instruction at the meteorological observatory), its facilities were available to serious amateur astronomers as well. See Ernest Mouchez, "Création de l'observatoire de Montsouris," *RMC* 54 (1877): 510–19, and Ernest Mouchez and Maurice Loewy, "Création de l'observatoire de Montsouris," *Annales du Bureau des longitudes et de l'Observatoire astronomique de Montsouris* 1, A (1877): 3–11. I am grateful to Guy Boistel for his guidance on these sources.

185. In fact, the base of expenditure was so low that some increase was not hard to achieve. In 1868, for example, the fifty-six faculties of the University (with their 408 professors) incurred a net deficit for the state of a mere 82,048 francs, most of their costs being recouped from student fees, mainly for examining, and other services. See Duruy, *Administration de l'instruction publique*, xx, note 1.

186. Jules Ferry's indignant address of 12 April 1870 to the Chamber of Deputies on the closure of the faculty gives a liberal perspective on the issues. See Paul Robiquet, ed., *Discours et opinions de Jules Ferry* (Paris, 1893–98), 1:305–10.

187. See Duruy's letter of 5 December 1874, originally published in *La Liberté*, in Duruy, *Notes et souvenirs*, 1:321–24.

188. This criticism of Duruy is alluded to in Dejob, "Réveil de l'opinion."

Chapter 4 · Science, Philosophy, and the Culture of Secularism

1. John William Draper, *History of the Conflict between Religion and Science* (London, 1875), and Andrew Dickson White, *A History of the Warfare of Science with Theology in Christendom*, 2 vols. (London, 1896).

2. Brooke, *Science and Religion: Some Historical Perspectives* (Cambridge, 1991), 5.

3. Ibid, 321.

4. William Coleman, *Georges Cuvier Zoologist: A Study in the History of Evolution Theory* (Cambridge, MA, 1964), 14–17 and 176–82.

5. Roger Hahn, *Pierre Simon Laplace 1749–1827: A Determined Scientist* (Cambridge, MA, 2005), 201–4.

6. These aspects of the lives of Ampère, Biot, and Cauchy emerge strongly from Claude-Alphonse Valson, *La Vie et les travaux d'André Marie Ampère*, new ed. (Lyon, 1897), 175–243; F. Lefort, "Un savant chrétien. J. B. Biot," *Le Correspondant* n.s., 36 (1867), 955–95; Valson, *La Vie et les travaux du baron Cauchy* (Paris, 1868), 1:108–21 and 1:169–213; and Bruno Belhoste, *Cauchy 1789–1857. Un mathématicien légitimiste au XIXe siècle* (Paris, 1985), 58–61 and 113–43.

7. Marcel de Serres, *De la Cosmogonie de Moïse comparée aux faits géologiques* (Paris, Strasbourg, and Nancy, 1838). Serres had taken up his chair in the faculty of science in Montpellier in 1811. By the time the third edition of his book appeared, in 1859, he was in his late seventies.

8. Serres's view embraced a developmental pattern, based on successive creations that brought the earth and its inhabitants to their present state. See the "Résumé" in Serres, *Cosmogonie de Moïse*, 218–28, and, in the third edition (2 vols., Paris, 1859), the much longer "Livre IV" concluding the second volume (376–88).

9. Serres, *De la création de la terre* (Paris, 1843), a restatement of Serres's view in which much of the lengthy preface (1–79) was devoted to answering Bonald's criticisms.

10. Lamartine, *Des destinées de la poésie* (Paris, 1834), 5–6.

11. Valson, *Vie du baron Cauchy*, 1:203–13.

12. Louis Liard, *L'Enseignement supérieur en France, 1789–1889 [1893]* (Paris, 1888–94), 2:244–47.

13. Janet, *La Crise philosophique. MM. Taine, Renan, Littré, Vacherot* (Paris, 1865), esp. 1–13.

14. Gabriel Monod, *Jules Michelet* (Paris, 1875), 41–45.

15. The contrast is made with particular vehemence in the essay on Taine in Janet, *Crise philosophique*, 93–135, esp. 98 and 134.

16. Janet had made the point in these terms three years earlier at the Sorbonne. See Janet, *Leçon d'ouverture du cours de philosophie prononcée à la faculté des lettres de Paris le 16 décembre 1862* (Paris, 1862), 18–19.

17. Janet, *Crise philosophique*, 7.

18. Ibid., 97.

19. Ravaisson's was one of the most substantial of the twenty-nine volumes of the collection, on which see chap. 3.

20. Ravaisson, *La Philosophie en France au XIXe siècle* (Paris, 1868), esp. 50–103.

21. Taine, *Les Philosophes français du XIXe siècle*, 2nd ed. (Paris, 1860), iii–iv.

22. Ibid., vi. As in the previous note, I cite the second edition, the first having been published without a preface.

23. Littré, *Auguste Comte et la philosophie positive* (Paris, 1863), i.

24. Ibid., esp. iii–iv, 498–515, and 662–81, where Littré described his break with Comte, a break that did nothing to diminish his affection and respect for a man he saw as "lit with the rays of genius" (681). Littré's engagement with Comtean positivism is a recurring theme in the collection of essays published as *Actes du colloque Littré. Paris, 7–9 octobre 1981 [Revue de synthèse*, 3rd ser., nos. 106–8] (Paris, 1982). See especially Ernest Coumet, "La philosophie positive d'E. Littré," 177–214, and Annie Petit, "Philologie et philosophie d'histoire," 215–43.

25. The articles, dating from 1844, 1849, and 1850, were republished as Littré, *Conservation, révolution et positivisme* (Paris, 1851), and I refer to them in this edition.

26. Janet, *Crise philosophique*, 113–22.

27. Littré, *Conservation, révolution et positivisme*, v–xxxii (Preface) and 169–328 ("Des progrès du socialisme").

28. Littré, *Paroles de philosophie positive* (Paris, 1859), 6–7.

29. *Oeuvres complètes d'Hippocrate*, ed. and trans. Emile Littré, 10 vols. (Paris, 1839–61).

30. Jean Mistler, *La Librairie Hachette de 1826 à nos jours* (Paris, 1964), 163–80. After having been commissioned to produce the dictionary in 1841, Littré had delivered the first part of his text in 1859. Printing began late in that year, and the work was completed with the publication of the last part of the second volume in 1872.

31. On Dupanloup's campaign, see François Lagrange, *Vie de Mgr Dupanloup, évêque d'Orléans, membre de l'Académie française*, 3 vols. (Paris, 1883–84), 2:407–17. A helpful study of Dupanloup's hostility to Littré's candidature in 1863 and his anger at Littré's eventual admission to the Académie in 1871 is Henri Welschinger, "L'élection de Littré à l'Académie française," *RDM* 6, no. 44 (1918): 394–422.

32. The attack, published as Dupanloup, *Avertissement à la jeunesse et aux pères de famille sur les attaques dirigées contre la religion par quelques écrivains de nos jours* (Paris, 1863) was widely read, independently of the election. It went through six editions in the year of its publication and was reprinted in 1868. Its main targets were not only Littré, Renan, and Taine but also (as an only slightly lesser evil) Renan's professorial colleague at the Collège de France, Alfred Maury.

33. Dupanloup, *L'Election de M. Littré à l'Académie française, suivi d'une réponse au Journal des débats* (Paris, 1872), 2.

34. Although Dupanloup's resignation was not accepted, he took no further part in the Académie's activities and gave the small allowance he received as an academician to charity. See Lagrange, *Dupanloup*, 3:246–47, and Emile Faguet, *Mgr Dupanloup. Un grand évêque* (Paris, 1914), 111–14.

35. *Journal des débats,* 3 January 1872. An unrepentant reply from Dupanloup was published on 5 January and reproduced in Dupanloup, *Election de M. Littré,* 27–32.

36. Moleschott, *Der Kreislauf des Lebens. Physiologische Antworten auf Liebig's "Chemische Briefe"* (Mainz, 1852); Büchner, *Kraft und Stoff. Empirisch-naturphilosophische Studien* (Frankfurt, 1855); Vogt, *Koehlerglaube und Wissenschaft. Eine Streitschrift gegen Hofraht R. Wagner in Göttingen* (Giessen, 1855). Vogt was replying to an address by the physiologist Rudolf Wagner, who in 1854 had argued, on moral rather than scientific grounds, that the existence of an immaterial soul was a necessary hypothesis. Without it, for Wagner, all noble thoughts became but "vain dreams."

37. Büchner, *Force et matière. Etudes philosophiques et empiriques de sciences naturelles. Mises à la portée de tout le monde,* trans. L. F. Gamper (Paris, 1863), and Moleschott, *La Circulation de la vie. Lettres sur la physiologie, en réponse aux Lettres sur la chimie, de Liebig,* trans. Emile Cazelles (Paris, 1865). In subsequent notes, I sometimes cite Büchner in the French edition, sometimes in the English edition (see note 42, below), and sometimes in both.

38. Among the responses, Janet's lengthy review of Büchner's *Force et matière* in the *Revue des deux mondes* in August and December 1863, also published as Janet, *Le Matérialisme contemporain. Examen du système du docteur Büchner* (Paris, 1864), stood out for its thoroughness.

39. Vogt, *Vorlesungen über den Menschen, seine Stellung in der Schöpfung und in der Geschichte der Erde,* 2 vols. (Giessen, 1863); translated as *Leçons sur l'homme, sa place dans la création et dans l'histoire de la terre* (Paris, 1865), with additional material that had not appeared in the German edition.

40. The expression "les Vogt et Cie" is quoted in Jacqueline Lalouette, *La libre pensée en France 1848–1940* (Paris, 1997), 161, from a disparaging comment by Charles Dollfus (a philosopher in the spiritualist tradition) to Auguste Nefftzer, 13 February 1857.

41. *La Pensée nouvelle. Science, lettres, arts, histoire et philosophie* succeeded the equally committed *Libre pensée* in 1867 but only lasted for one year.

42. Büchner, *Force and Matter: Empirico-philosophical Studies Intelligibly Rendered,* ed. J. Frederick Collingwood (London, 1864), xxvii. Collingwood included translations of the prefaces to the first, third, and fourth editions of *Kraft und Stoff.* The main text of his translation was of the eighth German edition.

43. Büchner, *Force and Matter,* 2. Cf. "Point de force sans matière—point de matière sans force" in Büchner, *Force et matière,* 2.

44. Büchner, *Force and Matter,* 9–15. Pages are the same in the French edition.

45. Büchner, *Force and Matter,* 239–58 (243–63 in the French edition). Cf. Harriet Martineau, *Autobiography* (London, 1877), 2:328–57, where Martineau recalled the gradual liberation of her spirit under the influence of the materialist writings of Henry George Atkinson in 1850 and 1851.

46. Büchner, *Force and Matter,* xiii.

47. Littré, "Préface d'un disciple," in Comte, *Cours de philosophie positive,* 2nd ed. (Paris, 1864), 1: xxvi–xxvii, and Leblais, *Matérialisme et spiritualisme. Etude de philosophie positive* (Paris, 1865), xix–xxii.

48. Pietro Corsi, *The Age of Lamarck: Evolutionary Theories in France 1790–1830,* trans. Jonathan Mandelbaum (Berkeley, CA, 1988), 88–103 and 136–45, and the significantly revised French translation, *Lamarck. Genèse et enjeux du transformisme 1770–1830,* trans. Diane Ménard (Paris, 2001), 116–32 and 170–79.

49. Cuvier, *Rapport historique sur les progrès des sciences naturelles depuis 1789, et sur leur état actuel* (Paris, 1810), 194. On the divergence between the views of Lamarck and Cuvier on spontaneous generation, see Corsi, *Age of Lamarck* (1988), 136–37 and 166, and *Lamarck* (2001), 170–71 and 202–3.

50. Büchner, *Force and Matter*, 63–88 (62–89 in the French edition). In later editions, Büchner recognized the declining position of spontaneous generation but continued to believe that generation of this kind might still be at work; see the third French edition, translated from the ninth German edition, Büchner, *Force et matière* (Paris, 1869), 136–41.

51. [Robert Chambers], *Natürliche Geschichte der Schöpfung des Weltalls, der Erde und der auf ihr befindlichen Organismen . . .,* trans. Carl Vogt (Braunschweig, 1851), 140n and, in the second edition (Braunschweig, 1858), 154n. For another expression of Vogt's agnosticism on the origin of living matter, cf. also his *Leçons sur l'homme,* 592.

52. For the similar assimilation in Britain, see James A. Secord, *Victorian Sensation: The Extraordinary Publication, Reception, and Secret Authorship of Vestiges of the Natural History of Creation* (Chicago, 2000), esp. 371 on Chambers's own concern about the tendency.

53. Pouchet's thesis, entitled *Essai sur l'histoire naturelle et médicale des solanées* and presented in the faculty of medicine in Paris on 21 December 1827, was dedicated to his father and to Flaubert, "mon premier maître." He then dedicated his first book, the *Traité élémentaire de zoologie, ou histoire naturelle du règne animal* (Rouen, 1832), to Blainville, "mon savant maître," whose system of zoological classification he used.

54. Pouchet, *Discours prononcé à la distribution des prix des écoles municipales de Rouen, en 1854* (Rouen, 1854).

55. Cf., for example, his contrast between the "unchanging wisdom" of God and the "minute power of our own intellect" in Pouchet, *Théorie positive de l'ovulation spontanée et de la fécondation des mammifères et de l'espèce humaine, basée sur l'observation de toute la série animale,* 1 vol. and 1 vol. of plates (Paris, 1847), 6–7.

56. Pennetier, "F.-A. Pouchet et son oeuvre," *PP,* 2nd ser., 17 (1876): 143.

57. Ibid.

58. Pouchet, *Théorie positive de la fécondation des mammifères, basée sur l'observation de toute la série animale* (Paris, 1842), 138.

59. As evidence of Pouchet's anxiety to win favor with the elite of French natural history, see the handwritten dedication of the copy of his thesis that he sent to Cuvier (BCMHN BB 335E): "Le baron Cuvier, Pair de France &&. Témoignage d'admiration par l'auteur. Pouchet." The fulsome dedication to Pierre Flourens of his *Théorie spontanée de l'ovulation spontanée* in 1847 reflected the same quest for recognition.

60. Flourens et al., "Rapport sur le prix de physiologie expérimentale pour l'année 1843," *CR* 20 (1845): 608–10.

61. See note 55 above.

62. Pouchet was defeated by Jacques-Armand Eudes-Deslongchamps in the election for one of two places in the zoology section on 10 December 1849 but was elected to the other vacancy a week later. See *CR* 29 (1849): 689 and 734.

63. Pouchet, "Note sur les proto-organismes végétaux et animaux dans l'air artificiel et dans le gaz oxygène," *CR* 47 (1858): 979–82 with a supplementary note (982–83) by Pouchet and Auguste Houzeau, professor of chemistry in the Ecole préparatoire à l'enseignement supérieur des sciences et des lettres in Rouen.

64. Milne Edwards, "Remarques sur la valeur des faits qui sont considérés par quelques naturalistes comme étant propres à prouver l'existence de la génération spontanée des animaux," *CR* 48 (1859): 23–29.

65. Pouchet, *Nouvelles expériences sur la génération spontanée et la résistance vitale* (Paris, 1864), vii–xv. Pennetier's indignation at Pouchet's treatment is evident in Pennetier, "Pouchet et

son œuvre" and "Un débat scientifique. Pouchet & Pasteur 1858–1868," *Actes du Muséum d'histoire naturelle de Rouen* 11 (1907): 5–55.

66. Farley and Geison, "Science, Politics and Spontaneous Generation in Nineteenth-Century France. The Pasteur-Pouchet debate," *Bulletin of the History of Medicine* 48 (1974): 161–98.

67. Geison, *The Private Science of Louis Pasteur* (Princeton, NJ, 1995), 321–22.

68. Farley, *The Spontaneous Generation Controversy from Descartes to Oparin* (Baltimore, MD, 1977), 47–55.

69. Pouchet, *Générations spontanées. Etat de la question en 1860* (Paris, 1861), 5.

70. Pasteur, "Mémoire sur les corpuscules organisés qui existent dans l'atmosphère: examen de la doctrine des générations spontanées," *Annales des sciences naturelles. Zoologie,* 4th ser., 16 (1861): 21–22.

71. Pasteur, "Mémoire sur la fermentation lactée," *Mém. Soc. Lille,* 2nd ser., 5 (1858): 13–26.

72. Pasteur, "Nouveaux faits pour servir à l'histoire de la levûre lactique," *CR* 48 (1859): 337–78.

73. "Lettre manuscrite de Pasteur à Pouchet," 28 February 1859, in *Œuvres de Pasteur*, ed. Pasteur Vallery-Radot (Paris, 1922–39), 2:628–30. The original is in the archives of the Muséum d'histoire naturelle de Rouen (MS 1023).

74. Described in Pasteur, "Mémoire sur les corpuscules organisés."

75. The list of publications on spontaneous generation between 1858 and 1869 in Pennetier, "Débat scientifique," 43–55, conveys the intensity of the debate. Pouchet's publications are also listed in Maryline Cantor [Cantor-Coquidé], *Pouchet, savant et vulgarisateur. Musée et fécondité* (Nice, 1992), 245–48.

76. *CR* 48 (1859): 535–36.

77. Pouchet, *Hétérogénie ou traité de la génération spontanée, basé sur de nouvelles expériences* (Paris, 1859), 126.

78. Ibid., 61–64 and 122–37; also the more succinct statement in his *Etat de la question en 1860*, 6–7, where Pouchet distanced himself from the "pretensions of the most nebulous ages of science." On the difficulty of determining Lamarck's precise stand on spontaneous generation, see Corsi, *Age of Lamarck*, 88–103, and *Lamarck*, 115–32.

79. Pouchet, *Hétérogénie*, 484–525.

80. Pouchet, *Les Créations successives et les soulèvements du globe. Lettres à M. Jules Desnoyers* (Paris, 1862), 9.

81. Pouchet, *Hétérogénie*, 98.

82. Ibid., 489.

83. Pouchet, *Nouvelles expériences*, xii; also Pouchet, *Hétérogénie*, ix, and *Notice sur les travaux de zoologie et de physiologie de M. F.-Archimède Pouchet* (Rouen, 1861), 5.

84. *CR* 55 (1862): 785.

85. Pasteur's victory, for his "Corpuscules organisés" (1861), was announced in the report on the competition in Bernard et al., "Prix Alhumbert pour l'année 1862," *CR* 55 (1862): 977–79.

86. A sentiment that Pouchet expressed in his *Nouvelles expériences*, esp. xiii–xv.

87. C. E. Alix, "Eloge de Nicolas Joly," *Mémoires de l'Académie des sciences, inscriptions et belles-lettres de Toulouse*, n.s., vol. 3 (1891): 491–524, esp. 500–502. Pierre-Adolphe Daguin in physics, Edouard Filhol in chemistry, and Joly formed a triumvirate of outstanding lecturers who made the faculty of science a focus for the city's intellectual life during the Second Empire; see Ulysse Lala, "L'Enseignement supérieur de la physique à Toulouse pendant un demi-siècle (1832–1882)," *Revue des Pyrénées* 22 (1910): 284–93.

88. Adolphe Félix Gatien-Arnoult, *Discours prononcé dans la séance publique de l'Académie*

impériale des sciences, inscriptions et belles-lettres de Toulouse, le 31 mai 1862 (Toulouse, 1862), 7, and Pouchet, *Nouvelles expériences*, x.

89. Pouchet, Joly, and Musset, "Expériences sur l'hétérogénie exécutées dans l'intérieur des glaciers de la Maladetta (Pyrénées d'Espagne)," *CR* 57 (1863): 558–61. Further exchanges between the authors and Pasteur followed in *CR* 57 (1863): 724–26 and 842–45.

90. Pasteur, "Note sur les générations spontanées," *CR* 58 (1864): 21–22.

91. Balard, Brongniart, Dumas, Flourens, and Milne Edwards, "Rapport sur les expériences relatives à la génération spontanée," *CR* 60 (1865): 384–97. An account of the closing stages of the affair that sets Pouchet, Joly, and Musset in a far more favorable light is Victor Meunier, "Histoire de la commission des générations spontanées," in Meunier, *La Science et les savants en 1864* (Paris, 1865), 315–44.

92. Balard et al., "Rapport sur les expériences," 396: "In sum, the facts observed by M. Pasteur and disputed by MM. Pouchet, Joly and Musset are accurate in every respect."

93. A contrast that drew an editorial comment in *Cosmos,* 3rd ser., 3 (1868): 198–99.

94. Pouchet, *L'Univers. Les infiniment grands et les infiniment petits* (Paris, 1865). The book and its reception are discussed in Cantor, *Pouchet, savant et vulgarisateur*, 77–88.

95. See, for example, his preface, dated 15 November 1867, to Georges Pennetier, *L'Origine de la vie* (Paris, 1868), vii–xvii. Here Pouchet claimed that heterogeny had emerged victorious from the debate in Germany, Italy, Britain, and America, if not yet in France (or rather "in Paris," as he pointedly expressed it).

96. Pouchet to Noël, 23 September 1865, quoted in Noel, *Conférence sur F.-A. Pouchet* (Rouen, 1874), 15.

97. Pouchet's involvement in the work of the Ligue suggests a growing radicalization, bred of understandable bitterness, in his later years. See Pennetier, "Pouchet et son oeuvre," 130 and 135.

98. Ibid., 142.

99. Flaubert to Ernest Feydau, 5 August 1860, in *Flaubert. Correspondance,* Pléiade edition, ed. Jean Bruneau (Paris, 1973–2007), 3:101.

100. Flaubert to Pouchet, 9 January 1864, ibid., 3:370. Flaubert was writing to thank Pouchet for a copy of his *Nouvelles expériences sur la génération spontanée.*

101. See, in particular, the excoriating dismissal of the Académie des sciences in the dedication ("To the press") of Meunier, *Essais scientifiques,* vol. 1, *L'apostolat scientifique* (Paris, 1857), 5–6. Meunier's true "academy," at least since the death of Arago, had been the millions of his readers across the globe.

102. Meunier, "Histoire de la commission," 328. The "Histoire" was one of several articles in Meunier's *La Science et les savants en 1864* that described in great, though partial, detail the later stages of the debate. In the autumn of 1865 Meunier was still not ready to let the matter rest; see Meunier, *La Science et les savants en 1865. 2e année, deuxième semestre* (Paris, 1866), 264–313, for the notes and papers that he submitted to the Académie des sciences between the autumn of 1865 and January 1866.

103. See, for example, his extended discussions of both Pouchet's and Pasteur's evidence in *ASI* 4 (1861): 247–56, and 5 (1861): 186–97. It was at the end of the latter contribution (197) that he expressed his weary resignation: "So on it meanders, vague and inconclusive, a debate that goes its way without advancing, a discussion that continues without shedding light."

104. Michelet to Joly, 16 January 1863, Bibliothèque interuniversitaire de Toulouse MS 212 (43).

105. Noel, *Les Générations spontanées* (Paris, 1865), 31–32.

106. Dumesnil, *L'Immortalité* (Paris, 1861), 172–81. Dumesnil dedicated the book to Noël.

107. Pasteur, "Des générations spontanées," *RCS* 1 (1863–64): 259.

108. Ibid., 259 and 260. Cf. Pasteur's very similar stance when, in the following year, he declared his preference for Charles Robin, rather than Henri de Lacaze-Duthiers, as a candidate for election to the Académie; see discussion later in this chapter.

109. Farley, *Spontaneous Generation Controversy*, 48–55.

110. On the views of Broca and Georges Pouchet, see chap. 6.

111. Robert E. Stebbins, "France," in Thomas F. Glick, ed., *The Comparative Reception of Darwinism* (Austin, TX, 1972), 122, notes five French reviews of the English edition of the *Origin*.

112. Sudre, "Des origines de la vie et de la distinction des espèces dans l'ordre animé," *Revue européenne* 2, no. 10 (1860): 582–605 and 820–37. Sudre is best known for his attack on the utopian revolutionary ideals of 1848, in his *Histoire du communisme ou réfutation historique des utopies sociales* (Paris, 1848).

113. It is relevant to the misreading of the *Origin* that Lamarck's work was by no means ignored c. 1860. The point emerges strongly from Corsi, *Age of Lamarck*, esp. 265–68, and Corsi, *Lamarck*, esp. 325–27. See also Goulven Laurent, *La Naissance du transformisme. Lamarck entre Linné et Darwin* (Paris, 2001), 123–30, and, for typical assimilations of Darwin's transformist doctrines to Lamarck's, Sudre, "Origines de la vie," 601, and Antoine Laurent Apollinaire Fée, *Le Darwinisme ou examen de la théorie relative à l'origine des espèces* (Paris, 1864), 5.

114. Dominique-Alexandre Godron, *De l'espèce et des races dans les êtres organisés et spécialement de l'unité de l'espèce humaine* (Paris, 1859), esp. 1:332–34, and 2:420–24.

115. Laugel, "Une nouvelle théorie d'histoire naturelle. L'Origine des espèces," *RDM*, 2ᵉ période, 26 (1860): 644–71.

116. Isidore Geoffroy Saint-Hilaire, *Histoire naturelle générale des règnes organiques* (Paris, 1854–62), 3:522–23. The comment appeared in the book's last chapter, reconstructed from notes left on the author's death in 1861.

117. Conry, *L'Introduction du darwinisme en France au XIXe siècle* (Paris, 1974), esp. 15–28.

118. Darwin, *De l'origine des espèces ou des lois du progrès chez les êtres organisés. Traduit en français sur la troisième édition avec l'autorisation de l'auteur par Clémence-Auguste Royer* (Paris, 1862), v–lxiv. On this and other translations of the *Origin*, see Gérard Molina, "Le savant et ses interprètes," in Patrick Tort, ed., *Darwinisme et sociéte* (Paris, 1992), 361–86, and Joy Harvey, "Darwin in a French Dress: Translating, Publishing, and Supporting Darwin in Nineteenth-Century France," in Eve-Marie Engels and Thomas F. Glick, eds., *The Reception of Charles Darwin in Europe* (London, 2008), 354–74.

119. On Royer's life and the circumstances of her engagement with the *Origin*, I draw on Geneviève Fraisse, *Clémence Royer. Philosophe et femme de sciences,* 2nd ed. (Paris, 2002), and Joy Harvey, *"Almost a Man of Genius": Clémence Royer, Feminism, and Nineteenth-Century Science* (New Brunswick, NJ, 1997), esp. 62–79.

120. Claparède, "M. Darwin et sa théorie de la formation des espèces," *Revue germanique française et étrangère* 16 (1861): 523–59.

121. Royer expressed the comprehensiveness of her scientism most fully in Royer, *Origine de l'homme et des sociétés* (Paris, 1870). In a detailed study of human origins, the book drew on a characteristically radical mix of Darwinian theory, spontaneous generation, and polygenism, presented as an antidote to the religious dogma, woolly thinking, and prejudice that had beset the subject since Rousseau's bold (but, for Royer, poorly founded) reflections on early man.

122. Royer, "Préface," in Darwin, *Origine des espèces* (1862), lxiv.

123. Darwin to Quatrefages, 11 July [1862], in *The Correspondence of Charles Darwin*, ed. Frederick Burkhardt et al. (Cambridge, 1985–2009), 10:313–15.

124. Claparède to Darwin, 6 September 1862, ibid., 10:398–400.

125. Altered to "des lois de transformation des êtres organisés" in the second edition of the translation (Paris, 1866).

126. Royer, "Préface" (1862), lxii.

127. Sudre, "Origines de la vie," 601. Sudre's review covered Pouchet's *Hétérogénie* as well as Royer's translation.

128. Ibid., 585.

129. In the second edition, Darwin, *Origine des espèces* (Paris, 1866), 95n, Royer explained that she had bowed to majority opinion by abandoning her use of "élection" in favor of the more literal translation of "sélection."

130. Flourens, *Examen du livre de M. Darwin sur l'origine des espèces* (Paris, 1864), 65.

131. Ibid., 32.

132. Ibid., 169–70. The importance that Flourens attached to the defeat of spontaneous generation is reflected in his decision to conclude his book with a dismissive two-page statement on the subject. The chapter ended categorically: "Spontaneous generation, therefore, does not exist. To continue discussing it is to misunderstand the question."

133. Janet, *Matérialisme contemporain*, 82–83 and 166–79.

134. Henri Boissonnot, *Le Cardinal Meignan* (Paris, 1899), 184–90.

135. Meignan, *Le Monde et l'homme primitif selon la Bible* (Paris, 1869), xi. For a discussion of Meignan's position, see Harry W. Paul, *The Edge of Contingency: French Catholic Reaction to Scientific Change from Darwin to Duhem* (Gainesville, FL, 1979), 24–40.

136. Meignan, *Le Monde et l'homme primitif*, 251–64.

137. Ibid., 173–74.

138. Ibid., 174–94.

139. "Program," *La Science et la foi. Journal religieux, scientifique et littéraire* 1, no. 1 (22 Dec. 1864): 1. By April 1865, the journal had been renamed *La Gazette du clergé*, a title that more accurately reflected its status as a mouthpiece of clerical opinion.

140. Here, Robin's position resembled that of Meunier, who also remained unconvinced by Darwin's theory. Meunier, however, firmly rejected the criticisms of A. L. A. Fée, whose *Darwinisme* (1864) had attacked the theory on the grounds of its incompleteness and similarity to the ideas of Lamarck and Geoffroy Saint-Hilaire; see Meunier, *La Science et les savants en 1865 . . . . deuxième semestre*, 190–95.

141. On Robin, I draw on two excellent biographical sources: Georges Pouchet, "Charles Robin (1821–1885). Sa vie et son oeuvre," *JAPNP* 22 (1886): i–clxxxiv, and Victor Genty, *Un grand biologiste. Charles Robin (1821–1855)* (Lyon, 1931).

142. In fact, Donald G. Charlton, in his *Positivist Thought in France during the Second Empire* (Oxford, 1959), 72–83, argues that Bernard subscribed to a positivism more orthodox than that of Comte himself. While Rayer was never a declared positivist, Littré clearly regarded him as a sympathizer; see his obituary of Rayer in *PP* 1 (1867): 489–91.

143. Robin, "Sur la direction que se sont proposée en se réunissant les membres fondateurs de la Société de biologie pour répondre au titre qu'ils ont choisi," *CRSB* 1 (1849): i–xi.

144. Cf. Comte, *Cours de philosophie positive* (Paris, 1830–42), 3:384–603 (lessons 41–44).

145. Robin expressed the same ideal again and more fully almost twenty years later in the leading positivist journal *La Philosophie positive*. See Robin, "De la biologie, son objet et son but, ses relations avec les autres sciences, la nature et l'étendue du champ de ses recherches, ses moyens d'investigation," *PP* 1 (1867): 78–101, 212–32, and 393–412, esp. 81–85.

146. In the heyday of the dinners, Modeste Magny's restaurant in the rue Contrescarpe-

Dauphine (renamed in 1867 the rue Mazet) was one of the most frequented in Paris. See Robert Baldick, *Dinner at Magny's* (London, 1971).

147. His activities are vividly recorded in the Goncourt Journal. See Edmond and Jules Goncourt, *Journal. Mémoires de la vie littéraire* (Paris: Fasquelle and Flammarion, 1956), esp. vol. 2, on the years 1864–78.

148. The gatherings at Magny's moved to Paul Brébant's restaurant in the boulevard Poissonnière in 1869.

149. An important vehicle for this aspect of Robin's program was the *Journal de l'anatomie et de la physiologie normales et pathologiques*, which he founded in 1864 and edited until his death in 1885, for most of the time with his friend Charles Brown-Séquard.

150. Paul Labarthe, *Nos Médecins contemporains* (Paris, 1868), 122–23.

151. Taine to Sainte-Beuve, 14 August 1865, in *H. Taine. Sa vie et sa correspondance* (Paris, 1902–7), 2:320–21. In Taine's words, Milne Edwards and Quatrefages regarded Robin as "too close to M. Littré." Lacaze-Duthiers was the natural choice for Milne Edwards, with whom he had worked as an assistant.

152. In his letter of 26 August 1865 to the princess, reproduced in Genty, *Charles Robin*, 47–48, Sainte-Beuve asked her to exert her influence with two of the Académie's most senior members, Antoine Serres and Chevreul (both now approaching eighty). In her receptiveness to liberal thought, the princess remained close to the emperor (whom she had once intended to marry), but she was estranged from the loyally Catholic Empress Eugénie. Her cultivation is reflected in Augustin Thierry's presentation of her as "Notre Dame des arts." On Sainte-Beuve's approach to Pasteur, see note 154 below.

153. Labarthe, *Nos Médecins contemporains*, 124–25. Of fifty-two votes cast, thirty went to Robin, twenty to Lacaze-Duthiers.

154. Pasteur to Saint-Beuve, 22 November 1865, in *Correspondance de Pasteur 1840–1895* (Paris, 1940–51), 2:213–14. Pasteur's comment ("The philosophical school to which he belongs is of no consequence to me") expressed (accurately, in my view) his refusal to allow extraneous considerations to color his scientific judgment of Robin's qualities.

155. Duruy, *Notes et souvenirs (1811–1894)* (Paris, 1901), 1:360, and Ernest Lavisse, *Un Ministre. Victor Duruy* (Paris, 1895), 100. Duruy's letter of 25 November 1866 to the vice rector of the academy of Paris urging vigilance with respect to both professors and students is in *Enseignement supérieur devant le Sénat. Discussion extraite du Moniteur. Avec préface et pièces à l'appui* (Paris, 1868), 45–46.

156. *Enseignement supérieur devant le Sénat.*, 27–29.

157. Ibid., 30–53. Also reported in *MU*, 28 March 1868.

158. The debates, originally published in issues of *MU* between 20 and 24 May 1868 (with an interesting record of the votes cast), were quickly republished in *Enseignement supérieur devant le Sénat*, together with a strongly anticlerical preface (1–26), unsigned but almost certainly by the historian Ernest Lavisse, who at the time was secretary to Duruy. They also appeared (with some lively commentaries) in numerous issues of the *JGIP* 38 between 28 May and 20 August 1868.

159. Nysten, *Dictionnaire de médecine, de chirurgie, de pharmacie, des sciences accessoires et de l'art vétérinaire*, 11th ed. (Paris, 1858), 54. Cf. the even more uncompromising version in the twelfth edition (Paris, 1865): "A term in biology that conveys the various functions of the brain and encephalic innervations, that is to say the perception of both external objects and inner sensations . . . this definition follows from current scientific orthodoxy, which excludes the possibility of a property or force without matter and of matter without a property or force, while always maintaining complete ignorance of the nature of force and matter" (55). Although the

wording of the article continued to be modified in later editions, the tone remained unchanged; see, for example, the fourteenth edition (Paris, 1878), 54.

160. Nysten, *Dictionnaire de médecine*, 11th ed. (1858), 533; 12th ed. (1865), 550.

161. *Enseignement supérieur devant le Sénat*, 172.

162. Ibid.

163. Nysten, *Dictionnaire de médecine*, 11th ed. (1858), 1138–39; 12th ed. (1865), 1213.

164. *Enseignement supérieur devant le Sénat*, 190–91.

165. Duruy's response to the thesis, which bore the title "Étude médico-psychologique du libre arbitre humain," is discussed in Chaix d'Est-Ange's report in *Enseignement supérieur devant le Sénat*, 49–52.

166. Bonnechose's position was uncompromising: "We must not be surprised if young people, after assimilating these doctrines, use them as a weapon in supporting socialism and fomenting revolutions." See *Enseignement supérieur devant le Sénat*, 181.

167. Ibid., 190–91.

168. The voting was eighty-four to thirty-one against the proposal to end the state's monopoly in higher education, and eighty to forty-three in favor of rejecting the accusations against the faculty.

169. Ferrière, *Essai sur le libre arbitre* (Paris, 1865).

170. Ferrière, *Le Darwinisme* (Paris, 1872), 6. I refer here to the longer of two publications by Ferrière with the same title. The abbreviated popular version of *Le Darwinisme* appeared six years later in the series "Bibliothèque utile" and sold for 60 centimes, compared with 4.50 francs for the main work. Both versions were published by the progressive publishing house of Germer Baillière.

171. Ferrière, *Les Apôtres. Essai d'histoire religieuse d'après la méthode des sciences naturelles* (Paris, 1879).

172. Ferrière, *Les Erreurs scientifiques de la Bible* (Paris, 1891) and *Les Mythes de la Bible* (1893).

173. Paul, *Edge of Contingency*, 82–93.

174. "Rapport de Mgr d'Hulst, président de la commission d'organisation à la première assemblée générale," in *Congrès scientifique international des Catholiques tenu à Paris du 8 au 13 avril 1888* (Paris, 1889), 1: lxix.

175. Ibid., 1:cxx. Although only a quarter of the subscribers seem to have attended the congress, the number remains impressive.

176. See Maisonneuve's brief intervention, ibid., 2:606–8.

177. Smets declared himself a convert to the theory of evolution following three years of study undertaken at the request of his superior; ibid., 2:608.

178. Gaudry, *Les Enchaînements du monde animal dans les temps géologiques. Mammifères tertiaires* (Paris, 1878), "Introduction" and "Résumé," esp. 257–58. The caution of Gaudry's position is discussed in Stebbins, "France," 136–38.

179. For an expression of his agnosticism, see Gaudry, *Les Enchaînements du monde animal dans les temps géologiques. Fossiles primaires* (Paris, 1883), 293.

180. The chapter on Thomism and science in Paul, *Edge of Contingency,* 179–94, is an essential source.

181. Abbé [Paul] de Broglie, *Le Positivisime et la science expérimentale,* 2 vols. (Paris, 1880).

182. Ibid., 1:vi–xiv, for a succinct statement of de Broglie's position.

183. On Loisy's commitment to the modernization of Catholic thought throughout his long life, see his *Mémoires pour servir à l'histoire religieuse de notre temps*, 3 vols. (Paris, 1930–31). In the

modern literature on Loisy and the modernist cause, Harvey Hill, *Alfred Loisy and the Scientific Study of Religion* (Washington, DC, 2002), esp. chaps. 1–3, is particularly helpful.

184. Evidence of Loisy's sense of a sudden change in the intellectual tone within the Church is in Loisy, *Choses passées* (Paris, 1913), 303–79, and *Mémoires*, vol. 2, esp. 259–377. The passage I quote from *E supremi apostolatus* appears in Loisy, *Mémoires*, 2:265, accompanied by Loisy's commentary on its significance.

185. In his last work, published posthumously, Quatrefages maintained that monogenism (with its core doctrine of the peopling of the earth through migrations) had triumphed, leaving few true polygenists; see Quatrefages, *Les Emules de Darwin* (Paris, 1894), 1:2–3. For Quatrefages, even polygenists had abandoned "true" polygenism by according a far greater importance to migration than Bory de Saint-Vincent and other pioneers of the polygenist position had once been willing to do.

186. In the congresses of Catholic scientists, as in the Catholic *Revue des questions scientifiques*, Nadaillac could be sure of finding receptive audiences. See, for example, Nadaillac, "Les découvertes préhistoriques et les croyances chrétiennes," in *Congrès scientifique international des catholiques . . . 1888*, 2:761–71, and (a paper presented, in his eightieth year, to the fourth international congress of Catholic scientists in Fribourg) "Unité de l'espèce humaine prouvée par la similarité des conceptions et des créations de l'homme," *RQS* 42 (1897): 415–48.

187. Mortillet, *Le Préhistorique. Antiquité de l'homme* (Paris, 1883), 8.

188. Milne Edwards, *Rapport sur les progrès récents des sciences zoologiques en France* (Paris, 1867), 420–32, esp. 420 and 425–29.

189. Stebbins, "France," 134.

190. Quatrefages, *Charles Darwin et ses précurseurs français* (Paris, 1870).

191. Ibid., 374.

192. Quatrefages, *Emules de Darwin*, 2:288.

193. In the absence of a single instance of a demonstrated transformation of a species, Blanchard remained unconvinced, preferring the traditional pattern of successive creations. See his "Origine des êtres," in Blanchard, *La Vie des êtres animés* (Paris, [1888]), 75–289, esp. 281–89; also the preface, vii–xv.

194. The new translations, both put out by the Reinwald publishing house, bore the title *L'Origine des espèces au moyen de la sélection naturelle ou la lutte pour l'existence dans la nature.* They were seen, not least by Darwin, as more faithful to the original than Royer's. Nevertheless, Royer's translation was reissued with revised prefatory material four times during her lifetime (in 1866, 1870, 1882, and 1901) and twice more after her death (1918 and 1932). See Fraisse, *Clémence Royer*, 168.

195. Giard, "Laboratoire de zoologie maritime à Wimereux (Pas-de-Calais)," in *AFAS, 3ᵉ session. Lille 1874*, 79.

196. Albert Dastre, "Une fondation de la ville de Paris à la faculté des sciences. La chaire d'évolution des êtres organisés," *RIE* 16 (1888): 521–36.

197. Donald G. Charlton, *Secular Religions in France 1815–1870* (London, 1963), esp. 13.

198. Ibid., 96–125.

199. Ibid., 155–99.

200. Ibid., 65–95.

201. Ibid., 38–64.

202. Ibid., 82–83.

203. "Foreword," in Gertrud Lenzer, ed., *Auguste Comte and Positivism: The Essential Writ-*

ings (Chicago, 1975), xv. Cf. the not dissimilar but carefully nuanced view of Andrew Wernick, who warns against the tendency to overstate the extent of the "real theoretical rupture" between the two phases of Comte's intellectual life; see Wernick, *Auguste Comte and the Religion of Humanity: The Post-Theistic Program of French Social Theory* (Cambridge, 2001), esp. 22–27.

204. Comte, *Catéchisme positiviste, ou sommaire exposition de la religion universelle, en onze entretiens systématiques entre une femme et un prêtre de l'humanité* (Paris, 1852), xxiii.

205. Comte characterized Clotilde de Vaux in this way in Comte, *Catéchisme positiviste*, xxii.

206. Comte, *Discours sur l'ensemble du positivisme, ou exposition sommaire de la doctrine philosophique et sociale propre à la grande république occidentale* (Paris, 1848), 315–93.

207. Quoted as the opening of the preface to Comte, *Catéchisme positiviste*, v. My translation is that of the English edition: *The Catechism of Positive Religion Translated from the French of Auguste Comte by Richard Congreve*, 2nd ed. (London, 1883), 1.

208. Comte, *Système de politique positive, ou traité de sociologie, instituant la religion de l'humanité* (Paris, 1851–54), 4:86–159 and 4:399–414. For a richly documented study of the religion of humanity, with special though not exclusive reference to France, see Jean-Claude Wartelle, *L'Héritage d'Auguste Comte. Histoire de "l'église positiviste" 1849–1946* (Paris, 2001).

209. A typical statement of the special role that women had in the religion of humanity is in the preface to Comte, *Catéchisme positiviste*, xxii–xxxix, from which I take all the phrases cited in this paragraph. See also Comte, *Système de politique positive*, 4:411–13.

210. For Comte, the "final regeneration" would only be possible when four sequential "revolutions" were completed. The first was the "philosophical" revolution, carried through by the philosophers honored in the eleventh month (Descartes); then came the "bourgeois" revolution, with its rejection of the inherited rights of the nobility, followed by the "proletarian" revolution of 1789. Only the "feminine revolution" remained to be completed.

211. Comte, *Catéchisme positiviste*, 193–204 (part 2, sixth interview), and *Système de politique positive*, 4:123–30.

212. On Audiffrent's role in the positivist movement and his differences with Laffitte, see Wartelle, *Héritage d'Auguste Comte*, esp. 113–19 and 140–50, and A.-M. Latour's entry on Audiffrent in *DBF*, vol. 4, cols. 383–85.

213. Although religious observances in themselves were not banned under the empire, they became suspect if there was any suggestion of a group involvement in politics or public affairs.

214. Wartelle, *Héritage d'Auguste Comte*, 123–26.

215. Recorded in a typed list in the Maison d'Auguste Comte, Archives, 4 (17).

216. The main sources on the support of positivists to the costs of the movement are the annual circulars that Laffitte prepared for distribution to all contributors. The figures I give are from the *Trente unième circulaire adressée à chaque coopérateur du libre subside institué par Auguste Comte pour le sacerdoce de l'humanité*, dated 2 January 1879.

217. Terence R. Wright, *The Religion of Humanity: The Impact of Comtean Positivism on Victorian Britain* (Cambridge, 1986), 73–88.

218. Comte, *Catéchisme positiviste*, 205–10, and *Système de politique positive*, 4:156.

219. On the purchase of the house and the inauguration of the chapel, see *Religion de l'humanité. Apostolat positiviste. Circulaire adressée aux occidentaux qui ont contribué à racheter, pour être consacrée au culte de l'Humanité, la maison où est morte Clotilde, rue Payenne n.5, à Paris. Rio, le 27 Descartes 51/117 (31 novembre 1905)*. The mistaken identity of the house was later the subject of exchanges in the *Mercure de France* no. 836 and 840 (1933): 477–84, 745–51, and 763–64.

220. Audiffrent, *Le Positivisme des derniers temps. Discours lus à la rue Jacob* (Paris, 1880), 3.

221. *Testament d'Auguste Comte avec les documents qui s'y rapportent . . . Correspondance avec Mme de Vaux . . . .* 2nd ed. (Paris, 1896), 4, and, in the English edition, *Confessions and Testament of Auguste Comte. And His Correspondence with Clotilde de Vaux*, ed. Albert Crompton (Liverpool, 1910), 473.

222. "Enseignement positiviste public et gratuit. Programme. Cours de géométrie professé par M. P. Laffitte," *RO* 2 (1879): 173–88.

223. Important sources on the schism of the 1870s include Audiffrent, *Circulaire exceptionnelle adressée aux vrais disciples d'Auguste Comte* (Paris, 1886) and *M. Laffitte et l'exécution testamentaire d'Auguste Comte* (Paris, 1903).

224. Harry W. Paul, "Scholarship and Ideology: The Chair of the General History of Science at the Collège de France, 1892–1913," *Isis* 67 (1976): 378–82; Annie Petit, "L'héritage du positivisme dans la création de la chaire d'histoire générale des sciences au Collège de France," *RHS* 48 (1995): 535–47; and, for Comte's letter of 28 October 1832, *Auguste Comte. Correspondance générale et confessions*, ed. Paulo E. de Berrédo Carneiro et al. (Paris, 1973–90), 1:406–8.

225. Emile Corra, "La philosophie des fêtes," *RPI* 122 (1910): 113–23.

226. Emile Corra, "Inauguration du nouveau siège social de la Société positiviste," *RPI* 126 (1914): 60–61.

227. Wartelle, *Héritage d'Auguste Comte*, 188–209.

228. Agathon [Henri Massis and Alfred de Tarde], *Les jeunes gens d'aujourd'hui. Le goût de l'action—La foi patriotique—Une renaissance catholique—Le réalisme politique* (Paris, 1913).

Chapter 5 · Science for All

1. The classic study of French "scientific poetry" remains Casimir Alexandre Fusil, *La Poésie scientifique de 1750 à nos jours. Son élaboration—Sa constitution* (Paris, 1918).

2. Ricard, *La Sphère, poëme en huit chants, qui contient les élémens de la sphère céleste et terrestre, avec des principes d'astronomie physique* (Paris, 1796).

3. Ibid., 317–475.

4. Delille, *Les trois règnes de la nature, avec des notes par M. Cuvier, de l'Institut, et autres savants*, 2 vols. (Paris, 1808). For a study of Delille, see Edouard Guitton, *Jacques Delille (1738–1813) et le poème de la nature en France de 1750 à 1820* (Paris, 1974). On the *Trois règnes*, see also Fusil, *Poésie scientifique*, 64–72.

5. Delille, "Discours préliminaire" in Delille, *Trois règnes de la nature*, 17.

6. The poem had the distinction of being published in full in the proceedings of the annual public meeting of the Institut de France: *Recueil des discours prononcés dans la séance publique annuelle de l'Institut royal de France . . . . 24 avril 1825* (BL 733.g.13 [5]). On its warm reception, see the publisher's preface to Daru, *L'Astronomie. Poème en six chants* (Paris, 1830), i.

7. Daru, *Astronomie*, iii–iv.

8. Ibid., viii.

9. Daru, "Fragment d'un poème sur l'astronomie," in *Institut royal de France. Séance publique annuelle des quatre académies, du mardi 24 avril 1827* (Paris, 1827), 61–81 (BL 733.g.13 [7]).

10. *Astronomicon libri sex. L'astronomie, poème en six chants, de feu M. le comte P. Daru*, trans. Clair-Louis Rohard Meru (Paris, 1839).

11. Daru was the author of an effusive twelve-page *Epître à Jacques Delille* (Paris, 1801), in verse, with the usual explanatory notes characteristic of the genre.

12. Guitton, *Jacques Delille*, 562–66.

13. [Anon.], "France," *Le Globe* 6 (24 Nov. 1827): 1.

14. [Anon.], "A tout le monde," *Le Magasin pittoresque* 1 (1833): 1.

15. Marie-Laure Aurenche, *Edouard Charton et l'invention du* Magasin pittoresque *(1833–1870)* (Paris, 2002), esp. 147–53.

16. This declaration of the *Magasin pittoresque*'s scope and intentions appears in a two-page notice, dated 31 December 1833 but untitled and unsigned, in the bound volume of the first year's issues, ii.

17. [Arnoult], "Prospectus," *L'Institut. Journal des académies et sociétés scientifiques de la France et de l'étranger* 1 (1833): 2.

18. *Echo du monde savant* 1, no. 4 (24 Apr. 1834): 13.

19. Cuvier et al., *Dictionnaire des sciences naturelles, dans lequel on traite méthodiquement des différens êtres de la nature . . . par plusieurs professeurs du Jardin du Roi, et des principales écoles de Paris* (Strasbourg, 1816–30), 1:vii–viii.

20. Jean-Baptiste Bory de Saint-Vincent et al., *Dictionnaire classique d'histoire naturelle par Messieurs Audouin . . . et Bory de Saint-Vincent* (Paris, 1822–31), 1:vii–viii.

21. On the *Dictionnaire classique,* see Pietro Corsi, *The Age of Lamarck: Evolutionary Theories in France, 1790–1830,* trans. Jonathan Mandelbaum (Berkeley, CA, 1988), 218–22, and Lamarck, *Genèse et enjeux du transformisme, 1770–1830,* trans. Diane Médard (Paris, 2001), 268–73.

22. Biot published his reflections on the effect of the *Comptes rendus* and of other facets of the Académie's openness in the *Journal des savants* in 1837 and 1842. See especially *JS* (Feb. 1837): 80–81, and (Nov. 1842): 641–42 and 654–55.

23. *JS* (Feb. 1837), 80, and *JS* (Nov. 1842): 652–53.

24. *JS* (Nov. 1842): 651.

25. Biot, *Mélanges scientifiques et littéraires* (Paris, 1858), 2:292. The phrase appeared in a note, added for the *Mélanges,* in which Biot described how by 1858 his worst fears about trends in the Académie had been realized.

26. [Libri], "Lettres à un Américain sur l'état des sciences en France," *RDM,* 4th ser., 21 (1840): 789–818, esp. 796–812.

27. Ibid., 794.

28. Ibid., 793–94 and 799–805.

29. Arago, "Ampère. Biographie lue par extraits en séance publique de l'Académie des sciences, le 21 août 1839," in *Œuvres [complètes] de François Arago,* ed. J.-A. Barral (Paris and Leipzig, 1854), 2:1–116.

30. [Libri], "Lettres à un Américain," 808–9.

31. Villemain, *Souvenirs contemporains d'histoire et de littérature* (Paris, 1854–55), 1:468.

32. Hugues Chabot, "Jacques Babinet, un savant vulgarisateur," in Jean Dhombres, ed., *Aventures scientifiques. Savants en Poitou-Charente du XVIe au XXe siècle* (Poitiers, 1995), 17–29.

33. Babinet, "Avis au lecteur," in Babinet, *Etudes et lectures sur les sciences d'observation et leurs applications pratiques* (Paris, 1855–68), 5:xi (1858).

34. The phrase is from the unpaginated dedication to Pouchet in Meunier, *La Science et les savants en 1865, 2ᵉ année, 2ᵉ semestre* (Paris, 1866).

35. Meunier, "Deux poids et deux mesures ou la politique de M. Pasteur," *Cosmos,* 3rd ser., 1 (1867–68): 1–3.

36. Pouchet, "Nécessité de réformer l'enseignement supérieur," *Cosmos,* 3rd ser., 1 (1867–68): 9–11.

37. On van Tieghem's unusual career, see the entry on him in Philippe Jaussaud and Edouard-Raoul Brygoo, eds., *Du Jardin au Muséum en 516 biographies* (Paris, 2004), 513–15.

38. Moigno presented himself as extending the pedagogical work of Cauchy in the introduc-

tions to the first two volumes of his *Leçons de calcul différentiel et de calcul intégral, rédigées d'après les méthodes et les ouvrages publiés ou inédits par M. A.-L. Cauchy*, 3 vols. (Paris, 1840–46).

39. On Moigno's life, I draw on the obituary by his friend the abbé Henri Valette, "Mort de M. l'abbé F. Moigno," *Cosmos-Les mondes*, 3rd ser., 1 (1884), 443–456b [*sic*], and the autobiographical notice in Moigno, *Les Splendeurs de la foi. Accord parfait de la révélation et de la science, de la foi et de la raison* (Paris, 1877–82), 4:1–25 (1879).

40. Moigno, *Traité de télégraphie électrique* (Paris, 1849).

41. The demise of *Cosmos* during the Franco-Prussian war allowed Moigno to reappropriate the title four years later. In launching a third series of *Cosmos-Les Mondes* in 1882, Moigno insisted on the continuity with the original *Cosmos* going back for thirty years; see L'abbé Moigno et ses collaborateurs, "A nos lecteurs," *Cosmos-Les Mondes*, 3rd ser., 1 (1882): 1–2.

42. *Cosmos*, 1 (1852): ii.

43. Ibid.

44. Ibid., i. Cf. Moigno's description of the character of the journal in his editorial in the May 1853 issue: "*Cosmos* thinks, speaks, judges, criticizes, encourages, blames, reprimands, corrects, praises"; *Cosmos*, 2 (1852–53), 531.

45. On the spiritual principles underlying Hirn's science, see Faidra Papanelopoulou, "Gustave-Adolphe Hirn (1815–90). Engineering thermodynamics in mid-nineteenth-century France," *BJHS* 39 (2006): 249–53.

46. On the Actualités scientifiques series of books, see later in this chapter. Moigno, *La Foi et la science. Explosion de la libre pensée en août et septembre 1874* (Paris, 1875), viii.

47. John William Draper, *Les Conflits de la science et de la religion* (Paris, 1875). The speed with which Draper's book was translated (in the same year as it had appeared in English) is striking.

48. Moigno, *Les Splendeurs de la foi* (Paris, 1877–82), 1:i–iv: "A notre très cher fils FRANÇOIS MOIGNO, chanoine de Saint-Denis. Léon XIII, Pape."

49. Ibid., vol. 1, appendices, 69–76, where Pius IX's encyclical is reproduced as "Encyclique addressée à tous les cardinaux, archevêques et évêques de France par notre Saint Père le Pape Pie IX. 21 mars 1853."

50. Ibid., 4:937–1259 ("Vérités absolues des livres saints"), and vol. 5 (devoted entirely to "Le miracle au tribunal de la science").

51. Ibid., 2:512–29.

52. *Cosmos. Revue des sciences et de leurs applications*, n.s., 1 (1885): 3–7.

53. On Flammarion's life, I draw mainly on his *Mémoires biographiques et philosophiques d'un astronome* (Paris, 1911).

54. Ibid., 210–11.

55. Ibid., 210.

56. Ibid., 211.

57. The imagery appears in Flammarion, *Astronomie populaire. Description générale du ciel* (Paris, 1880), 2.

58. Flammarion, *Dieu dans la nature* (Paris, 1867), xviii–xix. The italics are Flammarion's. Reynaud expounded his doctrines most fully in his *Philosophie religieuse. Terre et ciel* (Paris, 1854).

59. For a late and particularly virulent statement of his opposition to materialism ("a false, incomplete, and inadequate doctrine") and the positivism that fostered it, see Flammarion, *La Mort et son mystère* (Paris, 1920), 1:33–56.

60. Flammarion, *La Pluralité des mondes habités, étude où l'on expose les conditions d'habitabilité des terres célestes, discutées au point de vue de l'astronomie et de la physiologie* (Paris, 1862).

61. Ibid., 32.

62. Ibid., 42.

63. Flammarion, *Les Terres du ciel. Description astronomique, physique, climatologique, géographique des planètes qui gravitent avec la terre autour du soleil et de l'état probable de la vie à leur surface*, 1st ed. (Paris, 1877), 436–38, and *Terres du ciel*, 14th ed. (Paris, 1884), 193–94. I cite both editions, since although Flammarion's speculations about the conditions and forms of life on the different planets changed little, the presentation was significantly altered.

64. Flammarion, *La Planète Mars et ses conditions d'habitabilité. Synthèse générale de toutes les observations: climatologie, météorologie, aréographie, continents, mers et rivages, eaux et neiges, saisons, variations observées . . .* (Paris, 1892–1909), 1: 591.

65. Flammarion, *Terres du ciel* (1877), 497–99, and *Terres du ciel* (1884), 621–22.

66. Flammarion, *Terres du ciel* (1877), 369–72, and *Terres du ciel* (1884), 525–28.

67. "La vie dans l'infini," the final chapter of Flammarion, *Terres du ciel* (1877), 581–96, and *Terres du ciel* (1884), 753–69.

68. Flammarion, "Les autres mondes," *Almanach astronomique Flammarion* 1 (1884), 131–35 (131).

69. For a statement of this belief, see the "Avertissement" in *La Pluralité des mondes habités* (17th ed., Paris, 1872). The text appeared in numerous later editions as well.

70. Flammarion, *Mémoires d'un astronome*, 224–25.

71. Hermès [Camille Flammarion], *Des forces naturelles inconnues. A propos des phénomènes produits par les frères Davenport et par les médiums en général. Etude critique* (Paris, 1865).

72. A good contemporary account of the powers of the Fox sisters is in Louis Figuier, *Histoire du merveilleux dans les temps modernes* (Paris, 1860), 4:221–41.The eldest of the sisters, who bore the married name of Fish, was the daughter of Mrs. Fox by an earlier marriage.

73. For contemporary accounts, see Thomas Low Nichols, *A Biography of the Brothers Davenport. With Some Account of the Physical and Psychical Phenomena Which Have Occurred in their Presence, in America and Europe* (London, 1864) and Robert Cooper, *Spiritual Experiences, including Seven Months with the Brothers Davenport* (London, 1867).

74. Hermès [Flammarion], *Forces naturelles inconnues*, esp. 5–13 and 136–44.

75. Kardec expressed his favorable view of the brothers in a note preceding his appendix to the French translation of Nichols, *Brothers Davenport. Phénomènes des frères Davenport et leurs voyages en Amérique et en Angleterre*, trans. Madame Bernard Desrosne (Paris, 1865), 323.

76. Kardec, *Le Livre des esprits, contenant les principes de la doctrine spirite* (Paris, 1857); reproduced in the best modern edition, Kardec, *Livre des esprits* (Paris: Editions Trajectoire, 1998). On Kardec's life and work, see Thomas A. Kselman, *Death and the Afterlife in Modern France* (Princeton, NJ, 1993); John Warne Monroe, *Laboratories of Faith: Mesmerisim, Spiritism, and Occultism in Modern France* (Ithaca, NY, 2008), chaps. 3 and 4; and Sofie Lachapelle, *Investigating the Supernatural: From Spiritism and Occultism to Physical Research and Metaphysics in France, 1853–1931* (Baltimore, MD, 2011), 7–33.

77. Flammarion, *Forces naturelles inconnues*, 2 vols. continuously paginated (1907, 1917, and 1921), and *L'Inconnu et les problèmes psychiques*, 2 vols. (Paris, 1917).

78. Kardec's motto for the spiritist movement—"Charity our sole salvation"—also expressed Flammarion's ideals and those of his wife (a vehement champion of international disarmament) to perfection.

79. Babinet, "Les tables tournantes et les manifestations prétendues naturelles, considérées au point de vue des principes qui servent de guide dans les sciences d'observation," in Babinet, *Etudes et lectures*, 2:9–56 (1856).

80. For example, in the long footnote in Flammarion, *Terres du ciel* (1884), 181–82.

81. Flammarion, *Forces naturelles inconnues* (1907 ed.), 44–48.

82. Richet, "Expériences de Milan," *Annales des sciences psychiques* 3 (1893): 1–31. Flammarion's fullest account of Palladino's séances is in his *Forces naturelles inconnues* (1907 ed.), 89–260. On Palladino and the context of her activities, see (in a rich modern literature) Christine Blondel, "Eusapia Palladino. La méthode expérimentale et la 'diva' des savants," in Bernadette Bensaude-Vincent and Christine Blondel, eds., *Des Savants face à l'occulte 1870–1940* (Paris, 2002), 143–71; Monroe, *Laboratories of Faith,* 199–250 ; and Lachapelle, *Investigating the Supernatural,* 59–112.

83. Flammarion, *Forces naturelles inconnues* (1907 ed.), 259–60.

84. Ibid., 310.

85. Flammarion, *Lumen. Récits de l'infini. Histoire d'une comète dans l'infini* (Paris, 1873). Flammarion continued to work on *Lumen,* later editions of which show significant modifications.

86. Flammarion, *Rêves étoilés* (Paris, 1886), *Uranie* (Paris, 1889), and *Stella* (Paris, 1897).

87. Flammarion, *Lumen,* "Quatrième récit."

88. Flammarion, *Uranie,* 365.

89. A point made in Brian Stableford's introduction to his modern English edition of *Lumen* (Middletown, CT, 2002), xxvi.

90. On Figuier's life, see the obituary of him in his own annual publication, *ASI* 38 (1894; published 1895): vii–xii.

91. The pamphlet *Exposé des titres scientifiques de Mr Louis Figuier, candidat à l'Académie de médecine* (1856) summarizes Figuier's early career and publications, the first of which appeared in 1844.

92. Figuier, *Exposition et histoire des principales découvertes scientifiques modernes,* 4 vols. (Paris, 1851–57). Although the fourth volume was delayed until 1857, the whole work was in a fifth edition by 1858 and a sixth by 1862.

93. Figuier, *Exposition et histoire,* 1:i–iii, in a preface dated "Paris, 1er juillet 1851."

94. See above, note 72. The fourth volume is devoted to a critical examination of turning tables and mediums. The earlier volumes trace the history of superstition and beliefs in the extraordinary and the supernatural since antiquity.

95. Figuier, *Histoire du merveilleux,* vol. 3, esp. 238–69, and, for an account of Puységur's claims and the persistence of doctrines of animal magnetism in the postrevolutionary period, Monroe, *Laboratories of Faith,* 67–76.

96. Figuier, *Histoire du merveilleux,* 4:307–23.

97. Ibid., 4:345–48.

98. Cahagnet, *Magnétisme. Arcanes de la vie future dévoilés,* 3 vols. (Paris, 1848–54). The account of the exchanges with Swedenborg is on 3:205–61.

99. Baron du Potet de Sennevoy, "Bibliographie," *Journal du magnétisme* 8 (1849): 27.

100. See, for example, Cahagnet, *Révélations d'outre-tombe par les esprits Galilée, Hippocrate, Franklin, etc.* (Paris, 1856).

101. Cahagnet, *Magnétisme,* 3:318–34.

102. Figuier, *Histoire du merveilleux,* 4:308.

103. Figuier, *Le Lendemain de la mort ou la vie future selon la science* (Paris, 1872), 1–7, comprising the introduction to the book. This and subsequent references are to the fourth French edition. The words and phrases that I cite in English are from the English translation cited below, in note 108.

104. Figuier, *Lendemain de la mort,* 73–76.

105. Ibid., 79–82 and 105–30. Figuier conjectured that the planetary ether might be composed of hydrogen in a highly rarefied state; ibid., 68.

106. Ibid., 131–36.

107. Ibid., 185–202, esp. 186.

108. Figuier, *The Day after Death. Or, Our Future Life According to Science* (Paris, 1872).

109. Figuier, *Lendemain de la mort*, 27–28.

110. Figuier, *La Terre avant le déluge* (Paris, 1863), 361–65; *L'Homme primitif* (Paris, 1870): 26–36; and *Les Races humaines* (Paris, 1872), esp. 1–18. Figuier adopted d'Halloy's distinction between the five races of humanity rather than Quatrefages's division into three. Like d'Halloy and Quatrefages, Figuier defined the races by the color of their skin: white, yellow, brown, red, and black and saw the differences as having emerged through the influence of climate on the original, white Caucasian race. He could not explain how the races had diversified to acquire such distinctive characteristics while remaining members of the same species. But, in an obvious reference to the polygenists, he insisted that the lack of an explanation was no reason for following "the obsession of today's savants with wanting to explain everything"; see Figuier, *Races humaines,* 10.

111. Figuier, *Homme primitif,* 31–36. Cf. the English translation of the book, where "this strange genealogy" (p. 31) became "this strange, repugnant genealogy"; see Figuier, *Primitive Man: Revised Translation* (London, 1870), 30. Quatrefages expressed his rejection of a simian ancestry for the human race in his *Rapport sur les progrès de l'anthropologie* (Paris, 1867), 252.

112. Figuier, *Terre avant le déluge,* 361, as also (in a slightly modified passage) in the ninth edition (Paris, 1883), 451.

113. See, for example, Figuier, *Terre avant le déluge,* 9th ed. (1883), 449–63.

114. Ibid., 361.

115. Figuier, *Races humaines,* 5–18.

116. Desnoyers, "Note sur les indices matériels de la coexistence de l'homme avec l'Elephas meridionalis dans un terrain des environs de Chartres, plus anciens que les terrains de transport quaternaires des vallées de la Somme et de la Seine," *CR* 56 (1863): 1073–83.

117. Figuier, *Homme primitif,* 21. Some of the most influential of Figuier's contemporaries, among them Elie de Beaumont and the professor of geology at the Sorbonne Edmond Hébert, shared his skepticism. But subsequent discoveries (of stone axes in close proximity to the remains of an extinct quadruped) by the abbé Louis Bourgeois near Pontlevoy (Eure-et-Cher) in 1867 soon tipped the balance in favor of Desnoyers. For an indignant account of the lingering resistance to Desnoyers's conclusions, see Victor Meunier, *Les Ancêtres d'Adam. Histoire de l'homme fossile* (Paris, 1875), 257–74. Later in the book, Meunier presented the resistance as yet another manifestation of the refusal on the part of official science to give Boucher de Perthes (a member of "that noble race of free savants") his due; see page 280.

118. Parville, *Causeries scientifiques. Découvertes et inventions. Progrès de la science et de l'industrie. Première année* (Paris, 1862), v.

119. The most elaborate version of Arago's lectures appeared posthumously as Arago, *Astronomie populaire,* ed. J.-A Barral, 4 vols. (Paris and Leipzig, 1854–57). Comte published his lectures, which he inaugurated in 1830, as *Traité philosophique d'astronomie populaire* (Paris, 1844).

120. "Conférence," in Pierre Larousse, *Grand dictionnaire universel du XIXe siècle,* 17 vols. (1866–79), vol. 16, Supplement (1877), 590.

121. The estimate, by Louis Michel, is reproduced from *L'International* in Charles Louandre, "Les conférences libres," *JGIP* 34 (1865): 62.

122. Scoutetten, *Histoire chronologique des lectures publiques et des conférences* (Paris, 1867), 29–48. On Poncelet's lectures, see chap. 1.

123. Pierre Vinçard, "Associations pour l'enseignement gratuit des ouvriers," *L'Ami des sciences. Journal du dimanche* 6 (1860): 113–15; originally published in the Genevan newspaper *L'Espérance.* The article by Vinçard (who had himself attended lectures of the Association polytech-

nique) recorded the speech by the Minister of Public Instruction, Rouland, commending science as offering both utility and "relaxation for the mind." For the text of Rouland's address and other speeches, see *MU*, 23 January 1860.

124. Several of the seven series of collected lectures that were published between 1860 and 1867, as *Association polytechnique. Entretiens populaires publiés par Evariste Thevenin*, appeared in the "Bibliothèque des chemins de fer."

125. Henri Haran, "Enseignement populaire," in Eugène Lacroix, ed., *Etudes sur l'Exposition de 1867. Annales et archives de l'industrie au XIXe siècle* (Paris, 1868), 6, 114–45, esp. 128.

126. Deschanel, *Conférences*, 43, and [Anon.], "Inauguration des conférences du boulevard des Capucines," *RCL* 5 (1867–68): 186–88.

127. Louis Simonin, "Conférences du boulevard des Capucines," *RCL* 5 (1867–68): 356–60; Deschanel, *Conférences*, 32–34 and 42–46; Francisque Sarcey, *Souvenirs d'âge mûr* (Paris, 1892), 4–75.

128. Quoted in Bruno Béguet, "La science mise en scène. Les pratiques collectives de la vulgarisation au XIXe siècle," in Béguet, ed., *La Science pour tous. Sur la vulgarisation scientifique en France de 1850 à 1914* (Paris, 1990), 135.

129. Flammarion, *Mémoires d'un astronome*, 345–46. *Les Merveilles célestes. Lectures du soir* (Paris, 1865) was a mainly straightforward expository text appropriate to the Hachette "Bibliothèque des merveilles" (see discussion later in the chapter), though with two more speculative final chapters (including one on the plurality of worlds) on the "philosophical aspect of creation."

130. Deschanel, *Conférences*, 55, and Francisque Sarcey, "Conférences du boulevard des Capucines (séance de réouverture)," *RCL* 6 (1868–69): 19–24.

131. "Conférence de M. Fremy," in Meunier, *La Science et les savants en 1866, 3e année* (Paris, 1867), 109–11. See also "Conférences publiques données dans la salle du Conservatoire de musique," in *ASI* 11 (1866): 458–61.

132. *Cosmos* 1 (1852): 3–12. *Cosmos*, in fact, was the successor to a short-lived journal launched by de Montfort: *La Lumière. Revue de la photographie. Beaux-arts, héliographie, sciences* (1850–51).

133. *Cosmos* 2 (1852–53): 529–33.

134. Moigno, "Royal Panopticon des Sciences," *Cosmos* 4 (1854): 38–40, composed largely of an extract from Gilberto Govi's article, "Le Panopticon de Londres," *L'Illustration* 22 (17 Dec. 1853), 407–8.

135. See the admiring account by Victor Fournel, reproduced from *Le Français*, 6 December 1892, in Moigno, *Splendeurs de la foi*, 4:20–25.

136. Moigno, *Enseignement de tous par les projections. Les sciences, les industries, les arts enseignés et illustrés par quatre mille cinq cents photographies sur verre* (Paris, n.d.), iii; also the preface to the catalogue in *Cosmos-Les mondes*, 3rd ser., 2 (1882): 1–4. The figure of 50,000 is cited in Béguet, "Science mise en scène," 136, from a document in AN 18* III 137.

137. *Exposé de la situation de l'Empire présenté au Sénat et au Corps législatif. Février 1867* (Paris, 1867), 219.

138. *Exposé de la situation de l'Empire . . . Novembre 1867* (Paris, 1867), 170.

139. Duruy to the rector of the academy of Montpellier, 6 April 1864, in *Circulaires et instructions officielles relatives à l'instruction publique* (Paris, 1863–1902), 6:100–101.

140. Prefect of the Bas-Rhin to the Minister of Public Instruction, 2 December 1862, AD Bas-Rhin 1 TP/SUP.2.

141. Prefect of the Bas-Rhin to the Minister of Public Instruction, 9 May 1866, AD Bas-Rhin 1 TP/SUP.2.

142. On this and other examples of Duruy's readiness to intervene in defense of order and

decency, see the critical comment and accompanying extracts from the daily press in Louandre, "Conférences libres."

143. Meunier, *La Science et les savants en 1864* (Paris, 1865), 242–57. The success of Jamin's lecture was such that it was repeated three days later, to a similarly rapturous reception; see *MS* 6 (15 Mar. 1864): 281–84.

144. Meunier, *La Science et les savants en 1864*, 244.

145. Ibid., 251.

146. Henri Ernest Valette, "Soirée scientifique de l'Observatoire," *Cosmos-Les mondes,* 3rd ser., 1 (1882): 425–32, and "La grande soirée annuelle de l'Observatoire," *L'Astronomie* 1 (1882): 72–73 (the latter unsigned but almost certainly by Flammarion as editor of the journal).

147. Valette, "Soirée scientifique," 432.

148. Parville, *Causeries scientifiques. Première année*, 98.

149. On the origin of most of the pieces in these volumes, see Jacques Babinet, "Avis au lecteur," in Babinet, *Etudes et lectures*, 5:ix–xii (1858).

150. Flammarion, *Etudes et lectures sur l'astronomie*, 9 vols. (Paris, 1867–80). See 1:viii, for Flammarion's respectful reference to the pioneering role of Babinet and the support of Babinet's (and now his) publisher Gauthier-Villars.

151. The *Année scientifique et industrielle* continued after Figuier's death in 1894 under the editorship of Emile Gautier, who tempered its severity by introducing illustrations, including photographs. The war put an end to the publication, as it did to a latecomer to the genre, Max de Nansouty's *Actualités scientifiques*, nine volumes of which appeared between 1904 and the year of his death, 1913.

152. Liais, *L'Espace céleste et la nature tropicale. Description physique de l'univers d'après des observations personnelles faites dans les deux hémisphères* (Paris, 1865), 478–96.

153. Eugène Bouvard, "Nouvelles tables d'Uranus," *CR* 21 (1845): 524–25.

154. Liais, *Espace céleste*, 490.

155. Ibid., passim but esp. 555–87.

156. Figures recorded at the beginning of most of Flammarion's publications.

157. Gaston Tissandier, "L'enseignement supérieur en France," *La Nature* 1 (1873): 1–2.

158. Gaston Tissandier, "Notre quatorzième année," *La Nature* 14, 1ᵉʳ semestre (1886): 1–2 (1).

159. Florence Colin, "Les revues de vulgarisation scientifique," in Béguet, *Science pour tous*, 84.

160. A revealing testimony to Masson's standing is the memorial volume published shortly after his death in 1900: *A la mémoire de Georges Masson, libraire, éditeur, président de la Chambre de commerce de Paris . . . Né à Paris le 6 janvier 1839, mort à Paris le 6 janvier 1900* (BnF Rés. Ln27 49201).

161. See the discussion on the *Talisman* and the *Travailleur* later in the chapter.

162. A complete set, though with some specimens missing, is in the British Library (BL Tab.1800.c.1).

163. Emile Deyrolle fils, *L'Enseignement des sciences naturelles dans les écoles primaires* (Paris, [1871]).

164. *PNE* 1, no. 62 (15 Oct. 1872): 250.

165. Ibid.

166. The collection, belonging to a noted private collector, F. Monchicourt, was advertised in *Le Naturaliste* 1 (1879–81): 2.

167. *PNE* 1, no. 13 (1 Jan. 1870): 52.

168. *Les Etoiles et les curiosités du ciel. Description complète du ciel visible à l'œil nu et de tous les objets célestes faciles à observer; supplément de l'Astronomie populaire* (Paris, 1882), 671–85.

169. Flammarion, *Astronomie populaire*, 835.

170. Flammarion, *Etoiles et curiosités du ciel*, 684.

171. Flammarion, *Atlas céleste comprenant toutes les cartes de l'ancien atlas de Ch. Dien, rectifié, augmenté et enrichi . . . par Camille Flammarion* (Paris, 1877). Dien had published the first edition of his *Atlas* in 1864. In this third edition, Flammarion incorporated the fruits of new observations, especially of double stars, that he and others had made since 1864. Its price, 45 francs in its bound form or 40 francs for the loose sheets, made it an item that only dedicated observers were likely to buy.

172. Flammarion, *Etoiles et curiosités du ciel*, 685n.

173. The texts of all the plays are reproduced, each separately paginated and with a brief introduction and record of performances, in Figuier, *La Science au théâtre. Drames* (Paris, 1889) and *La Science au théâtre. Comédies* (Paris, 1889).

174. See the introduction to "Les six parties du monde," in Figuier, *La Science au théâtre. Drames*, vi–vii.

175. Figuier, "Préface," in *La Science au théâtre. Comédies*, v–xx, esp. xiii, where Figuier criticizes Claude Bernard and Michel Chevreul for their hostility to the popularization of science.

176. The aquarium's popular success was not matched by its scientific performance. The high mortality rate of the fish provoked the criticism that aesthetic considerations had been given undue priority. See *L'Aquarium du Trocadéro en 1900* (Paris, 1900), 7.

177. Georges Mareschal, "L'Aquarium d'eau de mer," *La Nature* 19, no. 2 (1900): 252–53.

178. [Anon.], "Observatoire populaire du Trocadéro, à Paris," *La Nature* 9 (1880–81): 13–14, and J. B., "Observatoire populaire du Trocadéro," *La Science populaire* 1 (1880–81): 367, 430–31, and 582–83. My comments draw mainly on the publicity brochures, correspondence, and newspaper clippings in the file on the observatory, AN F^{17} 2755.

179. *L'Astronomie* 2 (1883): 1. Cf. the figures in the same period of 15,000 for *La Nature* and 17,000 for the popular illustrated weekly, *L'Illustration*.

180. On the founding of the society, see [Anon., presumably Flammarion], "La Société astronomique de France" 6 (1887): 406–13.

181. [Anon., presumably Flammarion], "L'observatoire de Juvisy," *L'Astronomie* 6 (1887): 321–30; also Philippe de la Cotardière and Patrick Fuentes, *Camille Flammarion* (Paris, 1994), 162–64.

182. The Hôtel des sociétés savantes occupied the largely rebuilt Hôtel Panckoucke, following the building's refurbishment in 1887. By 1892, it housed thirty-two societies, with access to a large communal lecture theater and individual premises adjusted to each society's needs and financial circumstances. Although the Ministry of Public Instruction and the city of Paris backed the venture, the management committee for the Hôtel, chaired by Charles Gariel (see chap. 6), sought to raise a capital of 800,000 francs in shares of 500 francs each, in addition to the 200,000 francs that had been spent by 1892. Printed and manuscript material concerning the Hôtel is in AN F^{17} 3023.

183. Georges Fournier, "Un demi-siècle à l'observatoire de la Société astronomique de France," *L'Astronomie et Bulletin de la Société astronomique de France* 64 (1950): 325–40, and 386–94.

184. For a rapturous contemporary account of the new gallery of zoology, see Gaston Tissandier, "Nouvelles galeries de zoologie du Muséum d'histoire naturelle de Paris," *La Nature* 17, no. 2 (1889): 311–14.

185. Gaston Tisserand, "Conservatoire des arts et métiers. Nouvelle galerie des constructions civiles," *La Nature* 12, no. 1 (1884): 66–67; Georges Mareschal, "Les nouvelles galeries du Conser-

vatoire des arts et métiers à Paris," *La Nature* 19, no. 2 (1891): 39–42; and Béguet, "Le Conservatoire, Sorbonne et Musée de l'industrie vivante," in Béguet, *Science pour tous*, 148–50.

186. François Caron and Christine Berthet, "Electrical Innovation: State Innovation or Private Initiative? Observations on the 1881 Paris Exhibition," *History and Technology* 1 (1984): 307–18.

187. Henri Filhol, "Explorations sous-marines. Voyage du 'Talisman,'" *La Nature* 12, no. 1 (1884), 198–202.

188. A helpful overview of these series is Valérie Tesnière, "Diffuser la science," in Frédéric Barbier et al., eds., *Le Livre et l'historien. Etudes offertes en l'honneur du Professeur Henri-Jean Martin* (Geneva, 1997), 779–93.

189. The advertisement for the series in Achille Cazin, *La Spectroscopie* (Paris, 1878) lists sixty-six titles. In fact, the early volumes, beginning with William Huggins, *Analyse spectrale des corps célestes*, trans. François Moigno (Paris, 1866), did not bear the general title Actualités scientifiques, but they were incorporated in the series retrospectively from 1869.

190. Aurenche, *Edouard Charton*, 409–44.

191. Béguet, "Le livre de vulgarisation scientifique," in Béguet, *Science pour tous*, 50–70 (69), and Jean Mistler, *La Librairie Hachette de 1826 à nos jours* (Paris, 1964), 121–36 and 297–323.

192. Wurtz, *La Théorie atomique* (Paris, 1879). It is a reflection on the undoctrinaire character of the Bibliothèque scientifique internationale that both Wurtz and Berthelot, the archenemy of atomism, could have books in the same series. Berthelot's *Synthèse chimique* was published in 1876.

193. Valérie Tesnière, *Le Quadrige. Un siècle d'édition universitaire 1860–1968* (Paris, 2001), 83.

194. Ibid., 48–52 and 82–83, where Tesnière estimates that in 1883 alone 85,000 copies of titles in the series were printed for sale in France, with average printings of about 1,800 per volume. See also, for a similar point, Tesnière, "Diffuser la science," 783.

195. On the Baillières' internationalism, see the essays in Danielle Gourevitch and Jean-François Vincent, eds., *J.-B. Baillière et fils, éditeurs de médecine. Actes du colloque international de Paris (29 janvier 2005)* (Paris, 2006) and Josep Simon, "The Baillières. The Franco-British Book Trade and the Transit of Knowledge," in Robert Fox and Bernard Joly, eds., *Echanges franco-britanniques entre savants depuis le XVIIe siècle / Franco-British Exchanges in Science since the Seventeenth Century* (London, 2010), 243–62.

196. For an account of the Bibliothèque de philosophie scientifique, with statistics concerning the series, see Elisabeth Parinet, *Librairie Flammarion 1875–1914* (Paris, 1992), 235–43.

197. Foucault, *Mesure de la vitesse de la lumière. Etude optique des surfaces. Mémoires de Léon Foucault* (Paris, 1913).

198. Early successes in the series included two collective volumes on methodology in various scientific and scholarly disciplines and a translation of Wilhelm Ostwald's *Die Energie*. See P. Félix Thomas et al., *De la méthode dans les sciences* (Paris, 1909); Benjamin Baillaud et al., *De la méthode dans les sciences* (Paris, 1911); and Ostwald, *L'Energie*, trans. E. Philippi (Paris, 1910).

199. Béguet, "Livre de vulgarisation scientifique," 69.

200. Béguet, "La vulgarisation scientifique en France de 1850 à 1914. Contexte, conceptions et procédés," in Béguet, *Science pour tous*, 14–15, and Colin, "Revues de vulgarisation scientifique," ibid., 74–75.

201. Berthelot, *Les Origines de l'alchimie* (Paris, 1885), v and vi.

202. Colin, "Revues de vulgarisation scientifique," 75 and 83–84.

203. See, for example, Bernadette Bensaude-Vincent, "Florilège des societes industrielles," in *Le Livre des expositions universelles 1851–1989* (Paris, 1983), 276–86, and Paolo Brenni, "Dal Crys-

tal Palace al Palais de l'optique. La scienza alle esposizioni universali, 1851–1900," *Memoria e ricerca* 17 (2004): 35–63, esp. 54–56.

Chapter 6 · The Public Face of Republican Science

1. René Vallery-Radot, *La Vie de Pasteur* (Paris, 1900), 355–58. The sum corresponded roughly to the annual salary for the chair at the Sorbonne from which Pasteur's resigned in 1875.

2. August Wilhelm Hofmann, "Jean-Baptiste-André Dumas," *MS*, 3rd ser., 10 (1880): 416–19.

3. A detailed study of the scientific contribution to the French war effort is Maurice P. Crosland, "Science and the Franco-Prussian War," *SSS* 6 (1976): 185–214.

4. On this and the other more bizarre proposals that emanated from the patriotic clubs, see Crosland, "Science and the Franco-Prussian War," 196–97, and Melvin Kranzberg, *The Siege of Paris, 1870–1871: A Political and Social History* (Ithaca, NY, 1950), 88–90.

5. The work of this committee and the associated role of the Société des ingénieurs civils (which continued to hold meetings despite the siege) are described in *Société des ingénieurs civils. Séance d'installation du 7 juillet. Discours prononcés par M. Vuillemin et M. Yvon Villarceau* (Paris, 1871), esp. Vuillemin's account (6–7).

6. Berthelot, *Sur la force de la poudre et des matières explosives* (Paris, 1872), 7n and 32n.

7. A contemporary account in this vein was G. Grimaux de Caux, *L'Académie des sciences pendant le siège de Paris* (Paris, 1871).

8. Comte d'Hérisson [Maurice d'Irisson] *Journal d'un officier d'ordonnance (juillet 1870–février 1871)* (Paris, 1896), 253–63.

9. Auguste-Alexandre Ducrot, *Le Défense de Paris (1870–1871)*, (Paris, 1875–78), 1:185–86, and Henri Azeau, *Les Ballons de l'espoir* (Paris, 1987).

10. Louis Figuier's account in *ASI* (1870–71): 1–15 is especially vivid.

11. Dagron was the author of a *Traité de photographie microscopique* (Paris, 1864), essentially a sales brochure for the apparatus available from his Parisian workshop.

12. Mommsen, "Letter to the Editor of the Milanese Paper 'Il Secolo,' Inserted in the Number for the 20th of August, 1870," in Mommsen et al., *Letters on the War between Germany and France* (London, 1871).

13. Strauss, *Krieg und Friede. Zwei Briefe an Ernest Renan nebst dessen Antwort auf den ersten* (Leipzig, 1870), in which the two letters are reprinted. The first letter appeared in the *Augsburger Allgemeine Zeitung* of 18 August 1870 and, in French, in the *Journal des débats* of 15 September 1870. Renan replied in the *Journal des débats* of 16 September 1870; this reply was translated into German and published in Strauss's *Krieg und Friede*. Although Strauss's second letter appeared soon afterward, in the *Augsburger Allgemeine Zeitung* of 2 October 1870, Renan did not reply to it until almost a year later, for the reason I mention in note 21, below.

14. Strauss, *Krieg und Friede*, 58–59.

15. Vogt, *Carl Vogt's politischer Briefe an Friedrich Kolb* (Keil, 1870), containing twelve letters from Vogt to the liberal writer and political figure Georg Friedrich Kolb dating from between 10 and 24 October 1870. The letters were translated into French by Alfred Marchand, an editor of *Le Temps*, and published in the Swiss journal *Le Courrier du commerce* (in the German-language edition of which they had originally appeared) in 1871. For an example of warm French support for Vogt's stand against German predatoriness with regard to Alsace and parts of Lorraine, see François Garrigou, "Les lettres de Karl Vogt," *RPL*, 2nd ser., 1 (1871): 236–38.

16. Du Bois-Reymond, *Über den deutschen Krieg. Rede am 3 August 1870 . . . gehalten* (Berlin, 1870); translated into French as Du Bois-Reymond, "La guerre de 1870," *RCL* 7 (1870): 658–68 and into English as Du Bois-Reymond, *The German War*, trans. Gustav Solling (Berlin, 1870).

17. Du Bois-Reymond, "De l'organisation des universités," *RCS* 7 (1870): 322–35. Du Bois-Reymond had delivered the address in Berlin in October 1869.

18. The bitterness of French feeling is conveyed in Alfred Marchand, *Le Siège de Strasbourg 1870. La bibliothèque—La cathédrale,* 2nd ed. (Paris, 1870), esp. 46–53, and the detailed account of the destruction of the library by the French rector of the academy of Strasbourg, Jules-Sylvain Zeller, incorrectly printed as T. Zeller, ibid., 123–37.

19. Stark, *Die psychische Degeneration des französischen Volkes, ihr pathologischer Charakter, ihre Symptome und Ursachen. Ein irrenärtzlicher Beitrag zur Völkerpathologie* (Stuttgart, 1871).

20. [Anon], "Dé la dégénérescence intellectuelle des Allemands," *RPL* 1, no. 1 (1871): 182–86.

21. Renan, "Nouvelle lettre à M. Strauss," in Renan, *La Réforme intellectuelle et morale* (Paris, 1871), 196–208. This reply to Strauss's second letter was delayed, since Renan was unaware of the letter until after the armistice of Feb. 1871.

22. Quatrefages, "La race prussienne," *RDM* 2ᵉ période 91 (1871): 647–69. The article subsequently appeared in book form, with minor changes, as Quatrefages, *La Race prussienne* (Paris, 1871).

23. Virchow, "Les crânes finnois et esthoniens comparés aux crânes des tombeaux du nord-est de l'Allemagne," *RS,* 2nd ser., 2 (1872): 313–18, followed immediately in the same journal by "Observations de M. de Quatrefages," 318–20. Virchow's text had originally appeared in German in *Berliner Gesellschaft für Anthropologie, Ethnologie und Urgeschichte. Sitzung vom 10. Februar 1872* (Berlin, 1872), 8–18.

24. Virchow, "Crânes finnois," 318.

25. Virchow, *Die Aufgabe der Naturwissenschaften in dem neuen nationalen Leben Deutschlands. Rede, gehalten in der zweiten allgemeinen Sitzung der 44. Versammlung deutscher Naturforscher und Aerzte zu Rostock am 22. September 1871* (Berlin, 1871), 7–8.

26. Ibid., 33–34. Cf. an earlier, more explicit statement in Virchow, *Ueber die nationale Entwicklung und Bedeutung der Naturwissenschaften* (Berlin, 1865), 25–26, where Virchow invoked Protestantism as a cause of the strength of Britain and the Netherlands as scientific nations.

27. The work occupied more than a third of the volume of essays that appeared in 1871 as Renan, *Réforme intellectuelle et morale.*

28. For a flavor of the views that Renan expressed before 1870, see his preface to Renan, *Questions contemporaines* (Paris, 1868), i–xxxi.

29. Renan, *Réforme intellectuelle et morale,* 97–98.

30. Ibid., 124.

31. Vallery-Radot, *Vie de Pasteur,* 269–70 and 280–81.

32. *Bulletin de la Société d'acclimatation,* 2nd ser., 8 (1871): 63–65 (general meeting of the society, 27 Jan. 1871).

33. *BAM* 36 (1871): 130–33 and 144–48 (meetings of 7 and 14 Mar. 1871) and *Mémoires de l'Académie des sciences, belles-lettres et arts de Clermont-Ferrand* 13 (1871): 459–60 (meeting of 2 Mar. 1871).

34. *BAM* 36 (1871): 132 and 145.

35. Quatrefages, "La science et le patrie" [presidential address], in *AFAS, 1ère session. Bordeaux 1872,* 36–41 (38).

36. Pouchet, "Rapport sur une mission en Allemagne pour étudier les collections d'anatomie comparée," *RIE* 1 (1881): 486–503.

37. Ibid., 488.

38. *Flaubert. Correspondance,* Pléiade edition, ed. Jean Bruneau et al. (Paris, 1973–2007), 4:63–65.

39. "Lettre à la jeunesse," in Zola, *Roman experimental* (Paris, 1880), 97.

40. Pasteur, "Pourquoi la France n'a pas trouvé d'hommes supérieurs au moment du péril," in Pasteur, *Quelques réflexions sur la science en France* (Paris, 1871), 25–50.

41. *CR* 72 (1871): 237–39 (6 Mar. 1871) and 261–69 (13 Mar. 1871).

42. Antoine-Joseph Yvon Villarceau, "Note sur la destruction du cercle méridien no. 2 de Rigaud par les incendiaires de la Commune," *CR* 72 (1871): 611–12 (29 May 1871), and [Anon.], "L'Académie des sciences pendant l'armistice et la Commune," *RS*, 2nd ser., 1, no. 1 (1871): 37–47.

43. On the history of the society, see Hélène Gispert, *La France mathématique. La Société mathématique de France (1872–1914)* (Paris, 1991), esp. 11–213.

44. *Bulletin de la Société mathématique de France*, 1 (1872–73): 8.

45. Gispert, *France mathématique*, 24, where the contrast between research and the diffusion of mathematics is drawn.

46. Ibid., 166–67. Gispert also notes the equally significant reduction in the proportion of former *polytechniciens* among the membership. This fell from 58 percent in 1874 to 29 percent in 1914. By 1914, 72 percent of the members were pursuing academic careers, almost half of them in faculties of science.

47. Marcel Brillouin, "Les débuts de la Société française de physique," in *Livre du cinquante-naire de la Société française de physique* (Paris, 1925), 5–18, a text that includes Antoine-Jules Lissajous's recollections of the founding of the society.

48. Ibid., 8–9n. On d'Almeida's rise and Bertin's self-imposed marginalization, see also Daniel Jon Mitchell, "Gabriel Lippmann's Approach to Late-Nineteenth Century French Physics" (D.Phil. thesis, University of Oxford, 2009), 38–40.

49. Robert Fox, "The *Savant* Confronts His Peers: Scientific Societies in France, 1815–1914," in Robert Fox and George Weisz, eds., *The Organization of Science and Technology in France, 1808–1914* (Cambridge and Paris, 1980), 269–79, esp. table 2 on p. 275. Although the membership of the Société chimique was only three-quarters that of the Société française de physique, its income and capital were significantly larger.

50. *RS*, 2nd ser., 1: 1.

51. *JPTA*, 1 (1872–73): 5–6, in an untitled preface. On d'Almeida, see notice by Edmond Bouty, "Notice sur la vie et les travaux de J.-Ch. d'Almeida," *JPTA* 9 (1880): 425–34.

52. *AFAS, 1ère session. 1872 Bordeaux*, 41.

53. Figures in the society's centenary history, *Centenaire de la Société philomathique de Bordeaux 1808–1909* (Bordeaux, [1909]), convey the success of the courses offered in the building, where 2,192 pupils were enrolled by 1875.

54. Wurtz used the expression in addressing the inaugural general assembly of the promoters of the AFAS in Paris on 22 April 1872: *Association française pour l'avancement des sciences. Documents et informations diverses* 1 (20 July 1872), 6.

55. *AFAS, 1ère session. Bordeaux 1872*, 42.

56. *CSF, 29e session. Bordeaux 1861* 1: 62–70.

57. Ducrost had been curé of Solutré for fifteen years and was to become the first holder of the chair of geology when the Catholic university in Lyon established its faculty of science in 1877.

58. Broca, "Les troglodytes de la Vézère," in *AFAS, 1ère session. Bordeaux 1872*, 1199–1237, esp. (for Broca's interpretation) 1231–37. For discussions of Broca's polygenism and aspects of his anthropology relevant to this discussion, see Claude Blanckaert, *Paul Broca et l'anthropologie française (1850–1900)* (Paris, 2009), 209–53.

59. *AFAS, 5e session. Clermont-Ferrand 1876*, 554 (in a comment on a paper by Francisco María Tubino on the races of the Iberian peninsula).

60. Broca, "Troglodytes de la Vézère," 1237.

61. "Composition du Conseil supérieur de l'instruction publique. Amendement Paul Bert," in Bert, *Discours parlementaires. Assemblée nationale. Chambre des députés 1872–1881* (Paris, 1882), 1–39.

62. Bonnetty, "A nos abonnés. Sur la situation actuelle de la société chrétienne," *APC,* 6th ser., 2 (1870–71): 85–92. On Bonnetty, see the obituary in *Polybiblion. Revue bibliographique universelle. Partie littéraire,* 2nd ser., 9 (1879), 454–55.

63. Gaume, *Le Ver rongeur des sociétés modernes ou le paganisme dans l'éducation* (Paris, 1851) and *La Question des classiques ramenée à sa plus simple expression* (Paris, 1852).

64. Gaume, *Où en sommes-nous? Etude sur les évènements actuels. 1870 et 1871* (Paris, 1871), 365–77. Bonnetty conveyed his approval of Gaume's analysis, along with lengthy extracts from *Où en sommes-nous?* in "Où en sommes-nous? Etude sur les évènements actuels," *APC,* 6th ser., 2 (1870–71): 377–86, where Gaume's denunciation of free-thinking teachers is quoted (382).

65. See Parisis, *La Vérité sur la loi de l'enseignement* (Paris, 1850) for a particularly strong statement.

66. The gulf that separated Dupanloup from Bonnetty and Gaume on this question is evident from Gaume's *Lettres à Monseigneur Dupanloup, évêque d'Orléans sur le paganisme dans l'éducation* (Paris, 1852), responding to Dupanloup's *De l'éducation,* which began publication in 1850. For Dupanloup's views, see his *De l'éducation,* 5th ed. (Paris, 1861), 1:316–34 (book 5, chap. 6).

67. The far from overwhelming vote in the National Assembly in favor of the measure (316 to 266) reflects the keenness of the debate.

68. The most informative contemporary account, written from the informed but partial perspective of the universities' outstanding champion, is Maurice d'Hulst, *Les dix premières années des facultés libres. Rapport présenté à l'Assemblée générale des catholiques, le 26 mai 1886* (Paris, 1886).

69. Alfred Baudrillart, *Vie de Mgr d'Hulst,* 2 vols. (Paris, 1912–14), esp. vol. 1, 355–527.

70. See the addresses that d'Hulst delivered between 1881 and 1889, in his *Mélanges oratoires* (Paris, 1891), 2:219–422.

71. D'Hulst, *L'Empoisonnement de la science. Discours prononcé à Rouen au congrès des catholiques de Normandie le 23 novembre 1883* (Paris, 1883).

72. Ibid., 4.

73. A view that d'Hulst expressed with particular clarity in a speech to the Catholics of the Nord, *Le Rôle scientifique des facultés catholiques. Discours prononcé au congrès des catholiques du Nord à Lille, le 14 novembre 1883* (Paris, 1883).

74. *Observations présentées à MM. les sénateurs et députés au nom des principes et des intérêts de la science par le corps enseignant de l'Université catholique de Lille* (Lille, 1879), 18. The *Observations* were drawn up by the teaching staff at Lille, under the chairmanship of the rector, Edouard Hautcoeur, to protest against the proposed restrictions on the Catholic universities that came into force in 1880.

75. Donnadieu, *Université catholique de Lyon. Organisation du service de la zoologie à la faculté des sciences* (Paris, 1879).

76. Albert Lacroix, *Notice historique sur Albert-Auguste Lapparent lue dans la séance publique annuelle du 20 décembre 1920* (Paris, 1920), 16–17.

77. Charles Hardy, "Georges Lemoine," *Bulletin de la Société des sciences historiques et naturelles de l'Yonne* 77 (1923): 5–41, and L. J. Olmer, "Georges Lemoine," *RQS,* 4th ser., 3 (1923): 5–32.

78. Harry W. Paul, *From Knowledge to Power: The Rise of the Science Empire in France, 1860–1939* (Cambridge, 1985), 231.

79. André Girard, *Emile-Hilaire Amagat. Grand physicien français et fidèle citoyen de Saint-Satur (2 janvier 1841–14 février 1915* (Sancerre, 1941), 16n.

80. On the Béchamps *père et fils* and their difficulties at the Lille faculty, see Paul, *Knowledge to Power*, 231–35.

81. Blondel, "Branly face à l'innovation technique. Un cas d'espèce ?" *RHS* 46 (1993): 40–43. For the description of Branly's laboratory (presumably written by Branly himself), see *Les Installations des sciences physiques et naturelles à l'Université catholique de Paris* (Paris, 1879), 3–10.

82. Jeanne Terrat-Branly, *Mon père Edouard Branly* (Paris, 1941), 165–81 and 302–5; Gabriel Pelletier and J. Quinet, *Edouard Branly* (Paris, 1962), 114–18; and Philippe Monod-Broca, *Branly 1844–1940. Au temps des ondes et des limailles* (Paris, 1990), 328–31.

83. Terrat-Branly, *Mon père Edouard Branly*, 86; also (with a slightly different quotation) Monod-Broca, *Branly*, 138. Gariel had studied at the Ecole polytechnique, the Ecole des ponts et chaussées, and the faculty of medicine in Paris, and he now taught at both the Ecole des ponts et chaussées and the faculty. Closely involved in the founding of the AFAS, he served for thirty-five years as its secretary general.

84. Terrat-Branly, *Mon père Edouard Branly*, 56 and 73–75.

85. For more than three of these years, in the 1890s, Branly was responsible for physics in the clinic of the Ecole de psychologie, founded in the rue Saint-André-des-Arts by the pioneer of hypnotism and psychotherapy, Edgar Bérillon. See René Lacroix, *Le Docteur Bérillon 1859–1948. Un homme—un caractère—une oeuvre* (Paris, 1949), 23–25, and the plate (showing Branly at work in the clinic) facing page 33.

86. The episode is described, with correspondence between Curie and her sympathizers, in Karin Blanc, *Marie Curie et le Nobel* (Uppsala, 1999), 29–59. See also Terrat-Branly, *Mon père Edouard Branly*, 205–19, and Monod-Broca, *Branly*, 275–92. Branly's personal file in the archives of the Académie des sciences contains material on the election, including a glowing report on his work by Violle. For the vote, see *CR* 153 (1911): 170.

87. *CR* 147 (1908): 960 and 1460.

88. The fullest account of Branly's failure to receive the Nobel Prize in 1909 (as in 1904 and 1915, when he was also nominated) is in Monod-Broca, *Branly*, 250–56.

89. A brief mention in Pelletier and Quinet, *Edouard Branly*, 104, resurrected the story. Following repetition in the *Presse médicale*, the permanent secretaries of the Académie clarified the position in a letter of 7 January 1965 to the journal's editor, Roger Veylon. The correspondence is in Branly's file in the archives of the Académie.

90. See, in addition to d'Hulst's addresses (cited in note 70), those that Louis Baunard delivered in his capacity as rector of the Université catholique of Lille, in Baunard, *Université de Lille. Facultés catholiques. Vingt années de rectorat. Discours de rentrée et annexes* (Paris, 1909).

91. D'Hulst, *Dix premières années*, 8; also in *Bulletin de la Société générale d'éducation et d'enseignement* (1886), 407–19.

92. D'Hulst, *L'Institut catholique de Paris. Son caractère—son but—son importance—son organisation—ses résultats—ses besoins* (Paris, 1893), 7.

93. Paul, *Knowledge to Power*, 226. The numbers reported in *ED* 86 (1905) show that in virtually all the faculties in 1903–4 between a quarter and a half of the students were enrolled for the PCN. For the impact of the PCN on student numbers from 1893, see later in this chapter.

94. Paul, *Knowledge to Power*, esp. 235–44.

95. Ibid., and André Grelon, "Les enseignements techniques à Lille et dans sa région," in André Grelon and Françoise Birck, eds., *Des ingénieurs pour la Lorraine XIXᵉ–XXᵉ siècles* (Metz, 1998), 331–52, esp. 344–47. Indispensible on the foundation and early years of the Catholic facili-

ties in Lille is Emile Lesne, *Histoire de la fondation de l'Université catholique de Lille (1875–1877)* (Lille, 1927). I have also drawn on the *Livret des facultés catholiques de Lille,* published in 1902 with the support of the association of former students.

96. Lesne, *Fondation de l'Université catholique de Lille,* 85–105. See also "La Maternité Sainte-Anne" and "La Maison-Dieu" in Baunard, *Vingt années de rectorat,* 41–46.

97. *Livret des facultés catholiques de Lille,* 128.

98. The depth of Catholic resentment at the damagingly competitive attitudes of the ministerial authorities emerges strongly in *La Faculté de médecine et de pharmacie de l'Université catholique de Lille. Historique des difficultés qui précédèrent sa fondation* (Lille and Paris, May 1879).

99. Maurice Cesbron, a professor in the faculty of law at the Université catholique de l'Ouest in Angers, made the point with reference to his own discipline. Cesbron noted that whereas 40 percent of the pupils in secondary education were not in state schools, only 4 percent of the nation's law students were enrolled in Catholic faculties. See Charles-François Saint-Maur, "Nos universités catholiques. Leurs raisons d'être," in *Université catholique de l'Ouest. Livre du cinquantenaire 1875–1925. Travaux jubilaires offerts par les professeurs* (Angers, 1925), 28.

100. Carbonnelle, "L'aveuglement scientifique," *RQS* 1 (1877): 5–53. In contrast, Carbonnelle took a somewhat lenient attitude to Darwin's theory, seeing nothing in evolution by natural selection that set it against religion.

101. Ernest Lebon, *Gabriel Lippmann. Biographie, bibliographie analytique des écrits* (Paris, 1910), 2–3. The initiative appears to have come from two of Lippmann's strongest backers at the Ecole normale supérieure, Bertin and the director Ernest Bersot; see Mitchell, "Gabriel Lippmann's approach," 40.

102. *RSS,* 5th ser., 5 (1873): 204–25.

103. Simon's classic statement of the importance of prioritizing primary education (along with pleas for the promotion of the education of girls and for greater freedom of state control) is his book *L'Ecole* (Paris, 1874).

104. Simon, *La Réforme de l'enseignement secondaire* (Paris, 1874), 301–15.

105. Cf. the membership lists for December 1871 and December 1875 in the *Bulletin de la Société de géographie* 6th ser., 2 (1871) and 6th ser., 10 (1875). Membership continued to grow through to the First World War, standing at just under 2,000 in 1910. See *Société de géographie. Liste des membres. Janvier 1910,* published as a separately paginated supplement to *La Géographie. Bulletin de la Société de géographie,* 21 (1910).

106. The *Revue de géographie,* launched in 1877 as a monthly of at least eighty pages per issue, differed from the more popular journals in its resolutely scientific tone and in offering no illustrations, apart from the fine foldout maps that accompanied most issues.

107. The multiple objectives of the champions of geography are helpfully discussed in Vincent Berdoulay, *La Formation de l'école française de géographie (1870–1914)* (Paris, 1981), esp. 45–75, on the colonial movement, supported most importantly in the *Revue de géographie.* On Drapeyron's contribution and the difficulties arising from his position as a teacher of history and geography in the secondary system, see ibid., 160–63.

108. On the poor quality of French maps in the Franco-Prussian war, see also the comments of the center-left journalist and politician Ernest Picard in a letter of 17 December 1876 to Ludovic Drapeyron, in *RG* 1 (1877): 7–8.

109. Drapeyron, "De la transformation de la méthode des sciences politiques par les études géographiques," *RG* 1 (1877): 11–43, esp. 11, in the first issue of a journal that he had been instrumental in founding, following the international geographical congress of 1875 in Paris.

110. A reflection on the war that conveys Monod's moderation is his *Allemands et Français. Souvenirs de campagne. Metz—Sedan—La Loire* (Paris, 1872).

111. Monod, "Du progrès des études historiques en France depuis le XVIe siècle," *RH* 1 (1876): 5–38, esp. 26–38.

112. Boissier, *Cicéron et ses amis. Etude sur la société romaine au temps de César* (Paris, 1865), 22–23.

113. Paris, *Haut enseignement historique et philologique*, 15–16.

114. Thibaudet, *La République des professeurs* (Paris, 1927).

115. My interpretation of the reform movement in the 1880s and 1890s draws in particular on Antoine Prost, *Histoire de l'enseignement en France 1800–1967* (Paris, 1968), 223–44, and George Weisz, *The Emergence of Modern Universities in France, 1863–1914* (Princeton, NJ, 1983), chaps. 3–9.

116. The report and legislation creating the PCN and the closely related restructuring of medical education are in Beauchamp, *Lois et règlements*, 6:273–300. Modern studies of the effect of the PCN on scientific and medical studies include Paul, *Knowledge to Power*, 115 and 164–66, and Weisz, *Emergence of Modern Universities*, esp. 178, 183–84, and 358–59.

117. Louis Dulieu, *La Faculté des sciences de Montpellier de ses origines à nos jours* (Avignon, 1981), 21–24. For a contemporary account, see Henri Rouzaud, *Les Fêtes du VIe centenaire de l'université de Montpellier* (Montpellier and Paris, 1891), 37–45.

118. Louis Joubin, *La Faculté des sciences de Rennes* (Rennes, 1900), 91–103, and Jean-Yves Veillard, *Rennes au XIXe siècle. Architectes, urbanisme et architecture* (Rennes, 1978), 424–25.

119. The sense of continued privation is conveyed in the rector's speech at the university's annual inaugural ceremony for 1900–1901: *Université de Rennes. Rentrée solennelle des facultés de droit, des sciences & des lettres . . . 29 novembre 1900* (Rennes, 1901), 27. In the rector's words, "The faculty of science is rising majestically. It is being fitted out with instruments and taking in students, one category of whom, that of future doctors, was not foreseen at the time when the plans were laid. As a result, the building is becoming too small even before it is completed." The future doctors to whom the rector referred were candidates working for the PCN.

120. Georges Rayet, "Histoire de la faculté des sciences de Bordeaux (1838–1894)," *Actes Acad. Bordeaux,* 3rd ser., 59 (1897): 55.

121. Berthelot, *Science et philosophie* (Paris, 1905), i–ii.

122. Shinn, "The French Science Faculty System, 1808–1914. Institutional Change and Research Potential in Mathematics and the Physical Sciences," *HSPS* 10 (1979): 273–332, esp. 309–10 and 328.

123. Frédéric Blancpain, "La création du CNRS. Histoire d'une décision 1901–1939," *Bulletin de l'Institut international de l'administration publique* 32 (1974): 93–143.

124. Craig Zwerling, "The Emergence of the Ecole normale supérieure as a Centre of Scientific Education in the Nineteenth Century," in Fox and Weisz, *Organization of Science and Technology*, 31–60 (36–44). On the mathematical careers of former *normaliens,* see Gispert, *France mathématique,* passim but esp. 66–68 and 114.

125. Karady, "Educational Qualifications and University Careers in Science in Nineteenth-Century France," in Fox and Weisz, *Organization of Science and Technology*, 111.

126. The gradual eclipsing of literary figures by scientists in the last quarter of the nineteenth century is noted in Christophe Charle, *Naissance des "intellectuels" 1880–1900* (Paris, 1990), 28–38.

127. Christiane Sinding, "Claude Bernard and Louis Pasteur: Contrasting Images through Public Commemorations," in Pnina G. Abir-Am and Clark A. Elliott, eds., *Commemorative Prac-*

tices in Science: Historical Perspectives on the Politics of Collective Memory [*Osiris*, vol. 14] (1999), 61–85, esp. 63–64.

128. I use the list of state funerals in the appendix to Avner Ben-Amos's case study of the lavish funeral accorded to Victor Hugo: "Les funérailles de Victor Hugo. Apothéose de l'évènement spectacle," in Pierre Nora, ed., *Les Lieux de mémoire*, vol. 1, *La République* (Paris, 1984), 516–18.

129. Chevreul, *Distractions d'un membre de l'Académie . . . lorsque le roi de Prusse . . . assiégeait Paris de 1870 à 1871* (Paris, 1871). Chevreul made his protest before the Académie des sciences on the day after the bombardment. See *CR* 72 (1871): 41.

130. Bert, "Les travaux de Claude Bernard," in *L'Oeuvre de Claude Bernard. Introduction par Mathias Duval. Notices par E. Renan, Paul Bert et Armand Moreau* (Paris, 1881), 39–87 (86–87).

131. Gerald Geison's observation about Pasteur's lack of interest in religion and politics is convincing; see Geison, *The Private Science of Louis Pasteur* (Princeton, NJ, 1995), 42–43. Nevertheless Pasteur lived as a Catholic and died (as Vallery-Radot states) with a crucifix in his hand.

132. *Séance de l'Académie française. Discours de Pasteur. Réponse de Renan* (Paris, 1882), 3–26 (18).

133. Ibid., 29–54.

134. *1822–1892. Jubilé de M. Pasteur (27 décembre)* (Paris, 1893).

135. Christine Sinding, "La grande année Pasteur. Echec du contre culte?" in Pnina G. Abir-Am, ed., *La Mise en mémoire de la science. Pour une ethnographie historique des rites commémoratifs* (Paris, 1998), 289–310.

136. On this aspect of the changing character of the universal exhibitions, see chap. 5.

137. Picard, *Ministère du commerce, de l'industrie et des colonies. Exposition universelle internationale de 1889 à Paris. Rapport général* (Paris, 1891–92), vol. 10, "Pièces annexes," 438–39.

138. The attendances and other statistics are most easily consulted in Brigitte Schroeder-Gudehus and Anna Rasmussen, *Les Fastes du progress. Le guide des expositions universelles 1851–1992* (Paris, 1992).

139. *Ministère du commerce, de l'industrie, des postes et des télégraphes. Exposition universelle internationale de 1900 à Paris. Rapports du jury international. Classe 87—Arts chimiques et pharmacie. Rapport de M. A. Haller,* 2 vols. (Paris, 1901–2) and Cornu, "Introduction," in *L'Industrie française des instruments de précision. 1901–1902. Catalogue publié par le Syndicat des constructeurs en instruments d'optique & de précision* (Paris, 1901–2; facsimile reprint, Paris: Alain Brieux, 1980), v–xii.

140. Henri de Parville, *L'Exposition universelle*, 2nd ed. (Paris, 1890), 58.

141. Picard, *Le Bilan d'un siècle (1801–1900)* (Paris, 1906–7), 1:ii.

142. Ibid., 1:233.

143. Ibid., 1:235.

144. The classic contemporary accounts of the building of the tower are Picard, *Exposition de 1889. Rapport général*, 2:263–317, and Eiffel, *La Tour de trois cents mètres. Texte* (Paris, 1900); also the companion volume of plates the same year, published as Eiffel, *La Tour de trois cents mètres. Planches.*

145. Lockroy, Poubelle, and Eiffel had signed the agreement, incorporating a subsidy of 1.5m francs from public funds, on 8 January 1887; see Picard, *Exposition de 1889. Rapport général*, 2:267–68 and Eiffel, *Tour de trois cents mètres. Texte*, 6.

146. Picard, *Exposition de 1889. Rapport général*, 2:268–69, and Eiffel, *Tour de trois cents mètres. Texte*, 6–9.

147. Picard, *Exposition de 1889. Rapport général*, 2:268, and Eiffel, *Tour de trois cents mètres. Texte*, 6.

148. Coppée, "La tour Eiffel," in *Œuvres de François Coppée. Poésies 1886–1890* (Paris, 1891).

149. [Anon.], "Une tour de 300 mètres à Paris," *Cosmos,* n.s., 1 (1885): 23. The republican associations of the conception of the tower are explored in Miriam R. Levin, *Republican Art and Ideology in Late Nineteenth-Century France* (Ann Arbor, 1986), 41–45.

150. See the extract from Lockroy's letter to Alphand, in Picard, *Exposition de 1889. Rapport général,* 2:270–71, and Eiffel, *Tour de trois cents mètres. Texte,* 7.

151. Eiffel, *Tour de trois cents mètres. Texte,* 229.

152. Exemplified in Robida, *La Vieille France. Textes, desseins et lithographies,* 4 vols. (Paris, 1890–93), devoted respectively to Normandy, Brittany, Touraine, and Provence.

153. Robida, *Le Vingtième siècle* (Paris, 1883), and Robida, *Le Vingtième siècle. La vie électrique* (Paris, 1890).

154. Robida, *Le XIXe siècle* (Paris, 1888), 221.

155. *Le Vingtième siècle* sold at 25 francs, *La Vie électrique* at 20 francs. My comment on the print runs is supported only by the relative scarcity of the volumes in libraries today. The prices of these luxurious and aesthetically appealing productions contrast with those of the cheaper pocket editions (poorly illustrated and costing 3.50 francs in both cases).

156. Villiers de l'Isle-Adam, *L'Eve future* (Paris, 1886), esp. 87–105 (chaps. 2–5).

157. Richet, *Dans cent ans* (Paris, 1892), 147–81.

158. Ibid., 26.

159. Richet, "La Paix et la guerre," *Revue philosophique* 59 (1905): 113–32 and 252–70 (270).

160. Chesneaux, *The Political and Social Ideas of Jules Verne,* trans. Thomas Wikeley (London, 1972), 34. See also the same author's later work, *Jules Verne. Un regard sur le monde* (Paris, 2001), 58.

161. Verne, *Hector Servadac. Voyages et aventures à travers le monde solaire,* 2 vols. (Paris, 1877), part 2, chap. 8.

162. Verne published roughly half of his sixty or so novels initially as episodes in the *Magasin illustré d'éducation et de récréation,* a biweekly magazine that his publisher Jules Hetzel launched as family reading in 1864.

163. "Le récit de la science. Conversation avec Michel Serres," in Jean Demerliac, with Michel Serres and Jean-Yves Tadié, *L'Odyssée Jules Verne* (Paris, 2005), 50.

164. Verne, *Paris au XXe siècle* (Paris, 1994). On pages 15–16 of his preface to this first edition of the then recently discovered manuscript, Gondolo della Riva reproduced extracts from a draft letter (of late 1863 or early 1864) in which Hetzel rejected Verne's book.

165. Painlevé, "La philosophie de Marcellin Berthelot," *La Revue du mois* 3 (1907): 513.

166. Berthelot, *Science et philosophie* (Paris, 1886), *Science et morale* (Paris, 1897), *Science et education* (Paris, 1901), and *Science et libre pensée* (Paris, 1901).

167. *Cinquantenaire scientifique de M. Berthelot. 24 novembre 1901* (Paris, 1902), 11.

168. Ibid., 76.

169. Berthelot, *Science et libre pensée,* i.

170. Berthelot, *Science et morale,* 40. On the dinner, see Jacques, *Berthelot,* 209–18, and the unsigned "Banquet offert à M. Berthelot," *RS,* 4th ser., 3 (1895): 466–74, which includes a selection of the speeches at the dinner, including Berthelot's.

171. Berthelot, *Science et morale,* 41–42.

172. Ibid., 19–25.

173. The words are from Richet's address at the Saint-Mandé dinner, in "Banquet offert à M. Berthelot," 472.

174. Lefèvre's *La Philosophie* (Paris, 1879), esp. 443–44.

175. Lefèvre, *La Religion* (Paris, 1892), especially the excoriating "Conclusion. Regard en arrière et en avant," 552–73.

176. Ibid., 572.

177. Ibid., 573.

178. Brunetière, *La Science et la religion. Réponse à quelques objections* (Paris, 1895), 12–17 and 36.

179. Ibid., 10–25.

180. Ibid., 25–41.

181. "Cardinals" because of their religious allegiance, not their formal clerical rank, and "green" because of the green uniform of members of the Institute.

182. Lemaître, *Les Contemporains. Etudes et portraits littéraires* (Paris, 1898), 1:217–48, esp. 1:217 and 1:248.

183. Barbey d'Aurevilly, *Les Philosophes et les écrivains religieux* (Paris, 1860), 293–308 (294). Though unnumbered, this was the first volume of the multivolume *Les Oeuvres et les hommes.* Renan's treatment can be found in ibid., 125–51. For Barbey d'Aurevilly's opinion on Taine's book, see Barbey d'Aurevilly, *Les Oeuvres et les hommes (troisième série). Les philosophes et les écrivains religieux* (Paris, 1899), 330. The chapter on Taine had originally appeared as one of the first reviews of Taine, *De l'intelligence,* 2 vols. (Paris, 1870), in the Catholic newspaper *Le Constitutionnel* 55, no. 110 (20 Apr. 1870).

184. Bourget, *Le Disciple* (Paris, 1889), i–xii.

185. Ibid., xi.

186. Brunetière, "A propos du *Disciple,*" *RDM* 3ᵉ période 94 (1889): 214–26.

187. France, "P. Bourget, *Le Disciple,*" *Le Temps,* 23 June 1889. The responses of Brunetière and France to *Le Disciple* are discussed in Michel Mansuy, *Un Moderne. Paul Bourget. De l'enfance au Disciple* (Paris, 1960), 502–13.

188. See the autobiographical notice in *Oeuvres de L. Ackermann. Ma vie—premières poesies philosophiques* (Paris, 1885), xvi

189. Ackermann, *Contes et poésies* (Paris: Hachette, 1863), 253–54. Donald G. Charlton, *Positivist Thought in France during the Second Empire 1852–1870* (Oxford, 1959), 166–89, offers a helpful introduction to Ackermann's poetry and its positivist context.

190. Prudhomme, "Le Zénith. Aux victimes de l'ascension du ballon *Le Zénith,*" in *œuvres de Sully Prudhomme. Poésies (1868–1878)* (Paris, 1884).

191. Prudhomme, "Lucrèce. De la nature des choses. 1ᵉʳ livre. Préface," in *Œuvres de Sully Prudhomme. Poésies (1876–1879)* (Paris: Alphonse Lemerre, 1884).

192. Prudhomme, *Que sais-je? Examen de conscience. Sur l'origine de la vie terrestre* (Paris, 1896). The volume brought together the articles previously published in *La nouvelle revue,* a bimonthly literary and political journal with center-left republican leanings.

193. Richet, "L'effort vers la vie et la théorie des causes finales," *RS,* 4th ser., 10 (1898): 1–7. The article, Prudhomme's letters, and a response from Richet are reproduced in Prudhomme and Richet, *Le Problème des causes finales* (Paris, 1902), and I rely on this source for the discussion that follows.

194. Prudhomme, "Le crédit de la science," *RS,* 4th ser., 18 (1902): 545–48.

195. See his preface to Jules Michelet, *Bible de l'humanité* (Paris: Calmann-Lévy, 1899), i–xliii (being a late edition of a work first published by Michelet in 1864).

196. Danielle Fauque, "Edouard Grimaux (1835–1900) et l'affaire Dreyfus," in Hélène Gispert, ed., *Par la science, pour la patrie. L'Association française pour l'avancement des sciences, 1872–1914. Un projet politique pour une société savante* (Paris, 2002), 330–32.

197. A decision taken at the highest reaches of the Ecole polytechnique in agreement with the Minister of War had obliged Grimaux to take immediate retirement earlier in 1898. His support for Dreyfus was seen as incompatible with good order in the school and hence sufficient reason for his not appearing again before a student audience.

Conclusion

1. Pasteur, "Pourquoi la France n'a pas trouvé d'hommes supérieurs au moment du péril," in Pasteur, *Quelques réflexions sur la science en France* (Paris, 1871), 25–40 (29).

2. Duhem, "Usines et laboratoires," *Revue philomathique de Bordeaux et du Sud-Ouest* 2 (1899): 385–400 (387).

3. Quoted in Mary Jo Nye, *Science in the Provinces: Scientific Communities and Provincial Leadership in France, 1860–1930* (Berkeley, CA, 1986). Similar vituperation ran through Bouasse's *Bachot et bachotage. Etude sur l'enseignement en France* (Toulouse, 1910). See, for example, page 255, where (as part of his attack on the Académie) Bouasse advanced the characteristically unqualified judgment that "it is no secret that French science is in a state of unqualified decline."

4. Bouasse, *Vision et reproduction des formes et des couleurs, vision mono et binoculaire, illusions d'optique, photométrie, théorie des couleurs, photographie, procédés d'impression* (Paris, 1917), v–xxv (introduction entitled "L'inorganisation du travail scientifique en France").

5. Brigitte Schroeder-Gudehus, "Division of Labour and the Common Good: The International Association of Academies, 1799–1914," in Carl Gustaf Bernhard, Elisabeth Crawford, and Per Sorbom, eds., *Science, Technology and Society in the Time of Alfred Nobel* [Nobel Symposium 52] (Oxford, 1982), 3–20.

6. Fernand Papillon, "Cours de philosophie chimique fait au Collège de France par M. Ad. Wurtz," *MS* 6 (1864): 481.

7. Corsi, *The Age of Lamarck: Evolutionary Theories in France 1790–1830* (Berkeley, CA,1988), 236–39, and *Lamarck. Genèse et enjeux du transformisme 1770–1830*, trans. Diane Médard (Paris, 2001), 291–94.

8. Daniel Jon Mitchell, "Gabriel Lippmann's Approach to Late-Nineteenth Century Physics" (D.Phil. thesis, University of Oxford, 2009), 30–38.

9. This was precisely the criticism made by Emile Alglave in comparing French examinations unfavorably with those of Germany, which reflected a far greater emphasis on higher education as a preparation for research. See his editorial of 13 November 1869 in *RCS* 6 (1868–9): 785.

10. As Pestre notes, the weakness was relieved to some extent by the emergence of Henri Poincaré as a major figure from the 1880s. But, as he argues, it continued to mark French physics, especially theoretical physics, long after the First World War; see Pestre, *Physique et physiciens en France, 1918–1940* (Paris, 1984), esp. 31–65 and 104–46.

11. Forman, Heilbron, and Weart, "Physics *circa* 1900. Personnel, Funding, and Productivity of the Academic Establishments," *HSPS* 5 (1975): 11, and Shinn, "The French Science Faculty System, 1808–1914: Institutional Change and Research Potential in Mathematics and the Physical Sciences," *HSPS* 10 (1979): 302–28. Shinn, however, does observe a decline in productivity between 1900 and 1914. Forman, Heilbron, and Weart arrive at a figure of 2.5 publications a year for French physicists, compared with 3.2 a year for physicists in Germany.

12. See, in addition to their comments on Lille cited in chap. 6, Paul, *From Knowledge to Power: The Rise of the Science Empire in France, 1860–1939* (Cambridge, 1985), 221–50, esp. 235–44 (on the initiatives in Lille) and Grelon, "L'ingénieur catholique et son rôle social," in Yves Cohen and Rémi Baudouï, eds., *Les Chantiers de la paix sociale (1900–1940)* (Fontenay/Saint-Cloud, 1995), 167–83.

13. Nye, *Science in the Provinces*, 117–94.

14. Cf. ibid., 186, where Nye discusses the disappointment that Grignard, now in his sixties and after a quarter of a century in his chair, expressed in 1933 at the failure of provincial universities to establish themselves as centers for research and intellectual life comparable with Paris.

15. Herivel, "Aspects of French Theoretical Physics in the Nineteenth Century," *BJHS* 3 (1966–67): 130–31.

16. Moutier, *Eléments de thermodynamique* (Paris, 1872), v–vi. The contrast between the coolness of *la science officielle* toward the new energy physics (exemplified by Victor Regnault) and the interest among certain provincial *savants* (notably Marc Seguin and Hirn) is well drawn in Michel Cotte, "Les apports de Marc Seguin à la naissance de la thermodynamique," in Helge Kragh, Geert Vanpaemel, and Pierre Marage, eds., *History of Modern Physics: Proceedings of the XXth Congress of the History of Science. Actes du colloque de Liège, juillet 1997* (Turnhout, 2002), 125–32, and Faidra Papanelopoulou, "Gustave-Adolphe Hirn (1815–90): Engineering Thermodynamics in Mid-Nineteenth-Century France," *BJHS* 39 (2006): 231–54.

17. Michel Atten, "L'oeuf ou la poule. Science et tenchniques de l'é lectricité en France (1850–1900), *Réseaux. Communication—Technologie—Société* 11, 61 (1993), 113–23 (a reference I owe to Andrew Butrica), and "La physique en souffrance, 1850–1914," in Belhoste, Dahan-Dalmedico, and Picon, eds., *La Formation polytechnicienne 1794–1994* (Paris, 1994), 233–38; also (more fully) in Atten, "Théories électriques en France, 1870–1900. La contribution des mathématiciens, des physiciens, et des ingénieurs à la construction de la théorie de Maxwell" (Doctoral thesis, Ecole des hautes études en sciences sociales, 1992). As Atten indicates, Raynaud and Vaschy, who were appointed to relatively junior teaching posts at Polytechnique in 1882, probably exerted less influence than Mercadier, who began as a teaching assistant in 1876 and then moved on to positions as a leaving examiner in 1880 and director of studies in 1882.

18. Danielle Fauque, "De la réception de la théorie atomique en France sous le Second Empire et au début de la IIIe République," *AIHS* 53 (2003): 64–112.

19. Rocke, *Nationalizing Science: Adolphe Wurtz and the Battle for French Chemistry* (Cambridge, MA, 2001), 402.

20. Béhal had taught the atomic theory since the late 1880s in lectures that he gave as a supplement to the school's core program. I am grateful to Danielle Fauque for pointing out Béhal's difficulties to me. The evidence is in a letter from Charles Friedel to Scheurer-Kestner, 26 November 1893, Bibliothèque nationale et universitaire de Strasbourg MS 5982–83 (pp. 250–51).

21. Jules Marcou, *De la science en France. Première partie* (Paris, 1869), 209–324. On Marcou as a critic of the state of science in France, see also chap. 3.

22. Camille Limoges, "The development of the Muséum d'histoire naturelle, c.1800–1914," in Robert Fox and George Weisz, eds., *The Organization of Science and Technology in France, 1808–1914* (Cambridge and Paris, 1980), 211–40.

23. Gilpin, *France in the Age of the Scientific State* (Princeton, NJ, 1968), 77–123; Ben-David, "The Rise and Decline of France as a Scientific Center," *Minerva* 8 (1970): 160–79; and the almost identical text in Ben-David, *The Scientist's Role in Society: A Comparative Study* (Englewood Cliffs, NJ, 1971), 88–107.

24. During Pasteur's time as director of scientific studies between fifteen and twenty students a year were admitted to the scientific section of the Ecole normale supérieure, compared with roughly ten times that number going to Polytechnique, few of whom found their way into academic life.

25. I believe that the point about size stands, even though the contrast may have been somewhat exaggerated, as Peter Lundgreen has warned. See Lundgreen, "The Organization of Science

and Technology in France: A German Perspective," in Fox and Weisz, *Organization of Science and Technology*, 326–27.

26. The point is discussed, with reference to the 1850s and 1860s, in Atten, "La reine mathématique et sa petite soeur," in Bruno Belhoste, Hélène Gispert, and Nicole Hulin, eds., *Les Sciences au lycée. Un siècle de réformes des mathématiques et de la physique en France et à l'étranger* (Paris, 1996), 45–54.

I give the essential references in the notes and offer a complete online bibliography of my sources, both primary and secondary, at http://rfox.linacre.ox.ac.uk. This allows me to be briefer than would otherwise have been possible in the following commentary, in which I also make no attempt to replicate the full bibliographies that exist in many of the works cited below, notably (as starting points) those mentioned under General Secondary Sources.

General Secondary Sources

As wide-ranging histories of education at all levels in postrevolutionary France, there is still no rival to Antoine Prost, *Histoire de l'enseignement en France 1800–1967* (Paris, 1968; 2nd ed., 1970), and Françoise Mayeur, *De la Révolution à l'école républicaine*, vol. 3 of Louis-Henri Parias, ed., *Histoire générale de l'enseignement et de l'éducation en France*, 4 vols. (Paris, 1981).

Among general studies that relate more specifically to science, I have drawn heavily on George Weisz, *The Emergence of Modern Universities in France, 1863–1914* (Princeton, NJ, 1983); Harry W. Paul, *From Knowledge to Power. The Rise of the Science Empire in France, 1860–1939* (Cambridge, 1985); Nicole Hulin-Jung, *L'Organisation de l'enseignement des sciences. La voie ouverte par le Second Empire* (Paris, 1989); and the essays in Robert Fox and George Weisz, eds., *The Organization of Science and Technology in France, 1808–1914* (Cambridge and Paris, 1980; reissued, with corrections, 2009), which includes a comprehensive, though now inevitably dated bibliography on pp. 333–41.

Two general themes to which I return at different points in this book are the diffusion of the printed word and the relations between secular and religious traditions in French culture. Both have engendered a rich secondary literature, much of which I have used as a general background resource and therefore not always cited in the specific references that appear in the notes. A number of items in this literature call for notice here.

Any work that touches on the printed word has to take account of the four volumes of the *Histoire de l'édition française*, coordinated by Henri-Jean Martin and Roger Chartier (Paris, 1982–86), especially (for my purposes) vol. 3, *Le Temps des éditeurs. Du romantisme à la Belle Epoque*. On scientific books and periodicals, Paul, *From Knowledge to Power*, 251–87 (chap. 7) offers a good introduction. More focused (on the subject of chemistry textbooks) are the essays in Bernadette Bensaude-Vincent, Antonio García-Belmar, and José Ramón Bertomeu Sánchez, eds., *L'Émergence d'une science des manuels. Les livres de chimie en France (1789–1852)* (Paris, 2003). With regard to popular science, the secondary literature is growing encouragingly. A broadly cast history of scientific writing for the popular market is Daniel Reichvarg and Jean Jacques, *Savants et ignorants*.

Une histoire de la vulgarisation des sciences (Paris, 1991).More succinct but valuable as a guide to the providers and consumers of popular science is Bernadette Bensaude-Vincent's article "A Public for Science: The Rapid Growth of Popularization in Nineteenth Century France," *Réseaux* 3 (1995): 75–92. Apart from studies of individual popularizers, which I cite where appropriate, collective volumes have proved an effective way of tackling an inexhaustible subject. Bruno Béguet, ed., *La Science pour tous. Sur la vulgarisation en France de 1850 à 1914* (Paris, 1990) provides an informative and attractively presented overview. More detailed studies can be found in a special issue (on "Sciences pour tous") of *Romantisme. Revue de la Société des études romantiques et dix-neuvièmistes* 65 (1989), edited by Bernadette Bensaude-Vincent; here see, among a number of articles relevant to this book, Fabienne Cardot's on the scientific theater of Louis Figuier (59–67), a subject also treated in Reichvarg and Jacques, *Savants et ignorants,* 237–46, esp. 242–44. The collection of essays in Bernadette Bensaude-Vincent and Anne Rasmussen, eds., *La Science populaire dans la presse et l'édition. XIXe et XXe siècles* (Paris, 1997) has a more international focus but also contains much of relevance to France. Finally, mention must be made of a classic discussion that broaches general issues of popularization in the compass of a single, helpfully comparative article. This is Susan Sheets-Pyenson's "Popular Science Periodicals in Paris and London: The Emergence of a Low Scientific Culture, 1820–1875," *AS* 42 (1985): 549–72.

As with the history of the printed word, the references on the relations between secular and religious modes of thought that I give in the notes convey an incomplete impression of the richness of the secondary literature. One source that calls for special mention is Jacqueline Lalouette's history of French free thought between the mid-nineteenth century and the Second World War, *La libre pensée en France 1848–1940* (Paris, 1997). More specifically on the relations between science and religion, Georges Minois, *L'Eglise et la science. Histoire d'un malentendu,* 2 vols. (Paris, 1990–91) is a fine study; see especially the section expressively entitled "La guerre (XIXe–début du XXe siècle)," in vol. 2, 184–301. In a work covering almost a century and a half from the end of the ancien régime, Adrien Dansette's *Histoire religieuse de la France contemporaine,* 2 vols. (Paris, 1948–51) does not cover individual topics in detail, but it remains a good general account. With reference to the Second Empire, Jean Maurain's enduringly valuable *La Politique ecclésiastique du Second Empire de 1852 à 1869* (Paris, 1930) stands the test of time remarkably well. Maurain's account (855–64) of the Senate debates of 1868 on materialism, for example, is particularly well informed and perceptive. This said, there is no substitute for reading the debates themselves, assembled (following their publication in the *Moniteur universel,* the *JGIP,* and elsewhere) in *L'Enseignement supérieur devant le Sénat. Discussion extraite du Moniteur* (Paris, 1868).

Official Publications

In notes, I generally refer to the texts of decrees and other legislative documents in the form in which they appear in Arthur Marais de Beauchamp et al., *Recueil des lois et règlements sur l'enseignement supérieur comprenant les décisions de la jurisprudence et les avis des conseils de l'instruction publique et du Conseil d'Etat,* 7 vols. (Paris, 1880–1915). Cited in this book as Beauchamp, *Lois et règlements,* the work reproduces legislative and other administrative documents for the years from 1789 to 1914. It is available online via the collection of digitalized material at the Bibliothèque nationale de France, Paris, Gallica, except at the moment for volumes 2 and 6. An alternative and in many respects more reliable and complete source for the period that it covers (1802–1900) is *Circulaires et instructions officielles relatives à l'instruction publique. Publication entreprise par ordre de S. Exc. le Ministre de l'instruction publique [et des cultes],* 12 vols. (Paris, 1863–1902). Most of the documents in Beauchamp, *Lois et règlements* and the *Circulaires et instructions officielles* appeared first in issues of official or semiofficial publications for the relevant dates. See, in particular, *Bulletin administratif [du Ministère] de l'instruction publique* (145 vols., 1850–1932) and *Journal général [officiel] de l'instruction publique [et des cultes]* (54 vols., 1831–82), cited respectively as *BAIP* and *JGIP*. Among nonministerial publications, the *Revue internationale de l'enseignement* stands out as a rich source of information and critical commentary, with a reforming slant, on all aspects of higher education and research. The *RIE* was launched in 1881, as the journal of the Société de l'enseignement supérieur and the successor to the *Bulletin de l'enseignement supérieur* (1878–80).

These references indicate the sheer volume and importance of the official and quasi-official publications concerning education and research that appeared in the half century from the 1860s to the First World War. The 124 volumes of the *Enquêtes et documents relatifs à l'enseignement supérieur,* published at rather irregular intervals by the Ministry of Public Instruction between 1883 and 1929, are a mine of statistical and other information relating to all aspects of the administration of higher education and research and to all disciplines, scientific and non-scientific. Especially valuable for my purposes are the annual reports of the universities and faculties that appeared in this series. But other areas under the ministry's jurisdiction are well represented, notably by regular reports on provincial observatories. Among related, unofficial publications, the frequency with which I cite Albert Maire, *Catalogue des thèses de sciences soutenues en France de 1810 à 1890 inclusivement* (Paris, 1892) reflects its importance as a source of information on doctoral theses in science.

As Minister of Public Instruction from 1863 to 1869, Victor Duruy began the practice of publishing occasional reports and compilations of data relevant to the institutions under ministerial control. Of the substantial volumes concerning primary, secondary, and higher education that were published during and after Duruy's time at the ministry the most important for my purpose are *Statistique de l'enseignement supérieur, 1865–1868* (Paris, 1868); *Statistique de l'enseignement supérieur. Enseignement, examens, grades, recettes et dépenses, en 1876. Actes administratifs jusqu'en août 1878* (Paris, 1878) ; and *Statistique de l'enseignement supérieur. Enseignement, examens, grades, recettes et dépenses en 1886. Actes administratifs jusqu'en août 1888* (Paris, 1889). All of these works were published under the aegis of the Ministry of Public Instruction and bear the stamp of the strikingly efficient administrations of the Second Empire and early Third Republic. More "personal" in character were the collections of documents that ministers sometimes published as records of their own period in office. The most substantial and valuable of these is *L'Administration de l'instruction publique. De 1863 à 1869. Ministère de S. Exc. M. Duruy* (Paris, 1870), consisting mainly of speeches and administrative documents, though with an informative introductory essay by Duruy.

We are fortunate to possess a number of modern guides and collections of material drawn from official publications, usually with helpful introductions, commentaries, and further references. For the history of technical education, though with much material relevant to science, Thérèse Charmasson, Anne-Marie Lelorrain, and Yannick Rippa, *L'Enseignement technique de la Révolution à nos jours,* vol. 1, *De la Révolution à 1926* (Paris, 1987) is a model. Important for sources relating to science in secondary education (though with insights relevant to higher education as well) is *Les Sciences dans l'enseignement secondaire français. Textes officiels réunis et présentés par Bruno Belhoste avec la collaboration de Claudette Balpe et de Thierry Laport*, vol. 1, *1789–1914* (Paris, 1995).

Faced with the wealth of official and other publications and archival sources, anyone working on the history of research and education in nineteenth-century France needs guidance through what often appears an overwhelming quantity of material. Here, Thérèse Charmasson, ed., *Histoire de l'enseignement, XIXe–XXe siècles. Guide du chercheur* (Paris, 1986; 2nd, greatly enlarged edition, 2006) can be warmly recommended. So too can the multiauthored guide: Eric Brian and Christiane Demeulenaere-Douyère, eds., *Histoire et mémoire de l'Académie des sciences. Guide de recherches* (Paris, 1996), containing introductions to the archival resources of the Académie des sciences itself and of other archives and libraries, as well as brief essays on the history of the academy, a number of case-studies, mainly focused on the pre-nineteenth-century period, and a bibliography.

Universities and Faculties

The unified national university (variously Université royale, Université impériale, Université de France, or as I commonly refer to it, University) and the fifteen separate universities that were established in the later 1890s are prominent throughout the book, as they are in most studies of science in France in the period. The secondary literature is correspondingly abundant. Among works devoted specifically to the university system (including secondary as well as higher education), Paul Gerbod, *La Condition universitaire en France au XIXe siècle* (Paris, 1965) remains a classic work almost half a century on. Far briefer is the editors' introductory essay on "The Institutional Basis of French Science in the Nineteenth Century" in Fox and Weisz, *Organization of Science and Technology*, 1–28, much of which is devoted to the university system; in this essay, readers of the present book may find the table of the dates of foundation of the faculties of science and medicine and some related institutions (4–6) a useful source. See also my own "Science, the University, and the State in Nineteenth-Century France," in Gerald L. Geison, ed., *Professions and the French State, 1700–1900* (Philadelphia, 1984), 66–145, and the essays (on themes extending far beyond the sciences) in Christophe Charle and Régine Ferré, eds., *Le Personnel de l'enseignement supérieur en France aux XIXe et XXe siècles* (Paris, 1985).

A number of contemporary studies retain their value many decades after their publication. Preeminent among these is Louis Liard, *L'Enseignement supérieur en France, 1789–1889 [1893]*, 2 vols. (Paris, 1888–94), an essential source, albeit one to be used with caution, as the work of a far from unbiased champion of educational reform under the Third Republic. Other contemporary works of enduring value include Octave Gréard's *Education et instruction* (Paris, 1887), with its masterly vignette "L'enseignement supérieur à Paris en 1881" (1–78), and Liard's *Universités et facultés* (Paris, 1890), in which Liard invoked historical evidence in presenting the case for the establishment of independent universities that would have a high degree of autonomy while remaining under overall ministerial control.

For some thirty years now, Nicole Hulin has made a major contribution to our understanding of science in both higher and (especially) secondary education. In addition to her *Organisation de l'enseignement des sciences* (cited above), she has edited three collective volumes on aspects of the major reforms of 1902 in secondary education (reforms in which scientists were centrally involved with a view to promoting science as a core element of modern culture): *Physique et "humanités scientifiques." Autour de la réforme de l'enseignement de 1902. Etudes et documents* (Villeneuve d'Ascq, 2000); *Sciences naturelles et formation de l'esprit. Autour de la réforme de l'enseignement de 1902. Etudes et documents* (Villeneuve d'Ascq, 2002); and (with Hélène Gispert and Marie-Claire Robic) *Science et enseignement. L'exemple de la grande réforme des programmes du lycée au début du XXe siècle* (Paris, 2007). In her *L'Enseignement et les sciences. L'exemple français au début du XXe siècle* (Paris, 2005), Hulin brings together her interpretations of the period before 1914 but takes her account on into the 1930s.

By the 1980s, science in the provinces was beginning to attract greater attention among historians seeking to break with the traditional emphasis on the capital and with interpretations of research and teaching that tended to overlook the fine structure of geographical and cultural context and the diversity of French provincial life. The new focus was encapsulated and taken an important step forward in Mary Jo Nye, *Science in the Provinces: Scientific Communities and Provincial Leadership in France, 1860–1930* (Berkeley, CA, 1986). In this pioneering study, Nye drew attention explicitly to the intellectual vigor of the science faculties in Nancy, Grenoble, Lyon, Toulouse, and Bordeaux, identifying a particular vitality that characterized them between 1880 and 1910. Since Nye published her book, scholarly interest in institutions outside Paris has intensified, with encouraging results, some of which I cite below, in the section "Academies and societies." Engineering and applied science have attracted particular attention in this renewed focus on the provinces, notably in work by André Grelon and collaborative publications in which he has been involved, such as André Grelon and Françoise Birck, eds., *Des ingénieurs pour la Lorraine XIXe–XXe siècles* (Metz, 1998). The relations between science, industry, and locality that the authors in this volume treat are also the theme of Michel Grossetti, *Science, industrie et territoire* (Toulouse, 1995).The Catholic universities that were established in response to the law of 1875 granting freedom of higher education have been the subject of a relatively small but interesting literature. With regard to science, Paul, *From Knowledge to Power*, 221–50 (chap. 6) provides a good introduction. René Aigrain, *Les Universités catholiques* (Paris, 1935) is a useful international survey, with an informative section (19–62) on the French universities. Among several important studies stimulated by the fiftieth anniversary of the 1875 law, see, in particular, Emile Lesne, *Histoire de la fondation de l'Université catholique de Lille (1875–1877)* (Lille, 1927) and the collective volume *Université catholique de l'Ouest. Livre du cinquantenaire 1875–1925. Travaux jubilaires offerts par les professeurs* (Angers, 1925), in which eyewitness accounts by Monsignor H. Pasquier, "Mes premiers souvenirs universitaires" and Louis Le Helloco, "Comment a vécu l'Université catholique de l'Ouest depuis cinquante ans" are especially helpful. The centenary of the Institut catholique in Paris in 1975 yielded a substantial volume of historical studies and recollections: *Institut catholique de Paris. Livre du centenaire 1875–1975* (Paris, 1975). The contribution by Pierre Pierrard on "La fondation de l'Université catholique et l'enseignement des disciplines profanes 1875–1914," on pages 25–59, has much to say about science (among other disciplines), as do the essays on Edouard Branly by Robert Debré and Henri Rollet, on pages 329–38.

The impediments that the Catholic institutes suffered after 1880 (when their right to describe themselves as universities was withdrawn) are well treated, along with sympathetic discussions of the institutes' achievements, in Edouard Lecanuet, *La Vie de l'église sous Léon XIII* (Paris, 1930), 296–317 and in Pierrard's "Fondation de l'université catholique." The writings of individual champions of Catholic higher education likewise provide rich insights into the difficulty of competing with the faculties under ministerial control. The *Mélanges oratoires* of the abbé Maurice d'Hulst (2 vols., Paris, 1891) convey the determination of a leading pioneer of the Institut catholique de Paris to establish an independent alternative to the state system, as do the individual addresses by him that I cite in chapter 6 and Alfred Baudrillart's *Vie de Mgr d'Hulst*, 2 vols. (Paris, 1912–14). Science had a central role in this struggle, and d'Hulst's contributions to the establishment of a distinctively Catholic approach to scientific education are treated well in the introduction to Francesco Beretta, ed., *Monseigneur d'Hulst et la science chrétienne. Portrait d'un intellectuel* (Paris, 1996), 7–123. The annual addresses that Louis Baunard delivered as rector of the Université catholique de Lille for twenty years beginning in 1887 convey the particular problems (and successes) in the predominantly Catholic region of the Nord; see Baunard, *Université de Lille. Facultés catholiques. Vingt années de rectorat. Discours de rentrée et annexes* (Paris, 1909).

Research Institutions, "Grandes Ecoles," and Higher Technical Education

The leading national research institutions have all been the subject of an abundant literature, some of it now quite old and much of it unduly celebratory. Only thirty years after its foundation, for example, the Muséum d'histoire naturelle, was described in detail in Joseph-Philippe-François Deleuze, *Histoire et description du Muséum royal d'histoire naturelle*, 2 vols. [continuously paginated] (Paris, 1823). Deleuze's work is an eyewitness account and hence is essentially a primary source. As such, it is important, as are the two reports of the 1850s on the museum, which inform much of what I say about the institution in chapter 3. For a modern reader, however, the best entry point on the history of the museum in the nineteenth century is Claude Blanckaert, Claudine Cohen, Pietro Corsi, and Jean-Louis Fischer, eds., *Le Muséum au premier siècle de son histoire* (Paris, 1997). Older works still dominate the literature on the Collège de France. For my purposes, in fact, nothing has replaced *Le Collège de France (1530–1930). Livre jubilaire composé à l'occasion de son quatrième centenaire par MM. A. Lefranc, P. Langevin . . . professeurs au Collège de France* (Paris, 1932), which has good chapters on the scientific chairs in the nineteenth century. The Observatiore de Paris likewise awaits a major work that will complement more finely focussed accounts and continue the history of the institution beyond the prerevolutionary period, well treated in Charles Wolf, *Histoire de l'Observatoire de Paris, de sa fondation à 1793* (Paris, 1902). The Bureau des longitudes, however, has been the subject of a lengthy study, published in successive volumes of the *Annuaire du Bureau des longitudes*. See Guillaume Bigourdan, "Le Bureau des longitudes. Son histoire et ses travaux, de l'origine (1795) à ce jour," *ABL* (1928): A1–72, (1929): C1–92, (1930): A1–110, (1931): A1–145, and (1932): A1–117.

From their earliest days, the "grandes écoles" have generally been well served by their historians, and we do not want for dependable histories of the Ecole polytechnique, Ecole normale supérieure, and Ecole centrale des arts et manufactures, the most important of which I cite in notes. In the 1990s, an already rich body of material was enhanced by the fruits of a number of conferences and long-term collaborative projects devoted to the history of individual institutions. The Ecole polytechnique was a major beneficiary of bicentenary celebrations in 1994. Two particularly important volumes resulted: Bruno Belhoste, Amy Dahan Dalmedico, and Antoine Picon, eds., *La Formation polytechnicienne 1794–1994* (Paris, 1994) and Bruno Belhoste, Amy Dahan Dalmedico, Dominique Pestre, and Antoine Picon, eds., *La France des X. Deux siècles d'histoire* (Paris, 1995). The bicentenary of the Conservatoire national des arts et metiers, also in 1994, stimulated the publication of *Les Cahiers d'histoire du CNAM*, an occasional publication, edited by André Grelon, five issues of which appeared between 1992 and 1996. And the bicentenary of the Société d'encouragement pour l'industrie nationale in 2001 was the occasion for the preparation of a wide-ranging volume of essays, edited by Serge Benoit, Gérard Emptoz, and

Denis Woronoff, *Encourager l'innovation en France et en Europe. Autour du bicentenaire de la Société d'encouragement pour l'industrie nationale* (Paris, 2006).

Academies and Societies

By far the largest primary and secondary literature in this category relates to the Académie des sciences and the Institut de France, of which (with the other national academies) the Académie des sciences became a constituent part in 1795. Maurice P. Crosland, *Science under Control. The French Academy of Science, 1795–1914* (Cambridge, 1992) stands as a reliable and detailed history of the academy in the nineteenth and early twentieth centuries. Since the publication of Crosland's book, further work has been greatly facilitated by Brian and Demeunlenaere-Douyère, *Histoire et mémoire de l'Académie des sciences*, an indispensable guide for researchers to which I refer above. On the Institut de France more generally, Léon Aucoc, *L'Institut de France. Lois, statuts et règlements concernant les anciennes académies et l'Institut de 1635 à 1889. Tableau des fondations. Collection publiée sous la direction de la Commission administrative centrale* (Paris, 1889) conveniently assembles the relevant administrative documents up to the 1880s. For basic biographical information about members of all the academies composing the Institute, good sources are Amable Charles, Comte de Franqueville, *Le premier siècle de l'Institut de France. 25 octobre 1795–25 octobre 1895*, 2 vols. (Paris, 1895–96), and Jean Leclant et al., eds., *Le second siècle de l'Institut de France, 1895–1995. Recueil biographique et bibliographique des membres, associés étrangers, correspondants français et étrangers des cinq académies*, 3 vols. (Paris, 1999–2005). Though less ambitious than these works and limited to the Académie des sciences, the *Index biographique de l'Académie des sciences du 22 décembre 1666 au 1er octobre 1978* (Paris, 1979) remains a valuable tool.

The proceedings of the Académie des sciences are generally well recorded and easily accessible. For meetings held between 1795 and 1835, the ten volumes of *Procès-verbaux des séances de l'Académie tenues depuis la fondation de l'Institut jusqu'au mois d'août 1835* (Hendaye, 1910–22) are a printed edition of proceedings held in the Archives of the Académie des sciences. From 1835 onward, the debates were published more fully, and with remarkable speed and efficiency, in the weekly *Comptes rendus hebdomadaires des séances de l'Académie des sciences*, now digitized in the Gallica collection of the Bibliothèque nationale de France.

The wider world of the scientific and other learned societies, as it existed toward the end of the July Monarchy, was comprehensively mapped in the *Annuaire des sociétiés savantes de la France et de l'étranger, publié sous les auspices du Ministère de l'instruction publique. Première année. 1846* (Paris, 1846). Although it was intended that the *Annuaire* should become an annual publication, only this volume ever appeared. Among later attempts to provide a record of the work of the societies (of all kinds, both scientific and nonscientific), le comte Achmet d'Héricourt, *Annuaire des sociétiés savantes de la France et de l'étrange*, 2 vols. (Paris, 1863–64), 1:1–251 (on France) is useful. But Eugène Lefèvre-Pontalis, *Bibliographie des sociétés savantes de la France* (Paris, 1887) is more thorough and takes the inventory of societies and their publications into the 1880s. Histories of individual societies abound, and I cite these where appropriate. Among more broadly cast studies, Daniel Roche led the way with his classic work on the prerevolutionary provincial academies, *Le Siècle des lumières en province. Académies et académiciens provinciaux, 1680–1789*, 2 vols. (Paris, 1978). Since then, nineteenth-century societies have begun to receive systematic attention, though not yet in the same degree of detail. Jean-Pierre Chaline, *Sociabilité et érudition. Les sociétés savantes en France. XIXᵉ et XXᵉ siècles* (Paris, 1998) stands out as a comprehensive survey of societies of all types. My own contributions, focused mainly on scientific societies, include "The Savant

Confronts His Peers: Scientific Societies in France, 1815–1914," in Fox and Weisz, *Organization of Science*, 241–82, and "Learning, Politics, and Polite Culture in Provincial France. The *Sociétés Savantes* in the Nineteenth Century," *Historical reflections / Réflexions historiques* 7 (1980): 543–64.

France's academies and societies at all levels and in all disciplines were the subject of a major conference at the one hundredth Congrès national des sociétés savantes in Paris in 1975. The two volumes of papers that resulted from the conference are a rich resource. In one of these, *Colloque interdisciplinaire sur les sociétés savantes. Actes du 100e congrès national des sociétés savantes (Paris, 1975)* (Paris, 1976), I have found Marcel Baudot, "Trente ans de coordination des sociétés savantes (1831–1861)," 7–40; Yves Laissus, "Les sociétés savantes et l'avancement des sciences naturelles. Les musées d'histoire naturelle," 41–67; and Philippe Pinchemel, "Le sociétés savantes et la géographie," 69–78, particularly helpful. The companion volume, *Les Sociétés savantes. Leur histoire. Actes du 100e congrès national des sociétés savantes. Paris, 1975. Section d'histoire moderne et contemporaine et Commission d'histoire des sciences et des techniques* (Paris, 1976), is devoted to more than thirty case studies, many containing information not available elsewhere. Much briefer and inevitably dated, but still valuable is the entry on "Société" in *La Grande encyclopédie. Inventaire raisonné des sciences, des lettres et des arts par une société de savants et de gens de lettres sous la direction de MM. Berthelot . . .* 31 vols. (Paris, 1886–1902), 30:130–65, esp. the section on "Sociétés savantes, littéraires et artistiques," 30:147–56, by L. Sagney.

The Comité des travaux historiques et scientifiques and its associated congresses and initiatives in research were mainly concerned with historical, linguistic, and literary scholarship. But numerous attempts were made to integrate science in their activities, as I observe in chapter 3. Hence Xavier Charmes, *Le Comité des travaux historiques et scientifiques (Histoire et documents)*, 3 vols. (Paris, 1886) is a more relevant source for the history of science, in particular for the relations between the national and independent structures for research and learning, than might appear at first sight.

The interactions between the country's scientific center and the periphery are a recurring theme of *The Savant and the State*, and it is encouraging to see attention being increasingly directed to that subject. A particularly successful initiative in this area was the collaborative project that led to an important volume of essays on the early history of the Association française pour l'avancement des sciences, published as Hélène Gispert, ed., *"Par la science, pour la patrie." L'Association française pour l'avancement des sciences (1872–1914). Un projet politique pour une société savante* (Rennes, 2002).

Biographical Sources

The slowness with which the publication of the *Dictionnaire de biographie française*, ed. J. Balteau et al., with 19 volumes completed to date (Paris, 1933–) proceeds is disappointing, although many of the articles that have appeared are excellent. But a number of more specialized biographical dictionaries are now available in particular areas. Among those on which I have drawn are Christophe Charle and Eva Telkès, *Les Professeurs du Collège de France. Dictionnaire biographique (1901–1939)* (Paris, 1988) and (by the same authors) *Les Professeurs de la faculté des sciences de Paris. Dictionnaire biographique (1901–1939)* (Paris, 1989); Claudine Fontanon and André Grelon, eds., *Les Professeurs du Conservatoire national des arts et métiers. Dictionnaire biographique 1794–1955*, 2 vols. (Paris, 1994); Philippe Jaussaud and Edouard-Raoul Brygoo, eds., *Du Jardin au Muséum en 516 biographies* (Paris, 2004); and Laurence Lestel, ed., *Itinéraires de chimistes 1857–2007, 150 ans de chimie en France avec les présidents de la SFC* (Paris, 2007). The website of the Société d'histoire de la pharmacie offers a useful set of brief biographies of pharmacists, including many whose interests extended to areas of chemistry beyond pharmacy; see www.shp-asso.org/index.php? PAGE=personnages.

It is encouraging to note the growing number of book-length biographies of both scientists and educational administrators. In the 1980s, the series "Un savant, une époque," published by Editions Belin under the general editorship of Jean Dhombres, broke important new ground with biographical studies of a consistently high standard. In the main text, I cite volumes in this series on Cauchy (by Bruno Behlhoste), Branly (by Philippe Monod-Broca), and Berthelot (by Jean Jacques). Biographies, of course, eventually become dated both in style and in the sources on which they draw. But one that has survived remarkably well is Maurice Daumas, *Arago 1786–1853. La jeunesse de la science*, first published in 1943 and then reissued in the Belin series, with revisions by Emmanuel Grison and an essay by Charles Gillispie, in 1987.

The coverage of individuals who might be thought to merit full-length up-to-date biographies remains patchy. However, biographical materials are relatively accessible, not least through the often rich personal files of academicians in the Archives of the Académie des sciences and through the file that every employee of the Ministry of Public Instruction has in the F^{17} series in the Archives nationales. These resources have been increasingly quarried for some years now, with some excellent results. Even when surveying only very recent publications, as I do here, I find it heartening to see how many of the most glaring gaps are beginning to be filled. Anne-Claire Déré and Gérard Emptoz filled one such gap with their *Autour du chimiste Louis-Jacques Thenard (1777–1857). Grandeur et fragilité d'une famille de notables au XIXe siècle* (Chalon-sur-Saône, 2008), the first biography of Thenard since Paul Thenard's *Le Chimiste Thenard 1777–1857* (Dijon, 1950). James Lequeux has returned to the rich subject of Arago in his *François Arago. Un savant généreux. Physique et astronomie au XIXe siècle* (Paris and Les Ulis, 2008). Lequeux has also risen to the challenge of writing a biography of Le Verrier, the complexity and sheer quantity of whose work, and perhaps the character of the man, appear to have deterred earlier biographers. This makes Lequeux's *Le Verrier. Savant magnifique et détesté* (Paris, 2009) especially welcome. No less welcome is the magisterial study of Broca by Claude Blanckaert, *Paul Broca et l'anthropologie française (1850–1900)* (Paris, 2009), although in this case the subject has been one of perpetual interest to historians of the human sciences; among earlier works, see, for example, Francis Schiller, *Paul Broca: Founder of French Anthropology, Explorer of the Brain* (Berkeley, 1979) and Philippe Monod-Broca, *Paul Broca. Un géant du XIXe siècle* (Paris, 2005).

Conferences and collaborative projects have long helped to focus attention on lives and achievements that have daunted or for other reasons escaped the attention of individual biographers. Two recent works to result from initiatives of this kind, both bearing on major figures in this book, are those devoted to Arcisse de Caumont and Charles Dupin: Vincent Juhel, ed., *Arcisse de Caumont (1801–1873). Erudit normand et fondateur de l'archéologie française* (Caen, 2004) and Carole Christen and François Vatin, eds., *Charles Dupin (1784–1873). Ingénieur, savant, économiste, pédagogue et parlementaire du Premier au Second Empire* (Rennes, 2009). Authoritative and thorough though these volumes are, they should not allow older biographical studies to be overlooked. In addition to the sources on Dupin that I cite in chapter 1, for example, Joseph Bertrand's "Eloge historique de P. C. F. Dupin," in Bertrand, *Eloges académiques* (Paris, 1890), 221–46, is an essential source. Like Dupin, Caumont was the subject of numerous contemporary accounts and obituaries, often published in the proceedings of local academies and societies, and the same is true of virtually all the independent savants and *érudits* whom I discuss in chapter 2.

Among nineteenth-century Ministers of Public Instruction, Victor Duruy stands out for his sensitivity to the needs of science and scholarship, and the biographical studies of him reflect his importance. They include, most recently, Jean-Charles Geslot, *Victor Duruy (1811–1894). Historien et ministre* (Villeneuve d'Ascq, 2009). But, to pursue my point about the continuing value of older sources, Jean Rohr, *Victor Duruy, ministre de Napoléon III. Essai sur la politique de l'instruction publique au temps de l'Empire libéral* (Paris, 1967), and Sandra Horvath-Peterson, *Victor Duruy & French Education* (Baton Rouge, 1984), remain important. Most other ministers of the nineteenth century have been the subject of at least one biography. Among ministers who figure prominently in this book, Hippolyte Fortoul has been particularly well served by Paul Raphael and Maurice Gontard, *Un Ministre de l'instruction publique sous l'Empire autoritaire. Hippolyte Fortoul 1851–1856* (Paris, 1975), as have the comte de Salvandy, by Louis Trenard, *Salvandy en son temps 1795–1856* (Lille, 1968) and Denis Frayssinous, by Antoine Roquette, *Monseigneur Frayssinous, Grand-Maître de l'Université sous la Restauration (1765–1841). Evêque d'Hermepolis ou le chant du cygnet du trône et de l'autel* (Paris, 2007).

Manuscripts and Other Archival Sources

In the notes, I give full references to archival sources that I have used. Here, I wish only to pay tribute to the high quality of the finding aids that exist for most departmental and municipal collections. Where departmental archives are concerned, the existence of a broadly uniform national system of classification greatly facilitates access to relevant documents. For sources in the Archives nationales, the published guides to the archives of the Ministry of Public Instruction are excellent. Of these guides, *Les Archives nationales. Etat général des fonds*, vol. 1, *1789–1940*, coordinated by Jean Favier (Paris, 1978) is the best entry point, and it remains so even in the era of easy online searching. But, for work on the sources relevant to the history of science and scholarship in the Archives nationales, especially in the inexhaustibly rich F^{17} series of personal and institutional files, two additional volumes are essential. These are Marie-Elisabeth Antoine and Suzanne Olivier, *Inventaire des papiers de la Division des sciences et lettres du Ministère de l'instruction publique et des services qui en sont issus (Sous-série F^{17})*. vol. 1 (Paris, 1975), and Marie-Elisabeth Antoine, *Inventaire des papiers de la Division des sciences et lettres du Ministère de l'instruction publique et des services qui en sont issus (Sous-série F^{17})*, vol. 2 (Paris, 1981). These works are also models of the archivist's art. Even then, however, there is no substitute for time spent in the catalogue room of the Centre d'accueil et de recherche des Archives nationales (CARAN) in the rue des Quatre-Fils in Paris.

Beyond national and departmental archives, I have encountered the same efficiency in cataloguing and conservation wherever I have worked. Mentioning just one archive is invidious. But my own research has benefited in particular from the resources of the Bibliothèque centrale of the Muséum d'histoire naturelle. Here, the richness and ease of use of the many hundreds of items in the correspondence of Jean-Baptiste Mougeot (on which my comments on Mougeot in chapter 2 draw heavily) are typical. Yves Laissus bears witness to this in his two contributions, "Les papiers du docteur J.-B. Mougeot (1776–1858). Un fonds important pour l'histoire des sciences naturelles en Lorraine," in *Comptes rendus du 103e congrès national des sociétés savantes. Nancy 1978. Section des sciences. Fascicule V. Histoire des sciences et des techniques. Médecine* (Paris, 1978), 233–40, and

"La correspondance de Jean-Baptiste Mougeot," in Gérald Guéry, ed., *Histoire naturelle des Vosges. Sur les pas de Jean-Baptiste Mougeot* (Nancy, 1999) : 34–36.

In writing this bibliographical note, I have been anxious not only to comment on a selection of my main sources but also to point to the wealth of material that remains to be exploited. Looking to the future, I am conscious of working in the early dawn of a world of unprecedented access to published and manuscript sources through such sites as "Œuvres et rayonnement de Jean-Baptiste Lamarck (1749–1829)" (www.lamarck.cnrs.fr), coordinated by Pietro Corsi, and "Ampère et l'histoire de l'électricité" (www.ampere.cnrs.fr), coordinated by Christine Blondel. The challenge for the historian of science in nineteenth-century France, therefore, is increasingly one of selection, both from an immense range of primary sources and from a secondary literature that has burgeoned, and burgeoned excitingly, especially over the past three decades.